# CALCULUS

By the same author:

**Multivariate Calculus with Linear Algebra,** Wiley, 1972.

# CALCULUS
with An Introduction to Vectors

## Philip C. Curtis, Jr.
University of California, Los Angeles

John Wiley and Sons, Inc., New York · London · Sydney · Toronto

Library of Congress Catalogue Card Number: 75-160212

ISBN   0-471-18990-1

Printed in the United States of America

10  9  8  7  6  5  4  3  2  1

# PREFACE

This book is designed for a one-year course in single variable calculus containing an introduction to vectors for students in the physical sciences, engineering, and mathematics. My objective has been to strike a balance between theory and applications that will allow the student to perceive the structure of the subject matter as well as to understand the many applications of the calculus to the physical world.

My approach with the theoretical material has been to give careful statements of definitions and theorems following appropriate motivation. I have followed several routes with respect to proofs. Elementary arguments, especially those of a geometrical nature, are included in the body of the text. More technical proofs are either given in starred sections or appendices, which may be omitted if desired. Also, proofs are often handled as a sequence of advanced exercises.

The book breaks naturally into three parts. The first three chapters form an introduction to differentiation and integration with the more elementary applications. The material is further developed in Chapters 4 through 7, ending with discussion of elementary differential equations. The third part begins with the introduction of vectors and these techniques are applied to the study of curves in the plane and space. Taylor polynomials and infinite series conclude the book. For schools on a quarter plan, each of these parts constitutes approximately one-quarter's work. For a course taught on a semester basis, Chapters 1 to 5 comprise the material for the first semester, and Chapters 6 to 11 comprise the material for the second.

In addition to some of the more technical proofs, the starred sections and appendices contain material that may be omitted without affecting the continuity of the treatment. I have chosen to discuss the conic sections in an appendix as well since analytic geometry is becoming more and more a course in the senior year of high school. However, this book does not require a prior course in analytic geometry. The necessary discussion of lines in the plane is contained in Chapter 1. The conics are then discussed after the techniques of curve sketching have been developed.

In the end, success in mastering the calculus depends on the problems that a student does. I have included numerous exercise sets with many routine problems as well as many of a more challenging nature. Theoretical material is often explored in the exercises. I frequently use the exercises to anticipate material to be discussed later. The chapter on differential equations is included primarily to give a much broader spectrum of problems involving integration than would otherwise be possible. Answers to all problems of a computational nature are included in the book.

Vector techniques are of great importance to students in the sciences. The introduction to these methods is presented after the derivative and integral have been fully developed. The treatment of curves in Chapter 9 is then designed to let the student master the vector techniques from the previous chapter. This material can then serve as a foundation for a vector approach to multivariate calculus in the student's second year.

I am grateful to my publisher, John Wiley, for their assistance and co-operation, and especially to Jack Hoey and Fred Corey, whose help has been invaluable. I also thank Elaine Stafford for her expert typing of the manuscript. I particularly thank my students at UCLA who have used the material in preliminary form and whose comments and criticisms I have found especially helpful.

Los Angeles, California                               Philip C. Curtis, Jr.
August 1971

# CONTENTS

# GLOSSARY OF SYMBOLS AND FORMULAS

# CALCULUS

# CHAPTER ONE

# INTRODUCTION

## 1.1 Set Terminology and Notation

The so-called "theory of sets" invented in the latter part of the nineteenth century by the German mathematician, Georg Cantor, has widely influenced mathematical development during the twentieth century and has provided a very useful terminology that is applicable to every branch of mathematics. At this point we shall not discuss the subject in depth but shall merely provide a few of the elementary ideas in an informal way in order to take advantage of the notation.

Mathematics as a deductive science must be developed from certain undefined concepts. Strictly speaking, "set," and "member of a set" fall into that category. Thus, when we give synonyms for the word "set," we are attempting to convey informally the idea involved and are not attempting to give a mathematical definition of the term. By a set we mean a collection of objects that is completely determined by those objects belonging to it. We shall use capital letters $A, B, C, \ldots$ to denote sets and small letters $x, y, z, \ldots$ to denote the object belonging to the set. Then $A = \{1,3,5,\ldots\}$ expresses the fact that $A$ is the set of odd positive integers. Each odd integer $x$ is a member, or belongs to, the set $A$. To express this fact, we write

$$x \in A.$$

If an object $x$ is not a member of the set $A$, then we write

$$x \notin A.$$

Thus, if $A$ is the set of odd positive integers, $2 \notin A$.

When we say that a set $A$ is completely determined by the objects it contains, this means that two sets $A$ and $B$ are the same if and only if they have the same members, and then we write

$$A = B.$$

The simplest way to specify a set is to list its members as shown in the example above. We shall always use braces $\{ \ldots \}$ for this purpose. If it is impossible to write down all members of a set, three dots as in $\{1,3,5, \ldots\}$, are added to indicate "and so on." Often, instead of listing the members, it is more convenient to specify a property that defines the set. Thus if $N = \{1,2,3, \ldots\}$ is the set of positive integers, the set $B = \{1,2, \ldots, 10\}$ can be written

$$B = \{x \in N : 1 \leqslant x \leqslant 10\}.$$

This is read: $B$ is the set of those numbers $x$ belonging to $N$ such that $x$ is greater than or equal to 1 and less than or equal to 10. The colon: stands for "such that." In general, if $P(x)$ is a property that is either true or false for each $x$ belonging to a set $S$, then the set $T$ of those members of $S$ for which the property is true would be written

$$T = \{x \in S : P(x) \text{ is true}\}.$$

If it is clearly understood to which set $S$ the property $P(x)$ applies, then we often write just

$$T = \{x : P(x) \text{ is true}\}.$$

Thus if $R$ is the set of all real numbers,

$$\{x \in R : x^2 - x = 0\} = \{0,1\},$$

or more briefly

$$\{x : x^2 - x = 0\} = \{0,1\}.$$

Similarly

$$\{x \in R : x > 0 \text{ and } x^2 = 25\} = \{5\}.$$

We could have written the left-hand side as

$$\{x > 0 : x^2 = 25\}$$

or

$$\{x : x > 0 \text{ and } x^2 = 25\}.$$

It often happens that a set has just one member as in the example above. In such cases, we distinguish between the set and its single member. Thus $5 \in \{5\}$, but we do *not* write $5 = \{5\}$. There are technical reasons for making this distinction which we shall not investigate. In nonmathematical situations this convention is also observed. For example, if $A$ is a class with one student $x$, then $x \in A$; but it is *not* true that $x = A$.

As long as the property $P(x)$ is either true or false for each member $x$ of the set $S$, we insist that

$$\{x \in S : P(x) \text{ is true}\}$$

always defines a set $T$. If $P(x)$ is always false for $x$ in $S$, then we are led to a set with no members, called the *empty set* or the *null set*. This set is written $\emptyset$. Thus

$$\{x \in R : x^2 = -1\} = \emptyset.$$

If $S$ and $T$ are sets and each member of $S$ is a member of $T$, then we say that $S$ is a subset of $T$, or that $T$ contains $S$. This is written

$$S \subset T \qquad \text{or} \qquad T \supset S.$$

Thus $S \subset T$ if $x \in S$ implies that $x \in T$. We do not exclude the possibility that $S = T$. If, however, $S \subset T$ and $S \neq T$, then we say $S$ is a *proper subset* of $T$. Note that the null set $\emptyset$ is a subset of every set. (Why?)

Last, we define three operations on sets. If $S$ and $T$ are sets, then by $S \cup T$ (read $S$ *union* $T$), we mean

$$\{x : x \in S \text{ or } x \in T\}.$$

Note the use of the word "or.". In mathematics the word "or" always is taken in the nonexclusive sense. That is, the word "or" always stands for either-or, or both. Thus if $x \in S \cup T$, it may happen that $x \in S$ and $x \in T$.

By $S \cap T$ (read $S$ *intersection* $T$), we mean

$$\{x : x \in S \text{ and } x \in T\}.$$

Thus $S \cap T$ is the totality of elements belonging to both $S$ and $T$. Finally we define the *complement of T in S*. This set, written $S - T$, is defined to be $\{x : x \in S \text{ and } x \in T\}$. If $S = \{1,2,3,5\}$ and $T = \{2,4,5,7\}$, then

$$S \cup T = \{1, 2, 3, 4, 5, 7\}$$
$$S \cap T = \{2, 5\}$$
$$S - T = \{1, 3\}.$$

These three operations on sets may be illustrated diagramatically in the following way. Let $S$, $T$ be those points in the plane inside the following figures (Figure 1). The shaded portion denotes the appropriate set. These figures are called *Venn diagrams*.

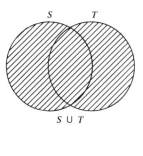

$S \cup T$

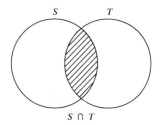

$S \cap T$

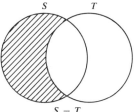

$S - T$

Fig. 1

**EXERCISES**

Let $A = \{1, 2, 4, 7\}$ $B = \{3, 4, 8\}$ $C = \{2, 7, 9\}$.

1. For $A, B, C$, as defined above, determine the following sets.
    (a) $(C \cap A) \cup (C \cap B)$
    (b) $(A \cap B) \cup C$
    (c) $(A \cup B) \cup (A \cap C)$
    (d) $(A \cup B) \cap (A - C)$
2. For the above sets $A, B, C$, are the following identities valid?
    (a) $C \cup (A \cap B) = (C \cup A) \cap (C \cup B)$
    (b) $A \cap (B \cup C) = A \cup (B \cap C)$
3. Determine all the subsets of

$$\{1, 2, 3\}.$$

There are eight in all.

4. Let $N = \{1, 2, 3, \ldots\}$ denote the set of positive integers, and let $R$ denote the set of all real numbers. Determine the following sets by listing the members.
    (a) $\{x \in N : x^2 + 1 \leqslant 17\}$
    (b) $\{x \in N : \pi x^2 = 1\}$
    (c) $\{x \in R : \pi x^2 = 1\}$
    (d) $\{x \in R : x^2 + x + 1 = 0\}$
5. Let # be an operation on sets. That is, if $S$ and $T$ are sets, $S \# T$ is also a set. For example, $\cup$, $\cap$, $-$ are operations on sets. The operation # is said to be

*associative* if, for each triple of sets $S$, $T$, $U$,

$$S \, \# \, (T \, \# \, U) = (S \, \# \, T) \, \# \, U.$$

The operation $\#$ is *commutative* if, for each pair of sets $S$ and $T$,

$$S \, \# \, T = T \, \# \, S.$$

Which of the operations $\cup$, $\cap$, $-$ on sets are associative? Which are commutative? Justify your answer. If the appropriate identity is incorrect, give an example for which it fails.

6. Let $A_1, \ldots, A_n$ be sets. The set $A_1 \cup \cdots \cup A_n$ is defined to be the set consisting of those elements belonging to at least one of the sets $A_k$. Give a similar definition for $A_1 \cap \cdots \cap A_n$.

7. To establish an identity for sets, such as

$$A \cup (A \cap B) = A$$

we proceed as follows: Let $X = A \cup (A \cap B)$.
To show $X = A$, it suffices to show that $A \subset X$ and $X \subset A$. Certainly $A \subset X$.
To show $X \subset A$, let $x \in X$. Then $x \in A$ or $x \in A \cap B$. But if $x \in A \cap B$, then $x \in A$. Hence $x \in X$ implies $x \in A$. This means $x \subset A$.

Establish the following identities for sets.
(a) $A \cap (A \cup B) = A$
(b) $A \cap (B \cup C) = (A \cap B) \cup (A \cap C)$
(c) $A \cup (B \cap C) = (A \cup B) \cap (A \cup C)$
(d) $A - (B \cup C) = (A - B) - C$
(e) $A - (B \cap C) = (A - B) \cup (A - C)$
(f) $(A \cup B) - C = (A - C) \cup (B - C)$

To show that an identity is false one must give an example of sets for which it fails. Thus the identity $A - (B \cup C) = (A - B) \cup (A - C)$ is in general false, for if $A = \{1, 2, 3\}$ $B = \{2, 4\}$ $C = \{3, 5\}$, then

$$A - (B \cup C) = \{1\},$$

and

$$(A - B) \cup (A - C) = \{1, 2, 3\}.$$

An example which illustrates that a given statement is not always true is called a *counter example* to that statement (Figure 2). Inspection of an appropriate Venn diagram is often helpful in deciding if an identity is true or false. For the above example the shaded portion is the desired set.

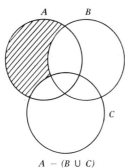 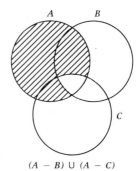

$A - (B \cup C)$                 $(A - B) \cup (A - C)$

Fig. 2

8. Prove or disprove the following identities for sets $R$, $S$, $T$. If the identity is false, provide a counter example.

(a) $R - (S \cap T) = (R - S) \cap (R - T)$
(b) $(R - S) - T = (R - S) \cup (R - T)$

(c) $(R \cup S) - T = (R-T) \cup (S-T)$
(d) $R - (S-T) = (R-S) - T$
(e) $(R \cap S) - T = (R-T) \cup (S-T)$

9. Show that there is only one null set. (If $\emptyset$ and $\emptyset'$ are null sets, show that $\emptyset \subset \emptyset'$ and $\emptyset' \subset \emptyset$.)

## 1.2 Real Numbers

The set of real numbers will play a central role in our development of the calculus. Here we shall informally discuss the axioms that describe the real numbers. These can be divided into three groups: arithmetic, order, and completeness. The axioms of arithmetic, or the field axioms, are the properties of addition and multiplication. We only list these and assume that the consequences are familiar to the student. The axioms of order are also familiar but are discussed in more detail. This discussion is presented because these axioms govern all calculations with inequalities, and these calculations are extremely important in our later work. The completeness axiom is probably unfamiliar and, since it occupies a central position in the development of the calculus, it will also be considered in detail.

1. *Field Axioms or Axioms of Arithmetic*

The set of real numbers $R$ is closed under the operations addition $+$, and multiplication $\cdot$. That is, for each pair of real numbers $x$ and $y$, $x+y$, and $x \cdot y$ are again real numbers. For all real numbers $x$, $y$, and $z$, these operations satisfy the following.

(a) The associative law: $(x+y)+z = x+(y+z)$ and $(x \cdot y) \cdot z = x \cdot (y \cdot z)$.
(b) The commutative low: $x+y = y+x$ and $x \cdot y = y \cdot x$.
(c) The distributive law: $x \cdot (y+z) = x \cdot y + x \cdot z$.

Each operation $+$, $\cdot$, has an identity, 0, 1 respectively, satisfying

$$0 + x = x \text{ and } 1 \cdot x = x \text{ for each real number } x.$$

Each real number $x$ has an additive inverse $-x$ satisfying

$$x + (-x) = 0.$$

Each real number $x \neq 0$ has a multiplicative inverse $x^{-1}$ satisfying

$$x \cdot x^{-1} = 1.$$

The familiar rules of arithmetic that follow from these axioms are assumed known and will be used without further justification.

2. *Order Axioms*

There is a subset $R^+$ of $R$, called the set of positive numbers, with the following properties.

(i) $1 \in R^+, 0 \notin R^+$.
(ii) If $x, y \in R^+$, then $x+y \in R^+$ and $x \cdot y \in R^+$.
(iii) If $x$ is a real number and $x \neq 0$, then $x \in R^+$ or $-x \in R^+$, but not both.

For real numbers $x$ and $y$, we write $x > y$ (read $x$ *is greater than* $y$) if $x - y \in R^+$. Equivalently, we may write $y < x$ (read $y$ *is less than* $x$). Thus, $x \in R^+$ if $x > 0$, and $-x \in R^+$ if $x < 0$. The set of negative numbers $R^-$, is defined to be the set $\{x \in R : -x \in R^+\}$. The following properties of the symbol $>$ are easily established.

1. If $x > y$ and $y > z$, then $x > z$.
2. If $x > y$, then $x + a > y + a$ for each real number $a$.
3. If $x > y$ and $a$ is positive, then $ax > ay$.
4. If $x > y$ and $a$ is negative, then $ay > ax$.

Property (4) expresses the well-known fact that if an inequality is multiplied by a negative number, the sense of the inequality is reversed.

The first three of these properties are left as exercises. To prove (4), suppose $x > y$. Then $x - y > 0$. If the number $a$ is negative, then $-a \in R^+$. Therefore, by (ii), $(-a)(x - y) > 0$. But

$$(-a)(x - y) = ay - ax.$$

Hence

$$ay > ax.$$

Properties (1) to (4) hold as well for the "less than" symbol $<$. Observe that (1) and (2) together imply

5. If $a < b$ and $x < y$, then $a + x < b + y$.

As an immediate consequence of the properties of $>$, we have the following classification of products $xy$ of real numbers.

(a) If $xy > 0$, then either $x > 0$ and $y > 0$, or $x < 0$ and $y < 0$. Conversely, if either $x > 0$ and $y > 0$, or $x < 0$ and $y < 0$, then $xy > 0$.

(b) If $xy < 0$, then either $x > 0$ and $y < 0$, or $x < 0$ and $y > 0$. Conversely, if either $x > 0$ and $y < 0$, or $x < 0$ and $y > 0$, then $xy < 0$.

**Example 1.**    Determine

$$\{x : x^2 + 2 > 3x\}.$$

This is equivalent to finding those numbers $x$ for which $x^2 - 3x + 2 > 0$. But $x^2 - 3x + 2 = (x - 2)(x - 1)$. By (b) above $(x - 2)(x - 1) > 0$ if

$$x - 2 > 0 \qquad \text{and} \qquad x - 1 > 0$$

or if

$$x - 2 < 0 \qquad \text{and} \qquad x - 1 < 0.$$

If $x - 2 > 0$ and $x - 1 > 0$, then $x > 2$ and $x > 1$. Hence $x > 2$. On the other hand if $x - 2 < 0$ and $x - 1 < 0$, then $x < 2$ and $x < 1$, which means $x < 1$. Thus

$$\{x : x^2 + 2 > 3x\} = \{x : x < 1\} \cup \{x : x > 2\}.$$

**Example 2.**    Determine

$$\left\{ x : \frac{x - 1}{x - 2} < 1 \right\}.$$

Again

$$\frac{x - 1}{x - 2} < 1 \qquad \text{if} \qquad \frac{x - 1}{x - 2} - 1 < 0.$$

But

$$\frac{x - 1}{x - 2} - 1 = \frac{x - 1 - x + 2}{x - 2} = \frac{1}{x - 2}.$$

Therefore, by (c) above,

$$\frac{1}{x - 2} < 0 \qquad \text{if} \qquad x - 2 < 0 \qquad \text{or} \qquad x < 2.$$

Therefore

$$\left\{ x : \frac{x - 1}{x - 2} < 1 \right\} = \{x : x < 2\}.$$

If $x > y$ or $x = y$, then we abbreviate this by writing $x \geqslant y$, (read $x$ is greater than or equal to $y$). Similarly, we abbreviate $x < y$ or $x = y$ by writing $x \leqslant y$, (which is read $x$ is less than or equal to $y$). The inequalities $>$, $<$ are called *strict* inequalities; $\geqslant$, $\leqslant$ are called *weak* inequalities.

The student should check that properties (1) to (4) hold for $\leqslant$ and $\geqslant$. Also, if $xy \geqslant 0$, then $x \geqslant 0$ and $y \geqslant 0$, or $x \leqslant 0$ and $y \leqslant 0$. If $xy \leqslant 0$, then $x \geqslant 0$ and $y \leqslant 0$, or $x \leqslant 0$ and $y \geqslant 0$. The real numbers $x$ satisfying $x \geqslant 0$ are called the *nonnegative* real numbers. Those that satisfy $x \leqslant 0$ are called *nonpositive*.

**Example 3.** Show that for all real numbers $x$, $x^2 - 2x + 5 > 0$. We know from elementary algebra that the polynomial $ax^2 + bx + c$ can be written in the form $a(x - x_0)(x - x_1)$ where $x_0$ and $x_1$ are real numbers if and only if the discriminant $b^2 - 4ac \geqslant 0$. In this case $b^2 - 4ac = 4 - 20 = -16 < 0$. Hence, we cannot write $x^2 - 2x + 5 = (x - x_0)(x - x_1)$ and, as a result, the technique of Example 1 is not applicable. Instead we observe that $x^2 - 2x + 5$ can be written as the sum of two quantities, both of which must be positive by the device of completing the square.

$$x^2 - 2x + 5 = x^2 - 2x + 1 + 4 = (x - 1)^2 + 4 \geqslant 4.$$

Thus $x^2 - 2x + 5$ has a minimum value of 4 when $x = 1$. Hence for all values of $x$, $x^2 - 2x + 5 > 0$.

## EXERCISES

1. Establish properties (1) to (3), (5) for the symbol $>$.
2. Establish properties (1) to (5) for $\geqslant$.
3. Establish properties (a), (b) for the strict inequalities $>$, $<$ respectively.
4. Establish properties (a), (b) for the weak inequalities $\geqslant$, $\leqslant$ respectively.
5. Determine the sets of numbers $x$ for which the following inequalities are valid.
    (*Note:* $a < x < b$ means $a < x$ and $x < b$).

   (a) $x^2 - 7x + 12 > 0$

   (b) $3x - x^2 - 2 > 0$

   (c) $x^2 + x + 1 < 0$

   (d) $x(x^2 - 1) > 0$

   (e) $1 < 2x - 3 < 4$

   (f) $-1 \geqslant 4x - 7 \geqslant -2$

   (g) $1 < x^2 - 4x + 1 < 2$

   (h) $\dfrac{1}{2x + 1} < -3$

   (i) $\dfrac{10 - x}{4x - 7} < 5$

   (j) $\dfrac{2x^2 - 1}{x + 2} < 3$

6. Show that for all values of $x$:

   (a) $x^2 + 2x + 9 > 0$

   (b) $2x^2 - 3x + 4 > 0$

   (c) $2x^2 - x > -2$

   (d) $x < x^2 + 4$

   (e) $5x - 2 < 4x^2$

   (f) $(x - 1)^2 > x - 4$

7. For what value of $x$ does $x^2 - 2x + 7$ take on its smallest value?
8. For what value of $x$ does $x - (x^2 + 1)$ take on its largest value?
9. Determine a condition on the constants $a$ and $b$ such that for all real numbers $x$

   $$x^2 + ax + b > 0.$$

10. Determine a condition on the constants $a$, $b$, $c$ such that for all real number $x$, $ax^2 + bx + c < 0$.

11. Is it possible to determine a constant $a$ such that for all values of $x$

    $$\frac{1}{ax + 1} > 1?$$

12. Determine a condition on the constants $a$ and $b$ such that for all values of $x$

    $$\frac{1}{ax + b} > 1.$$

### 1.3    Subsets of the Real Numbers

We shall denote the real numbers by $R$, and we shall always let

$$N = \{1, 2, 3, \ldots\}$$

denote the set of *positive integers* or whole numbers.

$$I = \{\cdots -2, -1, 0, 1, 2, \cdots\}$$

is the set of *integers*.

$$Q = \left\{\frac{p}{q} : p, q \in I, q \neq 0\right\}$$

is the set of *rational numbers* or common fractions. (We shall show in Section 1.5 that $Q$ is a proper subset of $R$.) A real number $x$ such that $x \notin Q$ is called *irrational*.

The set $\{x \in R : a < x < b\}$, which we write $(a, b)$, is called an *open* interval. The set $\{x \in R : a \leq x \leq b\}$, written $[a, b]$, is called a *closed* interval. Thus, for intervals the parenthesis sign ( is used if the end point is *not* included in the interval. The bracket symbol [ is used if the end point *is* included in the interval. Hence an interval is open if neither end point is included. An interval is closed if both end points are included. The intervals

$$\{x : a \leq x < b\} = [a, b)$$
$$\{x : a < x \leq b\} = (a, b]$$

are called half-open intervals.

We assume that the representation of the real numbers as points on a line is familiar to the student from plane geometry. Let us recall the construction (Figure 3). Fix a point $p_0$ on the line $l$ to serve as the origin, and a point

$$\cdots = p_{-2}p_{-1} = p_{-1}p_0 = p_0p_1 = p_1p_2 = \cdots$$

**Fig. 3**

$p_1$ to the right of $p_0$ to which we associate the number 1. Let $p_2$ be to the right of $p_1$ such that the segment $p_1p_2$ has the same length as $p_0p_1$. Associate the number 2 with $p_2$. Continuing in this way we have a representation for the nonnegative integers $\{0, 1, 2, \ldots\}$. If the analogous construction is carried out to the left we have the negative integers.

This construction may be extended in an easy way so as to identify each rational number $p/q$ with a unique point on the line $l$. We leave this construction to the exercises. However, we also assume that each real number $x$ corresponds to a unique point on the line, and also that to each point on the line there corresponds a unique real number. Thus, we may simplify our notation by writing $x$, rather than $p_x$ to denote the point on the line $l$ associated with the number $x$.

This representation gives us a means for "picturing" inequalities. Thus $x > y$ if the point $x$ lies to the right of the point $y$, and $x < y$ if $x$ lies to the left of $y$. Moreover a closed interval $[a, b]$ consists of all points on the line $l$ lying between $a$ and $b$ and including the points $a$ and $b$ (Figure 4). The open

**Fig. 4**   Geometric representation of the interval $[a, b]$.

interval $(a, b)$ is the set of points between $a$ and $b$ *excluding* the points $a$ and $b$.

**1.4   Absolute Values and the Triangle Inequality**

For each real number $x$ we define $|x|$, (read *the absolute value of x*) by the formula

$$|x| = x \quad \text{if} \quad x \geq 0$$
$$|x| = -x \quad \text{if} \quad x < 0.$$

Thus $|3| = 3$, $|-2| = -(-2) = 2$, and $|0| = 0$ etc. $|x|$ is often read the *magnitude* of $x$ or the *modulus* of $x$. We list some of the properties of $|x|$. The proofs of the easier statements are left as exercises.

**THEOREM 1.1**   *For each real number x,*

$$|x| \geq 0,$$

$$|x| = |-x|,$$

*and*

$$-|x| \leq x \leq |x|.$$

**THEOREM 1.2**   *For each pair of real numbers x and a,*

$$|x| \leq a \text{ if and only if} -a \leq x \leq a.$$

*Thus,* $\{x : |x| \leq a\}$ *is the closed interval* $[-a, a]$.

**Proof.**   We shall prove only Theorem 1.2 leaving Theorem 1.1 as an exercise. However, before beginning the argument, let us elaborate on the expression "if and only if." For statements $P$ and $Q$, the conditional statement "if $P$ then $Q$" means that $P$ implies $Q$ or that $Q$ follows as a consequence of $P$. The statement "$Q$, only if $P$" means that $Q$ implies $P$ or that $P$ follows as a consequence of $Q$. The statement "$P$ if and only if $Q$" is just a shorthand way of writing both of these statements together. Thus to verify "$P$ if and only if $Q$" we must prove that $P$ implies $Q$ and that $Q$ implies $P$. The phrase "if and only if" has the convenient abbreviation "iff."

To prove Theorem 1.2, we assume first that $|x| \leq a$. Then we must show that $-a \leq x \leq a$. In all cases, $|x| = x$ or $|x| = -x$. Hence, if $|x| \leq a$ and $|x| = x$, then $0 \leq x = |x| \leq a$. Consequently, $-a \leq x \leq a$. If $|x| \leq a$ and $|x| = -x$, then $0 \leq -x = |x| \leq a$. Multiplying by $-1$ we have $-a \leq x \leq 0 \leq a$. Thus, in both cases, $-a \leq x \leq a$.

Conversely, we must show that if $-a \leq x \leq a$, then $|x| \leq a$. But if $|x| = x$, then $x \leq a$ implies $|x| \leq a$. If $|x| = -x$, then $x \geq -a$ implies $-|x| \geq -a$. Multiplying by $-1$ we have $|x| \leq a$. This completes the proof.

If we note that $|x| = a$ if and only if $x = a$ or $x = -a$, then Theorem 1.2 has the following corollary.

**COROLLARY.**   *For each pair of real numbers x and a,* $|x| < a$ *iff* $-a < x < a$. *Thus,* $\{x : |x| < a\}$ *is the open interval* $(-a, a)$.

Using the same type of arguments we may derive the following result. We leave the proof as an exercise.

**THEOREM 1.3**   *For each pair of real numbers x and a, $|x| \geq a$ if and only if $x \geq a$ or $x \leq -a$. Thus*

$$\{x : |x| \geq a\} = \{x : x \geq 0\} \cup \{x : x \leq -a\}.$$

If we represent the real numbers as points on a line, then the sets $\{x : |x| \leq a\}$ and $\{x : |x| \geq a\}$ are represented as in Figures 5 and 6.

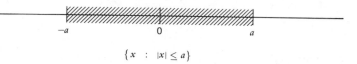

$$\{ x \ : \ |x| \leq a \}$$

**Fig. 5**

$$\{ x \ : \ |x| \geq a \}$$

**Fig. 6**

**THEOREM 1.4    (Triangle Inequality).**   *For each pair of real numbers x and y*

$$|x+y| \leq |x| + |y|.$$

**Proof.**   From Theorem 1.1

$$- |x| \leq x \leq |x| \text{ and} - |y| \leq y \leq |y|.$$

Adding these inequalities [property (5) of $\leq$] yields

$$-(|x| + |y|) \leq x + y \leq |x| + |y|.$$

Applying Theorem 1.2 we have

$$|x+y| \leq |x| + |y|.$$

**THEOREM 1.5**   *For each pair of real numbers x and y*

$$\pm (|x| - |y|) \leq \big||x| - |y|\big| \leq |x \pm y| \leq |x| + |y|.$$

**THEOREM 1.6**   *For each pair of real numbers x and y*

$$|xy| = |x| \cdot |y|.$$

The proofs of 1.5 and 1.6 are left as exercises at the end of the section. We illustrate the properties of $|x|$ with the following examples.

*Example 1.*   Determine

$$\{x : |2x+3| < 1\}.$$

By the corollary to Theorem 1.2, $|2x+3| < 1$ if and only if $-1 < 2x+3 < 1$. Subtracting 3 and dividing by 2 yields $-2 < x < -1$. Therefore

$$\{x : |2x+3| < 1\} = (-2, -1).$$

*Example 2.*   Determine

$$\{x : |x^2 - 3x + 2| < 2\}.$$

Again by the corollary to Theorem 1.2, $|x^2 - 3x + 2| < 2$ if and only if

$$-2 < x^2 - 3x + 2 < 2.$$

Hence, $\{x : |x^2 - 3x + 2| < 2\} = \{x : x^2 - 3x + 2 < 2\} \cap \{x : x^2 - 3x + 2 > -2\}$. First, $x^2 - 3x + 2 < 2$ iff $x(x - 3) = x^2 - 3x < 0$. This statement holds iff either $x > 0$ and $x - 3 < 0$ or $x < 0$ and $x - 3 > 0$. There are no values of $x$ satisfying the latter pair of inequalities. The first pair of inequalities will be satisfied if $0 < x < 3$. Second, $x^2 - 3x + 2 > -2$ iff $x^2 - 3x + 4 > 0$. This cannot be factored, so completing the square as before, we obtain

$$x^2 - 3x + 4 = x^2 - 3x + \tfrac{9}{4} + \tfrac{7}{4} = (x - \tfrac{3}{2})^2 + \tfrac{7}{4} \geqslant \tfrac{7}{4} > 0.$$

Therefore, for all values of $x$, $x^2 - 3x + 2 > -2$. Combining the two results, we get

$$\{x : |x^2 - 3x + 2| < 2\} = (0,3).$$

**Example 3.**   Show that if $|x - 1| < 1$, then $|x^2 - 1| < 3$. By Theorem 1.6 $|x^2 - 1| = |(x + 1)(x - 1)| = |x + 1||x - 1| < |x + 1|$ if $|x - 1| < 1$. We are finished if we can show that $|x - 1| < 1$ implies $|x + 1| < 3$. Now $|x - 1| < 1$ iff $-1 < x - 1 < 1$. Adding 2 yields $-3 < 1 < x + 1 < 3$ or

$$|x + 1| < 3.$$

Therefore, $|x^2 - 1| = |x + 1||x - 1| < 3$ if $|x - 1| < 1$.

**Example 4.**   Find a positive number $\delta$ so that $|x - 2| < \delta$ and $\delta < 2$ imply

$$|x^2 - 4| < 1.$$

Now $|x^2 - 4| = |x - 2||x + 2| < \delta|x + 2|$ if $|x - 2| < \delta$. But if $\delta < 2$, then this implies that

$$|x - 2| < 2$$

or

$$-2 < x - 2 < 2.$$

To estimate $|x + 2|$ we add 4 to this inequality, obtaining

$$2 < x + 2 < 6.$$

Therefore, if $|x - 2| < \delta < 2$, then $|x + 2| < 6$ and, consequently, $|x^2 - 4| < 6\delta$. If $6\delta < 1$ or $\delta < 1/6$, it follows that $|x^2 - 4| < 1$. Therefore, we have shown that if $\delta < 1/6$, and $|x - 2| < \delta$, then $|x^2 - 4| < 1$. Notice that the number $\delta$ is not unique. Any smaller positive number will do just as well.

## EXERCISES

1. Prove Theorem 1.1 by using the definition of $|x|$ and appropriate properties of inequalities.
2. Prove Theorem 1.3.
3. Prove Theorem 1.6.
4. Prove Theorem 1.5. There are several statements involved so we separate the proof into several parts.
   (a) To prove $|x - y| \leqslant |x| + |y|$, write $x - y = x + (-y)$ and apply Theorem 1.4
   (b) Prove $\pm (|x| - |y|) \leqslant ||x| - |y||$
   (c) To prove $|x| - |y| \leqslant |x - y|$, write $x = x - y + y$ and apply Theorem 1.4
   (d) Prove $|x| - |y| \leqslant |x + y|$
   (e) Prove $||x| - |y|| \leqslant |x \pm y|$
5. Determine the set of those numbers $x$ for which the following conditions are satisfied.
   (a) $|2 - 3x| < 1$
   (b) $|4x + 3| \geqslant -1$
   (c) $|2x - 1| = 1$
   (d) $|3x + 1| = -1$
   (e) $|x^2 - 1| \geqslant 8$
   (f) $|2x^2 + 3x - 2| < 1$
   (g) $|6x - 3x^2| \leqslant 4$
   (h) $|x^3 + x| \geqslant 2$

(i) $|x^4 - 4| < 1$

(j) $2|x+1|^2 - 9|x+1| + 4 = 0$

(k) $3|x-3|^2 - |x-3| - 4 = 0$

(l) $|x-1|^2 + 3|x-1| + 2 = 0$

6. Verify the following implications.

(a) $|x-2| < 1$ implies $|3x-6| < 3$

(b) $|x+2| < 1$ implies $|x^2-4| < 5$

(c) $|x-1| < 2$ implies $|x^2+x-2| < 10$

(d) $|x+1| < 1$ implies $|x^3-x| < 6$

(e) $|x| < a$ implies $|x^3+x^2| < a^3+a^2$

7. Verify the following implications for a fixed positive number $\delta$.

(a) $|x-3| < \delta$ implies $|3x-9| < 3\delta$

(b) $|x-1| < \delta < 1$ implies $|x^2-1| < 3\delta$

(c) $|x+2| < \delta < 2$ implies $|x^2-x-6| < 7\delta$

8. Find an appropriate positive number $\delta$ so that the following implications are satisfied.

(a) $|x-1| < \delta$ implies $|4x-4| < 1$

(b) $|x+1| < \delta$ and $\delta < 1$ imply $|x^2-1| < 1/2$

(c) $|x-1| < \delta$ and $\delta < 1$ imply $|2x^2-4x+2| < 1/10$

(d) $|x+2| < \delta$ and $\delta < 2$ imply $|x^2-4| < \epsilon$ where $\epsilon$ is an arbitrary positive number

9. In our discussion of the representation of real numbers, we gave an actual construction for the integers $\{\ldots -2, -1, 0, 1, 2, \ldots\}$. This could be accomplished by marking off on a fixed line with a compass, intervals of equal length. As a review of elementary geometry, let us construct the rational numbers on the line $l$ with a ruler and compass (Figure 7).

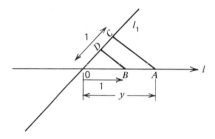

**Fig. 7**

First, for a positive integer $y$, we construct $1/y$ in the following manner. Draw a second line $l_1$ through the origin of the number line $l$ and mark off a unit distance $OC$.

If $OB$ and $OA$ on $l$ correspond to 1 and $y$ respectively, show that $OD$ corresponds to $1/y$ by exploiting the similarity of the triangles involved.

10. Convince yourself that this construction may be carried out by ruler and compass.

11. Figure out a corresponding construction for $x \cdot y$ if $x > 0$ and $y > 0$.

Since, for $x$ given, we can construct $-x$ (how?) we have now a method for constructing each rational number.

12. Give a ruler and compass construction for $\sqrt{2}$ and for $\sqrt{r}$ where $r$ is a positive rational number.

A logical question at this point is: Can the point on the line $l$ corresponding to each real number be constructed by ruler and compass? The answer is no. In fact $\sqrt[3]{2}$ cannot be constructed in this fashion. Indeed this is the arithmetic analogue of the famous geometrical result that it is impossible to construct by ruler and compass a cube having twice the volume of the unit cube. To prove this would take us too far afield. For a complete discussion of these matters, see *What Is Mathematics?* by R. Courant and H. Robbins, Oxford University Press. pp. 120–140.

**1.5 Completeness of the Real Numbers**

The field and order axioms that we have given so far are not sufficient to characterize the set of real numbers. Indeed, one cannot conclude from these axioms that there exists a solution to the simple quadratic equation $x^2 = 2$. It is clear that the field and order axioms hold for the rational numbers, but we shall show presently that there cannot exist a common fraction or rational number $x$ that is a solution to the equation $x^2 = 2$. However, if we consider the real numbers as corresponding to points along a line, then of course there does exist a point on that line whose distance $x$ from the origin satisfies $x^2 = 2$. Indeed a ruler and compass construction for that point is Exercise 12 of Section 1.4. Unfortunately not all points on a line, hence not all real numbers, can be constructed by the use of a ruler and compass.

What we need is an axiom for the real numbers that can be used in a wide variety of situations to assert the existence of a real number with certain desired properties. It is rather remarkable that the following postulate, which intuitively says that there are no gaps or holes in the real numbers, is sufficient. Before stating this completeness axiom we need some terminology. Let $S$ and $T$ be nonempty sets of real numbers. If for each $s \in S$ and $t \in T$ it follows that $s \leq t$, then we call the sets $S$ and $T$ *separated*. If, further, there is a number $m$ satisfying $s \leq m \leq t$ for all numbers $s \in S$ and $t \in T$, then we say that the number $m$ *separates* the sets $S$ and $T$. Intuitively, a number $m$ separates the sets $S$ and $T$ if all the numbers in $S$ are one one side of $m$ and all the numbers of $T$ are on the other. Thus $m$ is "in between" $S$ and $T$. The desired completeness axiom can now be stated as follows.

**Completeness Axiom for the Real Numbers**

> *Let $S$ and $T$ be nonempty sets of real numbers that are separated. Then there exists at least one real number $m$ that separates the sets $S$ and $T$.*

The power of this completeness or separation axiom is not immediately apparent. Its utility stems from the infinite variety of ways that the sets $S$ and $T$ may be specified. We shall use the axiom many times in the course of our work.

***Example 1.*** Let $S = \{x \in R : x < 2\}$, $T = \{x \in R : x > 2\}$. Clearly, $s \leq t$ if $s \in S$ and $t \in T$. Hence, $S$ and $T$ are separated. Indeed, the number 2 separates $S$ and $T$ since $s \leq 2 \leq t$ for each $s \in S$ and $t \in T$. Notice, however, that the number 2 does not belong to either $S$ or $T$ (Figure 8).

**Fig. 8** The number 2 separates $S$ and $T$.

***Example 2.*** Let $S = \{x \in R : x \leq 2\}$ and $T = \{x \in R : x \geq 2\}$. The sets $S$ and $T$ are still separated. Indeed, they are separated by the number 2, which in this case belongs to both $S$ and $T$. Thus separated sets $S$ and $T$ may overlap, but if they do, $S \cap T$ must contain only one point. (Why?)

***Example 3.*** There may be infinitely many numbers "sitting in between" two separated sets $S$ and $T$.

For example, if $S = \{x \in R : x \leq 2\}$ and $T = \{x \in R : x \geq 3\}$ then $S$ and $T$ are clearly separated. If $2 \leq y \leq 3$ then

$$s \leq y \leq t$$

for each $s \in S$ and $t \in T$, and thus every number $y$ between 2 and 3 separates $S$ and $T$. This is illustrated in Figure 9.

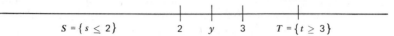

$S = \{s \leq 2\}$     2     $y$     3     $T = \{t \geq 3\}$

**Fig. 9**   If $2 \leqslant y \leqslant 3$, then $y$ separates $S$ and $T$.

The last example illustrates the fact that if there are two numbers $x$ and $y$ in between two separated sets $S$ and $T$, then the numbers $s \in S$ and $t \in T$ cannot be arbitrarily close together. This is clear because if

$$s \leqslant x < y \leqslant t$$

for each $s \in S$ and $t \in T$, then

$$t - s \geqslant y - x.$$

Thus the number $t$ and $s$ are farther apart than $y - x$. As a consequence we have an important condition which guarantees that there is exactly one number separating $S$ and $T$:

Suppose the sets $S$ and $T$ are separated. If for each positive number $\delta$, no matter how small, there are numbers $s \in S$ and $t \in T$ satisfying

$$t - s \leqslant \delta,$$

then there is *exactly* one number $m$ satisfying

$$s \leqslant m \leqslant t$$

for each $s \in S$ and $t \in T$.

The above examples are all quite trivial and do not involve the completeness axiom in an essential way since it is obvious from the start which numbers separate $S$ and $T$. A nontrivial and very important application of the completeness axiom is the following result. It asserts the existence of $\sqrt{2}$.

**THEOREM 1.7**   *There is a positive real number $x$ satisfying*

$$x^2 = 2.$$

**Proof.**   We produce $x$ as the number "in between" two separated sets $S$ and $T$. Let

$$S = \{x > 0 : x^2 < 2\}$$

and

$$T = \{x > 0 : x^2 > 2\}.$$

We assert that $S$ and $T$ are separated. These sets are clearly nonempty since $1 \in S$ and $2 \in T$. If $x \in S$ and $y \in T$, then we assert $x < y$. By definition   $x^2 < 2 < y^2$.   Therefore   $0 < y^2 - x^2 = (y - x)(y + x)$.   Since $x, y > 0$, this implies $y > x$. Thus $S$ and $T$ are separated.

Let $m$ separate the sets $S$ and $T$. We assert

$$m^2 = 2.$$

Certainly either $m^2 < 2$, $m^2 = 2$, or $m^2 > 2$. We shall show that the first and third alternatives are impossible. If $m^2 < 2$, then we claim that for a small enough positive number $\delta$, it will be true that

$$(m + \delta)^2 < 2.$$

**Fig. 10**

That this is possible is reasonably clear from Figure 10. However, to actually calculate how small $\delta$ must be in order for $(m+\delta)^2 < 2$ we proceed as follows. If $0 < \delta < 1$, then $\delta^2 < \delta$ and

$$(m+\delta)^2 = m^2 + 2m\delta + \delta^2 < m^2 + 2m\delta + \delta$$

$$= m^2 + (2m+1)\delta.$$

Thus, to prove $(m+\delta)^2 < 2$, it suffices to choose $\delta$ so that $m^2 + (2m+1)\delta < 2$. But now $m^2 + (2m+1)\delta < 2$ if

$$\delta < \frac{2-m^2}{2m+1}.$$

Hence, if

$$\delta < 1 \quad \text{and} \quad \delta < \frac{2-m^2}{2m+1},$$

it follows that $(m+\delta)^2 < 2$. On the other hand, it is impossible that $(m+\delta)^2 < 2$. If this were true, then

$$m+\delta \in S.$$

Now, by definition of $m$, $m \geq s$ for each $s \in S$. Hence, if $m+\delta \in S$, we must have $m \geq m+\delta$. This is clearly impossible since $\delta > 0$. Therefore the assumption that $m^2 < 2$ leads to a contradiction. Hence it is impossible that $m^2 < 2$. A similar line of reasoning shows that $m^2 > 2$ is also impossible. We leave the verification of this as an exercise. Since both the alternatives $m^2 < 2$ and $m^2 > 2$ cannot occur, it must follow that $m^2 = 2$, which was what we wanted to prove.

If $x^2 = 2$ and $x > 0$, then we denote $x$ by $\sqrt{2}$. The next result shows that this number cannot be a rational number. From this we conclude that there must be real numbers that cannot be written as the quotient of two integers. Such numbers we call *irrational*.

**THEOREM 1.8**   *If $x > 0$ and $x^2 = 2$, then $x$ is not a rational number.*

**Proof.**   Suppose $x^2 = 2$, and $x$ is a rational number. Then $x = p/q$, where $p$ and $q$ are integers. We may assume that $p$ and $q$ have no common factors. Now

$$x^2 = \frac{p^2}{q^2} = 2$$

implies $p^2 = 2q^2$. Therefore $p^2$ is even. We wish to conclude that $p$ is even. This follows since if $p$ were odd, $p^2$ would also be odd. Now if $p$ is even, then

$$p = 2r$$

for some integer $r$, and hence

$$p^2 = 4r^2.$$

Since $p^2 = 2q^2$, this implies that

$$q^2 = 2r^2,$$

or that $q^2$ is even. Therefore, $q$ is even, or $q = 2s$ for some integer $s$. Thus, if $x$ is rational, $p$ and $q$ must have 2 as a common factor. Since we have assumed that $p$ and $q$ have no common factor, we conclude that $x$ cannot be a rational number.

The fact that $\sqrt{2}$ is not rational was discovered by Greek geometers during the 5th century B.C. It was a very disconcerting discovery because one of the principal tenets of Pythagorean geometry had been that all natural phenomena could be described by whole numbers or their ratios. However, it was not until the 19th century that the concept of an irrational number was completely understood. The completeness axiom of this section was invented by the German mathematician Richard Dedekind, who published it in 1872 in a little book entitled *Continuity and Irrational Numbers*. An excerpt from this can be found in *The World of Mathematics* by James R. Newman, volume 1.

Exercise 16 shows that we may conclude from the arguments of Theorems 1.7 and 1.8 that the completeness axiom does not hold for the set of rational numbers. That is, there exist separated sets of rational numbers with no rational number separating them.

**EXERCISES**

Let $S$ and $T$ be sets of real numbers as specified below. Determine if these sets are separated. If they are not, explain why not. If $S$ and $T$ are separated, determine those numbers $m$ that separate $S$ and $T$.

1. $S = \{0, 2, 4, 6, \ldots\}$
   $T = \{1, 3, 5, \ldots\}$

2. $S = \{0, -1, -2, \ldots\}$
   $T = \{1, \frac{1}{2}, \frac{1}{3}, \ldots\}$

3. $S = \{x : |x-1| < 2\}$
   $T = \{x : |x-4| < 1\}$

4. $S = \{x : |x-2| < 1\}$
   $T = \{x : |x-4| < 2\}$.

5. $S = \{x : x^2 - x < 0\}$
   $T = \{x : x^2 - x > 0\}$.

6. $S = \{x : 1 - x^2 > 0\}$
   $T = \{x : x > 0 \text{ and } 1 - x^2 < 0\}$.

7. $S = \{x : x^2 - 5 < 0\}$
   $T = \{x : 11x - x^2 - 30 > 0\}$.

8. $S = \{x : x > 0 \text{ and } x^2 < 3\}$
   $T = \{x : x > 0 \text{ and } x^2 > 3\}$.

9. Show that in the proof of Theorem 1.7 it is impossible that $m^2 > 2$. (If $m^2 > 2$, then for a small enough positive number $\delta$, $(m-\delta)^2 > 2$. To determine how small to choose $\delta$, exploit the fact that

$$(m-\delta)^2 = m^2 - 2\delta m + \delta^2 > m^2 - 2\delta m.)$$

10. Show that there is a positive real number $x$ such that $x^2 = 3$. (Determine appropriate sets $S$ and $T$ and apply the argument of Theorem 1.7.)

11. Generalize the argument of Theorem 1.7 and show that for any positive real number $a$ there exists a real number $x$ satisfying $x^2 = a$.

12. Show that if $x^2 = 3$, then $x$ is not a rational number. (Replace the even-odd part of the argument of Theorem 1.8 by something that exploits the fact that a positive integer $n$ may be written $n = 3m$; $n = 3m + 1$ or $n = 3m + 2$ for some integer $m$.)

13. Show that if $x^2 = 5$, then $x$ is not a rational number. (This result may be generalized to the following: if $n$ is a positive integer and $x^2 = n$, then $x$ is irrational unless $n$ is the square of an integer. Thus the only rational solutions to the equation $x^2 = n$ occur when $n = 1, 4, 9, 16, \ldots$)

14. Let $S$ and $T$ be separated sets of real numbers. Show that the intersection of $S$ and $T$ is either empty or contains precisely one point.

15. For $k = 1, 2, 3, \ldots$ let $I_k = [a_k, b_k]$ be a closed interval. If $I_1 \supset I_2 \supset \cdots \supset I_n \supset \cdots$ prove that there always exist real numbers that are common to all of the intervals $I_k$. (Construct two separated sets of real numbers and apply the completeness axiom.)

16. Does the separation axiom hold for the set of rational numbers? That is, if $S$ and $T$ are separated sets of rational numbers, must there exist a rational number separating them?

## 1.6   Functions

One of the most important, if not the most important, concepts occurring in the study of the calculus is that of a function. Indeed, calculus may be thought of as an investigation of certain properties of functions. Elementary functions occur early in mathematics. For example, polynomial functions $f(x) = a_n x^n + a_{n-1} x^{n-1} + \cdots + a_0$ and exponential functions $f(x) = a^x$ occur in elementary algebra. Trigonometric functions are encountered in elementary geometry.

Functions also comprise the mathematical language used to describe the interaction of natural phenomena. For example, the position, velocity, and acceleration of a moving body may be expressed as a function of time. The profit from a business enterprise may be expressed as a function of the cost of raw materials, labor, taxation, and other economic factors. We wish a mathematical definition of function broad enough to include these and many other examples.

To phrase this definition we need one further notion from the elementary theory of sets, that is, the concept of the cartesian product of two sets $A$ and $B$.

**Definition.**   *If $A$ and $B$ are nonempty sets, then the cartesian product of $A$ and $B$, written $A \times B$, is the set of all ordered pairs $(a, b)$, where $a \in A$ and $b \in B$.*

*Symbolically then*

$$A \times B = \{(a, b) : a \in A, b \in B\}.$$

We use the expression "ordered pair" to stress the fact that $(a, b) = (a_1, b_1)$ if and only if $a = a_1$ and $b = b_1$. Thus $(1, 2) \neq (2, 1)$.

**Example 1.**   Let $A = \{0,1\}$, $B = \{1,2,3\}$. Then $A \times B = \{(0,1), (0,2), (0,3), (1,1), (1,2), (1,3)\}$ and $B \times A = \{(1,0), (2,0), (3,0), (1,1), (2,1), (3,1)\}$. If $A$ and $B$ are the same set, then we abbreviate $A \times A$ by $A^2$.

It is a familiar fact from coordinate geometry that the cartesian product $R \times R = R^2$, where $R$ is the set of real numbers, may be identified with the set of all points in the plane. This is accomplished by drawing two lines $l_1$, $l_2$ the first horizontal, the second vertical. We assume that the positive direction on the horizontal line is to the right, and the positive direction on the vertical line is up. The two lines $l_1$, $l_2$ form *coordinate axes* for the plane. The point of intersection is called the *origin*.

Now if $P$ is a point in the plane and $l$ a line, it is an axiom of elementary geometry that there is exactly one line through $P$ perpendicular to $l$. If we let $x$ be the number corresponding to the point of intersection of $l_1$ with the line through $P$ perpendicular to $l_1$, and $y$ be the number corresponding to the point of intersection of $l_2$ with the line through $P$ perpendicular to $l_2$, we have associated the ordered pair $(x, y)$ with the point $P$.

To each point $P$ there is one and only one pair $(x, y)$ and vice versa. The numbers $x$, $y$ are the *coordinates* of the point $P$. It is a time-honored tradition to denote the first or horizontal coordinate of $P$ by $x$ and the second or vertical coordinate of $P$ by $y$. The first coordinate of $P$ is called the *abscissa* of $P$ and the second coordinate the *ordinate*. The lines $l_1$, $l_2$ are called the *first and second coordinate axes*; or traditionally the $x$ and $y$ coordinate axes. For certain applications, it is more convenient to write $(x_1, x_2)$ instead of $(x, y)$, but for the moment we shall adhere to the traditional $x$, $y$ notation. This coordinate system for the plane is illustrated in Figure 11.

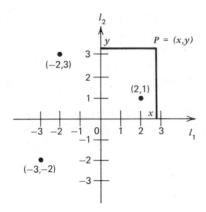

**Fig. 11**   Coordinate system for the plane.

We now give the definition of a function.

**Definition.**   *Let A and B be nonempty sets. Let f be a nonempty subset of the cartesian product $A \times B$. Then f is a function if there are no ordered pairs P and Q in f having the same first coordinate but different second coordinates.*

Thus if $f$ is a function, and $P = (x, y)$ is a point in $f$, then $P$ is the only point of $f$ having first coordinate $x$. The set of *first* coordinates of points in $f$ is called the *domain* of $f$. The set of *second* coordinates is called the *range* or *image* of $f$. The definition implies that for each $x$ belonging to the domain of $f$ there is one and only one point $y$ in the range such that $(x, y)$ is an ordered pair of $f$. It is customary to write this point $y = f(x)$. We then say that $y$ is the *value* of the function at the point $x$.

It is helpful to look at this definition of a function from a slightly different viewpoint. Since to each $x$ in the domain, there corresponds exactly one $y$ in the range that satisfies $y = f(x)$, we may think of the function as defined by a rule that assigns a unique value of $y$ for a given value of $x$. In fact, functions are usually defined by specifying this rule or method of correspondence.

**Example 2.**   Let $A = \{0,1,2\}$, $B = \{3,4,5\}$, then $f = \{(0,3), (1,5), (2,4)\}$ is a function with domain $A$ and range $B$.

**Example 3.**   If $A$ and $B$ are as in Example 1, then the set of ordered pairs $\{(0,4), (1,3), (2,5)(0,3)\}$ does not define a function, since there are two ordered pairs with the same first coordinate and different second coordinates.

**Example 4.**   Let $A$ be the set of calculus books in the United States, and let $N$ be the positive integers. Define $f$ to be the set of ordered pairs $(x, y) \in A \times N$ such that $y$ is the number of pages in the calculus book $x$. The domain of $f$ is all of $A$, whereas the range of $f$ is some proper subset of $N$.

**Example 5.**   Let $A$ be the set of all persons living in California. Let $f = \{(x, y) \in A^2 : y$ is the spouse of $x\}$. Assuming bigamy is impossible, then $f$ is a function.

The domain and range for $f$ coincide and consist of those married persons in California whose spouses also live in California.

**Example 6.**    Let $A$ be the set of girls at UCLA, $B$ the set of boys. Let $f = \{(x, y) \in A \times B : y$ is a boyfriend of $x\}$. This correspondence probably does not define a function since some girls may have many boyfriends.

If sets $A$ and $B$ are indicated by Venn diagrams, then sometimes it is convenient to represent a function with domain $D$ in $A$ and range $E$ in $B$ as in Figure 12.

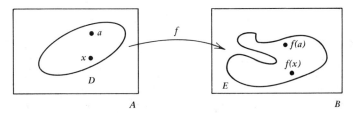

Fig. 12

This focuses attention on the fact that the value $f(x)$ is obtained by applying the rule $f$ to the point $x$. Thus $f$ can be thought of as a transformation that "maps" or transforms the points $x$ in $D$ into the points $f(x)$ in $E$.

We shall be primarily concerned in this book with functions $f$ such that the domain and the range of $f$ are subsets of the set of real numbers. Such a function is commonly called a *real valued function of a real variable*. A symbol standing for an arbitrary point in the domain of a function $f$ is often called the *independent variable* of $f$. A symbol standing for a point in the range is called the *dependent variable* of $f$.

When working with real valued functions, it is often easier to think of the function $f$ as the rule that assigns to each $x$ in the domain of $f$ the corresponding point in the range rather than to think of $f$ as the complete set of ordered pairs. Also, we shall try to keep our notation for functions as simple as possible. Thus the function $f = \{(x, y) \in R^2 : y = x^2 - 2x + 1\}$ will usually be written the function $y = x^2 - 2x + 1$ or the function $f(x) = x^2 - 2x + 1$. Whenever a function $f$ is defined by an equation in this way, the domain of $f$ is always taken to be the set of points $x$ for which the equation makes sense.

**Example 7.**    The domain of the function $f$ defined by the equation $y = (x - 1/x + 1)$ is the set of all real numbers different from $x = -1$. For if $x = -1$, then $x + 1 = 0$ and division by zero is impossible.

**Example 8.**    The domain of the function $f$ defined by $f(x) = \sqrt{1 - x^2}$ is the interval $[-1, 1]$. If $x$ lies outside this interval, $1 - x^2 < 0$ and the square root of a negative number cannot be defined as a real number.

If $f$ is a function, then $f(x)$ is the value of $f$ at the point $x$. Hence, $f(2), f(z)$, $f[(a + 1)/(a - 1)]$ are the respective values of $f$ at the points $2$, $z$, $(a + 1)/(a - 1)$ respectively. These values of $f$ are obtained by applying the function $f$ to the points $2$, $z$, $(a + 1)/(a - 1)$ respectively. If the function $f$ is given by a formula or expression involving $x$, then $f(2), f(z), f[(a + 1)/(a - 1)]$ are computed by substituting $2$, $z$, $(a + 1)/(a - 1)$ for $x$ everywhere that it occurs.

**Example 9.**    If $f(x) = \sqrt{x^2 + 1}$ then $f(2) = \sqrt{5}$, $f(z) = \sqrt{z^2 + 1}$ and $f[(a + 1)/(a - 1)] = \sqrt{[(a + 1)/(a - 1)]^2 + 1}$.

**EXERCISES**

1. Let $A = \{x \in R : 0 \leqslant x \leqslant 10\}$, $B = \{x \in R : 5 \leqslant x \leqslant 15\}$. Which of the following sets define a function with domain in $A$ and range in $B$. If such a function is not defined, explain why not.
   (a) $\{(0,5), (2,7), (3,12), (10,5)\}$
   (b) $\{(2,5), (3,7), (10,5), (2,8)\}$
   (c) $\{(0,6), (-1,10), (5,12), (3,7)\}$
   (d) $\{(0,11), (10,1), (9,8), (2,13)\}$

2. Determine the domains of the following functions.
   (a) $\{(x, y) : y = 2x + 3\}$
   (c) $\{(x, y) : y = \sqrt{4 - x^2}\}$
   (b) $\left\{(s, t) : t = \dfrac{2}{s - 4}\right\}$
   (d) $\{(u, v) : v = \sqrt{u^2 - a^2}\}$

3. Determine the ranges of the following functions.
   (a) $\{(u, v) : v = 2 - u\}$
   (b) $\{(x, y) : y = x^2 - 1\}$
   (c) $\{(t, s) : s = \sqrt{2 - t^2}\}$

4. Compute the domains and ranges, as subsets of the real numbers, of the functions defined by
   (a) $f(x) = x^3 + 1$
   (b) $g(x) = \sqrt{x - 1}$
   (c) $h(x) = \dfrac{1}{x^2 - 1}$

   Compute the values of each of the above functions at $x = 2$, $x = a$, $x = (y + a)/(y - a)$.

5. For real numbers $x$, define the function $f$ by the rule

   $$f(x) = x^2 + \sqrt{x - 1}.$$

   What is the domain of $f$? The range? Compute $f(0)$, $f(4)$, $f(a + b)$, $f(1 + h)$,

   $$\frac{f(1 + h) - f(1)}{h}, \quad \frac{f(y^2 + k) - f(y^2)}{k}.$$

6. Let $g(x) = \sqrt{(x + 1)/(x - 1)}$. What is the domain and range of $g$?

   Compute $g(2)$, $g[(x + 1)/(x - 1)]$, $g[(x - 1)/(x + 1)]$, $g(2 + h)$, $\dfrac{g(2 + h) - g(2)}{h}$,

   $\dfrac{g(a + h) - g(a)}{h}$.

7. Let $f$ and $g$ be functions. If for some $x$ belonging to the domain of $f$, $f(x)$ belongs to the domain of $g$, then $g(f(x))$ is a point in the range of $g$. For the functions $f$, $g$, $h$ of Exercise 4, determine the following quantities, or explain why this is impossible
   (a) $g(f(1))$
   (e) $h(g(f(1)))$
   (b) $f(g(1))$
   (f) $g(h(1))$
   (c) $g(f(-2))$
   (g) $h(f(g(2)))$.
   (d) $h(g(f(2)))$

8. Define sets $A$ and $B$ which are not sets of numbers. Give three examples of functions with domain $a$ subset of $A$ and range a subset of $B$.

9. Specify three correspondences between the set $A$ and $B$ of Exercise 8 that do not define functions and explain why.

## 1.7 Real-Valued Functions

Identifying the cartesian product $R \times R$ with the $x$, $y$ plane gives us a simple geometric interpretation of a real-valued function of a real variable.

**Definition.** *If $f$ is a nonempty subset of $R^2$, then $f$ defines a function if each vertical line intersects $f$ at most once.*

The set of first coordinates of points on $f$ is the domain of $f$. The set of second coordinates is the range of $f$. If the function $f$ is defined to be the set of points $(x, y)$ satisfying a given equation, then the domain of $f$ is the set of $x$ coordinates of pairs satisfying the equation. The range of $f$ is the set of $y$ coordinates.

The set of ordered pairs $(x, y)$ that satisfy a given equation is called the *graph* of the equation. Since the graph of a function $f$ is the set of ordered pairs $(x, y)$ that satisfy the given functional relationship, there is no distinction between a function and its graph. To use the phrase "sketch the graph of the function $f$" in place of "sketch the function $f$" is just a stylistic preference.

*Example 1.* $\{(x, y): y = 2x + 1\}$ defines a function. The graph (Figure 13) is a straight line through $(0, 1)$ with slope 2. The domain and range of this function are both the entire real line.

*Example 2.* The graph $\{(x, y): x^2 + y^2 = 1\}$ is a circle (Figure 14). This does not define a function since for $-1 < x < 1$ each vertical line through $x$ intersects $G$ twice.

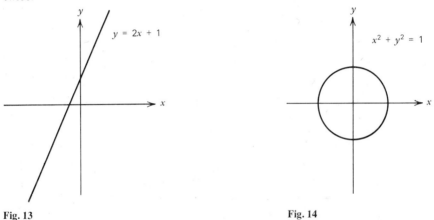

Fig. 13                                        Fig. 14

*Example 3.* $\{(x, y): y = |x|\}$ defines the absolute value function. The graph is the line $y = x$ if $x \geq 0$ and $y = -x$ if $x < 0$. The graph is sketched in Figure 15.

*Example 4.* Let $f = \{(x, y): y = (|x|/x)\}$. Applying the definition of $|x|$ we see that $y = 1$ if $x > 0$ and $y = -1$ if $x < 0$. Since division by zero is impossible, $x = 0$ is not a point of the domain of $f$. Hence the domain of $f$ is the set $\{x: x \neq 0\}$. The range is the two point set $\{-1, 1\}$. Figure 16 is a sketch of the graph.

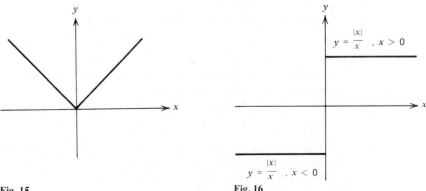

Fig. 15                          Fig. 16

*Example 5.*  Let $f = \{(x, y) : y = 6 \text{ if } 0 < x \leqslant 1$
$$y = 12 \text{ if } 1 < x \leqslant 2$$
$$y = 18 \text{ if } 2 < x \leqslant 3 \text{ etc.}\}.$$

Clearly each vertical line intersects $f$ at most once. The function $f$ is often called the postage stamp function since if $x$ is the weight in ounces of a letter, then $y = f(x)$ is the first class postage for the letter. It is an example of an important special class of functions called *step functions*. To indicate on the sketch of the graph (Figure 17) the value of $f$ at the points $x = 0, 1, 2, \ldots$ we draw a large dot.

*Example 6.*  Let $f = \{(x, y) : y = 1/(x - 1)\}$. Again each vertical line intersects $f$ at most once. Hence $f$ is a function. Since division by zero is impossible, the domain of $f = \{x : x \neq 1\}$. The range is $\{y : y \neq 0\}$. Figure 18 is a sketch of the graph.

*Example 7.*  The equation $y = \sqrt{x - 1}$ clearly defines a function $f$. Since the square root operation is only defined for $x \geqslant 0$, the domain of $f$ equals $\{x : x \geqslant 1\}$. The range is the set of nonnegative real numbers. The graph is sketched in Figure 19.

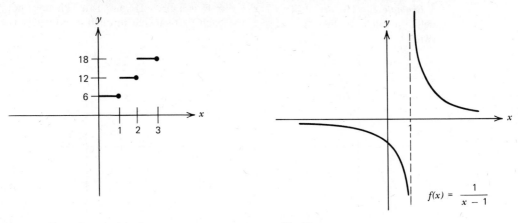

**Fig. 17**  Postage stamp function.

**Fig. 18**

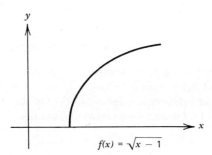

**Fig. 19**

*Example 8.*  A set of ordered pairs may define a function, but it may not be possible to draw an adequate sketch of this graph. Let

$$f = \{(x, y) : y = 1 \text{ if } x \text{ is rational}$$
$$y = 2 \text{ if } x \text{ is irrational}\}.$$

The domain of this function $f$ is the set of all real numbers, and $\{1, 2\}$ is the range. If we attempt to sketch the graph of this function, the best we can do is to draw two horizontal lines, one through $y = 1$, the other through $y = 2$. But this is obviously not the graph of the function $f$!

It is important to observe that the operations of addition and multiplication can be defined for real-valued functions. If $f$ and $g$ are real-valued functions with the same domain $D$, then for each $x \in D$ we may form

$$f(x) + g(x).$$

If we set $h(x) = f(x) + g(x)$, then this rule defines a new function $h$, called the *sum* of $f$ and $g$, and we write

$$h = f + g.$$

Thus

(1) $$(f+g)(x) = f(x) + g(x)$$

Similarly we define the *difference* $f - g$; *product* $f \cdot g$, and *quotient* $f/g$ by the formulas

(2) $$(f-g)(x) = f(x) - g(x)$$

(3) $$(f \cdot g)(x) = f(x) \cdot g(x)$$

(4) $$(f/g)(x) = \frac{f(x)}{g(x)} \text{ if } g(x) \neq 0.$$

If $f$ and $g$ have domains $D_1$ and $D_2$ that are not assumed to be the same, then the functions (1) to (4) may still be defined, with domain the intersection $D_1 \cap D_2$.

We note one final fact. Since functions are sets of ordered pairs, two functions $f$ and $g$ are equal if and only if they are defined by the same set of ordered pairs.

### EXERCISES

Consider the following sets of ordered pairs. Sketch the sets if possible and determine which sets are the graphs of functions. Specify the domain and range of each function.

1. $\{(x, y) : 2y + 3x = 1\}$.
2. $\{(x, y) : y - x^3 = 1\}$
3. $\{(x, y) : 2x^2 + y^2 = 1\}$.
4. $\{(x, y) : x^2 + y^2 = 1 \text{ and } y \geq 0\}$.
5. $\{(x, y) : y = |x|\}$.
6. $\{(x, y) : x + y \geq 1\}$.
7. $\{(x, y) : y = |2x - 1|\}$.
8. $\left\{(x, y) : y = \frac{x^2 - 1}{x + 1}\right\}$.

9. $\{(x, y) : y = \sqrt{4 - x^2}\}$.
10. $\{(x, y) : y^3 + x^3 = 1\}$.
11. $\{(x, y) : y = 1 - |x|\}$.
12. $\{(x, y) : y^2 + |x - 1| = 1\}$.
13. $\{(x, y) : y^2 + x^2 = -1\}$.
14. $\left\{(x, y) : y + \frac{1}{4 - x} = 0\right\}$.

15. Let $f(x) = |x - 1|$ and $g(x) = x$.
    Compute $f + g$, $f - g$, $2f - g$, and sketch the graphs.
16. Let $f(x) = |2x + 1|$ and $g(x) = 1 - 3x$.
    Compute $f + g$, $f - g$, $2f - 3g$, and sketch the graphs.
17. Let $f(x) = x^2$
    $$g(x) = \sqrt{x + 1}$$
    $$h(x) = \frac{x + 1}{x - 1}.$$
    Compute $f + g$, $g \cdot h$, $f - h$, $h/g$. What are the domains D of these functions?
18. Let $f(x) = \sqrt{x^2}$, $g(x) = |x|$. Are $f$ and $g$ the same function?
19. Define $\text{Sgn}(x) = 1$ if $x > 0$
    $$= 0 \text{ if } x = 0$$
    $$= -1 \text{ if } x < 0$$

Sgn is called the sign function, or the signum function. Let $f(x) = |x|/x$. Is Sgn $= f$?

20. Sketch the graphs of the following functions.

(a) $f(x) = \text{Sgn}\ (x^2 - 1)$

(c) $f(x) = \dfrac{x}{\text{Sgn}\ (x-1)}$

(b) $f(x) = \dfrac{\text{Sgn}\ (x-1)}{x}$

(d) $f(x) = \dfrac{x^2\ \text{Sgn}\ (x-1)}{\text{Sgn}\ (x-2)}$

21. Let $f(x) = x^2$, $g(x) = \sqrt{x+1}$, $h(x) = \dfrac{x+1}{x-1}$.

Compute the following quantities or explain why this cannot be done.

(a) $\dfrac{f(x+1) - f(x)}{g(x+1) - g(x)}$

(b) $f(g(1)) \cdot g(h(2))$

(c) $f(g(h(1))) \cdot g(f(2))$

(d) $(f(x+1) - g(x-1))^2$

(e) $g(h(-2)) \cdot f(g(2))$

## 1.8   Polynomials

If $a_0, \ldots, a_n$ are real numbers, then the function $f$ defined by

$$f(x) = a_n x^n + a_{n-1} x^{n-1} + \cdots + a_1 x + a_0$$

is called a *polynomial*. Its domain is the set of all real numbers. The numbers $a_0, \ldots, a_n$ are called the *coefficients* of $f$. The largest integer $n$ for which $a_n \neq 0$ is called the *degree* of the polynomial. It is assumed that the student is familiar with linear and quadratic polynomials. Our discussion of them will be brief. Polynomials of higher degree will be considered later after we have developed some of the machinery of the calculus. If $n = 0$, then

$$f(x) = a_0,$$

and $f$ is called a constant function. The graph of $f$ is, of course, the horizontal line through $a_0$ (Figure 20).

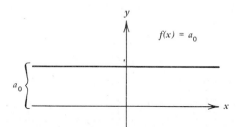

**Fig. 20**

If $n = 1$, $f(x) = a_1 x + a_0$ is called a *linear function*, and the graph of $f$ is a nonvertical straight line. The first coefficient $a_1$ is called the *slope* of the line, and the constant term $a_0$ is the value of $f$ at $x = 0$. The points at which the graph of $f$ intersects the $x$ and $y$ axes are called the $x$ and $y$ *intercepts* of the line. Clearly $a_0$ is the $y$ intercept. Setting $y = 0$ and solving for $x$ yields

$$x = \frac{-a_0}{a_1},$$

which is the $x$ intercept of the line.

Often we are given a line $l$ defined by certain geometric conditions, and we wish to determine the linear function $f(x) = a_1 x + a_0$ with this line as its graph. This amounts to exploiting the geometric conditions to determine the constants $a_0$ and $a_1$.

We do this by using the fact that a point $(x, y)$ will lie on the line if and only if $y = a_1 x + a_0$. This equation is called the equation for the line $l$ (Figure 21).

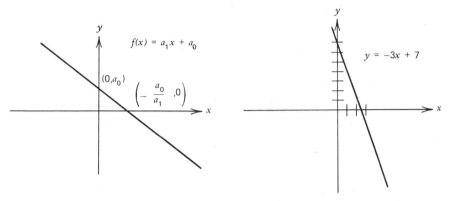

Fig. 21                                   Fig. 22

*Example 1.* Find an equation for the line passing through $(2, 1)$ and $(3, -2)$. Substituting these points for $(x, y)$ in the equation $y = a_1 x + a_0$ yields

$$1 = 2a_1 + a_0$$

and

$$-2 = 3a_1 + a_0.$$

Solving we get $a_1 = -3$, $a_0 = 7$.

Therefore the desired linear function is

$$f(x) = -3x + 7$$

The graph is sketched in Figure 22.

In general if $(x_0, y_0)$ and $(x_1, y_1)$ are two points on the graph of $f$, then

$$a_1 = \frac{y_1 - y_0}{x_1 - x_0}.$$

This yields the fact that $a_1$ is the tangent of the angle $\alpha$ that the line makes with the positive $x$ axis if this angle is measured in a counterclockwise direction from the $x$ axis (Figure 23).

Let $l_1$, $l_2$ be two intersecting nonvertical straight lines with slopes $a_1$, $a_1'$ as in Figure 24. Then $a_1 = \tan \alpha$ and $a_1' = \tan \alpha'$. If $\beta$ is the angle between $l_1$ and $l_2$ measured from $l_2$ to $l_1$, then $\beta = \alpha - \alpha'$, and

$$\tan \beta = \tan (\alpha - \alpha') = \frac{a_1 - a_1'}{1 + a_1 a_1'}.$$

Now $l_1$ and $l_2$ are perpendicular if $\beta = 90°$ or if $a_1 a_1' = -1$ in the above equation. Thus two nonvertical straight lines are perpendicular if and only if the product of the slopes equals $-1$.

Clearly two lines are parallel if and only if their slopes are equal. A line is vertical if and only if its equation is $x = a$. This equation does not define a function $y = f(x)$.

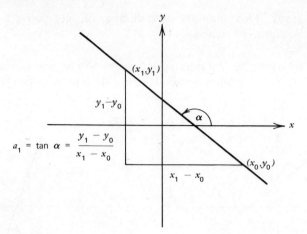

**Fig. 23**

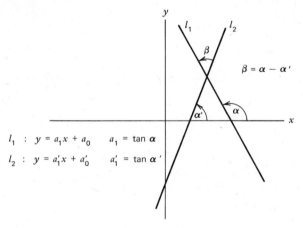

**Fig. 24**

If $n = 2$, the graph of the polynomial

$$f(x) = a_2 x^2 + a_1 x + a_0$$

is a *parabola*. It is easily shown that if $a_2 > 0$ there is a minimum point $(x_0, y_0)$ on the graph of $f$ and if $a_2 < 0$ there is a maximum point. This point is called the vertex of the parabola. To sketch the graph it usually suffices to locate the vertex and the intercepts (the points where the graph crosses the coordinate axes). The parabola opens up if $a_2 > 0$ and down if $a_2 < 0$. The vertex of a parabola may be determined by completing the square. We illustrate this by an example and leave the general argument as an exercise.

*Example 2.*    Determine the vertex of the graph of $f(x) = 2x^2 + 4x + 7$ and sketch the graph.

$$
\begin{aligned}
f(x) &= 2x^2 + 4x + 7 \\
&= 2x^2 + 4x + 2 + 5 \\
&= 2(x^2 + 2x + 1) + 5 \\
&= 2(x+1)^2 + 5 \geq 5.
\end{aligned}
$$

Therefore, there is a minimum to the graph of $f$ at the point with coordinates $x = -1$, $y = 5$. The point $(0, 7)$ is the $y$ intercept. Figure 25 is a sketch of the graph.

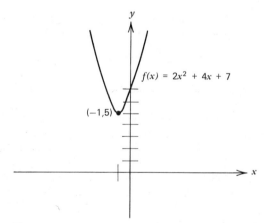

$f(x) = 2x^2 + 4x + 7$

$(-1, 5)$

**Fig. 25**

The vertical line through the vertex of the parabola is called the *axis* of the parabola. The graph of the parabola is symmetric with respect to this line (see Exercise 18, p. 28).

We close this section with an observation on function notation. We have defined a function $f$ as a collection of ordered pairs $(x, y)$ such that no two pairs from this collection have the same first coordinate and different second coordinates. For a given ordered pair $(x, y)$ belonging to this collection we write $y = f(x)$. Thus $f(x)$ is the value of the function $f$ at the point $x$. However, in place of the phrase "the function $f$" one often sees "the function $f(x)$" or "the function $y = f(x)$." This latter notation is somewhat inexact since it blurs the distinction between the function $f$ and its value at a fixed point $x$. This notation should cause no confusion, however, if the distinction between $f$ and $f(x)$ is kept clearly in mind.

**EXERCISES**

Determine the functions representing the following straight lines and sketch the graphs.

1. Line through $(1, -2)$ and $(2, 4)$.
2. Line through $(1, 3)$ perpendicular to the line $3x + 4y - 1 = 0$.
3. Line with $x$ and $y$ intercept $2, -4$ respectively.
4. Lines through $(1, 2)$ making an angle of $45°$ with $x$ axis.
5. Line through $(4, 2)$ parallel to line through $(1, 3)$ and $(-2, 5)$.
6. Line through $(2, 4)$ perpendicular to line through $(-2, 5)$ and $(1, 7)$.

Determine the functions representing the following parabolas and sketch the graphs.

7. Passing through $(1, 1)\,(2, -1)\,(0, 4)$.
8. Passing through $(0, 4)\,(3, -6)\,(2, -1)$.
9. With axis $x = 2$; minimum value $y = 5$; passing through $(3, 7)$.
10. With vertex $(1, -1)$ passing through $(2, 7)$.
11. Determine the vertices of the following parabolas and sketch the graphs
    (a) $y = 3x^2 + x - 1$
    (b) $y = -x^2 + 4x + 7$
12. If $f(x) = a_2 x^2 + a_1 x + a_0$, show that $f$ has a vertex at

$$\left(-\frac{a_1}{2a_2},\, a_0 - \frac{a_1{}^2}{4a_2}\right).$$

Show that if $a_2 > 0$, then this vertex is at the lowest point on the graph; if $a_2 < 0$, it is at the highest.

It is easily verified that the midpoint of the line segment determined by $(x_1, x_1)$, $(x_2, y_2)$ is given by

$$\left(\frac{x_1 + x_2}{2}, \frac{y_1 + y_2}{2}\right).$$

13. Determine an equation for the perpendicular bisector of the line segment joining $(1, 3)$ $(2, -1)$.
14. Determine an equation for the perpendicular bisector of the chord formed by the intersection of the line $2y + 3x - 16 = 0$ with the parabola $y = x^2 + 4x - 5$.
15. A line not parallel to the axis of a parabola can be said to be tangent to the parabola at the point $(x_0, y_0)$ if the line intersects the parabola only at the point $(x_0, y_0)$. Using the fact that a quadratic equation $ax^2 + bx + c = 0$ has one real root if and only if $b^2 - 4ac = 0$, determine an equation for the tangent line to the parabola $y = x^2$ at the point $(2, 4)$.
16. Determine an equation for the tangent line to the circle $x^2 + y^2 = 25$ at the point $(3, 4)$ if the same definition of tangent line is used.
17. If an object falls subject to the force of gravity, its height $y$ above the surface of the earth as a function of time is given by

$$y = at^2 + b,$$

for appropriate constants $a$ and $b$. If a ball is dropped from 1000 feet and after 3 seconds it has fallen 144 feet, when will it hit the earth?
18. Show that a parabola $y = a_2x^2 + a_1x + a_0$ is symmetric about its axis. [Let $x = a$ be the $x$ coordinate of the vertex of the parabola (Exercise 12). For each $x$ show that if $(a + x, y)$ lies on the parabola, then $(a - x, y)$ does also.]

## 1.9  Limits

One of the reasons that calculus is considered by some to be a difficult subject is that many of the deepest and most fundamental ideas occur at the very beginning. Foremost among these is the notion of a limit. This basic concept of limit underlies most of the machinery of the calculus. Indeed, this machinery was fully developed long before the definition of a limit had been formulated as it is at present. This formulation took two centuries to achieve, so it is not surprising that the full import of the idea is not immediately obvious.

Assume $f$ is a function defined on the interval $(a, b)$ except possibly at the point $x_0$, $a < x_0 < b$. Then we say that the function $f$ has the number $L$ as a limit at $x_0$ if the number $L$ can be approximated to any degree of accuracy by all of the values $f(x)$ provided only that $x$ is chosen sufficiently close to $x_0$. When this is true we write $\lim_{x \to x_0} f(x) = L$. This may be read either as "$L$ is the limit of $f$ at $x_0$" or "$L$ is the limit of $f(x)$ as $x$ approaches $x_0$." We shall clarify what we mean by "any degree of accuracy" and "chosen sufficiently close to $x_0$" presently but first let us consider an example.

Suppose $f(x) = x^2$. One would conjecture that the limit of $f$ at $x = 0$ is also zero. To demonstrate this we must show that 0 may be approximated by the numbers $f(x)$ as closely as desired provided that $x$ is chosen sufficiently close to 0. If $\epsilon$ measures the accuracy of our approximation, then how small must $x$ be chosen to guarantee that

$$|0 - f(x)| = |x^2| < \epsilon?$$

Clearly, if $|x| < \sqrt{\epsilon}$, then $|x^2| < \epsilon$. Hence we can approximate 0 by $f(x)$ within an error of $\epsilon$ whenever $|x|$ is chosen smaller than $\sqrt{\epsilon}$ (see Figure 26).

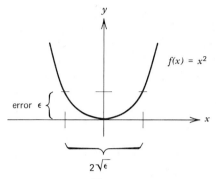

**Fig. 26** If $|x| < \sqrt{\epsilon}$, then $|f(x)| = |x^2| < \epsilon$.

Before proceeding further let us give the precise arithmetic definition of the limit of a function $f$ at a point $x_0$.

**Definition.** *Let $f$ be a function defined on the interval $(a, b)$ except possibly at the point $x_0$, $a < x_0 < b$. Then $\lim_{x \to x_0} f(x) = L$ if for each positive number $\epsilon$ a corresponding positive number $\delta$ can be found so that $L - \epsilon < f(x) < L + \epsilon$ whenever $x \neq x_0$ and $x_0 - \delta < x < x_0 + \delta$.*

Since $L - \epsilon < f(x) < L + \epsilon$ if and only if $|f(x) - L| < \epsilon$, we may reword our statement in the following way:

$$\lim_{x \to x_0} f(x) = L \text{ if to each } \epsilon > 0$$

there exists a corresponding $\delta > 0$ such that

(1) $$0 < |x - x_0| < \delta \text{ implies } |f(x) - L| < \epsilon.$$

Notice that in verifying the conditional statement (1) we only consider $x \neq x_0$, i.e. those $x$ for which $|x - x_0| > 0$, because the existence of a limit of $f$ at $x_0$ depends only on the behavior of those values of $f(x)$ for $x$ near $x_0$. In many examples the function $f$ will not be defined at $x_0$. We leave it as an exercise to verify that the operation of computing the limit of a function $f$ at a point $x_0$ is uniquely defined. That is, if $\lim_{x \to x_0} f(x) = L$ and $\lim_{x \to x_0} f(x) = M$, then $L = M$.

To verify the conditional statement (1), one usually analyzes what it means for $|f(x) - L|$ to be small, that is, what it means for $|f(x) - L|$ to be less than a fixed small positive number $\epsilon$. This analysis usually indicates how one must choose $\delta$ in terms of $\epsilon$. Once the recipe for the choice of $\delta$ has been determined, it must then be verified that for this choice of $\delta$ if

$$0 < |x - x_0| < \delta, \text{ then } |f(x) - L| < \epsilon.$$

Sometimes it is easy to see how to choose $\delta$ in terms of $\epsilon$, the point $x_0$, and perhaps other fixed constants. But this is by no means always true. It is relatively straightforward to verify that

$$\lim_{x \to 1} \frac{x^2 - 1}{x - 1} = 2.$$

However, we shall eventually prove such statements as

$$\lim_{x \to 0} \frac{\sin x}{x} = 1 \quad \text{and} \quad \lim_{x \to 0} \frac{\sin x - x}{x^3} = -\frac{1}{6}.$$

Why either of these statements should be true is by no means apparent. However, both can be proved by verifying the conditions in the definition of a limit. Success with these arguments, often called "$\epsilon$, $\delta$" arguments, comes only with experience. The student should not feel discouraged if it takes some time to develop this skill.

**Example 1.**   If for all $x$, $f(x) = c$, then for each point $x_0$

$$\lim_{x \to x_0} f(x) = c.$$

This follows immediately from the definition, and we leave the verification as an exercise.

**Example 2.**   Let $f$ be the linear function $f(x) = 3x - 4$. We assert

$$\lim_{x \to 2} f(x) = 2 = f(2).$$

For a given $\epsilon > 0$ we must pick a positive $\delta$ depending on $\epsilon$ so that whenever $0 < |x - x_0| = |x - 2| < \delta$ it follows that

$$|f(x) - L| = |3x - 4 - 2| < \epsilon.$$

But

$$|3x - 4 - 2| = |3x - 6| = 3|x - 2|.$$

Hence

$$3|x - 2| < \epsilon \text{ if } |x - 2| < \frac{\epsilon}{3}.$$

Therefore, if we choose $\delta$ satisfying $0 < \delta \leqslant \epsilon/3$, it follows that

$$0 < |x - 2| < \delta$$

implies

$$|3x - 6| = 3|x - 2| < 3\delta \leqslant \epsilon,$$

which was to be shown. The choice of $\delta$ in terms of $\epsilon$ is illustrated in Figure 27.

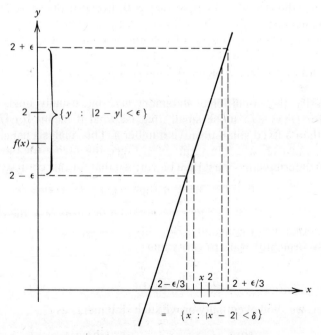

**Fig. 27**   If $\delta \leqslant \dfrac{\epsilon}{3}$, then $0 < |2 - x| < \delta$ implies $|f(x) - 2| < \epsilon$.

The result may be immediately extended to the following. If $f(x) = a_1 x + a_0$, then $\lim_{x \to x_0} f(x) = a_1 x_0 + a_0 = f(x_0)$. We leave the verification of this as an exercise.

The following simple consequence of the definition of a limit is often useful.

**THEOREM 1.9**  *Let $f$ and $g$ be defined on the interval $(a, b)$ except possibly at the point $x_0$, $a < x_0 < b$. Assume for $x \neq x_0$ that $f(x) = g(x)$. Then if $\lim_{x \to x_0} f(x) = L$, it follows that $\lim_{x \to x_0} g(x) = L$.*

We leave the proof as an exercise.

***Example 3.***  Let $f(x) = 3x + 1$.

If

$$g(x) = \frac{f(1+x) - f(1)}{x},$$

show that $\lim_{x \to 0} g(x) = 3$. The function $g$ is not defined at $x = 0$. But for $x \neq 0$.

$$g(x) = \frac{f(1+x) - f(1)}{x} = \frac{3(1+x) + 1 - 4}{x} = 3.$$

Since $\lim_{x \to 0} 3 = 3$, it follows that $\lim_{x \to 0} g(x) = 3$.

***Example 4.***  We have already shown that $\lim_{x \to 0} x^2 = 0$. It is true that $\lim_{x \to x_0} x^2 = x_0^2$. We shall verify it for $x_0 = 2$. To show that $\lim_{x \to 2} x^2 = 4$ we must verify that for a given $\epsilon > 0$ a corresponding $\delta > 0$ may be chosen so that $|x^2 - 4| < \epsilon$ whenever $0 < |x - 2| < \delta$. To see how to choose $\delta$ we estimate the size of $|x^2 - 4|$.

$$|x^2 - 4| = |x - 2||x + 2| \leq |x - 2|(|x| + 2)$$

by the triangle inequality. Again by the triangle inequality, $|x| - 2 < |x - 2|$. Hence, if $|x - 2| < 1$, then $|x| - 2 < 1$ and $|x| < 3$. Consequently

$$|x^2 - 4| \leq |x - 2|(|x| + 2) \leq 5|x - 2|.$$

Now $5|x - 2| < \epsilon$ if $|x - 2| < \epsilon/5$. Hence, if we choose $\delta \leq \epsilon/5$ *and* $\delta < 1$, it follows that

$$0 < |x - 2| < \delta \text{ implies } |x^2 - 4| \leq 5|x - 2| < \epsilon,$$

and we have shown that $\lim_{x \to 2} x^2 = 4$. Notice that in the above argument we have placed two restrictions on the choice of $\delta$. One is that $\delta \leq \epsilon/5$. The second is that $\delta < 1$. Indeed, in "$\epsilon, \delta$" arguments, $\delta$ may depend on finitely many fixed constants in addition to depending on $\epsilon$. This is no handicap because we are only interested in producing some value of $\delta$ for the given value of $\epsilon$. We are not interested in the largest value of $\delta$ that will work.

The above argument can be immediately extended to show that $\lim_{x \to x_0} x^2 = x_0^2$.

Let us now investigate what it means for $\lim_{x \to x_0} f(x) = L$ to be a false statement. Two things may happen. One is that $\lim_{x \to x_0} f(x)$ may be a number different from $L$. The other is that $f$ may not have a limit at $x_0$. We investigate two examples of the second of these possibilities.

***Example 5.***  Show that $\lim_{x \to 3} 1/(x - 3)$ does not exist.

This is intuitively clear from Figure 28 because if $|x - 3|$ is small, then $|f(x)| = 1/|x - 3|$ is large. In particular $f(x)$ is not close to a fixed number $L$. However, it is instructive to give a complete "$\epsilon, \delta$" proof of the fact that $\lim_{x \to 3} 1/(x - 3)$ does not exist. To show that $\lim_{x \to x_0} f(x)$ does not exist we must show that no number $L$ can

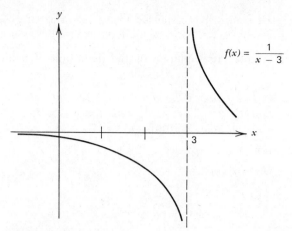

**Fig. 28**

satisfy the definition of a limit. This means that for each number $L$ we must produce some $\epsilon > 0$ such that no matter how small $\delta$ is chosen there are values of $x$ satisfying $0 < |x - x_0| < \delta$ but $|f(x) - L| \geq \epsilon$. In the case that $f(x) = 1/(x-3)$ and $x_0 = 3$ it is not hard to see how to choose $\epsilon$.

First, assume $L = 0$. Then we may take $\epsilon = 1$. For if $\delta < 1$ and $0 < |x - 3| < \delta$, it follows that $1/|x-3| > 1/\delta > 1 = \epsilon$. Hence $0$ cannot be the limit. Next assume $L \neq 0$. We assert that we may take $\epsilon = |L|$. To see this, note that if $\delta \leq 1/2|L|$ and $0 < |x - 3| < \delta$, then

$$\left| \frac{1}{|x-3|} - L \right| \geq \frac{1}{|x-3|} - |L| > \frac{1}{\delta} - |L| \geq 2|L| - |L| = |L| = \epsilon.$$

Hence a nonzero number cannot be the limit of $1/(x-3)$ at $x_0 = 3$. Thus $\lim_{x \to 3} 1/(x-3)$ does not exist.

**Example 6.** Show that $\lim_{x \to 0} |x|/x$ does not exist.

As in Example 5 one must show that for each real number $L$ there is a corresponding $\epsilon$ so that no matter how small $\delta$ is chosen there are points $x$, such that $0 < |x| < \delta$ but $|f(x) - L| \geq \epsilon$. It turns out that if we choose $\epsilon$ to be any fixed number less than one, then $\epsilon$ will have the desired property. This may be verified by inspecting the graph of $f$, Figure 29. We leave the details as an exercise.

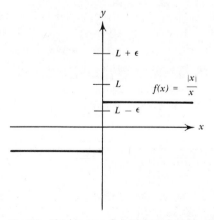

**Fig. 29**   If $L$ is any real number and $\epsilon < 1$, then for each $\delta > 0$ there exist $x$ satisfying

$$0 < |x| < \delta, \text{ but } \left| \frac{|x|}{x} - L \right| \geq \epsilon.$$

## EXERCISES

When verifying $\lim_{x \to x_0} f(x) = L$, show how $\delta$ is chosen in terms of $\epsilon$ and verify that $0 < |x - x_0| < \delta$ implies $|f(x) - L| < \epsilon$.

1. If $f(x) = c$ for all $x$, show that $\lim_{x \to x_0} f(x) = c$ for each real number $x_0$.

2. Show that $\lim_{x \to 2} 3x + 1 = 7$.

3. Show that $\lim_{x \to -1} 2 - x = 3$.

4. Show that $\lim_{x \to 2} 4 - 3x = -2$.

5. Show that $\lim_{x \to x_0} a_1 x + a_0 = a_1 x_0 + a_0$.

6. Prove Theorem 1.9.

7. Using Exercise 5 and Theorem 1.9 compute the following limits.

   (a) $\lim_{x \to 1} \dfrac{x^2 - 1}{x - 1}$

   (b) $\lim_{x \to 3} \dfrac{x^2 - 5x + 6}{x - 3}$

   (c) $\lim_{x \to 0} \dfrac{(1 + x)^2 - 1}{x}$

   (d) $\lim_{x \to 1} \dfrac{x^3 + x^2 - x - 1}{x^2 - 1}$

8. Compute $\lim_{x \to 0} \dfrac{f(x + 1) - f(1)}{x}$ if $f(x) = x^2$

9. Compute $\lim_{x \to 0} \dfrac{f(x + 2) - f(2)}{x}$ if $f(x) = 2x^2 - 1$

10. Show that if $x_0 \neq 0$, then $\lim_{x \to x_0} \dfrac{1}{x} = \dfrac{1}{x_0}$.

11. Show that if $f(x) = a_2 x^2 + a_0$, then $\lim_{x \to x_0} f(x) = f(x_0)$

12. Compute the following limits using Theorem 1.9 and previous exercises.

    (a) $\lim_{t \to 0} \dfrac{f(2 + t) - f(2)}{t}$ if $f(t) = t^3$

    (b) $\lim_{x \to 0} \dfrac{f(x) - f(2)}{x - 2}$ if $f(x) = 2x^3 - x$

    (c) $\lim_{s \to -1} \dfrac{f(s) - f(-1)}{s + 1}$ if $f(s) = 1 - 2s^3$

13. Show that $\lim_{x \to 2} \dfrac{1}{x - 2}$ does not exist.

14. Show that $\lim_{x \to 4} \dfrac{1}{2x - 8}$ does not exist.

15. Finish the demonstration that $\lim_{x \to 0} \dfrac{|x|}{x}$ does not exist.

16. Show that $\lim_{x \to 1} \dfrac{|x - 1|}{x - 1}$ does not exist.

17. Show that $\lim_{x \to 5} \dfrac{|2x - 10|}{x - 5}$ does not exist.

18. If $f(x) = |x|$, compute $\lim_{h \to 0} \dfrac{f(a + h) - f(a)}{h}$ for each $a$ for which this limit exists.

19. Show that $\lim_{x \to 2} \dfrac{1}{(x - 2)^2}$ does not exist.

20. Let $f(x) = \operatorname{Sgn}(x)$ (see p. 23). Compute $\lim_{x \to a} f(x)$ for each number $a$ for which this limit exists.

21. If $\lim_{x \to x_0} f(x) = L$ and $\lim_{x \to x_0} f(x) = M$, show that $L = M$.

22. If $\lim_{x \to x_0} f(x) = L$, show that $\lim_{x \to x_0} |f(x)| = |L|$. If $\lim_{x \to x_0} |f(x)| = |L|$ does it necessarily follow that $\lim_{x \to x_0} f(x) = L$?

23. Consider the following set of points in the plane

$$\left\{ (1, 1), \left( \tfrac{1}{2}, 0 \right), \left( \tfrac{1}{3}, 1 \right), \left( \tfrac{1}{4}, 0 \right), \left( \tfrac{1}{5}, 1 \right), \left( \tfrac{1}{6}, 0 \right), \dots \right\}.$$

Connect these points by straight lines and reflect the resulting set about the $y$ axis. This set forms the graph of a function $f$ with domain $\{x : 0 < |x| \leq 1\}$. Sketch the graph of this sawtooth function. What can you say about $\lim_{x \to 0} f(x)$?

24. Do exactly the same if the points now are

$$\left\{ (1, 1), \left(\frac{1}{2}, 0\right), \left(\frac{1}{3}, \frac{1}{3}\right), \left(\frac{1}{4}, 0\right), \left(\frac{1}{5}, \frac{1}{5}\right), \left(\frac{1}{6}, 0\right), \cdots \right\}.$$

Does $\lim_{x \to 0} f(x)$ exist? If so what is the limit?

### 1.10   Continuous Functions; Limit Theorems

In our discussion of limits in the previous section we showed that for many functions $f$,

$$\lim_{x \to x_0} f(x) = f(x_0).$$

In other words, the *limit* of $f$ at $x_0$ equals the *value* of $f$ at $x_0$. Such a function $f$ is called *continuous* at the point $x_0$. If $f$ is continuous at all points of an interval $(a, b)$, we say that $f$ is *continuous on the interval* $(a, b)$. The graph of a continuous function defined on an interval $(a, b)$ is characterized by the fact that it is unbroken and has no gaps. Thus Figure 30 is the graph of a continuous function.

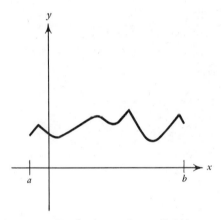

**Fig. 30**   Graph of a continuous function.

We saw in the previous section that the functions $f(x) = a_1 x + a_0$, $f(x) = |a_1 x + a_0|$, $f(x) = a_2 x^2 + a_0$ were all continuous. The set of continuous functions is an extremely important class of functions. The construction of these functions is greatly facilitated by the fact that the sum, difference, and product of continuous functions is also continuous. Furthermore, the quotient $f/g$ of continuous functions is continuous at each point $x$ for which $g(x) \neq 0$. These facts are immediate corollaries of the following fundamental theorem on the calculus of limits.

**THEOREM 1.10**   *Let $f$ and $g$ be defined on the interval $(a, b)$ except possibly at the point $x_0$, $a < x_0 < b$. Assume $\lim_{x \to x_0} f(x) = L$, and $\lim_{x \to x_0} g(x) = M$. Then*

(1)     $$\lim_{x \to x_0} cf(x) = cL \text{ for each real number } c$$

(2)     $$\lim_{x \to x_0} f(x) + g(x) = L + M$$

(3)     $$\lim_{x \to x_0} f(x) \cdot g(x) = L \cdot M$$

(4)
$$\lim_{x \to x_0} \frac{f(x)}{g(x)} = \frac{L}{M}, \text{ if } M \neq 0.$$

**COROLLARY.** *If $f$ and $g$ are continuous at the point $x_0$, then $f + g$, and $f \cdot g$ are continuous at $x_0$. If $g(x_0) \neq 0$, then $f/g$ is continuous at $x_0$.*

The proof of Theorem 1.10 is given in the appendix at the end of the section. It is not essential that the "$\epsilon, \delta$" details be mastered at this time. The theorem itself is very important, however, and will be used repeatedly. It follows from (2) and (3) and mathematical induction that a polynomial $p(x) = a_n x^n + a_{n-1} x^{n-1} + \cdots + a_0$ is continuous at each real number $x_0$. We leave the proof of this fact as an exercise in the next section after we have considered induction.

For the quotient $f/g$ of two functions $f$ and $g$, Theorem 1.10 states that the limit of the quotient is the quotient of the limits provided that the limit of the denominator is not zero. However, if $g(x)$ tends to zero as $x$ tends to $x_0$, then the quotient $f/g$ may or may not have a limit. If under this assumption $\lim_{x \to x_0} [f(x)/g(x)]$ does exist, then $\lim_{x \to x_0} f(x) = 0$ exists as well. The proof of this useful fact is a simple consequence of Theorem 1.10.

**THEOREM 1.11** *Let $f$ and $g$ be defined on the interval $(a, b)$, except possibly at $x_0$, $a < x_0 < b$.*
*Suppose $\lim_{x \to x_0} (f(x)/g(x)) = L$ and $\lim_{x \to x_0} g(x) = 0$. Then $\lim_{x \to x_0} f(x) = 0$.*

**Proof.** For $x \neq x_0$, $f(x) = (f(x)/g(x)) \cdot g(x)$. By Theorem 1.10 (3) $\lim_{x \to x_0} (f(x)/g(x)) \cdot \lim_{x \to x_0} g(x) = L \cdot 0 = 0$, and the proof is complete.

If $\lim_{x \to x_0} f(x) = 0 = \lim_{x \to x_0} g(x)$, then Theorem 1.10 yields no information on whether $\lim_{x \to x_0} (f(x)/g(x))$ exists or not. Some sort of algebraic manipulation with the fraction $f/g$ is necessary before Theorem 1.10 can be used.

Let us look now at what it means for a function not to be continuous at a point $x_0$ belonging to the domain of $f$. When this happens $f$ is said to be *discontinuous* at $x_0$ and we call the point $x_0$ a *discontinuity* of $f$. There are two possibilities. The first of these is that $f$ has no limit as $x \to x_0$. The second is that $f$ has a limit, but $\lim_{x \to x_0} f(x) \neq f(x_0)$.

*Example 1.* If $f$ is defined by

$$f(x) = |x|, x \neq 0$$

$$f(x) = 1, \quad x = 0,$$

then $f$ is not continuous at $x = 0$ since $\lim_{x \to 0} f(x) = 0$ but $f(0) = 1$. When $f$ has a limit at $x_0$ and $f$ is discontinuous at $x_0$, we say that $x_0$ is a *removable discontinuity* for $f$. This means that $f$ may be redefined at $x_0$ in such a way as to make $f$ continuous. In fact, we need only define $f(x_0)$ to be the limit of $f$ at $x_0$.
In Figure 31 if we redefine $f$ at 0 so that $f(0) = 0$, then $f$ is continuous at 0.

*Example 2.* Let $f(x) = ((1 + x)^2 - 1)/x$ if $x \neq 0$ and $f(0) = 0$. Determine if $f$ is continuous or discontinuous at $x = 0$. Since, for $x \neq 0$,

$$f(x) = \frac{(1 + x)^2 - 1}{x} = \frac{1 + 2x + x^2 - 1}{x} = 2 + x,$$

we see that $\lim_{x \to 0} f(x) = 2$. Since $f(0) = 0$, $f$ is discontinuous at 0 and indeed $x = 0$ is a removable discontinuity (see Figure 32).

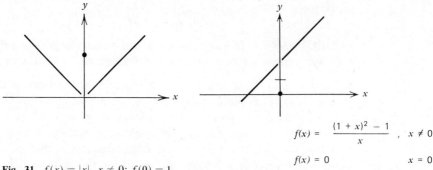

$$f(x) = \frac{(1 + x)^2 - 1}{x} \quad , \quad x \neq 0$$

$$f(x) = 0 \qquad\qquad x = 0$$

**Fig. 31** $f(x) = |x|$, $x \neq 0$; $f(0) = 1$ has a removable discontinuity at $x = 0$.

**Fig. 32** $f$ has a removable discontinuity at $x = 0$.

If, however, the function $f$ has no limit at $x_0$, then it is impossible to redefine $f$ at $x_0$ so that $f$ will be continuous at $x_0$. The functions $f(x) = 1/x$, $f(x) = |x|/x$ have no limit at $x = 0$. The sawtooth function of Exercise 23, p. 33 has no limit at the origin either. Hence, if either of these functions are defined at $x = 0$, this point must be a discontinuity of the function (see Figure 33).

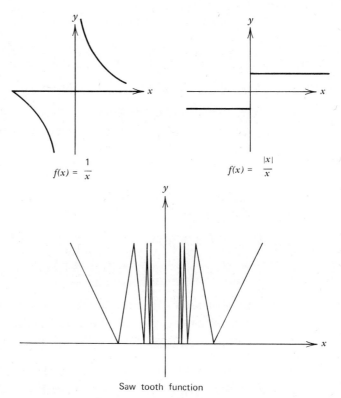

$$f(x) = \frac{1}{x} \qquad\qquad f(x) = \frac{|x|}{x}$$

Saw tooth function

**Fig. 33** Three functions that do not possess limits at $x = 0$.

It is often convenient to write $f(x) \to L$ as $x \to x_0$ in place of $L = \lim_{x \to x_0} f(x)$. The statement $f(x) \to L$ is read "$f(x)$ approaches $L$ or $f(x)$ tends to $L$." Thus "$L$ is the limit of $f$ at $x_0$" may be reworded as "$f(x)$ tends to $L$ as $x$ tends to $x_0$."

**EXERCISES**

1. Suppose $g(x) \to 0$ and $f(x) \to L \neq 0$ as $x \to x_0$. What can you say about

$$\lim_{x \to x_0} \frac{f(x)}{g(x)}?$$

2. Suppose $g(x) \to 0$ as $x \to x_0$ and $\lim_{x \to x_0} f(x)$ does not exist. What can you say about

$$\lim_{x \to x_0} \frac{f(x)}{g(x)}.$$

3. Give examples of functions $f$ and $g$ such that for some point $x_0$ $\lim_{x \to x_0} f(x) = 0 = \lim_{x \to x_0} g(x)$ and (1) $\lim_{x \to x_0} (f(x)/g(x))$ exists; (2) $\lim_{x \to x_0} (f(x)/g(x))$ does not exist.

4. Determine if the following limits exist. Compute them if they do.

   (a) $\lim_{x \to 1} 3x^2 - x + 4$

   (b) $\lim_{x \to 3} \dfrac{2x - 1}{x + 3}$      (g) $\lim_{x \to 1} \dfrac{x^2 - x - 2}{x^2 + x - 6}$

   (c) $\lim_{x \to 1} \dfrac{x + 1}{2x^2 - x + 1}$      (h) $\lim_{x \to 2} \dfrac{x^2 - 9}{x - 2}$

   (d) $\lim_{x \to 1} \dfrac{x^2 + x - 2}{x - 1}$      (i) $\lim_{x \to 1} \dfrac{x - 1}{x^2 - 2x + 1}$

   (e) $\lim_{x \to -1} \dfrac{x - 2}{x + 1}$      (j) $\lim_{x \to 4} \dfrac{x^2 - 5x + 4}{x^2 - 16}$

   (f) $\lim_{x \to 1} \dfrac{x^2 + 3x + 2}{x - 1}$      (k) $\lim_{x \to 2} \dfrac{x^3 - 2x^2 + 4x - 8}{x - 2}$

5. Determine the points of discontinuity for the following functions. If there are any discontinuities, determine which are removable and which are not. (Remember that in order for a function to be continuous or discontinuous, at a point $x$, the point $x$ must belong to the domain of the function.)

   (a) $f(x) = 2x^2 - 3x + 1$

   (b) $f(x) = \dfrac{x - 1}{x + 1}, f(-1) = 0$

   (c) $f(x) = \dfrac{x^2 + x - 2}{x - 1}, f(1) = 2$

   (d) $f(x) = \dfrac{x^2 - x - 2}{x^2 - 1}, f(1) = f(-1) = 1$

   (e) $f(x) = \dfrac{x^2 - 1}{x^2 + 1}$

   (The function $\mathrm{Sgn}(x)$ is defined on p. 23).

   (f) $f(x) = \mathrm{Sgn}(x^2 - 4)$

   (g) $f(x) = \mathrm{Sgn}\left(\dfrac{x - 1}{x + 2}\right)$

   (h) $f(x) = \mathrm{Sgn}(x - 2)^2$

   (i) $f(x) = \mathrm{Sgn}\dfrac{(x - 2)^2}{(x + 3)}$

   (j) $f(x) = \mathrm{Sgn}\dfrac{(x^2 - 1)}{(x - 2)^2}$

6. Define $f(x) = (x^n - 1)/(x-1)$, $x \neq 0$, $f(1) = a$. For what value of $a$ is $f$ continuous at $x = 1$?

7. Let $g(x) = (f(a+x) - f(a))/x$, $x \neq 0$. For what values of $a$ is it possible to define $g(0)$ so that $g$ is continuous at $x = 0$? When this is the case, determine $g(0)$ if

(a) $f(x) = x^2$

(b) $f(x) = 3x^2 - x$

(c) $f(x) = \dfrac{2}{x}$

(d) $f(x) = \dfrac{3}{x^2}$

(e) $f(x) = \dfrac{x-1}{x+1}$

(f) $f(x) = \dfrac{x+2}{3x-1}$

8. If $\lim_{x \to 0} (f(a+x) - f(a))/x$ exists, is $f$ continuous at the point $a$? Explain.

9. If $p$ is a polynomial, then it is not hard to show that $\lim_{x \to x_0} p(x) = 0$ if and only if $(x - x_0)$ is a factor of $p(x)$. Suppose $p$ and $q$ are polynomials. Assume $p(x) \to 0$ and $q(x) \to 0$ as $x \to x_0$. Determine a criterion in terms of the multiplicity of the factor $(x - x_0)$ that will establish whether or not

$$\lim_{x \to x_0} \frac{p(x)}{q(x)} \text{ exists.}$$

10. Let $f$ be the sawtooth function of Exercise 24, p. 34. Define $f(0) = 0$ so that $f$ is defined on the interval $[-1, 1]$. Is $f$ continuous on this interval? Explain.

# APPENDIX

### 1.11.*    The Well-Ordering Principle and the Principle of Induction

From time to time, we shall have occasion to establish identities or statements that are to be valid for all positive integers. For example

(1)
$$1 + 2 + \cdots + n = \frac{n(n+1)}{2}$$

(2)
$$1^2 + 2^2 + \cdots + n^2 = \frac{n(n+1)(2n+1)}{6}.$$

These identities can be easily checked for specific values of $n$. This, however, does not constitute a proof that the identities hold for all values of $n$. For example, consider the statement "For each positive integer $n$, $f(n) = n^2 - n + 41$ is a prime number." This is a true statement for $n = 1, 2, 3, \ldots, 40$ but fails for $n = 41$ since $f(41) = (41)^2$.

Usually the only way to establish an identity such as (1) or (2) is to appeal to one or the other of the following properties of the set of positive integers.

#### Well-Ordering Principle

*Each nonempty set of positive integers contains a smallest element.*

#### Principle of Induction

*Let $S$ be a subset of the positive integers $N$. If (i) $1 \in S$ and (ii) for each $k \in S$ it follows that $k + 1 \in S$, then $S = N$.*

We illustrate these two principles by proving the identities (1) and (2). We prove (1) by the well-ordering principle.

The statement $1+2+\cdots+n=(n(n+1))/2$ will be true for each $n \in N$ if

$$F=\left\{n \in N: 1+2+\cdots+n \neq \frac{n(n+1)}{2}\right\}$$

is empty. To prove $F$ is empty, we use an indirect argument. That is, we assume $F \neq \emptyset$ and arrive at a contradiction. If $F \neq \emptyset$, then by the well-ordering principle $F$ has a smallest element $k$. $k \neq 1$ because the identity is true for $n=1$. Therefore $k > 1$, and $k-1$ is a positive integer. Consequently the identity must be true for $n=k-1$ since $k$ is the smallest positive integer for which the identity is false. Therefore

$$1+2+\cdots+k-1=\frac{(k-1)(k-1+1)}{2}.$$

Adding $k$ to both sides yields

$$1+2+\cdots+k-1+k=\frac{(k-1)k}{2}+k=\frac{(k-1)k+2k}{2}=\frac{k(k+1)}{2}.$$

Therefore (1) is true for $n=k$, and this contradicts the choice of $k$. Hence $F$ has no smallest element. Therefore $F=\emptyset$, and the identity is valid for all positive integers.

Next we prove

$$1^2+2^2+\cdots+n^2=\frac{n(n+1)(2n+1)}{6}$$

for each positive integer $n$ by using the principle of induction. Let

$$T=\left\{n \in N: 1+2^2+\cdots+n^2=\frac{n(n+1)(2n+1)}{6}\right\}.$$

Assert $T=N$. First, $1 \in T$ since $(1(1+1)(2+1))/6=\frac{6}{6}=1$. Next we verify that $if\ k \in T$, then $k+1 \in T$. Suppose $k \in T$. Then

$$1+2^2+\cdots+k^2=\frac{k(k+1)(2k+1)}{6}.$$

Adding $(k+1)^2$ to both sides yields

$$1+2^2+\cdots+k^2+(k+1)^2=\frac{k(k+1)(2k+1)}{6}+(k+1)^2$$

$$=\frac{(k+1)[k(2k+1)+6(k+1)]}{6}$$

$$=\frac{(k+1)(k+2)(2k+3)}{6}.$$

Therefore, $1+2^2+\cdots+n^2=(n(n+1)(2n+1))/6$ if $n=k+1$. Hence $k+1 \in T$, and applying the principle of induction we infer that

$$T=N.$$

The statements to be verified may take the form of inequalities. For example, verify that for each integer $n$

$$(3) \qquad 1^3+2^3+\cdots+(n-1)^3 < \frac{n^4}{4}.$$

We establish this by using the well-ordering principle.

Let $F = \{n : 1^3 + \cdots + (n-1)^3 \geqslant n^4/4\}$. If $F \neq \emptyset$, $F$ has a smallest element $k$. This number $k \neq 1$ because

$$(1-1)^3 = 0 < \tfrac{1}{4}.$$

Hence, $k > 1$ and $k-1 \notin N$. Since $k-1 \notin F$, (3) is valid for $l = k-1$. Therefore

$$1^3 + \cdots + (l-1)^3 < \frac{l^4}{4}.$$

Adding $l^3$ to both sides yields

$$1^3 + \cdots + (l-1)^3 + l^3 < \frac{l^4}{4} + l^3.$$

But

$$\frac{l^4}{4} + l^3 = \frac{l^4 + 4l^3}{4} < \frac{l^4 + 4l^3 + 6l^2 + 4l + 1}{4} = \frac{(l+1)^4}{4}.$$

Therefore (3) is valid for $l+1 = k$. Hence $k$ is not the smallest member of $F$. This contradicts the choice of $k$. Hence $F$ is empty, and statement (3) is valid for each positive integer $n$.

The reason that the principle of induction and the well-ordering principle hold for the set $N$ of positive integers is really a matter of definition  Indeed, our informal definition that $N = \{1, 2, 3, \ldots\}$ assumes some form of the induction principle in order to give mathematical meaning to "and so on." If a precise "inductive" definition of $N$ is given, then the well-ordering principle may be derived as a consequence. A sketch of one way to do this is contained in Exercise 11.

**EXERCISES**

1. Prove that the following statements are valid for all positive integers $n$ by using the well-ordering principle.
   (a) $1 + 3 + 5 + \cdots + (2n-1) = n^2$
   (b) $(1+x)^n \geqslant 1 + nx$ if $x > -1$ (Bernoulli's inequality)
   (c) $\dfrac{n^4}{4} < 1^3 + \cdots + n^3$

2. Establish the validity of the following statements for each positive integer $n$ by using the principle of induction.
   (a) $1^3 + 2^3 + \cdots + n^3 = (1 + 2 + \cdots + n)^2$
   (b) $1^2 + 2^2 + \cdots + (n-1)^2 < \dfrac{n^3}{3} < 1^2 + \cdots + n^2$
   (c) $\dfrac{1}{1 \cdot 2} + \dfrac{1}{2 \cdot 3} + \cdots + \dfrac{1}{n(n+1)} = \dfrac{n}{n+1}$

3. Note that

$$\frac{1}{2} = 2 - \frac{3}{2}$$

$$\frac{1}{2} + \frac{2}{4} = 2 - \frac{4}{4}$$

$$\frac{1}{2} + \frac{2}{4} + \frac{3}{8} = 2 - \frac{5}{8}.$$

Guess a general identity for which the above are special cases and prove it by using either well ordering or induction.

4. Note that

$$\left(1-\frac{1}{2}\right) = \frac{1}{2}$$

$$\left(1-\frac{1}{2}\right)\left(1-\frac{1}{3}\right) = \frac{1}{3}$$

$$\left(1-\frac{1}{2}\right)\left(1-\frac{1}{3}\right)\left(1-\frac{1}{4}\right) = \frac{1}{4}.$$

Guess the general identity implied and prove it by using well ordering or induction.

5. Prove that if $x \neq 1$ and $n$ is a positive integer, then

$$1 + x + \cdots + x^n = \frac{1-x^{n+1}}{1-x}.$$

6. Prove by induction that for each positive integer $n$

$$1 + \frac{1}{2} + \frac{1}{3} + \cdots + \frac{1}{2^n} \geq 1 + \frac{n}{2}.$$

(Experiment with small values of $n$.)

7. For a fixed integer $k$ let $I_k = \{n \in I : n \geq k\}$. Formulate an appropriate induction principle for $I_k$ and prove it. [If $f(x) = k + 1 - n$, $n \in N$, then the range of $f$ is all of $I_k$. Furthermore, $n = m$ if and only if $f(n) = f(m)$, and $n < m$ if and only if $f(n) < f(m)$.]

8. Show that the well-ordering principle holds for $I_k$.

9. If $n \geq 0$ is an integer, prove using induction or the well-ordering principle that $\lim_{x \to a} x^n = a^n$.

10. Using Exercise 9 show that if $p$ is a polynomial and $a$ is a real number, then $\lim_{x \to a} p(x) = p(a)$. [Use induction or well ordering to prove that for each non-negative integer $n$

$$\lim_{x \to a} (a_n x^n + \cdots + a_0) = a_n a^n + \cdots + a_0.]$$

11.* To prove that the well-ordering principle holds for the set $N$ of positive integers we must give a precise definition of $N$. To do this we proceed as follows. Call a set $S$ of real numbers *inductive* if

(i)                                    $1 \in S$

and

(ii)                    whenever $x \in S$, then $x + 1 \in S$.

The set of all real numbers, $R$, is clearly an inductive set. We then define

$N = $ intersection of all inductive subsets of $R$.

The fact that the principle of induction holds for $N$ follows immediately from the definition. Prove that $N$ satisfies the well-ordering principle.

[Suppose $S \subset N$ has no smallest element. Let $T = \{n \in N : n < s \text{ for all } s \in S\}$. Apply the principle of induction to show that $T = N$. In the course of the proof you may use the "obvious" fact that if $n$ and $k$ are positive integers and $n < k + 1$ then $n \leq k$. For a "complete" argument you should demonstrate (by induction) this fact as well.]

## 1.12*   Proof of Theorem 1.10

We prove separately each assertion of Theorem 1.10.

(1)          If $\lim_{x \to x_0} f(x) = L$ and $c$ is constant, then $\lim_{x \to x_0} cf(x) = cL$.

***Proof.*** For a given $\epsilon > 0$ we must determine a recipe for choosing a corresponding $\delta > 0$ so that whenever $0 < |x - x_0| < \delta$ it follows that $|cf(x) - cL| < \epsilon$. If $c = 0$, the result is obvious, so assume $c \neq 0$. Now $|cf(x) - cL| = |c||f(x) - L|$, and $|c||f(x) - L| < \epsilon$ if $|f(x) - L| < \epsilon/|c|$. By assumption $\lim f(x) = L$. Hence for each $\epsilon > 0$ there is a corresponding $\delta > 0$ such that $0 < |x - x_0| < \delta$ implies $|f(x) - L| < \epsilon$. In particular for each $\epsilon > 0$ there is a corresponding $\delta > 0$ such that $0 < |x - x_0| < \delta$ implies $|f(x) - L| < \epsilon/|c|$. (Why?) But then $|cf(x) - cL| < \epsilon$, and we are finished.

(2)   If $\lim_{x \to x_0} f(x) = L$ and $\lim_{x \to x_0} g(x) = M$, $\lim_{x \to x_0} f(x) + g(x) = L + M$.

***Proof.*** For a given $\epsilon > 0$ we must pick $\delta > 0$ so that if $0 < |x - x_0| < \delta$ then $|f(x) + g(x) - L - M| < \epsilon$. Now $|f(x) + g(x) - L - M| \leq |f(x) - L| + |g(x) - M|$.

By hypothesis, positive numbers $\delta_1, \delta_2$ can be found so that

$$0 < |x - x_0| < \delta_1 \text{ implies } |f(x) - L| < \frac{\epsilon}{2}$$

and

$$0 < |x - x_0| < \delta_2 \text{ implies } |g(x) - M| < \frac{\epsilon}{2}.$$

Therefore, if $\delta$ is chosen so that $\delta < \delta_1$ *and* $\delta < \delta_2$, it follows that $0 < |x - x_0| < \delta$ implies $|f(x) - L| < \epsilon/2$ and $|g(x) - M| < \epsilon/2$. Hence if $0 < |x - x_0| < \delta$, then

$$|f(x) + g(x) - (L + M)| \leq |f(x) - L| + |g(x) - M|$$

$$< \frac{\epsilon}{2} + \frac{\epsilon}{2} = \epsilon,$$

and we are finished.

(3)   If $\lim_{x \to x_0} f(x) = L$ and $\lim_{x \to x_0} g(x) = M$, then $\lim_{x \to x_0} f(x) \cdot g(x) = L \cdot M$.

**Proof.** To find $\delta$ so that $0 < |x - x_0| < \delta$ implies $|f(x)g(x) - LM| < \epsilon$ we must provide an estimate on the size of $|f(x)g(x) - LM|$. However by the triangle inequality

$$|f(x)g(x) - LM| = |f(x)g(x) - Lg(x) + Lg(x) - LM|$$

$$\leq |f(x) - L||g(x)| + |L||g(x) - M|.$$

First choose $\delta_1$ so that $0 < |x - x_0| < \delta_1$ implies

$$|L||g(x) - M| < \frac{\epsilon}{2}$$

*and* $|g(x) - M| < 1$. (Why can this be done?) If $|g(x) - M| < 1$, then $|g(x)| < |M| + 1$ and

$$|f(x) - L||g(x)| < |f(x) - L|(M + 1).$$

Now $\delta_2$ may be chosen so that $0 < |x - x_0| < \delta_2$ implies

$$|f(x) - L|(M + 1) < \frac{\epsilon}{2}.$$

(Why?) Hence, if $\delta < \delta_2$ *and* $\delta < \delta_1$, then $0 < |x - x_0| < \delta$ implies

$$|f(x)g(x) - LM| \le |f(x) - L||g(x)| + |L||g(x) - M|$$
$$< |f(x) - L|(|M| + 1) + |L||g(x) - M|$$
$$< \frac{\epsilon}{2} + \frac{\epsilon}{2} = \epsilon.$$

This completes the proof.

We first prove a special case of (4).

(4′)    If $\lim\limits_{x \to x_0} f(x) = L$ and $L \ne 0$, then $\lim\limits_{x \to x_0} \dfrac{1}{f(x)} = \dfrac{1}{L}$.

**Proof.**    Again the problem is given $\epsilon > 0$, choose $\delta > 0$ so that

$$0 < |x - x_0| < \delta \text{ implies } \left| \frac{1}{f(x)} - \frac{1}{L} \right| < \epsilon.$$

Now

$$\left| \frac{1}{f(x)} - \frac{1}{L} \right| = |L - f(x)| \frac{1}{|f(x)||L|}.$$

Since $L \ne 0$ by hypothesis, there is a $\delta_1 > 0$ such that $0 < |x - x_0| < \delta_1$ implies $|f(x) - L| < L/2$. (Why?) If this holds, the triang e inequality then implies

$$|L| - |f(x)| \le |f(x) - L| < \frac{|L|}{2}.$$

Hence

$$|f(x)| > \frac{|L|}{2},$$

and

$$|L - f(x)| \frac{1}{|f(x)||L|} < \frac{2}{|L|^2} |L - f(x)|.$$

Since $L = \lim\limits_{x \to x_0} f(x)$, we may choose $\delta_2$ so that

$$0 < |x - x_0| < \delta_2 \text{ implies } \frac{2}{|L|^2} |L - f(x)| < \epsilon.$$

Hence, if $\delta < \delta_1$ and $\delta < \delta_2$, then

$$0 < |x - x_0| < \delta \text{ implies } \left| \frac{1}{f(x)} - \frac{1}{L} \right| < \frac{2}{|L|^2} |L - f(x)| < \epsilon.$$

This completes the proof.

Statements (3) and (4′) together imply that if $\lim\limits_{x \to x_0} f(x) = L$ and $\lim\limits_{x \to x_0} g(x) = M$, then

$$\lim\limits_{x \to x_0} f(x)/g(x) = \lim\limits_{x \to x_0} f(x) \cdot \lim\limits_{x \to x_0} \frac{1}{g(x)} = L/M \text{ if } M \ne 0.$$

This is statement (4).

# CHAPTER TWO

# THE DERIVATIVE

## 2.1 Motion

The problem of providing a mathematical model for the motion of a particle along a line is an extremely old one. It had its origin in Greek geometry during the 5th century B.C. and a successful solution to the problem did not come until the invention of the calculus over 2000 years later. Position, velocity, and acceleration are words that are used to describe the motion of the particle. What, however, are their appropriate mathematical definitions?

It is certainly reasonable to assume the existence of a position function $s = f(t)$ that measures the distance of the particle from a fixed origin at time $t$. Thus at each value of time $t$, the position of the particle $s = f(t)$ is uniquely determined. The problem comes with the velocity. Is it possible to define velocity as a function of time? In other words, is it possible to associate with each value of $t$ a number $v$ that measures the velocity of the particle at time $t$?

Over an interval of time $[t_0, t_1]$ the notion of velocity causes no problem. Indeed, if $s_1 = f(t_1)$, $s_0 = f(t_0)$ denote the positions of the particle at $t = t_1$ and $t = t_0$, respectively, then the average value of the velocity over the interval $[t_0, t_1]$ is given by the quotient

(1)
$$\frac{s_1 - s_0}{t_1 - t_0} = \frac{f(t_1) - f(t_0)}{t_1 - t_0}.$$

For example, if the position function is defined by $s = f(t) = t^2 + t$, then the average velocity over the interval $[1, 2]$ is

$$\frac{f(2) - f(1)}{2 - 1} = \frac{6 - 2}{1} = 4.$$

The average velocity over the interval $[1, t]$ is

$$\frac{f(t) - f(1)}{t - 1} = \frac{t^2 + t - 2}{t - 1}.$$

However, formula (1) cannot be used to define the velocity at a point $t_0$. For if we set $t_1 = t_0$, then both the numerator and denominator in (1) vanish. In order to get around this difficulty we must use the notion of a limit.

**Definition.** *If $s = f(t)$ denotes the position of the particle at time $t$, then the velocity $v$ of the particle at time $t$ is defined to be*

$$\lim_{t_1 \to t} \frac{f(t_1) - f(t)}{t_1 - t}$$

*if this limit exists.*

Thus, in the previous example if the position of the particle at time $t$ is described by the function $f(t) = t^2 + t$, then the average velocity over an interval $[t, t_1]$ is

$$\frac{t_1^2 + t_1 - (t^2 + t)}{t_1 - t} = \frac{(t_1 - t)(t_1 + t + 1)}{t_1 - t} = (t_1 + t + 1).$$

Hence the velocity at the point $t$ equals $\lim_{t_1 \to t} (t_1 + t + 1) = 2t + 1$.

Calculations with quotients $(f(t_1) - f(t))/(t_1 - t)$ are usually simplified if we make a change of variable. Let $t_1 - t = h$. Then $t_1 = t + h$ and

$$\frac{f(t_1) - f(t)}{t_1 - t} = \frac{f(t + h) - f(t)}{h}.$$

Hence

$$\lim_{t_1 \to t} \frac{f(t_1) - f(t)}{t_1 - t} = \lim_{h \to 0} \frac{f(t + h) - f(t)}{h}.$$

Thus, if the position of the particle at time $t$ is described by $s = f(t)$, the velocity at time $t$ is defined to be

$$\lim_{h \to 0} \frac{f(t + h) - f(t)}{h}.$$

The particle is said to have a velocity at each point $t$ for which

$$\lim_{h \to 0} \frac{f(t + h) - f(t)}{h}$$

exists. If this limit does not exist at the point $t$, then the velocity of the particle does not exist at that point. Indeed, we may define a velocity function $v$ by the formula

$$(2) \qquad v(t) = \lim_{h \to 0} \frac{f(t + h) - f(t)}{h}.$$

The domain of this function will be the set of points for which the limit in (2) exists.

**Example 1.**   A particle moves along a line so that its position at time $t > 0$ is given by the function $f(t) = 1/t$. Determine the average velocity over an interval $[t, t + h]$. Determine the velocity function $v(t)$. The average velocity over the interval $[t, t + h]$ is given by

$$\frac{f(t + h) - f(t)}{h} = \frac{1/(t + h) - (1/t)}{h} = \frac{t - t - h}{h(t + h)t} = -\frac{1}{(t + h)t}.$$

The velocity function is

$$v(t) = \lim_{h \to 0} \frac{f(t + h) - f(t)}{h} = \lim_{h \to 0} \frac{-1}{(t + h)t} = -\frac{1}{t^2}.$$

The last equality follows from Theorem 1.10 (4). Notice that for $t > 0$ the velocity $v(t)$ is negative. This means that if the line is vertical and the positive direction along the line is up, then the particle is moving down. This fact is clearly illustrated by the graph of the function $f(t) = 1/t$ (Figure 1).

When physical phenomena are described by functions, the function notation is often "abused" in the following way. In place of writing $s = f(t)$ with three different letters, $f$ standing for the function, $t$ standing for a point in the

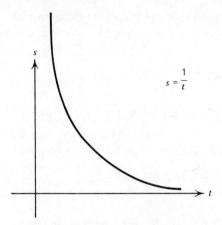

$s = \dfrac{1}{t}$

**Fig. 1**   As $t$ increases, $s = 1/t$ moves down.

domain of $f$, and $s$ standing for a point in the range, it is very common notation to write $s = s(t)$ or $v = v(t)$ where now the same letter is used for both
the function and a point in the range. It is imprecise notation, but convenient
since fewer letters are being used. If $s = s(t)$, then the value of the function
$s$ at points $t_0$, $t_1$ would then be written $s_0 = s(t_0)$ and $s_1 = s(t_1)$.

## EXERCISES

1. If $s(t) = t - 3t^2$ measures the position of a particle at time $t$, determine the
   average velocity over the intervals $[1, 2]$, $[2, 3]$, $[3, 4]$. What is the velocity
   at $t = 1, 2, 3, 4$? Determine the velocity function $v(t)$.
2. If $s(t) = t + (1/t)$ measures the position of a particle at time $t$, determine the
   average velocity over the intervals $[1, 2]$, $[2, 3]$, $[a, a+h]$. Determine the
   velocity function $v(t)$.
3. If $s(t) = t^2 - 1$ measures the position of a particle at time $t$, determine the
   average velocity on the interval $[a, a+h]$ and determine the velocity function.
   At what point does the velocity vanish?
4. If $s(t) = 3t - t^3$, determine the velocity function $v(t)$. At which points $t$ is the
   velocity zero?
5. Show that if the position of a particle is given by $s(t) = at + b$, then the velocity
   is constant with value A. It is a fact, which we cannot prove at the present time,
   that if the velocity of a particle is known to be equal to a constant $a$, then
   $v(t) = at + b$ where $b$ is constant
6. If the velocity of a particle is positive on an interval $(a, t_0)$ and negative on an
   interval $(t_0, b)$, or the opposite, then the velocity changes sign at the point $t_0$, and
   the sense of the motion is reversed. Determine the points, if any, where the
   velocity changes sign for the following position functions $s(t)$.
   (a) $s(t) = t - t^2$
   (b) $s(t) = t^3$
   (c) $s(t) = t^3 - 3t$
7. If $v(t)$ is the velocity of a particle moving along a line, then the *average acceleration* of the particle over the interval $[t, t+h]$ is given by $(v(t+h) - v(t))/h$.
   The *acceleration* $a(t)$ of the particle at the point $t$ is defined by

   $$a(t) = \lim_{h \to 0} \frac{v(t+h) - v(t)}{h}.$$

   The function $a(t)$ defined by this formula is called the acceleration function of
   the particle. Determine the acceleration functions for the following position
   functions $s(t)$.

(a) $s(t) = 2t - 1$

(c) $s(t) = \dfrac{1}{t}$

(b) $s(t) = t + 3t^2$

(d) $s(t) = \dfrac{2}{t-1}$

8. A particle moves so that its position at time $t$ is given by $s(t) = at^2 + bt + c$. Show that the acceleration is constant and equals $2a$. The converse of this statement is also true, but we cannot prove it at the present time: if a particle moves with constant acceleration $2a$, then the position function

$$s(t) = at^2 + bt + c$$

where $b$ and $c$ are constant.

9. If an object is dropped and falls under the action of gravity, its distance from the surface of the earth at time $t$ is given by the function

$$s(t) = at^2 + b$$

where $a$ and $b$ are constant. If a ball is dropped from 100 feet (at $t = 0$) and if two seconds later it has fallen 64 feet, what is the velocity in feet per second of the ball at time $t$? What is the velocity when the ball hits the earth?

10. If a projectile is fired from a given position, and we neglect friction, its height above the surface of the earth at time $t$ is given by

$$s(t) = at^2 + bt + c$$

for appropriate constants $a$, $b$, $c$. What is the velocity at time $t$? If a bullet is fired from a height of 1000 feet and one second later its position and velocity are 1184 feet and 168 feet/second, respectively, at what time does the bullet hit the earth? How high does the bullet go?

11. A particle moves along a line so that at time $t$, $s(t) = |t|$. What is the velocity $v(t)$ if $t > 0$, if $t < 0$? Does the particle have a velocity at $t = 0$?

## 2.2   Tangents

In the previous section we used the idea of a limit to solve the problem of defining the velocity of a moving particle. Indeed, if $s(t)$ measures the position of the particle at time $t$, then the velocity $v(t)$ at time $t$ is given by the formula

$$v(t) = \lim_{h \to 0} \frac{s(t+h) - s(t)}{h}.$$

Another problem for which the notion of a limit provides the solution is the question of defining the tangent line to the graph of a function $f$.

The tangent to a circle at a point $(x_0, y_0)$ can be defined as the line that touches the graph only at the point $(x_0, y_0)$. However, generalizing this definition to the graphs of functions causes difficulty. We certainly cannot use the definition that the tangent line at the point $(x_0, y_0)$ is the line that intersects the graph only at the point $(x_0, y_0)$. In Figure 2 the lines $l_2$ and $l_3$, which intersect the graph only at $(x_0, y_0)$, are obviously not the tangent line, whereas the line $l_1$, which perhaps should be the tangent, intersects the graph twice. There is the added difficulty of describing mathematically what it means for a line to "touch" the graph of a function at the point $(x_0, y_0)$. In fact, this is the fundamental problem.

To see how to proceed for an arbitrary function $f$, let us examine Figure 3. Let $P = (x_0, y_0)$ be a fixed point on the graph of $f$, and let $Q = (x_1, y_1)$, $R = (x_2, y_2)$ be points on the graph on either side of $P$. Let $l_1$ be the line through $Q$ and $P$ and $l_2$ be the line through $P$ and $R$. Now if the points $x_1$ and

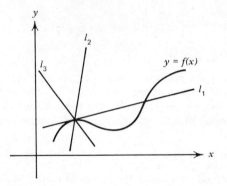

**Fig. 2**

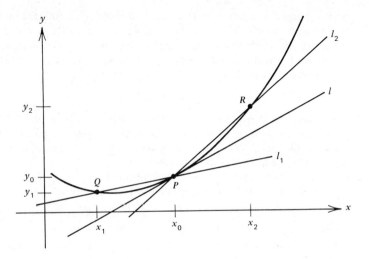

**Fig. 3**

$x_2$ are chosen close to $x_0$, the lines $l_1$, and $l_2$ in some sense approximate the line $l$. Furthermore, it is in accord with our intuitive idea of tangency that this line $l$ touches the graph at $P$ and should be taken to be the tangent line. The problem now is to translate these vague notions into something precise.

Now, if a line is not vertical, it has an equation

$$y = mx + b.$$

Since the tangent line to $f$ at $P = (x_0, y_0)$ must pass through $P$, we have that

$$y_0 = mx_0 + b.$$

This determines $b$ in terms of the slope $m$. Hence, we shall have solved our problem if we can give an appropriate geometric condition that will determine the slope $m$. Here is where we need the notion of a limit.

Referring again to Figure 3, let $m_1 = (y_1 - y_0)/(x_1 - x_0)$ be the slope of $l_1$ and $m_2 = (y_2 - y_0)/(x_2 - x_0)$ be the slope of $l_2$. It is intuitively clear that when we think of the lines $l_1$, $l_2$ approximating the line $l$, what we really mean is that the slope $m$ of $l$ is approximated by the slopes of $l_1$ and $l_2$. Thus, we are requiring that the slope $m$ of the tangent line can be approximated as well as we please by the slopes $m_1$, $m_2$ of the lines $l_1$, $l_2$ just so long as the points $x_1$, $x_2$ are chosen sufficiently close to $x_0$. This just says that $m$ is the limit of $m_1$ and $m_2$ as $x_1$, $x_2$ approach $x_0$. To make a formal definition let us write $x_1 = x_0 + h_1$ where $h_1 < 0$ and $x_2 = x_0 + h_2$ where $h_2 > 0$. Thus

$$m_1 = \frac{y_1 - y_0}{x_1 - x_0} = \frac{f(x_0 + h_1) - f(x_0)}{h_1}$$

and

$$m_2 = \frac{y_2 - y_0}{x_2 - x_0} = \frac{f(x_0 + h_2) - f(x_0)}{h_2}.$$

**Definition.**   *Let $(x, f(x))$ be a point on the graph of the function $f$. If*

$$\lim_{h \to 0} \frac{f(x+h) - f(x)}{h}$$

*exists, then $f$ is said to have a nonvertical tangent at the point $x$. The tangent to $f$ at $x$ is defined to be the line with slope*

$$m = \lim_{h \to 0} \frac{f(x+h) - f(x)}{h}$$

*which passes through the point $(x, f(x))$. This is illustrated in Figure 4.*

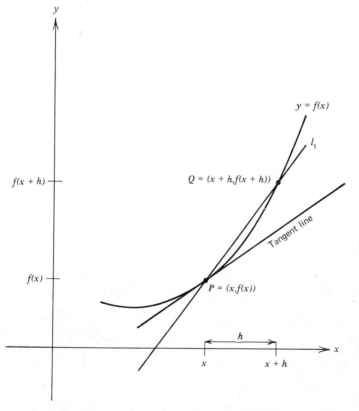

**Fig. 4**   $\dfrac{f(x+h) - f(x)}{h}$ = slope of line $l_1$. $m = \lim\limits_{h \to 0} \dfrac{f(x+h) - f(x)}{h}$ = slope = slope of tangent line.

   If $\lim\limits_{h \to 0} (f(x+h) - f(x))/h$ *fails to exist at the point $x$, then we say that $f$ does not have a tangent at $x$. More precisely, $f$ does not have a non-vertical tangent at the point $x$.*

**Example 1.**  Determine an equation for the tangent line to the function $f(x) = x^2$ at the point $(1, 1)$. If this tangent line exists it will have equation $(y-1)/(x-1) = m$ where

$$m = \lim_{h \to 0} \frac{f(1+h) - f(1)}{h}.$$

But

$$\frac{f(1+h) - f(1)}{h} = \frac{(1+h)^2 - 1}{h} = \frac{1 + 2h + h^2 - 1}{h} = 2 + h.$$

Hence

$$\lim_{h \to 0} \frac{f(1+h) - f(1)}{h} = 2.$$

The tangent line has equation

$$\frac{y-1}{x-1} = 2 \qquad \text{or} \qquad y = 2x - 1.$$

See Figure 5.

**Example 2.**  Determine the point $(x, y)$ on the graph of the parabola $y = x^2 + x + 1$ at which the slope of the tangent line $= 3$. Letting $f(x) = x^2 + x + 1$, the slope $m$ of the tangent line at the point $x$ is given by

$$m = \lim_{h \to 0} \frac{f(x+h) - f(x)}{h}.$$

But

$$\frac{f(x+h) - f(x)}{h} = \frac{(x+h)^2 + (x+h) + 1 - (x^2 + x + 1)}{h}$$

$$= \frac{2xh + h^2 + h}{h} = 2x + h + 1.$$

Hence

$$\lim_{h \to 0} \frac{f(x+h) - f(x)}{h} = 2x + 1.$$

Now $2x + 1 = 3$ if $x = 1$. Since $f(1) = 3$, $(1, 3)$ is the desired point.

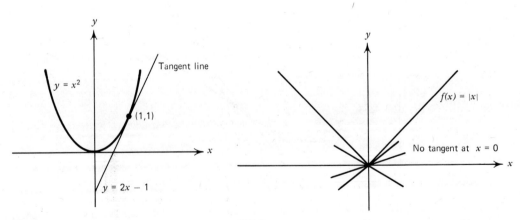

**Fig. 5**                              **Fig. 6**

**Example 3.**  Show that there is no tangent line to $f(x) = |x|$ at $x = 0$ (see Figure 6).

We must show that for $x = 0$ $\lim_{h \to 0} (f(0+h) - f(0))/h$ does not exist. But $(f(0+h) - f(0))/h = |h|/h$ and we have already observed (p. 32) that $\lim_{h \to 0} |h|/h$ does not exist since $|h|/h = 1$ if $h > 0$, and $|h|/h = -1$ if $h < 0$.

**EXERCISES**

1. Determine an equation for the tangent line to the parabola $y = 4x^2 + 4$ at the point $x = 2$.
2. Determine an equation for the tangent line to the parabola $y = 1 - x^2$ at the point $x = 2$.
3. For what value of $x$ does the tangent line to $f(x) = x^2 + 3x + 1$ at the point $x$ have slope $= 5$?
4. At which point $x$ does the tangent line to $f(x) = x^2 + x + 1$ pass through the origin?
5. For what values of $x$ does the tangent line to $f(x) = x^3 + 1$ at the point $x$ have slope $= 12$?
6. Show that the vertex of a parabola $y = a_2x^2 + a_1x + a_0$ (cf. Exercise 12, p. 27) is precisely the point at which the slope of the tangent line equals zero.
7. Locate the vertices of the following parabolas by using the results of Exercise 6.
   (a) $4y = 4x + 8 - x^2$
   (b) $x - y + x^2 = 4$
   (c) $3x + x^2 - y = 1$
8. The graph of the function $f(x) = 1/x$ is a rectangular hyperbola. Determine the slope of the tangent line to this hyperbola at the point $x = 2$.
9. Show that the tangent line to the rectangular hyperbola $xy = 1$ at the point $(x_0, y_0)$ has equation $y_0x + x_0y = 2$.
10. Show that at the point $(1, 1)$ the function

$$f(x) = 1 - |1 - x|$$

has no tangent line. Sketch the graph of this function.
11. Suppose

$$f(x) = \frac{|x|}{x} \text{ for } x \neq 0$$

$$= 0 \quad \text{if} \quad x = 0.$$

Does $\lim_{h \to 0} (f(x+h) - f(x))/h$ exist for $x = 0$? Explain your answer. Does $f$ have a tangent line at $x = 0$?
12. If $\lim_{h \to 0} (f(x+h) - f(x))/h$ exists, explain why $\lim_{h \to 0} f(x+h) = f(x)$.

## 2.3 The Derivative

Let $f$ be a real valued function defined on an interval $(a, b)$. The quotient $(f(x+h) - f(x))/h$ is called the *difference quotient of $f$ at $x$*. If the limit of the difference quotient exists as $h \to 0$, then $f$ is said to be *differentiable at the point $x$*. The function $f'$ defined by

(1) $$f'(x) = \lim_{h \to 0} \frac{f(x+h) - f(x)}{h}$$

is called the *derivative of $f$*. The domain of this function $f'$ is the set of points $x$ such that the limit in (1) exists. A function $f$ that is differentiable at each point of its domain is called a *differentiable function*.

The derivative of a function $f$ at the point $x_0$ is also called "the rate of change of $f$ at $x_0$." Indeed, we saw in Section 2.1 that if $f$ is the position function of a particle moving along a line, then $f'$ is the rate of change of the position function, or the velocity. Thus if $y = f(x)$, we say that $f'(x_0)$ is the rate of change of $f$ with respect to $x$ at the point $x_0$, or alternatively the rate of change of $y$ with respect to $x$ at the point $x_0$. Geometrically the existence of the derivative $f'$ at a point $x_0$ means that the graph of $f$ has a nonvertical tangent line at the point $x_0$. Indeed, the slope of this line is just

$f'(x_0)$, which is also called the slope of the function $f$ at the point $x_0$. Now if $f$ has a tangent at $x_0$, then the graph of $f$ must be unbroken at the point $x_0$. In other words $f$ is continuous at the point $x_0$. The proof of this important fact is very simple and follows immediately from Theorem 1.10.

**THEOREM 2.1**   *If the function $f$ is differentiable at the point $x$, then $f$ is continuous at $x$.*

**Proof.**   To show $f$ is continuous at $x$ we must show that $\lim_{h\to 0} f(x+h) = f(x)$. But for $h \neq 0$

$$f(x+h) - f(x) = \frac{f(x+h) - f(x)}{h} \cdot h.$$

Therefore, by Theorem 1.10 (4),

$$\lim_{h\to 0} [f(x+h) - f(x)] = \lim_{h\to 0} \frac{f(x+h) - f(x)}{h} \cdot \lim_{h\to 0} h$$

$$= 0.$$

It is important to note that the converse to this statement is not valid. That is, if a function is continuous at a point $x$, it is not necessarily differentiable there. The function

$$f(x) = |x|,$$

which we have examined repeatedly, is an example. $\lim_{h\to 0} f(0+h) = f(0) = 0$ yet as we showed on p. 32.

$$\lim_{h\to 0} \frac{f(0+h) - f(0)}{h} = \lim_{h\to 0} \frac{|h|}{h}$$

does not exist. See Figure 7.

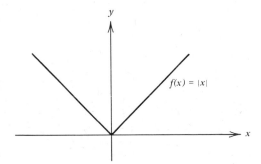

**Fig. 7**   $f(x) = |x|$ is continuous at $x = 0$ but not differentiable there.

Intuitively speaking, a function $f$ will be differentiable at $x_0$ if the graph is smooth at this point. If $f'(x_0)$ does not exist but $f$ is continuous at $x_0$, then there will be some sort of irregularity in the graph at the point $x_0$. For example, if there is a sharp corner or cusp in the graph as at the point $x_0$ in Figure 9, this means that $f'(x_0)$ does not exist. There are more complicated irregularities that may occur if $f'(x_0)$ does not exist. These are difficult to sketch accurately, and we shall defer the discussion of them to Chapter 5. Of course, if $f$ is not continuous at $x_0$, i.e., there is a break in the graph at that point, then $f$ is not differentiable at that point either. Figures 8, 9, and 10 illustrate these possibilities.

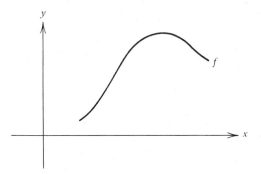

**Fig. 8** The function $f$ is differentiable at all points $x$.

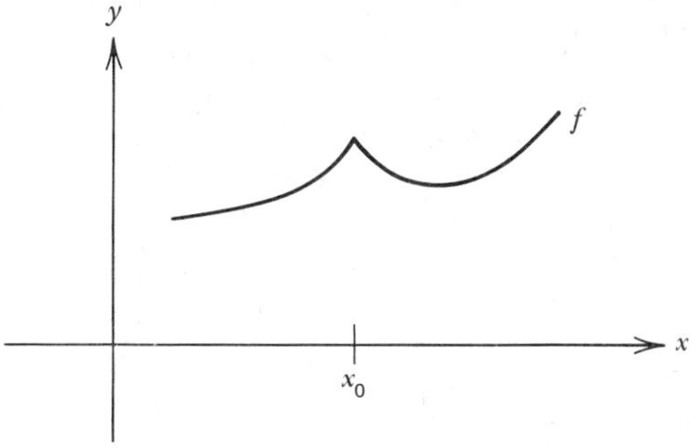

**Fig. 9** The function $f$ is continuous at the point $x_0$ but not differentiable there.

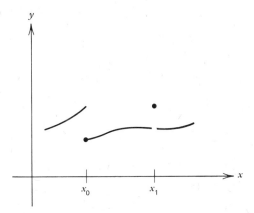

**Fig. 10**

In Figure 10 the function $f$ is discontinuous at $x_0$ since $\lim_{h \to 0} f(x_0 + h)$ does not exist. Therefore, $f$ is not differentiable there. At $x_1$, $\lim_{h \to 0} f(x_1 + h)$ exists, but the limit does not equal $f(x_1)$. The function is not continuous or differentiable at this point either. To indicate the value of $f$ at a discontinuity $x_0$ we draw a boldface dot as in Figure 10.

Since the derivative $f'$ is again a function, we may form the difference quotient of $f'$ at $x$, namely

$$\frac{f'(x+h) - f'(x)}{h}.$$

If $\lim_{h \to 0} (f'(x+h) - f'(x))/h$ exists, then $f$ is said to be twice differentiable at $x$. The function $f''$, defined by the formula

$$f''(x) = \lim_{h \to 0} \frac{f'(x+h) - f'(x)}{h},$$

is called the *second derivative* of $f$, which measures the rate of change of $f'$. In the case that $f$ is a position function of a particle, then $f''$ is the *acceleration*. (Exercise 7, p. 46). We shall postpone the discussion of the geometric significance of the second derivative until Chapter 5.

The process of computing derivatives, called *differentiation*, may be repeated, forming third, fourth, and higher derivatives, which we write

$f'''(x), f'^{v}(x)$, etc. The superscript is called the *order* of the derivative. Thus, $f'^{v}$ is the fourth-order derivative and $f^{(n)}$ the *n*th-order derivative. A function $f$ with derivatives of all orders is said to be *infinitely differentiable*. We shall show in the next section that polynomials are infinitely differentiable functions. Rational functions, or the quotients of polynomials, are also infinitely differentiable at each point at which the denominator fails to vanish.

In the next two sections we shall develop formulas that greatly facilitate the calculation of derivatives. However, the student should develop some facility for computing derivatives directly from the definition. For rational functions this calculation takes the following form.

1. Compute the difference quotient $(f(x+h)-f(x))/h$.
2. Simplify and cancel $h$ from the denominator.
3. Evaluate the resulting limit as $h \to 0$ by applying Theorem 1.10.

**Example 1.** If $f(x)=c, c$ a constant, show that $f'(x)=0$. Forming the difference quotient we get

$$\frac{f(x+h)-f(x)}{h} = \frac{c-c}{h} = 0.$$

Therefore, $\lim_{h\to 0} (f(x+h)-f(x))/h = 0$.

**Example 2.** If $f(x)=cx, c$ constant, show that $f'(x)=c$. Again for $h \neq 0$

$$\frac{f(x+h)-f(x)}{h} = \frac{c(x+h)-cx}{h} = c.$$

Therefore, $\lim_{h\to 0} (f(x+h)-f(x))/h = c$ as required.

**Example 3.** If $f(x)=1/(x+1)$, compute $f'(x)$. For $h \neq 0$ we have

$$\frac{f(x+h)-f(x)}{h} = \frac{\dfrac{1}{x+h+1} - \dfrac{1}{x+1}}{h}$$

$$= \frac{x+1-(x+h+1)}{(x+h+1)(x+1)\cdot h} = \frac{-1}{(x+h+1)(x+1)}.$$

Therefore

$$\lim_{h\to 0} \frac{f(x+h)-f(x)}{h} = \lim_{h\to 0} \frac{-1}{(x+h+1)(x+1)} = \frac{-1}{(x+1)^2}.$$

and, consequently, $f'(x) = -1/(x+1)^2$.

**Example 4.** Determine an equation for the tangent line to $f(x)=1/(x+1)$ at the point $(2, \frac{1}{3})$. The slope $m$ of the tangent line at $x$ is given by $f'(x)$. But by Example 3 if $f(x)=1/(x+1)$, $f'(x) = -1/(x+1)^2$. Hence $f'(2)=-\frac{1}{9}$. Therefore

$$\frac{y - \dfrac{1}{3}}{x-2} = -\frac{1}{9}$$

or

$$9y+x-5=0$$

which is an equation for the tangent line (see Figure 11).

**Example 5.** Determine the second derivative $f''(x)$ if $f(x)=1/(x+1)$. By definition

$$f''(x) = \lim_{h\to 0} \frac{f'(x+h)-f'(x)}{h}.$$

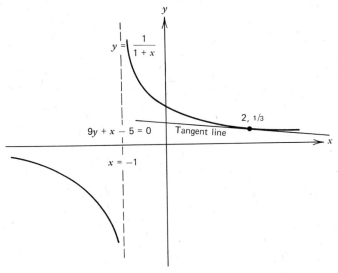

**Fig. 11**

From Example 3, $f'(x) = -1/(x+1)^2$. Hence, if $h \neq 0$,

$$\frac{f'(x+h) - f'(x)}{h} = \frac{-1/(x+h+1)^2 + 1/(x+1)^2}{h} = \frac{(x+h+1)^2 - (x+1)^2}{h(x+h+1)^2(x+1)^2}$$

$$= \frac{-2h(x+1) + h^2}{h(x+h+1)^2(x+1)^2} = \frac{-2(x+1) + h}{(x+h+1)^2(x+1)^2}.$$

Therefore

$$f''(x) = \lim_{h \to 0} \frac{-2(x+1) + h}{(x+h+1)^2(x+1)^2}$$

$$= \frac{-2(x+1)}{(x+1)^4} = \frac{-2}{(x+1)^3}.$$

***Example 6.*** Define a function $f$ as follows:

$$\begin{aligned}
f(x) &= 1 & 0 &\leqslant x \leqslant 1 \\
f(x) &= 2 & 1 &< x < 2 \\
f(x) &= 3 & x &= 2 \\
f(x) &= 2 & 2 &< x \leqslant 3 \\
f(x) &= x - 1 & 3 &< x \leqslant 4 \\
f(x) &= \frac{x^2}{8} + 1 & 4 &\leqslant x \leqslant 5.
\end{aligned}$$

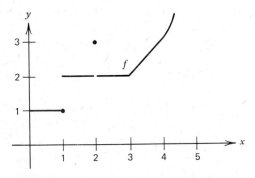

**Fig. 12**

Discuss the differentiability and continuity of $f$ at the points $x = 1, 2, 3, 4$ (see Figure 12).

$x = 1$:

$\lim_{x \to 1} f(x)$ (or $\lim_{h \to 0} f(1 + h)$) does not exist, so $f$ is not continuous and hence not differentiable at $x = 1$

$x = 2$:

$\lim_{x \to 2} f(x) = 2$ but $f(2) = 3$. Therefore $f$ is not continuous and hence not differentiable at $x = 2$.

$x = 3$:

$\lim_{x \to 3} f(x) = \lim_{h \to 0} f(3 + h) = 2 = f(3)$. Therefore $f$ is continuous at $x = 3$. However for $-1 < h < 0$

$$\frac{f(3 + h) - f(3)}{h} = \frac{2 - 2}{h} = 0,$$

and if $0 < h < 1$

$$\frac{f(3 + h) - f(3)}{h} = \frac{3 + h - 1 - 2}{h} = 1.$$

Therefore $\lim_{h \to 0} (f(3 + h) - f(3))/h$ does not exist, and $f$ is not differentiable at $x = 3$.

$x = 4$:

For $-1 < h < 0$, $(f(4 + h) - f(4))/h = (4 + h - 1 - 3)/h = 1$.
For $0 < h < 1$

$$\frac{f(4 + h) - f(4)}{h} = \frac{(4 + h)^2/8 + 1 - 3}{h} = \frac{h + (h^2)/8}{h} = 1 + \frac{h}{8} \to 1 \text{ as } h \to 0.$$

Therefore, $\lim_{h \to 0} (f(4 + h) - f(4))/h = 1$ and $f$ is differentiable at $x = 4$. Since differentiability implies continuity, $f$ is also continuous at $x = 4$.

## EXERCISES

1. For the following functions compute the first and second derivatives directly from the definition.

   (a) $f(x) = 1 - 3x$

   (b) $f(x) = 2x^2 + x$

   (c) $f(t) = t - 3t^2$

   (d) $g(s) = (s - 1)^2$

   (e) $h(s) = \dfrac{1}{4 - s}$

   (f) $g(s) = \dfrac{1}{2s + 1} + s$

   (g) $h(t) = t^3 + 1$

   (h) $p(u) = (u - 1)^3$

   (i) $p(r) = \dfrac{1}{r^2 - 1}$

   (j) $f(t) = \dfrac{t + 1}{t - 1}$

   (k) $g(s) = \dfrac{s}{s^2 + 1}$

   (l) $h(u) = \left(\dfrac{1}{1 + u}\right)^2$

2. Determine the equation of the tangent line to the function $f(x) = 2x^2 + x$ at the point $x = 1$.

3. Determine the equation of the tangent line to the function $f(t) = t - 3t^2$ at the point $t = 0$.

4. At what point $x_0$ is the tangent line to the function $g(x) = (x - 1)^2$ at $x_0$ perpendicular to the line $y = 2x$?

5. At what point $x_0$ is the tangent line to the function $f(x) = x^3 + 1$ at the point $x_0$ parallel to the line $y - 12x + 2 = 0$?

6. A line $l$ is normal to a function $f$ at a point $x$, if the line $l$ passes through the point $(x, f(x))$ and is perpendicular to the tangent line to $f$ at $x$. Determine equations for the normal lines to the following functions at the indicated points

(a) $f(x) = 1 - 2x^2$     $x = 2$

(b) $g(s) = \dfrac{1}{2+s}$     $s = -1$

(c) $h(t) = (t+1)^3$     $t = 1$

7. If $f(x) = x^n$, $n$ being a positive integer, show that $f'(x) = nx^{n-1}$. [By the binomial formula

$$(x+h)^n = x^n + nx^{n-1}h + \frac{n(n-1)}{2}x^{n-2}h^2 + \cdots + h^n.$$

Using this compute $((x+h)^n - x^n)/h$, simplify, and then compute the limit as $h \to 0$.]

8. Show that this formula is still valid if $n$ is a negative integer.

9. Determine if the following functions are continuous or differentiable at the point $x = 0$. Give reasons for your answers, and sketch the graph of the function.

(a) $f(x) = 2, x \leqslant 0$
       $= 1, x > 0$

(b) $f(x) = |x|, x \neq 0$
       $= 1, x = 0$

(c) $f(x) = x^2, x \leqslant 0$
       $= x^3, x > 0$

(d) $f(x) = x, x \leqslant 0$
       $= x^2, x > 0$

(e) $f(x) = x, x \leqslant 0$
       $= x - x^2, x > 0$

10. Let $f$ be differentiable at the point $x_0$. Let $l(x) = ax + b$ be the tangent line to $f$ at $x_0$. Then $d(x) = l(x) - f(x)$ is the vertical distance between $f$ and the tangent line at the point $x$. (Figure 13.) Show that $\lim_{x \to x_0} d(x)/(x - x_0) = 0$

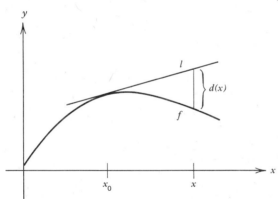

**Fig. 13**

11. Conversely, if there is a linear function $l(x) = ax + b$ with the property that

$$\lim_{x \to x_0} \frac{l(x) - f(x)}{x - x_0} = 0,$$

show that $f$ is differentiable at $x_0$ and that $y = l(x)$ is the tangent line to $f$ at $x_0$.

If $y = l(x)$ is any linear function passing through $(x_0, f(x_0))$ we may think of $l$ as approximating $f$ near $x_0$. The error of this approximation is the difference $l(x) - f(x)$ (Figure 13). The *relative error* is the quotient $(l(x) - f(x))/(x - x_0)$. Thus Exercises 10 and 11 show that $f$ is differentiable at $x_0$ if and only if there is a straight line passing through $x_0$ such that the relative error

$$\epsilon(x) = (l(x) - f(x))/(x - x_0)$$

tends to zero as $x \to x_0$. We shall explore the significance of this fact in Section 2.11.

**2.4 Differentiation Formulas**

If the functions $f$ and $g$ are differentiable at $x$, we wish to derive formulas for the derivative of the sum, difference, product, and quotient of $f$ and $g$. We summarize these below.

**THEOREM 1.9** *Let $f$ and $g$ be differentiable at $x$.*

(1) *If $c$ is constant and $h(x) = cf(x)$, then $h$ is differentiable at $x$, and*

$$h'(x) = cf'(x).$$

(2) *If $s(x) = f(x) + g(x)$, then $s$ is differentiable at $x$, and*

$$s'(x) = f'(x) + g'(x).$$

(3) *If $p(x) = f(x)g(x)$, then $p$ is differentiable at $x$, and*

$$p'(x) = f'(x)g(x) + f(x)g'(x).$$

(4) *If $q(x) = (f(x)/g(x))$ and $g(x) \neq 0$, then $q$ is differentiable at $x$, and*

$$q'(x) = \frac{g(x)f'(x) - f(x)g'(x)}{[g(x)]^2}.$$

*Proof.* To establish these results, we shall use Theorem 1.10 of Chapter 1, and we shall make repeated use of the fact that if $f$ is differentiable at $x$, then $f$ is continuous at $x$; that is, $\lim_{h \to 0} f(x+h) = f(x)$ (Theorem 2.1). We leave the proof of (1) as an exercise and begin with (2). Forming the difference quotient of $s$ we have

$$\frac{s(x+h) - s(x)}{h} = \frac{f(x+h) - f(x)}{h} + \frac{g(x+h) - g(x)}{h}.$$

Since the limit of a sum is the sum of the limits [Theorem 1.10(2)]

$$\lim_{h \to 0} \frac{s(x+h) - s(x)}{h} = \lim_{h \to 0} \frac{f(x+h) - f(x)}{h} + \lim_{h \to 0} \frac{g(x+h) - g(x)}{h}$$

$$= f'(x) + g'(x).$$

Hence, $s$ is differentiable at $x$ and

$$s'(x) = f'(x) + g'(x)$$

For (3), if $p(x) = f(x)g(x)$ then

$$\frac{p(x+h) - p(x)}{h} = \frac{f(x+h)g(x+h) - f(x)g(x)}{h}$$

$$= \frac{f(x+h)g(x+h) - f(x)g(x+h) + f(x)g(x+h) - f(x)g(x)}{h}$$

$$= \frac{f(x+h) - f(x)}{h} \cdot g(x+h) + f(x) \cdot \frac{g(x+h) - g(x)}{h}.$$

Now, by Theorem 1.10 (3) and Theorem 2.1

$$\lim_{h \to 0} \frac{p(x+h) - p(x)}{h} =$$

$$\lim_{h \to 0} \frac{f(x+h) - f(x)}{h} \cdot \lim_{h \to 0} g(x+h) + \lim_{h \to 0} f(x) \cdot \lim_{h \to 0} \frac{g(x+h) - g(x)}{h}$$

$$= f'(x)g(x) + f(x)g'(x).$$

Hence, $p$ is differentiable at $x$, and

$$p'(x) = f'(x)g(x) + f(x)g'(x).$$

To prove (4) we first prove the special case

(4')  *Let* $q(x) = 1/g(x)$, *and assume* $g(x) \neq 0$. *Then* $q$ *is differentiable at* $x$, *and*

$$q'(x) = -\frac{g'(x)}{[g(x)]^2}.$$

Forming the difference quotient for $q$ we have

$$\frac{q(x+h) - q(x)}{h} = \frac{1/g(x+h) - 1/g(x)}{h}$$

$$= -\frac{g(x+h) - g(x)}{h} \cdot \frac{1}{g(x) \cdot g(x+h)}.$$

Applying Theorem 1.10 (4') and the fact that $\lim_{h \to 0} g(x+h) = g(x)$ (Theorem 2.1) we conclude that

$$\lim_{h \to 0} \frac{q(x+h) - q(x)}{h} = -\lim_{h \to 0} \left[ \frac{g(x+h) - g(x)}{h} \right] \cdot \lim_{h \to 0} \frac{1}{[g(x) \cdot g(x+h)]}$$

$$= \frac{-g'(x)}{[g(x)]^2}.$$

Hence, $q$ is differentiable at $x$, and

$$q'(x) = -\frac{g'(x)}{[g(x)]^2}.$$

We leave as an exercise the verification that (4') combined with (3) yields (4).

The following examples illustrate the computation of derivatives by the use of Theorem 2.2. It is immaterial, of course, which symbols $f$, $g$, $h$, etc., are used to represent the function and which symbols $x$, $y$, $s$, $t$, etc., are used to represent points in the domain of the function.

*Example 1.*  Let $f(x) = x^2 = x \cdot x$. Compute $f'(x)$. Applying (3) we have

$$f'(x) = 1 \cdot x + x \cdot 1 = 2x.$$

*Example 2.*  Let $g(t) = t^2(3 + t)$. Compute $g'(t)$. Again, applying (3) we have

$$g'(t) = 2t(3 + t) + t^2 \cdot 1 = 3t^2 + 6t.$$

*Example 3.*  Let $h(s) = (s + 1)/(1 + s^2)$. Compute $h'(s)$. By (4)

$$h'(s) = \frac{(1 + s^2) \cdot 1 - (s + 1)(2s)}{(1 + s^2)^2} = \frac{1 - s^2 - 2s}{(1 + s^2)^2}.$$

*Example 4.*  If $n$ is a positive integer and $f(x) = x^n$, show that $f$ is differentiable and show that

$$(5) \qquad\qquad\qquad f'(x) = nx^{n-1}.$$

We verify this by induction. If $n = 1$, $f(x) = x$, and $f'(x) = 1$, the statement is true for $n = 0$. Assume the result for $n = k$.

We must prove that if $n = k + 1$, then $f(x) = x^n$ is differentiable and $f'(x) = nx^{n-1}$.

But if $f(x) = x^n = x^{k+1}$, then $f(x) = x^k \cdot x$ and by Theorem 2.2 (3), $f$ is differentiable. Moreover by the product formula (3)

$$f'(x) = kx^{k-1} \cdot x + x^k \cdot 1$$
$$= (k+1)x^k.$$

Applying the principle of induction we conclude the result for all positive integers $n$.

Formula (5) is valid for negative integers as well, provided that $x \neq 0$.

To see this, assume $k < 0$ and set $n = -k$. Therefore $f(x) = x^n = x^{-k} = 1/x^k$. Moreover by the quotient formula (4)

$$f'(x) = -\frac{kx^{k-1}}{x^{2k}} = \frac{-k}{x^{k+1}} = nx^{n-1}.$$

If $n = 0$, then $f(x) = x^n = 1$ and $f'(x) = 0$. Hence the power function $f(x) = x^n$ is differentiable for each integer $n$ and $f'(x)$ is given by the *power formula* (5). We state this important resultant as a theorem.

**THEOREM 2.3**   *Let $n$ be an integer, and set $f(x) = x^n$. Then $f$ is differentiable and*

(5) $$f'(x) = nx^{n-1}.$$

*If $n < 0$, it is assumed that $x \neq 0$.*

If we use the notation

$$(f + g)(x) = f(x) + g(x),$$
$$(f \cdot g)(x) = f(x) \cdot g(x)$$

and

$$\left(\frac{f}{g}\right)(x) = \frac{f(x)}{g(x)}$$

to denote the sum, product, and quotient of the functions $f$ and $g$, then formulas (2), (3) and (4) of Theorem 2.2 become

(2) $$(f + g)' = f' + g'$$

(3) $$(f \cdot g)' = f' \cdot g + f \cdot g'$$

and

(4) $$\left(\frac{f}{g}\right)' = \frac{gf' - fg'}{g^2}.$$

**EXERCISES**

Differentiate the following functions using the formulas of Theorems 2.2 and 2.3.

1. $f(x) = 3x - 2$.

2. $h(x) = 3x^2 + 4x + 1$.

3. $g(t) = 4t^5 - \dfrac{2}{t^3}$.

4. $f(x) = (2x + 5)^2$.

5. $g(p) = \dfrac{3p + 1}{2p - 4}$.

6. $p(s) = s^2(2s - 1)$.

7. $m(t) = \dfrac{1}{4t - 7}$.

8. $f(y) = \dfrac{y^2 + 4}{1 - 2y^3}$.

9. $g(x) = \dfrac{x^2}{x^2 - 1}$.

10. $f(y) = \dfrac{y + 1}{y - 1}$.

11. $h(s) = \dfrac{1}{2s^2 - s}$.

12. $p(t) = \dfrac{1 - t^2}{t^2 + 1}$.

13. $l(t) = \dfrac{7t^5 - t^3}{t+1}$.

14. $f(t) = (t+1)(2t^2 - 4)$.

15. $m(y) = \dfrac{1+y}{(2-y)(3+y)}$.

16. $f(t) = (2-t)^2(3+t)$.

17. $g(z) = \dfrac{(4z-3)(z^2)}{(2z+1)(3z-2)}$.

18. $g(s) = \dfrac{3s^2 + s - 1}{s+2}$.

19. $m(u) = \dfrac{u^5 + 4u}{u-3}$.

20. $h(x) = 2(x+1)(x-2)(x+3)$.

21. $f(s) = s(s^2 - 1)^2$.

22. $p(t) = \dfrac{3t^2 - 1}{4t^5}$.

23. $f(x) = \dfrac{(2x-1)^2}{x^6}$.

24. Establish formula (1) of Theorem 2.2.

25. Let $a$ and $b$ be constants and assume $f$ and $g$ are differentiable at $x$. If $h(x) = af(x) + bg(x)$, show that $h$ is differentiable at $x$ and show that $h'(x) = af'(x) + bg'(x)$.

26. Finish the proof of Theorem 2.2 by showing that (3) and (4′) imply (4).

27. Let $f$ be differentiable at $x$ and assume $n$ is a nonnegative integer. Show that $g(x) = [f(x)]^n$ is differentiable at $x$ and show that $g'(x) = n[f(x)]^{n-1} \cdot f'(x)$. (Adapt the argument in example 4.)

28. If $f$ is differentiable at $x$ and $n$ is a negative integer show that $g(x) = [f(x)]^n$ is differentiable at $x$ if $f(x) \neq 0$ and furthermore $g'(x) = n[f(x)]^{n-1}f'(x)$.

29. Using the general power formula of Exercises 27 and 28 differentiate the following functions.

(a) $f(x) = (3x+1)^5$

(b) $g(x) = (2x^2 - 1)^6$

(c) $p(t) = (t^2 + 3t)^8$

(d) $m(s) = \dfrac{1}{(2s+3)^5}$

(e) $g(x) = \dfrac{4}{(2x^2 + 3)^6}$

(f) $f(x) = \dfrac{(2x-3)^4}{(3x+2)^3}$

(g) $p(t) = \dfrac{(4t^2 + 3t)^5}{(1 + 1/t)^4}$

(h) $m(t) = \dfrac{(1 - 1/t)^3}{(t + 4/t^3)^6}$

(i) $f(x) = \dfrac{(4x-1)^5(2x+3)^4}{(5-2x)^3}$

(j) $p(t) = \dfrac{(2t+4)^5(2-3t)^2}{(t+1)^4(4-t)^3}$

## 2.5  Composition of Functions

If $f$ and $g$ are real valued functions, then we may use the arithmetic operations to form new functions $f+g$, $f \cdot g$, $f/g$ from the given ones. Another extremely important device for constructing new functions is the operation of function composition. Before giving the formal definition let us illustrate the idea with an example.

Let $P = (x_0, y_0)$ be a point in the plane and let $x = f(t)$ be the position function of a particle moving along the $x$ axis.

What is the distance from the point $P$ to the particle as a function of time? The distance $d$ between two points in the plane with coordinates $(x, y)$ and $(x_0, y_0)$ is defined to be

$$d = \sqrt{(x - x_0)^2 + (y - y_0)^2}.$$

Hence, if the point $(x, y)$ lies on the $x$ axis, we may write the distance $d$ as a function of the $x$ coordinate of the point, and we have

(1) $$d(x) = \sqrt{(x - x_0)^2 + y_0^2}.$$

If the point $P = (x_0, y_0)$ lies on the $x$ axis as well, then we have

$$d(x) = \sqrt{(x - x_0)^2} = |x - x_0|$$

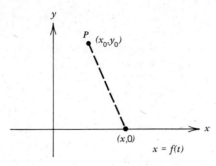

**Fig. 14**

Now for each value of $t$ the distance between the moving particle and the point $P$ can be written as a function of time by substituting $f(t)$ for $x$ in equation (1) (see Figure 14). If we denote this function by $g$, we have

$$g(t) = \sqrt{(f(t) - x_0)^2 + y_0^2}$$

This function $g$ is called the composition of $d$ and $f$. For example, if $f(t) = t^2 + 1$, then

$$g(t) = \sqrt{(t^2 + 1 - x_0)^2 + y_0^2}.$$

We now give the formal definition of the composition of two functions.

**Definition.**   *Let $f$ and $g$ be functions. The composition of $g$ and $f$, written $g \circ f$ is the set of ordered pairs $(x, z)$ such that for some $y$ belonging to the domain of $g$*

$$y = f(x)$$

*and*

$$z = g(y).$$

Thus, to obtain $z$ from $x$ we first apply $f$, obtaining $y = f(x)$. Then we apply $g$, obtaining

$$z = g(y) = g(f(x)).$$

We may abbreviate the definition by writing

$$g \circ f = \{(x, z) : z = g(f(x))\}.$$

Hence

$$(g \circ f)(x) = g(f(x)).$$

The notation $g \circ f$ for composition should not be confused with $g \cdot f$ which is the product of $g$ and $f$.

**Example 1.**   If $f(x) = 1 + x^2$ and $g(x) = x^{10}$, then $(g \circ f)(x) = (1 + x^2)^{10}$. Note $(f \circ g)(x) = 1 + (x^{10})^2 = 1 + x^{20}$. Thus $g(f(1)) = (g \circ f)(1) = 2^{10}$, whereas $f(g(1)) = (f \circ g)(1) = 2$. This example shows that function composition is not commutative. Thus $f \circ g \neq g \circ f$.

If we think of a function $f$ as a transformation or mapping that transforms the point $x$ into the point $f(x)$, then $g \circ f$ transforms the point $x$ into the point $g(f(x))$. The arrow notation for function introduced in Figure 12 (Chapter 1, p. 00), illustrates this idea quite well (see Figure 15).

**Example 2.**   Let $f(x) = 1 - x^2$, $g(x) = \sqrt{x}$. Determine $g \circ f$ and specify its domain.

$$(g \circ f)(x) = g(f(x)) = \sqrt{1 - x^2}.$$

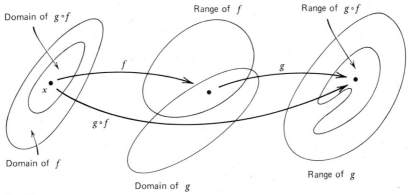

**Fig. 15**

In order for $y = 1 - x^2$ to lie in the domain of the square root function, we must have $y \geq 0$. Therefore, the domain of

$$(g \circ f)(x) = \sqrt{1 - x^2} = \{x : 1 - x^2 \geq 0\} = \{x : -1 \leq x \leq 1\}.$$

Note that in general the domain of $g \circ f$ is the set of those points $x$ belonging to the domain of $f$ such that $f(x)$ belong to the domain of $g$.

It is an important fact, and one that is easy to prove, that the composition of two continuous functions is continuous. This is our next result.

**THEOREM 1.9** *Let $f$ and $g$ be real valued functions defined on the intervals $(a, b)$ and $(c, d)$ respectively. Let $x_0 \in (a, b)$ and assume $y_0 = f(x_0) \in (c, d)$. If $f$ is continuous at $x_0$ and $g$ is continuous at $y_0$, then $g \circ f$ is continuous at $x_0$.*

**Proof.** We must show that $\lim_{x \to x_0} g(f(x)) = g(f(x_0))$. This means that we must show that for each $\epsilon > 0$ there is a corresponding $\delta > 0$ such that whenever $|x - x_0| < \delta$, it follows that $|g(f(x)) - g(f(x_0))| < \epsilon$. But since $g$ is continuous at $y_0 = g(f(x_0))$, for each $\epsilon > 0$ there is a corresponding $\delta_1$ such that whenever $|y - y_0| < \delta_1$ it follows that $|g(y) - g(y_0)| < \epsilon$. But now by the continuity of $f$ at $x_0$, there is a number $\delta$ such that whenever $|x - x_0| < \delta$ it follows that $|f(x) - f(x_0)| < \delta_1$. This implies that $|g(f(x)) - g(f(x_0))| < \epsilon$. This shows that $\lim_{x \to x_0} g(f(x)) = g(f(x_0))$.

The choice of $\delta_1$ in terms of $\epsilon$ and then $\delta$ in terms of $\delta_1$ for two functions $f$ and $g$ is illustrated in Figure 16.

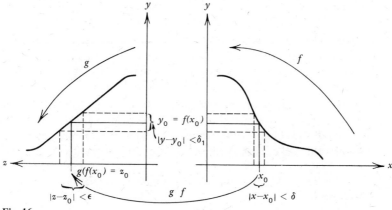

**Fig. 16**

**EXERCISES**

1. A particle moves along the $y$ axis such that at time $t$, $y = 2t - 1$. Write a function $f$ that expresses the distance from the particle to the point $(1, 3)$ as a function of time. Determine $g$ and $h$ such that $f = g \circ h$.

2. A particle moves along the line $y = 3x + 2$ such that at time $t$, $x = 2t^2 - 4$. Express the distance of the particle from the $x$ axis as a function of time. What are functions $f$ and $g$ such that this distance function equals $g \circ f$?

3. A particle moves along the parabola $4y = 3x^2 - 6x + 2$ such that at time $t$, $x = t^2 + 4$. Write a function that expresses the cube of the distance from the particle to the $x$ axis as a function of time. Express this function as the composition of three functions $f$, $g$, $h$.

4. Let $f(x) = 2x + 3$, $g(x) = x^2 - 4$. Compute $(f \circ g)(x)$, $(g \circ f)(x)$, $(f \circ f)(x)$, and $(g \circ g)(x)$. Evaluate each of these functions at $x = 1$.

5. Let $f(x) = 1/(2x - 1)$, $g(x) = x^2$. Compute $f \circ g$ and $g \circ f$. Determine the domain of each of these functions.

6. If $f(x) = 1/(3x - 4)$, determine $f \circ f$. What is the domain of this function?

7. Let $f(x) = 1/(x - 1)$, $g(x) = \sqrt{1 - x}$. Determine $f \circ g$ and $g \circ f$ and their respective domains.

8. Let $f(x) = 1/(1 - x^2)$, $g(x) = x^2$. Compute $(f \circ g)(x)$, $(g \circ f)(x)$, $(f \circ g)(2)$, $(g \circ f)(2)$. Specify the domain of $f \circ g$ and $g \circ f$. Compute $f \circ (g \circ f)$, $g \circ (f \circ g)$.

9. If $f(x) = \sqrt{1 - x}$ and $g(x) = |x - 1|$, compute $f \circ g$, $g \circ f$ and specify the domains of these functions.

10. Let $f(x) = 1/(x - 1)$, $g(x) = 3x + 1$. Write $f \circ g$ and determine the domain of this function. Compute

$$\lim_{h \to 0} \frac{(f \circ g)(x + h) - (f \circ g)(x)}{h}.$$

11. Let $f(x) = 1/(x^2 - 1)$, $g(x) = 2x - 1$. Write $f \circ g$ and determine the domain of this function. Compute.

$$\lim_{h \to 0} \frac{(f \circ g)(x + h) - (f \circ g)(x)}{h}.$$

12. Show that the operation of function composition is associative That is $f \circ (g \circ h) = (f \circ g) \circ h$ for each triple of functions $f$, $g$, $h$. As a result we may omit the parentheses and write

$$f \circ g \circ h = f \circ (g \circ h) = (f \circ g) \circ h$$

## 2.6   The Chain Rule

Suppose a particle moves along the line $y = 3x$ in such a way that at time $t$, the $x$ coordinate of the position of the particle equals $2t$. Thus $x = 2t$, and since $x$ is a linear function of $t$ the rate of change of $x$ with respect to $t$ equals 2. Similarly if $y = 3x$, then the rate of change of $y$ with respect to $x$ equals 3. What is the rate of change of the $y$ coordinate of the particle with respect to time? Substituting $x = 2t$ in $y = 3x$ yields $y = 6t$. Hence the rate of change of $y$ with respect to $t$ equals 6. This is the product of the two rates of change. In this example we have formed the composition $g(f(t))$ where $f(t) = 2t$ and $g(x) = 3x$. Moreover if $h = g \circ f$, then the derivative $h'$ is the product of the derivative of $f$ with the derivative of $g$. This is a special case of a general result for the composition of two functions $g$ and $f$ called the chain rule. Indeed if $h(x) = g(f(x))$, then the chain rule asserts that $h'(x)$ is the product of $g'$ evaluated at $f(x)$ and $f'(x)$. That is,

(1)                               $h'(x) = g'(f(x))f'(x).$

The chain rule enables us to differentiate immediately the composition of two differentiable functions. Before proving the result, let us compute a few examples.

**Example 1.** Let $f(x) = x^2 - 1$, $g(x) = x^3$. Then $h(x) = g(f(x)) = (x^2 - 1)^3$. Now $f'(x) = 2x$ and $g'(x) = 3x^2$. Hence, $g'(f(x)) = 3(x^2 - 1)^2$, and by the chain rule formula (1)

$$h'(x) = g'(f(x))f'(x) = 3(x^2 - 1)^2 2x.$$

We could, of course, have derived this formula by successive applications of the product formula for derivatives. If $g(x) = x^{25}$, though, this would be very tiresome. However, the derivative of $h(x) = g(f(x)) = (x^2 - 1)^{25}$ can be computed virtually by inspection from the chain rule. Indeed $g'(x) = 25x^{24}$. Hence, $g'(f(x)) = 25(x^2 - 1)^{24}$, and applying (1) we have

$$h'(x) = 25(x^2 - 1)^{24} 2x.$$

The chain rule formula (1) may be iterated for three or more functions. Thus, if

$$p(x) = h(g(f(x))),$$

then

$$p'(x) = h'(g(f(x))) \cdot g'(f(x)) \cdot f'(x).$$

**Example 2.** If $g(x) = 1 + x^{10}$, $h(x) = x^5$, and $f(x) = x + (1/x)$, then

$$p(x) = h(g(f(x))) = \left(1 + \left(x + \frac{1}{x}\right)^{10}\right)^5.$$

But $f'(x) = 1 - (1/x^2)$, $g'(x) = 10x^9$, and $h'(x) = 5x^4$.
By the chain rule

$$p'(x) = \underbrace{h'(g(f(x)))} \cdot \underbrace{g'(f(x))} \cdot \underbrace{f'(x)}$$

$$= 5\left(1 + \left(x + \frac{1}{x}\right)^{10}\right)^4 \cdot 10\left(x + \frac{1}{x}\right)^9 \cdot \left(1 - \frac{1}{x^2}\right).$$

**THEOREM 2.5** *Let the functions $f$ and $g$ be defined on the intervals $(a, b)$ and $(c, d)$, respectively. If $f$ is differentiable at $x$, $a < x < b$, and $g$ is differentiable at $y = f(x)$, where $c < y < d$, then $g \circ f$ is differentiable at $x$ and*

(1)
$$(g \circ f)'(x) = g'(f(x)) \cdot f'(x) = g'(y) \cdot f'(x).$$

**Proof.** Let $p(x) = g(f(x))$. Form the difference quotient

(2)
$$\frac{p(x+h) - p(x)}{h} = \frac{g(f(x+h)) - g(f(x))}{h}.$$

A straightforward way to show that (2) has a limit as $h \to 0$ and to derive formula (1) would be the following. Let $y = f(x)$ and set $k = f(x+h) - f(x)$. Then $f(x+h) = y + k$. Multiply numerator and denominator of (2) by $f(x+h) - f(x)$ obtaining

(3)
$$\frac{g(f(x+h)) - g(f(x))}{h} = \frac{g(y+k) - g(y)}{f(x+h) - f(x)} \cdot \frac{f(x+h) - f(x)}{h}$$

$$= \frac{g(y+k) - g(y)}{k} \cdot \frac{f(x+h) - f(x)}{h}.$$

Since $f$ is differentiable, $f$ is continuous. Hence, $\lim_{h \to 0} f(x+h) - f(x) = 0$, or $k \to 0$ as $h \to 0$. Since $g$ is assumed to be differentiable at $y$, it appears that

(4)
$$\lim_{h \to 0} \frac{g(y+k) - g(y)}{k} = \lim_{k \to 0} \frac{g(y+k) - g(y)}{k} = g'(y).$$

Hence, by Theorem 1.10 (3),

$$\lim_{h \to 0} \frac{p(x+h)-p(x)}{h} = \lim_{h \to 0} \frac{g(y+k)-g(y)}{k} \cdot \lim_{h \to 0} \frac{f(x+h)-f(x)}{h}$$

$$= \lim_{k \to 0} \frac{g(y+k)-g(y)}{k} \cdot f'(x)$$

$$= g'(y) \cdot f'(x) = g'(f(x)) \cdot f'(x).$$

which is the required formula (1).

However, if we attempt to justify (4), we run into the following problem. The device of multiplying the right-hand side of (3) by $(f(x+h)-f(x))/(f(x+h)-f(x))$ is only legitimate if $k = f(x+h)-f(x) \neq 0$, and this cannot always be guaranteed. To get around this difficulty we use the following technique, which is useful in other similar situations. Define:

(5)
$$\begin{cases} u(k) = \dfrac{g(y+k)-g(y)}{k} - g'(y) \text{ if } k \neq 0 \\ \quad = 0 \text{ if } k = 0 \end{cases}$$

Now $\lim_{k \to 0} u(k) = 0$, since $\lim_{k \to 0} (g(y+k)-g(y))/k = g'(y)$. Hence $u$ is continuous at $k = 0$. Since $k \to 0$ as $h \to 0$ we may conclude from Theorem 2.4 that $\lim_{h \to 0} u(k) = \lim_{k \to 0} u(k) = 0$.

Next we note that for all values of $k$ we have

$$g(y+k)-g(y) = [u(k)+g'(y)] \cdot k.$$

Hence

$$\frac{p(x+h)-p(x)}{h} = \frac{g(y+k)-g(y)}{h} = [u(k)+g'(y)]\frac{k}{h}.$$

Applying Theorem 1.10 (3) and Theorem 2.4 we infer that

$$\lim_{h \to 0} \frac{p(x+h)-p(x)}{h} = \lim_{h \to 0} [u(k)+g'(y)] \cdot \lim_{h \to 0} \left[\frac{k}{h}\right]$$

$$= \lim_{k \to 0} [u(k)+g'(y)] \cdot \lim_{h \to 0} \left[\frac{k}{h}\right]$$

$$= g'(y) \cdot f'(x) = g'(f(x)) \cdot f'(x).$$

Thus $p = g \circ f$ is differentiable at $x$ and the chain rule formula (1) is valid.

As an application of the chain rule we shall extend the power formula for differentiation (p. 60) to rational exponents as well as integral exponents. Thus, we wish to show that if

$$f(x) = x^{m/n}, \text{ then}$$

(6)
$$f'(x) = \frac{m}{n} x^{(m/n)-1}$$

First let us briefly discuss the function

$$f(x) = x^{1/n}, x \geq 0.$$

For $n = 1, 2, 3, \ldots$ the graphs of the equation $y = x^n$, $x \geq 0$ are given in Figure 17.

It appears from the sketch that for each $y \geq 0$ there is an $x \geq 0$ such that $y = x^n$. In other words, the range of the function $f(x) = x^n$ is the set of all

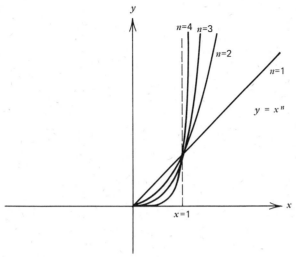

**Fig. 17**

nonnegative numbers. This is, of course, true, but to prove it one must exploit both the continuity of $f(x) = x^n$ and the completeness of the real numbers. Let us take this fact for granted now.

If for each $y \geq 0$ there exists an $x \geq 0$ such that $y = x^n$, then this value of $x$ certainly is unique. Hence the set of ordered pairs $(y, x)$ defines a function $f$. This function is called the $n$th-root function and we write $x = f(y) = y^{1/n}$. To sketch the graph of $f$ in the usual way with its domain along the $x$ axis we need only reflect Figure 17 about the line $y = x$. This amounts to interchanging the roles of $x$ and $y$.

Let us assume now that for each $x > 0$, the function $f(x) = x^{1/n}$ is differentiable. This can be demonstrated by an "$\epsilon, \delta$" argument but the result is apparent from the way the graph of $f$ has been constructed (Figure 18).

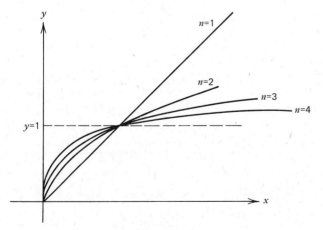

**Fig. 18** Graphs of the functions $f(x) = x^{1/n}$, $x \geq 0$, for $n = 1, 2, 3, 4$.

A nonhorizontal tangent line to the graph of $y = x^n$ is reflected into a non-vertical tangent line to the graph of $y = x^{1/n}$.

The formula for $f'(x)$ may now be easily derived by using the chain rule.

If $f(x) = x^{1/n}$, $x > 0$, then

(7)
$$f(x)^n = x.$$

If we set $h(x) = f(x)^n$, then $h$ is the composition of $g(x) = x^n$ and $f$. Hence

$$h(x) = g(f(x)),$$

and by the chain rule

$$h'(x) = g'(f(x))f'(x).$$

On the other hand $h(x) = x$. Therefore

$$1 = h'(x) = g'(f(x)) \cdot f'(x)$$

or

$$1 = nf(x)^{n-1} \cdot f'(x).$$

Substituting $f(x) = x^{1/n}$ and solving for $f'(x)$, we obtain

$$f'(x) = \frac{1}{n}[x^{1/n}]^{(1-n)} = \frac{1}{n}x^{(1/n)-1}$$

which is a special case of (6).

Next we consider $f(x) = x^{m/n}$, where $m$, $n$ are positive integers. From elementary algebra $x^{m/n} = [x^{1/n}]^m = [x^m]^{1/n}$. Hence $f(x) = x^{m/n}$ implies

(8)
$$[f(x)]^n = x^m.$$

If we differentiate (8) and apply the chain rule to the left-hand side, we obtain

$$n[f(x)]^{n-1}f'(x) = mx^{m-1}.$$

Substituting $f(x) = x^{m/n}$ and solving for $f'(x)$, we get

$$f'(x) = mx^{m-1} \cdot \frac{1}{n}x^{-(m/n) \cdot n-1}$$

$$= \frac{m}{n}x^{m-1-m+(m/n)}$$

$$= \frac{m}{n}x^{(m/n)-1}.$$

This is the desired formula (6).

Now if $f(x) = x^{-m/n}$, then $f(x)^n = x^{-m}$. Exactly the same argument as the above shows that $f'(x) = -(m/n)x^{-(m/n)-1}$. Hence under the assumption that $f(x) = x^{1/n}$ is differentiable we have shown that for each rational number $k$ if $f(x) = x^k$, $x \geqslant 0$, then $f'(x) = kx^{k-1}$. Hence the power formula for differentiation holds for all rational exponents. A direct proof that $f(x) = x^{1/n}$ is indeed a differentiable function is contained in Exercises 27 and 28 at the end of this section.

**Example 3.**  Compute $f'$ if $f(x) = \sqrt{1+x^2}$. If $h(x) = 1+x^2$, $g(x) = \sqrt{x}$, then $h'(x) = 2x$, $g'(x) = \frac{1}{2}x^{-1/2}$. Hence

$$f(x) = g(h(x)),$$

and

$$f'(x) = g'(h(x))g'(x)$$

$$= \frac{1}{2}(1+x^2)^{-1/2} \cdot 2x$$

$$= \frac{x}{(1+x^2)^{1/2}}.$$

*Example 4.*   Differentiate

$$h(x) = \left(1 + \frac{x+1}{\sqrt{x}}\right)^{3/4}.$$

If $f(x) = 1 + (x+1)/\sqrt{x}$ and $g(x) = x^{3/4}$, then $h(x) = g(f(x))$. Hence

$$h'(x) = g'(f(x)) \cdot f'(x)$$

$$= \frac{3}{4} f(x)^{(3/4)-1} \cdot f'(x)$$

$$= \frac{3}{4}\left[1 + \frac{x+1}{\sqrt{x}}\right]^{-1/4} \cdot \left[\frac{\sqrt{x} - \frac{1}{2}(x+1)/\sqrt{x}}{x}\right]$$

$$= \frac{3}{4}\left(1 + \frac{x+1}{\sqrt{x}}\right)^{-1/4} \cdot \frac{x-1}{x^{3/2}}.$$

We note one final fact. If $n$ is an odd positive integer, then the function $f(x) = x^{1/n}$ is defined for all real numbers $x$. This is clear since, in this case, for each real number $y$ there exists a unique number $x$ satisfying $y = x^n$. Hence, the ordered pairs $(y, x)$ define a function. The graph of $y = x^{1/n}$ is obtained as before by reflecting the graph of $y = x^n$ about the line $y = x$ (Figures 19 and 20).

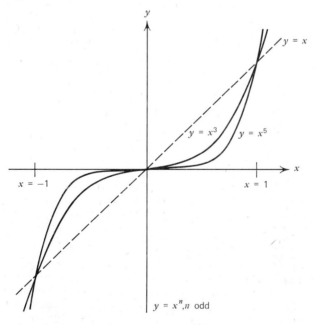

Fig. 19

The power formula (6) is valid in this case as well. We summarize this discussion of the function $f(x) = x^{m/n}$ in the following result.

**THEOREM 2.6**   *Let $f(x) = x^{m/n}$. If $n$ is even, we assume that $x \geqslant 0$. Then, for each $x \neq 0$, $f$ is differentiable at $x$ and*

$$f'(x) = \frac{m}{n} x^{(m/n)-1}.$$

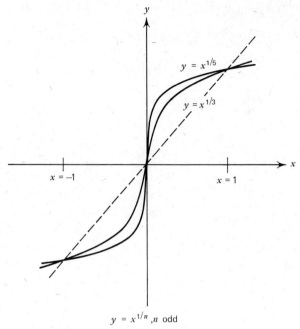

$$y = x^{1/n}, n \text{ odd}$$

**Fig. 20**

## EXERCISES
Differentiate the following functions using the chain rule where necessary.

1. $f(x) = \sqrt{1+3x}$.
2. $f(x) = (3+x^2)^5$.
3. $g(t) = \sqrt[3]{3+4t+2t^2}$.
4. $p(s) = (1+3s)^5$.
5. $g(x) = \left(\dfrac{x-1}{x+1}\right)^{1/2}$.
6. $f(x) = (2x+\sqrt{x})^3$.

7. $q(r) = (1+\sqrt{r})^{10}$.
8. $g(t) = \sqrt{\dfrac{2t-1}{2t+1}}$.
9. $f(x) = (4-x+3x^2)^3$.
10. $g(x) = \dfrac{1}{(3x-5)^5}$.

Find $f'(x)$ for the following functions $f$.

11. $f(x) = (x^2-1)\sqrt{1+x^2}$.
12. $f(x) = \dfrac{1}{(3x-2)^2}$.
13. $f(x) = \dfrac{\sqrt{1-x^2}}{\sqrt[3]{1+x^2}}$.
14. $f(x) = (1+x^2-x^4)^{1/5}$.
15. $f(x) = \dfrac{\sqrt{x}-\sqrt{1+x}}{\sqrt{x}+\sqrt{1+x}}$.
16. $f(x) = (48-x^2)^{-1}$.
17. $f(x) = (2+3x^2-\sqrt{x})^{-3}$.
18. $f(x) = \sqrt{1+\sqrt{1+\sqrt{x}}}$.

19. $f(x) = \left(\dfrac{x^2-1}{x^2+1}\right)^{3/4}$.
20. $f(x) = \left(2+\dfrac{1}{x}-\sqrt{x}\right)^{-3/2}$.
21. $f(x) = ((1+\sqrt{x})^2+1)^{3/2}$.
22. $f(x) = \dfrac{(x^2-1)^{1/2}(4-x)^2}{(2x+1)^3}$.
23. $f(x) = \sqrt{x^2 \cdot \dfrac{1-\sqrt{x}}{1+\sqrt{x}}}$.
24. $f(x) = \left(\left(x-\dfrac{1}{x}\right)^3+\sqrt{x}\right)^{-1/2}$.

25. Sketch the graph of the function $f(x) = x^{-1/n}$ when $n = 1, 2, 3, 4$. Determine the domain of $f$ for the various choices of $n$.
26. Verify the power formula for $f(x) = x^{-m/n}$, $m, n$ being positive integers, by applying the chain rule to the equation $f(x)^n = x^{-m}$.
27.* To show that $f(x) = x^{1/n}$ is differentiable we must first prove that $f$ is continuous.

(a) Prove this for $x = 0$ by showing $\lim\limits_{x \to 0} x^{1/n} = 0$.
(The choice of $\delta$ for a given $\epsilon$ is easy in this case.)
(b) Prove continuiuty at $x = 1$ by showing $\lim\limits_{x \to 1} x^{1/n} = 1$.

[To estimate $|1 - x^{1/n}|$ use the identity $(1 - x) = (1 - x^{1/n})(1 + x^{1/n} + \cdots + x^{(n-1)/n})$ valid if $x \geq 0$. Hence if $x \geq 0$, $|1 - x| = |1 - x^{1/n}| \cdot (1 + x^{1/n} + \cdots + x^{(n-1)/n}) \geq |1 - x^{1/n}|$.]

(c) Prove that $x^{1/n}$ is continuous at $x = a$ if $a \neq 0$ by using part (b) and the fact that

$$x^{1/n} = \left(\frac{x}{a}\right)^{1/n} \cdot a^{1/n}.$$

(Apply the product theorem for limits and the fact that $x \to a$ implies $y = (x/a) \to 1$.)

28*. Prove that $f(x) = x^{1/n}$ is differentiable if $x \neq 0$ and derive the formula for $f'(x)$. (Let $y = x^{1/n}$, $y + k = (x + h)^{1/n}$. Then $x = y^n$, $x + h = (y + k)^n$ and

$$\frac{f(x+h) - f(x)}{h} = \frac{(x+h)^{1/n} - x^{1/n}}{h} = \frac{y + k - y}{x + h - x}$$

$$= \frac{k}{(y+k)^n - y^n} = 1 \bigg/ \frac{(y+k)^n - y^n}{k}.$$

Now as $h \to 0$, $k \to 0$ (why?), and hence

$$1 \bigg/ \frac{(y+k)^n - y^n}{k} \to \frac{1}{n y^{n-1}}.$$

(Why?) From this conclude that $f'(x) = (1/n)x^{(1/n)-1}$.

## 2.7  Other Notations for the Derivative

The invention of the derivative is usually ascribed to Sir Isaac Newton during 1665–1666 and independently to the German mathematician and philosopher Gottfried Wilhelm von Leibniz in 1673–1676. The modern definition that we have presented evolved gradually. The statement that

(1) $$f'(x) = \lim_{h \to 0} \frac{f(x+h) - f(x)}{h}$$

occurs first in the work of Bolzano early in the 19th century. In place of $h$ he used the symbol $\Delta x$.

Virtually every writer on the calculus during the 18th and 19th centuries introduced his own variation on the notation. The notation $f'(x)$, $f''(x)$, $f'''(x)$, etc., to denote successive derivatives of $f$ at the point $x$ was introduced in 1797 by the French mathematician Lagrange. This notation is used primarily to stress the fact that from the function $f$ a new function $f'$ has been constructed via the formula

$$f'(x) = \lim_{h \to 0} \frac{f(x+h) - f(x)}{h}.$$

Leibniz' own notation for the derivative was $dy/dx$ where $y = f(x)$. Indeed letting $\Delta x = h$ and

$$\Delta y = f(x + \Delta x) - f(x)$$

then

$$\frac{dy}{dx} = \lim_{\Delta x \to 0} \frac{\Delta y}{\Delta x}.$$

Higher order derivatives $f''(x),\ldots,f^{(k)}(x)$ are written

$$\frac{d^2y}{dx^2},\ldots,\frac{d^ky}{dx^k}.$$

The Leibniz notation is still widely used today because of its simplicity in calculations and applications. In particular the formula for the chain rule becomes very simple. Indeed, if $z = g(y)$ and $y = f(x)$, then setting $h(x) = g(f(x))$ we have $z = h(x)$. Moreover

$$\frac{dz}{dx} = h'(x) = g'(y)f'(x) = \frac{dz}{dy}\frac{dy}{dx}.$$

Another virtue of the $dy/dx$ notation for derivatives is that it focuses attention on the independent variable of the function $f$. For example, if

$$y = (x^2 - a^3)^5,$$

then

$$\frac{dy}{dx} = 5(x^2 - a^3)^4 \cdot (2x).$$

Thus it is assumed that $a$ is constant. Perhaps $a$ is to be thought of as the independent variable and $x$ is to be fixed. Then

$$\frac{dy}{da} = 5(x^2 - a^3)^4(-3a^2).$$

A disadvantage of the $dy/dx$ notation for derivatives is that we have lost sight of the point at which the derivative is to be evaluated. To write $f'(a)$ one must write

$$\frac{dy}{dx}_{x=a} \quad \text{or} \quad \frac{dy}{dx}(a).$$

Other common notations for $\dfrac{dy}{dx}$ and $f'$ are $y'$ and $\dfrac{df}{dx}$. Successive derivatives are then written $y'',\ldots,y^{(k)}$, and

$$\frac{d^2f}{dx^2},\ldots,\frac{d^kf}{dx^k}.$$

The differentiation operation itself is often indicated by $D$ or $\dfrac{d}{dx}$. Thus if $y = f(x)$,

$$f' = Df = \frac{d}{dx}(f)$$

$$f'' = D^2f = \frac{d^2}{dx^2}(f)$$

$$(f+g)' = D(f+g) = \frac{d}{dx}(f+g).$$

Using these notations for the differentiation operation, the product formula $(fg)' = f'g + fg'$ becomes

$$D(fg) = (Df)\cdot g + f\cdot(Dg)$$

or

$$\frac{d}{dx}(fg) = \frac{d}{dx}(f)g + f\cdot\frac{d}{dx}(g).$$

For quotients we have

$$D\left(\frac{d}{g}\right) = \frac{g\,Df - f\,Dg}{g^2},$$

and

$$\frac{d}{dx}\left(\frac{f}{g}\right) = \frac{g\dfrac{d}{dx}(f) - f\dfrac{d}{dx}(g)}{g^2}.$$

**EXERCISES**

1. If $y = x - \dfrac{1}{\sqrt{x-1}}$, determine $\dfrac{dy}{dx}, \dfrac{d^2y}{dx^2}$.

2. If $s = \sqrt{\dfrac{1}{1+t^2}}$, determine $\dfrac{ds}{dt}, \dfrac{d^2s}{dt^2}$.

3. If $f(x) = (1 + (1 + x^2)^2)^{1/2}$, determine $Df$, $D^2f$, $(D^2f)\,(0)$.

4. If $u = \dfrac{\sqrt{v - w}}{v + w}$, compute $\dfrac{du}{dv}$ and $\dfrac{du}{dw}$.

5. If $y = 2x^2 - x$, show that $x\dfrac{dy}{dx} = 2y + x$.

6. If $y = 2x^3 - x + 1$, show that $y'' = xy'''$.

7. If $y = \sqrt{4x^2 + 1}$, show that $yy' - 4x = 0$.

8. If $y = \sqrt{x^2 + ax}$, show that $x^2 + y^2 = 2xyy'$.

9. If $y = \sqrt{ax^2 + bx}$, show that $y'' = \dfrac{-b^2}{4y^3}$.

## 2.8 Implicitly Defined Functions and Implicit Differentiation

We have emphasized that the unit circle

(1) $$G = \{(x, y) : x^2 + y^2 = 1\}$$

does not define $y$ as a function of $x$ since for each point $x$, $-1 < x < 1$, the vertical line through $x$ intersects $G$ twice (Figure 21). Nevertheless there are

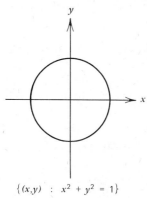

$\{(x,y) \; : \; x^2 + y^2 = 1\}$

**Fig. 21**

functions $f$ defined on the interval $-1 \leqslant x \leqslant 1$ such that for each value of $x$ in this interval

(2) $$x^2 + [f(x)]^2 = 1.$$

Indeed, for a function $f$ to satisfy (2) all that has to be required is that for each

$x$ in the interval $[-1, 1]$

$$f(x) = \sqrt{1-x^2}$$

(3)        or

$$f(x) = -\sqrt{1-x^2}.$$

This choice may be effected in infinitely many ways. However, if we wish $f$ to be continuous on $[-1, 1]$, then there are only two possibilities. Either $f(x) = \sqrt{1-x^2}$ for each $x$ in $[-1, 1]$ or $f(x) = -\sqrt{1-x^2}$ for each $x$ in $[-1, 1]$. See Figure 22.

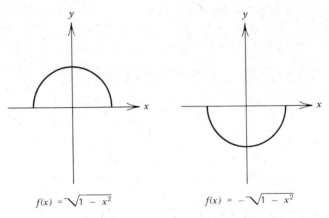

$$f(x) = \sqrt{1-x^2} \qquad\qquad f(x) = -\sqrt{1-x^2}$$

**Fig. 22**   Two continuous functions defined on $[-1, 1]$ that satisfy (1).

If a function $y = f(x)$ defined on some interval $(a, b)$ satisfies equation (1) for all $x$ in this interval, then $f$ is said to be *implicitly defined* by (1). We also say that $f$ is an *implicit solution* to (1). Thus $f$ is an implicit solution to (1) if for each $x$ in $(a, b)$ either $f(x) = \sqrt{1-x^2}$ or $f(x) = -\sqrt{1-x^2}$. If we remove the ambiguity and specify for each value of $x$ which sign is to be taken, then $f$ is said to be an *explicit solution* to (1).

**Example 1.**   Determine two explicit solutions $y = f(x)$ to the equation

(4)                    $$xy^2 + y - 1 = 0.$$

Any function $f$ defined on an interval $(a, b)$ and which satisfies $x(f(x))^2 + f(x) - 1 = 0$ for each $x \in (a, b)$ is an *implicit solution* to (4).

Solving (4) by the quadratic formula, this means that for $x \neq 0$ either

$$f(x) = \frac{-1 + \sqrt{1+4x}}{2x}.$$

or

$$f(x) = \frac{-1 - \sqrt{1+4x}}{2x}.$$

Two explicit solutions $f_1, f_2$ to (4) are defined by setting

$$f_1(x) = \frac{-1 + \sqrt{1+4x}}{2x} \quad x \geq -\tfrac{1}{4}, \, x \neq 0$$

and

$$f_2(x) = \frac{-1 - \sqrt{1+4x}}{2x} \quad x \geq -\tfrac{1}{4}, \, x \neq 0.$$

Now for many equations in $x$ and $y$, in fact most, it may be awkward or impossible to give an explicit formula for the functions $f$ that satisfy the given equation.

*Example 2.* Consider the equation

(5) $$x = y^3 + 3xy^2 + \frac{y}{x} - 4.$$

The graph of this equation

$$G = \left\{ (x, y) : x = y^3 + 3xy^2 + \frac{y}{x} - 4 \right\}$$

is not empty because $(1, 1) \in G$. Moreover, there are functions $y = f(x)$ that satisfy (5) but the explicit determination of any of these functions would be quite tedious.

We now formulate these ideas precisely. Let $F$ be a real valued function defined on a set in the $(x, y)$ plane. Let

$$G = \{ (x, y) : F(x, y) = 0 \}$$

be the graph of the equation $F(x, y) = 0$ and assume that $G$ is not empty.

If $f$ is a function defined on an interval $(a, b)$ such that for all $x$ in $(a, b)$ $F(x, f(x)) = 0$, then $f$ is said to be *implicitly defined* by the equation

$$F(x, y) = 0,$$

and we say that the function $f$ is an *implicit solution* to this equation.

Conditions on $F$ may be given which will insure the existence of such solutions $f$ and which furthermore will guarantee that $f$ is differentiable. This problem is very important, but we are not in a position to consider it at the present time. However, if we *assume* the existence of a differentiable function $y = f(x)$ satisfying

$$F(x, y) = F(x, f(x)) = 0,$$

the chain rule gives us a method for determining the derivative $f'$ explicitly as a function of $x$ and $y$.

*Example 3.* Determine $f'$ as a function of $x$ and $y$ if $y = f(x)$ is an implicit solution to the equation

(6) $$xy^2 + y - 1 = 0.$$

Since $f$ satisfies (6) we have

$$x[f(x)]^2 + f(x) - 1 = 0.$$

Differentiating and applying the chain rule we get

$$1 \cdot [f(x)]^2 + 2xf(x)f'(x) + f'(x) = 0.$$

Solving for $f'(x)$ we have

$$f'(x) = -\frac{f(x)^2}{1 + 2xf(x)} = -\frac{y^2}{1 + 2xy}.$$

The notation is simplified if we write $y'$ for $f'(x)$.

*Example 4.* Determine $y'$ if $y = f(x)$ is differentiable and is defined implicitly by

(7) $$x = y^3 + 3xy^2,$$

Differentiating (7) and applying the chain rule yields

$$1 = 3y^2 y' + 3y^2 + 6xy y'.$$

Solving for $y'$ we have

$$y' = \frac{1 - 3y^2}{3y^2 + 6xy}.$$

If $y = f(x)$ is an implicit solution to the equation $F(x, y) = 0$, the process of determining $y'$ as a function of $x$ and $y$ by differentiating the equation $F(x, y) = 0$ is called *implicit differentiation*.

**Example 5.** Determine the point $(x, y)$ on the circle $x^2 + y^2 = 1$ at which the tangent line is parallel to the line $2y - x + 1 = 0$.

One answer to this problem is to solve explicitly for functions $y = f(x)$ which satisfy the equation of the circle, differentiate, and then set the derivative equal to the slope of the line. Using implicit differentiation we proceed as follows. Since $x^2 + y^2 = 1$, if we differentiate and apply the chain rule, we obtain $2x + 2yy' = 0$, or $y' = -x/y$. Setting $y'$ equal to the slope of the line we have $y' = -x/y = \frac{1}{2}$ or $y = -2x$. Substituting this value of $y$ in the equation for the circle we get $x^2 + 4x^2 = 1$ or $x = \pm 1/\sqrt{5}$. Hence the desired points $(x, y)$ are $(1/\sqrt{5}, -2/\sqrt{5})$ and $(-1/\sqrt{5}, 2/\sqrt{5})$. (See Figure 23.)

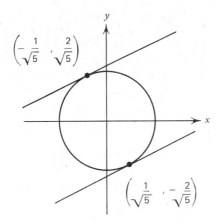

**Fig. 23**

**Example 6.** If $2x^2 + 4y^2 = 1$, determine $y''$ as a function of $x$ and $y$. Differentiating implicitly we have

(8) $$4x + 8yy' = 0.$$

If we differentiate (8) remembering that both $y$ and $y'$ are functions of $x$, we obtain

(9) $$4 + 8(y')^2 + 8yy'' = 0.$$

From (8), $y' = -x/2y$. Solving (9) for $y''$ we have

$$y'' = -\frac{4 + 8(y')^2}{8y} = -\frac{4 + (2x^2/y^2)}{8y} = -\frac{1}{8y^3}.$$

We emphasize again that nothing in the above discussion *asserts the existence* of a differentiable function $y = f(x)$ satisfying

$$F(x, y) = F(x, f(x)) = 0.$$

We always *assume* that this is the case. Even in simple cases such as

(10) $$y^n - x = 0$$

the existence of such a function is by no means obvious. In our discussion of (10) in Section 2.6, we tacitly *assumed* the existence of the function $f(x) = x^{1/n}$, which satisfies (10). However, to *prove* that this function exists we must appeal to the completeness axiom in precisely the same way as was done in the proof that $\sqrt{2}$ exists. Then, to assert that $x^{1/n}$ is *differentiable* we must argue as in Exercises 27 and 28 of Section 2.6.

It may happen that there are no functions $f$ defined implicitly by the equation $F(x, y) = 0$. Then, of course, it is meaningless to talk of $f'$.

**Example 7.** Let $f$ be an implicit solution to

(11) $$x^2 + y^2 + 1 = 0.$$

Determine $f'$ as a function of $x$ and $y$. Differentiating (11) implicitly we obtain

$$2x + 2yy' = 0$$

or

$$y' = \frac{-x}{y}.$$

This, however, is meaningless because there are no points $(x, y)$ satisfying (11)!

**EXERCISES**

Assuming the existence of a differentiable function $y = f(x)$ defined implicitly by the following equations, compute $y' = f'(x)$ as a function of $x$ and $y$ by implicit differentiation.

1. $3x^2 + 4y^2 = 1$.

2. $2x^2 - y^2 = 4$.

3. $x = 4y^2 + y + 1$.

4. $x^2 + xy^3 - y = 0$.

5. $\sqrt{x + y^2} - y = x$.

6. $6(xy)^3 - y^5 - x = 0$.

7. $\dfrac{x+y}{x-y} - 1 = x$.

8. $\dfrac{y-a}{y+a} = \dfrac{x+a}{x-a}$.

9. $\left(\dfrac{x-y}{1+x}\right)^{1/3} = 1 - (x-y)^{3/5}$.

10. $\sqrt{x+y} - \sqrt{x-y} = x$.

Find $y'$, $y''$ if

11. $x^2 + xy + y^2 = 1$.

12. $2x^2 - 4y^2 = 1$.

13 $\dfrac{x+y}{x-y} = 1/y$.

14. Find the slope $y'$ to the parabola $x = 4y^2$ at the point $(16, 2)$.

15. Find the slope $y'$ to the ellipse $x^2/a^2 + y^2/b^2 = 1$ at a point $(x_0, y_0)$.

16. Find the equation of the tangent line to the hyperbola $x^2 - y^2 = 1$ at the point $(\sqrt{5}, 2)$.

17. Find the equation of the tangent line to the graph of $x = y^3 - y$ at the point $(6, 2)$.

18. Show that the tangent lines to the graphs of the equations $3y = 2x + x^4y^3$ and $2y + 3x + y^5 = x^3y$ are perpendicular at the origin.

19. Show that the equation of the tangent line to the ellipse

$$\frac{x^2}{a^2} + \frac{y^2}{b^2} = 1$$

at the point $(x_0, y_0)$ is given by $x_0 x/a^2 + y_0 y/b^2 = 1$.

20. Show that the equation of the tangent line to the parabola $y^2 = 4px$ at the point $(x_0, y_0)$ is given by

$$y_0 y = 2p(x + x_0).$$

21. Differentiate implicitly each of the following equations obtaining $y' = f(x, y)$. For which of the equations is this a meaningful formula?
   (a) $x^2 + 2x + y^2 = 0$
   (b) $x^2 - 2x + y^2 + 2 = 0$
   (c) $2x^2 - 12x + y^2 + 11 = 0$

22. Criticize the following statement
   For $-1 \leqslant x \leqslant 1$ we define the function $f$ by the following formula.
   $f(x) = \sqrt{1 - x^2}$ if $x$ is a rational number;

$f(x) = -\sqrt{1-x^2}$ if $x$ is an irrational number.

The function $y = f(x)$ is a solution to the equation $x^2 + y^2 = 1$. At each point $(x, y)$ on the unit circle the derivative $y' = f'(x)$ is given by $y' = -x/y$.

23. Let $f_1(x) = \dfrac{-1 + \sqrt{1+4x}}{2x}$, $x \geqslant -\tfrac{1}{4}$, $x \neq 0$

$f_2(x) = \dfrac{-1 - \sqrt{1+4x}}{2x}$, $x \geqslant -\tfrac{1}{4}$, $x \neq 0$

be two explicit solutions to the equation $xy^2 + y - 1 = 0$. (See Example 1, p. 74) Which of these functions can be defined to be continuous at $x = 0$? Explain.

24. The graph of

(12)                                    $y = x^3 - x$

is given in Figure 24.

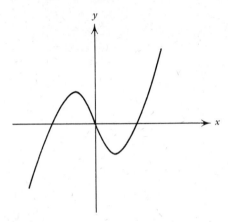

**Fig. 24**

Sketch the graph of $x = y^3 - y$. Indicate graphically how to define a function that is an implicit solution to (12) for all values of $x$. Define graphically three such functions. Convince yourself that no such function can be continuous for all values of $x$. Why?

## 2.9   Primitives

For a function $f$, the physical interpretation of the derivative $f'(x)$ is the rate of change of $f$ at the point $x$. As we have seen, if

$$s = s(t)$$

measures the position of a particle moving along a straight line as a function of time, then

$$v = v(t) = s'(t)$$

measures the velocity or rate of change of position. Differentiating again we obtain the acceleration, the rate of change of velocity, as a function of time.

Suppose now we have two particles $p_1$ and $p_2$ moving along the same line, which at time 0 have the same position. Let $s_1(t)$, $s_2(t)$ be the respective position functions and let $v_1(t)$, $v_2(t)$ be the respective velocities. Suppose, in addition for all values of $t$ the velocities $v_1(t)$ and $v_2(t)$ are equal. Then does it follow that the positions of the particles must always be the same? One's immediate response to this equation is to answer, yes, of course. Indeed, any other answer would completely violate one's intuitive concept of motion and velocity.

However, let us see if we can justify this on purely mathematical grounds. First, if $v_1(t) = v_2(t)$, then $v_1(t) - v_2(t) = 0$. Since

$$v_1(t) - v_2(t) = s_1'(t) - s_2'(t) = (s_1 - s_2)'(t),$$

we are asserting that $(s_1 - s_2)'(t) = 0$ for all $t$. Now we know that if $f$ is a constant function, then its derivative vanishes. Let us assume for the moment that the converse of this statement is true. That is, if $f'(t) = 0$ for all $t$ in some interval, then $f$ is constant on this interval. We assert that *if* this converse statement is true, then we have answered our question about the position functions $s_1$ and $s_2$. Suppose

$$s_1(t) - s_2(t) = \text{constant} = k$$

By assumption for $t = 0$,

$$s_1(0) = s_2(0).$$

Hence, $k = s_1(0) - s_2(0) = 0$, and for all values of $t$,

$$s_1(t) = s_2(t).$$

Hence the position of the particles is always the same  Let us state as a theorem the mathematical result we have used.

**THEOREM 2.7   (Vanishing Derivative Theorem).** *Let $f$ be a function such that for all $x$ in the interval $(a, b)$*

$$f'(x) = 0,$$

*then $f(x)$ is constant on that interval.*

**COROLLARY 2.8**   *If $f'(x) = g'(x)$ on the interval $(a, b)$, then*

$$f(x) - g(x) = constant$$

*for all $x$ in that interval.*

Since it is so easy to prove that the derivative of a constant function is zero, one might expect that the proof of the "Vanishing Derivative Theorem" would be equally simple. This is far from the case, however. In fact, we must postpone the proof of this theorem to Chapter 5. Nevertheless, the theorem is a cornerstone of the work we do next, and we shall use the result freely.

**Definition.**   *If the function $f$ is given, and the function $F$ satisfies*

$$F'(x) = f(x)$$

*for all values of $x$ in the domain of $f$, then $F$ is called a primitive or an antiderivative of $f$.*

Corollary 2.8 asserts that on an interval $(a, b)$, if $F$ is one primitive of $f$, then *all* primitives of $f$ are given by the family of functions $F + C$ where $C$ is a constant. The process of finding primitives for a given function $f$ is sometimes called *antidifferentiation*. We shall investigate this problem in some detail in later chapters. For the moment let us solve the problem in a few special cases.

Consider first $f(x) = x^k$ where $k$ is a rational number. Now $g(x) = x^{k+1}$ implies that $g'(x) = (k+1)x^k$. Hence if $k+1 \neq 0$ and we set

$$F(x) = \frac{x^{k+1}}{k+1},$$

then
$$F'(x) = \frac{k+1}{k+1} x^k = x^k.$$

Thus if $k \neq -1$, $F(x) = x^{k+1}/(k+1)$ is a primitive for $f(x) = x^k$. By Corollary 2.8 all primitives for $x^k$ must be of the form $x^{k+1}/(k+1)+c$. The differentiation formulas (1) and (2) of Theorem 2.2 can be translated immediately into statements about primitives.

**THEOREM 2.9**   *Let F be a primitive for f and G be a primitive for g, and let c, d be real numbers. Then cF + dG is a primitive for cf + dg.*

*Proof.*   We need only to verify that
$$(cF + dG)' = cf + dg.$$
But
$$(cF + dG)' = cF' + dG'$$
by (1) and (2) of Theorem 2.2. Since $F' = f$, $G' = g$ by hypothesis, we have
$$(cF + dG)' = cf + dg.$$

We can extend this result to $n$ functions by applying induction. In particular, if $f$ is a polynomial and
$$f(x) = a_n x^n + a_{n-1} x^{n-1} + \cdots + a_0,$$
then
$$F(x) = \frac{a_n x^{n+1}}{n+1} + \frac{a_{n-1}}{n} x^n + \cdots + a_0 x$$
is a primitive for $f$. Furthermore, each primitive for $f$ is given by $F + C$ where $C$ is a constant.

*Example 1.*   Find all primitives $F$ for the function
$$f(x) = 3x^2 - 5x + 3.$$
*Solution:*
$$F(x) = \frac{3x^3}{3} - \frac{5x^2}{2} + 3x + C.$$
$$= x^3 - \frac{5}{2} x^2 + 3x + C,$$
where $C$ is a constant.

*Example 2.*   Find all antiderivatives for the function
$$f(x) = 4x^2 + 3\sqrt{x} + \frac{1}{2}(1/\sqrt{x^3}).$$
since
$$f(x) = 4x^2 + 3x^{1/2} + \frac{1}{2} x^{-3/2}$$
all antiderivatives for $f$ take the form
$$F(x) = \frac{4}{3} x^3 + \left(3/\frac{3}{2}\right) x^{3/2} + \frac{1}{2}\left(1/\left(-\frac{1}{2}\right)\right) x^{-1/2} + C$$
$$= \frac{4}{3} x^3 + 2x^{3/2} - x^{-1/2} + C.$$

If one specific primitive for $f$ is desired, then a value must be assigned to the constant $C$.

***Example 3.*** Determine a primitive $F$ for $f(x) = 5x^4 - 6x + 4$ which satisfies $F(1) = 5$. All primitives for $f$ are of the form

$$F(x) = \frac{5}{5}x^5 - \frac{6}{2}x^2 + 4x + C$$

$$= x^5 - 3x^2 + 4x + C.$$

If we require that $F(1) = 5$, then this determines the constant $C$. Indeed $5 = F(1) = 1 - 3 + 4 + C$ implies $C = 3$. Hence, $F(x) = x^5 - 3x^2 + 4x + 3$ is the desired primitive for $f$.

The determination of the constant $C$ in a primitive for $f$ may depend on physical information.

***Example 4.*** A ball is dropped from a height of 64 feet. Falling under the action of gravity, its velocity is $-32t$ feet per second. When does the ball hit the earth?
*Solution* If $v(t)$ is the velocity of the ball, then

$$v(t) = -32t.$$

If $s(t)$ is the height of the ball at time $t$,

$$s'(t) = -32t.$$

Therefore

$$s(t) = -\frac{32t^2}{2} + C = -16t^2 + C.$$

At $t = 0$

$$s(0) = 64 = C.$$

Therefore, $s(t) = 64 - 16t^2$, and $s(t) = 0$ if $0 = 64 - 16t^2$ or $t = \pm 2$. We are only concerned with $s(t)$ for $t \geqslant 0$. Hence, we infer that the ball hits the earth when $t = 2$ seconds.

## EXERCISES

1. Determine the set of all primitives for the following functions $f$.

(a) $f(x) = 3x^2 + 1$

(b) $f(x) = 5x^4 + 3x^2 - 2$

(c) $f(x) = x^5 - 7x^2 + 3$

(d) $f(x) = (x^2 - 5x + 1)^2$

(e) $f(x) = \frac{2}{x^3} - \frac{4}{x^5}$

(f) $f(x) = 4x^3 - \frac{7}{x^3} + 3\sqrt{x}$

(g) $f(x) = 4\sqrt{x} - \sqrt{x^3}$

(h) $f(x) = 2x^{2/3} - 4x^2 + 5x^{-1/5}$

(i) $f(x) = 5 + \frac{12}{\sqrt{x^5}} - \frac{2}{\sqrt[3]{x^7}}$

(j) $f(x) = 6x - 7x^3 + \frac{10}{x^{1/2}} + \frac{1}{x^5}$

2. Determine primitives $F$ for the following functions $f$, which in addition satisfy $F(a) = b$

(a) $f(x) = 2x - 1$ $\quad\quad (a, b) = (1, 4)$

(b) $f(x) = x^2 - 4x + 1$ $\quad (a, b) = (1, 2)$

(c) $f(x) = 3x^4 + 5x^2 - 6$ $\quad (a, b) = (0, 7)$

(d) $f(x) = \frac{2}{x^2} - x + \frac{4}{x^5}$ $\quad (a, b) = (1, -3)$

(e) $f(x) = 3\sqrt{x} - \frac{2}{x^3}$ $\quad (a, b) = (2, 4)$

(f) $f(x) = x^{3/2} - 4x^{-1/2}$ $\quad (a, b) = (1, -1)$

(g) $f(x) = x^2 - \frac{2}{x^{3/2}}$ $\quad (a, b) = (1, 4)$

(h) $f(x) = 3x^2 - \frac{4}{\sqrt[3]{x^4}}$ $\quad (a, b) = (2, 1)$

3. If $f'(x) = x(x-1)(x-2)$ and $f(0) = 1$, determine $f$.

4. If $f'(x) = \left(x + \dfrac{1}{x}\right)^2$ and $f'(1) = 0$, determine $f$.

5. If $f''(x) = \left(1 - \dfrac{1}{x^2}\right)^2$ and $f'(1) = f(1) = 0$, determine $f$.

6. If $f''(x) = (2x + \sqrt{x})^3$ and $f'(1) = f(0) = 1$, determine $f$.

7. At each point $x$ the slope of a function $f$ is given by $a(x-1)^2$ for some constant $a$. If the $x$ intercept of the graph is 2 and the $y$ intercept is 4, determine the function $f$.

8. The acceleration of a car at a drag race is proportional to the square of the time. If one second after the start of the race the car has traveled 10 feet, how long does it take the car to travel one half of a mile?

9. If an object falls near the surface of the earth and resistance due to air is neglected, then the acceleration of the object is constant. If distance is measured in feet and time in seconds then this constant has the value 32. A ball is dropped from a height of 100 ft. When does it hit the earth?

10. An archer fires an arrow straight up in the air with an initial velocity of 50 feet per second. How high does the arrow go and when does it hit the earth?

11. A boy on the top of a building that is 100 ft high, throws a rock straight down. It hits the ground in one second. How fast did he throw it?

12. Let $v$ be the velocity of an object falling only under the influence of gravity. After the object has fallen $s$ units, the velocity $v$ is given by

$$v^2 = 2gs$$

where $g$ is the acceleration due to gravity. Derive this formula.

13. What constant acceleration will enable a runner to sprint the 100 yard dash in 10 seconds?

14. The chain rule asserts that if $f(x) = g(h(x))$, then $f'(x) = g'(h(x))h'(x)$. Hence if we are given a function $f$ that can be written $f(x) = g'(h(x))h'(x)$, then all primitives $F$ for $f$ can be written

$$F(x) = f(g(x)) + c.$$

Using this observation determine all primitives for the following functions.

(a) $2(2x+1)^5$

(b) $-(1-x)^3$

(c) $\sqrt{1+x}$

(d) $\dfrac{2}{(2x+1)^3}$

(e) $2(1+2x)^{-3/2}$

(f) $\sqrt{1+2x}$

(g) $\dfrac{1}{(1-x)^3}$

(h) $x\sqrt{1+x^2}$

## 2.10 Applications of the Chain Rule

Consider an equation in $x$ and $y$ such as

(1) $$x^3 + (x-y)^3 + y^2 = 1.$$

If $y$ is to be thought of as a function of $x$, i.e., $y = f(x)$, then, as we have seen, we may compute $y' = dy/dx$ by differentiating (1) with respect to $x$ obtaining

$$0 = \frac{d}{dx}(1) = \frac{d}{dx}(x^3 + (x+y)^3 + y^2)$$

$$= \frac{d}{dx}(x^3) + \frac{d}{dx}(x-y)^3 + \frac{d}{dx}(y^2)$$

$$= 3x^2 + 3(x-y)^2\left(1 - \frac{dy}{dx}\right) + 2y\frac{dy}{dx}.$$

Hence

$$\frac{dy}{dx} = -\frac{3(x^2 + (x-y)^2)}{2y - 3(x-y)^2}.$$

Often, $x$ and $y$ are to be thought of as functions of a third variable $t$. Then differentiating (1) with respect to $t$ and applying the chain rule we have

(2) $$0 = \frac{d}{dt}(1) = \frac{d}{dt}(x^3 + (x-y)^3 + y^2)$$

$$= 3x^2\frac{dx}{dt} + 3(x-y)^2\left(\frac{dx}{dt} - \frac{dy}{dt}\right) + 2y\frac{dy}{dt}.$$

If we denote $dx/dt$, $dy/dt$ by $x'$ and $y'$ respectively, then (2) becomes

(3) $$0 = 3x^2x' + 3(x-y)^2(x' - y') + 2yy'.$$

Thus, if two functions $x = f(t)$ and $y = g(t)$ are related by equation (1), then we may differentiate (1) to obtain an equation relating the rates of change $x'$ and $y'$. Indeed, if $x$, $y$, $x'$ are specified for some value of $t$, then $y'$ may be computed from (3). A problem such as this is called a problem in *related rates*.

**Example 1.** A particle is moving along the parabola $y = 4x^2$. Let $dx/dt = y'(t)$ and $dy/dt = y'(t)$ denote the rates of change of the $x$ and $y$ coordinates. If $x'(t) = 4$ at the point $(2, 16)$, determine $y'(t)$ at this point. Differentiating the equation $y = 4x^2$ with the respect to $t$, we obtain

$$\frac{dy}{dt} = 8x\frac{dx}{dt}.$$

Hence, if $dx/dt = 4$ when $x = 2$, then $dy/dt = 64$.

**Example 2.** A conical tank, with vertex at the bottom is being filled at the rate of one cubic foot per minute. If at the top the tank has a radius of 5 feet and a height of 10 feet, how fast is the water level rising when the depth of water is 5 feet?

*Solution.* Referring to Figure 25, the volume $V$ of water in the tank equals $\frac{1}{3}\pi r^2 h$ where $h$ is the height, and $r$ the radius at the surface. The quantities $V$, $r$, and $h$ are functions of time.

Hence, differentiating the equation $V = \frac{1}{3}\pi r^2 h$ with respect to time, we obtain

(4) $$1 = \frac{dV}{dt} = \frac{2}{3}\pi rh\frac{dr}{dt} + \frac{\pi}{3}r^2\frac{dh}{dt}$$

which relates the rates of change of the radius, $dr/dt$, and the height of the water, $dh/dt$. Exploiting the similarity of triangle $pab$ with triangle $pcd$ (Figure 25 ), we see that

$$\frac{r}{h} = \frac{5}{10} \quad \text{or} \quad r = \frac{1}{2}h.$$

Hence, $dr/dt = 1/2(dh/dt)$. Substituting in (4) these values of $r$ and $dr/dt$, we obtain

$$1 = \frac{2\pi}{3}\left(\frac{h}{2}\right)h\left(\frac{1}{2}\frac{dh}{dt}\right) + \frac{\pi}{3}\frac{h^2}{4}\frac{dh}{dt} = \frac{\pi}{4}h^2\frac{dh}{dt}.$$

Thus, if $h = 5$, $dh/dt = 4\pi/25$. Hence, the water level is rising at a rate of $4\pi/25$ feet per second when there is 5 feet of water in the tank.

It may often be necessary to determine primitives or antiderivatives in related rate problems.

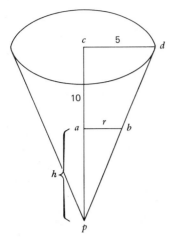

Fig. 25

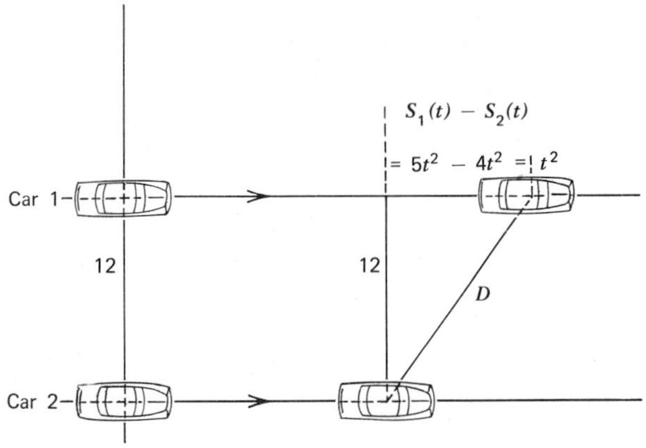

Fig. 26

*Example 3.* At the start of a drag race two cars are 12 feet apart (Figure 26). The two cars race parallel to each other along straight tracks with constant acceleration. If the acceleration of the first car is 10 feet per second per second and the acceleration of the second is 8 feet per second per second, determine how fast the distance between the cars is changing 6 seconds after the start of the race.

*Solution.* Let $a_1(t) = 10$, $a_2(t) = 8$ be the accelerations of the two cars. Let $v_1(t)$, $v_2(t)$ be the respective velocities; and $s_1(t)$, $s_2(t)$ be the respective positions. Then $a_k(t) = v'_k(t)$ and $v_k(t) = s'_k(t)$ for $k = 1, 2$. Since $v_1(0) = v_2(0) = s_1(0) = s_2(0) = 0$, $v_1(t) = 10t$, $v_2(t) = 8t$, $s_1(t) = 5t^2$, and $s_2(t) = 4t^2$. Therefore, at time $t$, car 1 is $5t^2 - 4t^2 = t^2$ feet ahead of car 2. Let $D(t)$ be the distance between the cars. By the Pythagorean theorem

$$[D(t)]^2 = 12^2 + (s_1(t) - s_2(t))^2.$$

Therefore, at time $t$,

$$[D(t)]^2 = 12^2 + t^4.$$

Differentiating with respect to $t$ and applying the chain rule, we obtain

$$2D(t)D'(t) = 4t^3.$$

Therefore, $D'(t) = 2t^3/(D(t))$. When $t = 6$, $D(6) = (12^2 + 6^4)^{1/2} = 12\sqrt{10}$. Therefore

$$D'(6) = 36/\sqrt{10} \text{ feet per second.}$$

### EXERCISES

1. A particle moves along the curve $y^2 = x^3$. Determine $x'$ at the point $(1, 1)$ if $y' = 4$.
2. A particle slides down the curve $y^2 = x^3$ under the force of gravity with no friction. Thus, at time $t$, $y''(t) = -16$ feet per second. If at $t = 0$ the particle is at rest at the point $(10, 10\sqrt{10})$, determine $x'(1)$.
3. A particle moves along the unit circle $x^2 + y^2 = 1$. If at the point $(\frac{3}{5}, \frac{4}{5})$, $x'(t) = -1$, determine $y'(t)$. If in addition $x''(t) = 1$, determine $y''(t)$.
4. A particle moves along the graph of the function $y = f(x)$. At each point $t$ such that $x'(t) \neq 0$, show that the quotient $y'(t)/x'(t)$ equals the slope of the graph
5. A ladder 15 feet long stands against a wall. If the base of the ladder slides away from the wall at 2 feet per second, how fast is the top of the ladder falling when the base of the ladder is 5 feet from the wall? How fast is the midpoint falling at the same time? Does the top of the ladder fall down the wall at a constant rate?

6. At noon a ship leaves port sailing due south at a rate of 10 knots. At 1 P.M. a second ship leaves port sailing due west at 20 knots. How fast is the distance between the ships increasing at 3 P.M.?

7. A policeman standing 100 feet from a right angle intersection sees an automobile accelerate through the intersection at a constant rate of 10 feet per second per second. If the velocity of the automobile is 25 feet per second when it passes through the intersection, how fast is the distance between the automobile and the policeman changing 5 seconds later?

8. A baseball diamond is a square, 90 feet on each side. A batter lines a grounder down the third base line at a rate of 50 feet per second. How fast is the distance between the ball and first base changing as the ball passes over third base? If the ball is batted at the pitcher's mound, how fast is the distance between the ball and first base changing as the ball passes second base?

9. A particle on the rim of a hemispherical bowl 4 feet in diameter slides down the bowl under just the force of gravity with no friction. How fast is the distance from the particle to the bottom of the bowl changing when the particle is equidistant between the rim and the bottom of the bowl?

10. A birdbath is in the form of a hemispherical bowl with a 2-foot radius. If the bowl is being filled with water at a constant rate of $\frac{1}{2}$ cubic foot per minute, how fast is the water level rising when it overflows? How fast is the water level rising when half of the bowl has been filled? (The volume of a section of a sphere of height $h$ is $\pi h^2 (r - (h/3))$ where $r$ is the radius of the sphere.)

11. A man 6 feet tall is walking away from a lamppost 15 feet high at a constant rate of 13 feet per second. At what rate is the tip of his shadow receding from the lamppost when he is 6 feet from the lamppost?

12. A spherical ball 2 inches in diameter is lowered at a constant rate $r$ into a high circular cylinder 4 inches in diameter, which is partially filled with water. How fast is the water level rising in the cylinder when the ball is half submerged?

## 2.11  Approximation by Linear Functions

It is often necessary to approximate complicated functions by the use of simpler ones. Often just to compute a function value $f(x)$ we must resort to an approximation of the function $f$. The simplest approximations are afforded by linear functions. We shall investigate in this section how well an arbitrary function $f$ can be approximated by a linear function. A preliminary discussion of this problem is contained in Exercises 10 and 11, p. 57.

To this end, let $f$ be a function defined on $(a, b)$. Let $l(x)$ be a nonvertical straight line passing through the point $(x_0, f(x_0))$, $a < x_0 < b$. Then $l(x) = f(x_0) + m(x - x_0)$ where $m$ is the slope of the line. If we use $l(x)$ to approximate $f(x)$ for $x$ near $x_0$, then the error $\epsilon(x)$ in this approximation is

$$\epsilon(x) = l(x) - f(x).$$

The quantity $\epsilon(x)$ is the vertical distance from $f(x)$ to $l(x)$. (See Figure 27.)

In order for $l$ to be a usable approximation to $f$ the error $\epsilon(x)$ must tend to zero as rapidly as possible as $x$ tends to $x_0$. Indeed $\lim_{x \to x_0} \epsilon(x) = 0$ means that $f$ is continuous at $x_0$. This is clear since

$$\lim_{x \to x_0} (l(x) - f(x)) = f(x_0) - \lim_{x \to x_0} m(x - x_0) - \lim_{x \to x_0} f(x)$$

$$= f(x_0) - \lim_{x \to x_0} f(x).$$

Hence, $\lim_{x \to x_0} \epsilon(x) = 0$ if and only if $\lim_{x \to x_0} f(x) = f(x_0)$.

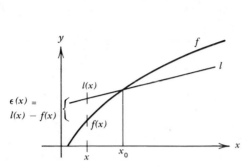

**Fig. 27**

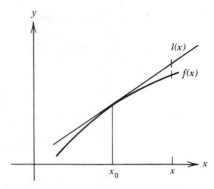

**Fig. 28** The tangent line $l(x) = f(x_0) + f'(x_0)(x-x_0)$ is the best linear approximation to $f$ near $x_0$.

Now we ask, for what choice of the slope $m$ is the linear approximation $l(x)$ the best, in the sense that the error $\epsilon(x)$ tends to zero as fast as possible as $x \to x_0$. A reasonable guess is that we should take $m = f'(x_0)$. Indeed, for that choice of $m$,

$$\lim_{x \to x_0} \frac{\epsilon(x)}{x - x_0} = 0.$$

Let us state the result precisely.

**THEOREM 2.10**   Let $l(x) = f(x_0) + m(x-x_0)$ be a linear approximation to $f$ near the point $x_0$ and set $\epsilon(x) = l(x) - f(x)$. Then

$$\lim_{x \to x_0} \frac{\epsilon(x)}{x - x_0} = 0$$

if and only if $f$ is differentiable at $x_0$ and we choose $m = f'(x_0)$.

**Proof.** The verification of this is extremely simple. Since $\epsilon(x) = l(x) - f(x)$

$$\frac{\epsilon(x)}{x - x_0} = \frac{f(x_0) + m(x - x_0) - f(x)}{x - x_0}$$

$$= \left( m - \frac{f(x) - f(x_0)}{x - x_0} \right).$$

Hence

$$\lim_{x \to x_0} \frac{\epsilon(x)}{x - x_0} = 0$$

if and only if

$$m = \lim_{x \to x_0} \frac{f(x) - f(x_0)}{x - x_0} = f'(x_0).$$

This completes the proof.

Notice that if $f$ is *not* differentiable at $x_0$, then this theorem asserts that it is *impossible* to approximate $f$ near $x_0$ by a linear function $l$ so that

$$\lim_{x \to x_0} \frac{\epsilon(x)}{x - x_0} = 0.$$

On the other hand if $f$ is differentiable at $x_0$ and has a nonvertical tangent there, then any line through $(x_0, y_0)$ has an error of approximation $\epsilon(x)$

satisfying $\lim_{x \to x_0} \epsilon(x) = 0$. The tangent line is the only one for which

$$\lim_{x \to x_0} \epsilon(x)/(x - x_0) = 0.$$

(See Figure 28.)

The function $f'(x_0)(x - x_0)$ as a function of $x - x_0$ is called the *differential of f at $x_0$*, and we write this function $df_{x_0}$ or just $df$ if the point $x_0$ is understood. Thus

$$df(x - x_0) = f'(x_0)(x - x_0)$$

or setting $x - x_0 = h$ we have

(1)
$$df(h) = f'(x_0)h.$$

Thus the differential of $f$ at $x_0$ is the linear function passing through the origin with slope $f'(x_0)$. The derivative $f'(x_0)$ is sometimes called the *differential coefficient of f at $x_0$*, but we shall not use this terminology. If we have written $y = f(x)$, then it is customary to use $dx$ rather than $h$ in (1) to denote the independent variable of the differential. In that case we write $dy$ in place of $df(h) = df(dx)$. Hence, (1) becomes

(2)
$$dy = f'(x_0)dx.$$

**Example 1.** Compute the differential of $f(x) = \sqrt{x}$ at $x_0 = 2$. Since $f'(x_0) = \frac{1}{2}(1/\sqrt{x_0})$, we have for $x_0 = 2$, $df(h) = (1/2\sqrt{2})h$. Setting $h = dx$, this becomes

$$dy = \frac{1}{2\sqrt{2}}dx.$$

For $x$ close to $x_0$, the linear function $f(x_0) + f'(x_0)(x - x_0)$ provides an approximation to $f(x)$. It is convenient to use the symbol "$\approx$" to denote approximate equality. Hence, we write

(3)
$$f(x) \approx f(x_0) + f'(x_0)(x - x_0)$$

if $x$ is close to $x_0$.

**Example 2.** Estimate $\sqrt{4.1}$ by using the differential approximation (3). If $f(x) = \sqrt{x}$, set $x = 4.1$ and $x_0 = 4$. Then $x - x_0 = .1$ and by (3)

$$\sqrt{4.1} \approx \sqrt{4} + \frac{1}{2\sqrt{4}}(.1)$$

$$= 2 + 0.0250 = 2.0250.$$

The correct value to four decimal places is 2.0248.

If we wish to estimate $f(x) - f(x_0)$, then from (3) we have

(4)
$$f(x) - f(x_0) \approx f'(x_0)(x - x_0).$$

Hence, for $x$ close to $x_0$, $df(x - x_0) = f'(x_0)(x - x_0)$ is an approximation to the difference $f(x) - f(x_0)$. If we write $\Delta x = x - x_0$ and $\Delta f = f(x) - f(x_0)$, then (4) becomes

$$\Delta f \approx f'(x_0)\Delta x.$$

Thus, if $x_0$ is changed by a small amount $h$ or $\Delta x$, the corresponding change $\Delta f$ in $f$ is approximately $f'(x_0)\Delta x$.

**Example 3.** The radius of a sphere is increased from one unit to 1.05 units. Determine approximately the corresponding change in the volume. If $r$ is the radius and $V$ is the volume of the sphere, $V = \frac{1}{3}\pi r^3$. Hence $V'(r) = \pi r^2$. For a small change $\Delta r$ in the radius, the corresponding change $\Delta V$ in the volume is approximately

$$V'(r)\Delta r = \pi r^2 \Delta r.$$

In this example $r = 1$ and $\Delta r = 0.05$. Setting $\pi \approx 3.07$ we have

$$\Delta V \approx 3.07(1)0.05 = 0.1535 \text{ cubic units.}$$

**EXERCISES**

1. For the following functions $f$ determine the differential $df_{x_0}(h)$ at the indicated point $x_0$.

(a) $f(x) = 3x^2 + x$      $x_0 = 2$

(b) $f(x) = \sqrt{x^2 + 1}$      $x_0 = 3$

(c) $f(x) = \dfrac{1}{\sqrt[3]{4x - 2}}$      $x_0 = 1$

(d) $f(x) = \sqrt{\dfrac{x-1}{x+1}}$      $x_0 = a$

(e) $f(x) = \dfrac{x^2 - 1}{\sqrt{x+1}}$      $x_0 = a$

(f) $f(x) = \dfrac{\sqrt[3]{x^2 + 1}}{x - 1}$      $x_0 = a$

2. Determine the best linear approximations $l(x)$ to the following functions $f$ near the given points $x_0$.

(a) $f(x) = 4x^2 - 4$      $x_0 = 2$

(b) $f(x) = 3x^5 - \dfrac{5}{x^2}$      $x_0 = 1$

(c) $f(x) = \sqrt{x^2 + x}$      $x_0 = 2$

(d) $f(x) = \sqrt{\dfrac{x^2 + 1}{x^2 - 1}}$      $x_0 = 4$

(e) $f(x) = \sqrt[3]{\dfrac{x}{x^2 + 1}}$      $x_0 = 1$

3. Compute approximations to the following quantities by using an appropriate differential approximation

(a) $(10.1)^5$

(b) $\sqrt[3]{27.05}$

(c) $\sqrt[5]{32.1}$

(d) $\sqrt[7]{124}$

(e) $\sqrt[4]{265}$

4. Let $f(x) = |x^2 - 1|$. Let $l(x)$ be a linear approximation to $f(x)$ near the point $x_0$. For which values of $x_0$ is it possible to construct an approximation satisfying

$$\lim_{x \to x_0} \frac{l(x) - f(x)}{x - x_0} = 0.$$

5. Let $f(x) = \sqrt[3]{x}$. For which values of $x_0$ is it possible to construct a linear approximation $l(x)$ such that

$$\lim_{x \to x_0} \frac{l(x) - f(x)}{x - x_0} = 0.$$

6. A machinist is manufacturing cubes that are 2 inches on each side. If an error of 0.01 inch is made in the side of the cube, determine an approximation to the error produced in the volume and in the surface area.

7. Steel spheres with a radius of 4 inches are to have a volume accurate to $10^{-4}$ cubic inches. Approximately how much error can be allowed in the radius?

8. A hemispherical bowl 2 feet deep is half full of water. If the water level is raised one tenth of an inch, what is the approximate change in the volume? (See Exercise 10, p. 85).

CHAPTER THREE

# THE INTEGRAL

### 3.1  Area

The problem of determining the area of plane regions with curved boun-
daries goes back to the Greek school of geometry. The method of exhaustion, probably invented by Eudoxus in the 4th century B.C., was the first
technique for attacking such problems. Eudoxus used this method to prove
that areas of circles were proportional to the square of their diameters. The
idea was further refined by Archimedes, who used it to derive the formula
for the area of a parabolic segment. This area problem continued to occupy
mathematicians until the time of Newton and Leibniz, when the development of the calculus provided a solution to the problem. Although the
technique of the calculus provided answers to a vast number of "area
problems," the precise definition of the concept of area was not discovered
until the 19th century after the arithmetic formulation of the central ideas
of the calculus had been given.

The problem of area is twofold. First, a careful definition of what is
meant by the area of an arbitrary set in the plane *must be given.* Mathematically it is not satisfactory to take the position that a region such as
that in Figure 1 obviously has a well-defined area, with the only problem
being to compute it. After an appropriate definition of area has been given,
techniques must be developed for its computation. It is here that the
methods of the calculus provide the essential tool.

Let us consider informally how to define the area $A$ of a region $T$ such as
in Figure 1. For a set $T$ in the plane, let $A(T)$ denote its area, *if this can be
defined.* First, if $R$ is a rectangle with base $b$ and height $h$ (Figure 2), we
define

(1) $$A(R) = b \cdot h.$$

If we consider points $P$ and line segments $l$ as degenerate rectangles, then
it follows from (1) that

$$A(P) = 0 \qquad \text{and} \qquad A(l) = 0.$$

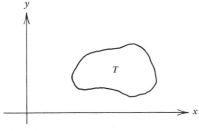

Fig. 1

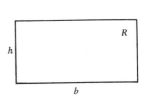

Fig. 2

Suppose $R_1$ and $R_2$ are two rectangles that do not overlap, i.e., $R_1 \cap R_2 = \emptyset$; then we define

(2) $$A(R_1 \cup R_2) = A(R_1) + A(R_2)$$

See Figure 3.

Formula (2) should hold in slightly more general situations. Indeed if the rectangles $R_1$ and $R_2$ overlap at most on a portion of an edge, then (2) should hold. Let us give this situation a name. If sets $S_1$, and $S_2$ satisfy $S_1 \cap S_2 = \emptyset$, then we say the sets $S_1$ and $S_2$ are *disjoint*. If rectangles $R_1$ and $R_2$ intersect at most on a portion of an edge, we say $R_1$ and $R_2$ are *almost disjoint* (Figure 4). Extending (2) by induction we have the following.

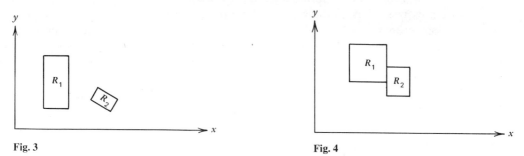

Fig. 3                                             Fig. 4

**Definition.**    *If G is the union of a finite number of rectangles $R_1, \ldots, R_n$ such that each pair $R_i$ and $R_j$ are almost disjoint, then*

(3)    $$A(G) = A(R_1 \cup \cdots \cup R_n) = A(R_1) + A(R_2) + \cdots + A(R_n).$$

For lack of a better name, a finite collection of pairwise, almost disjoint rectangles, the edges of which are parallel, will be called a grid (Figure 5). We do not exclude the possibility that some of these rectangles are degenerate; that is, they may be points or line segments. Thus the area of a grid is the sum of the areas of the constituent rectangles.

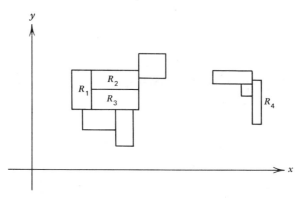

**Fig. 5**    A grid G

For a more general set $T$ we proceed as follows. First we assume that $T$ may be enclosed in a rectangle $R$. Such a set $T$ is called *bounded*. Next we construct a grid $G$ by dividing the base and the altitude of the rectangle $R$ into finitely many pieces, as in Figure 6. Let $R_i$ denote the subrectangles of

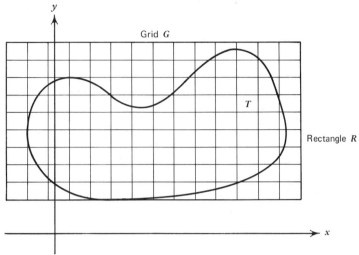

Fig. 6

this grid. We call this grid a *partition* of the rectangle $R$. From this partition $G$ we construct two auxilliary grids $G_1$ and $G_2$.

The grid $G_1$, which we call the inner grid, is the union of all the rectangles $R_i$ that are totally contained in $T$.

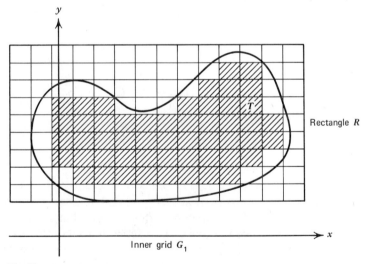

Fig. 7

For the set $T$, this is Figure 7. The grid $G_2$, which we call the outer grid, is the union of all those rectangles $R_i$, each of which contains a portion of the set $T$ (Figure 8).

Since each rectangle $R_i$ of the inner grid $G_1$ is also a rectangle of the outer grid $G_2$, the area, $A(G_1)$ of the inner grid must be less than or equal to the area $A(G_2)$ of the outer grid. Thus

$$A(G_1) \leq A(G_2).$$

Furthermore, if $T$ is to have an area $A(T)$, then we require that

(4) $$A(G_1) \leq A(T) \leq A(G_2).$$

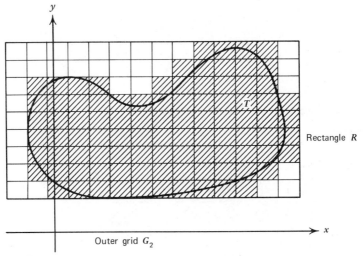

**Fig. 8**

If there is *precisely one number A* with the property that

(5) $$A(G_1) \leq A \leq A(G_2)$$

for each choice of inner grid $G_1$ and outer grid $G_2$, then we say that $T$ has an area $A(T)$ and we define $A(T)$ to be that unique number $A$.

If there is more than one number $A$ satisfying (5) for each choice of inner and outer grids $G_1$ and $G_2$, then we say that $T$ has no area.

Now if $G_1$ is an inner grid associated with one partition of $R$ and $G_2$ is an outer grid associated with another, then it is easy to show (see Exercise 8, p. 95) that

$$A(G_1) \leq A(G_2).$$

Hence, the two sets of numbers

$$\{A(G_1) : G_1 \text{ an inner grid for } T\} = U$$

$$\{A(G_2) : G_2 \text{ an outer grid for } T\} = V$$

are separated (Chapter 1, p. 13). Thus by the completeness axiom for the real numbers there must exist at least one real number $A$ separating the sets $U$ and $V$. If this number $A$ is unique, then we say that $T$ has area equal to $A$.

On p. 14 we gave conditions on two separated sets of real numbers $U$ and $V$ which guarantees that there exists a unique number separating the two sets. For $U$ and $V$ this condition is just the requirement that for each positive number $\epsilon$ no matter how small, an inner grid $G_1$ and outer grid $G_2$ can be found such that

$$A(G_2) - A(G_1) < \epsilon.$$

Thus the number $A$ separating $U$ and $V$ will be unique if and only if $A$ may be approximated arbitrarily well from below by the area of inner grids and arbitrarily well from above by the areas of outer grids.

*Example 1.*   It is a bit hard to imagine an example of a set $T$ for which no area can be assigned. However, we may construct such an example in the following way. Let $f$ be the function defined on the interval [0, 1] that satisfies

$$f(x) = 2 \text{ if } x \text{ is a rational number};$$

$$f(x) = 1 \text{ if } x \text{ is an irrational number}.$$

Then we set

$$T = \{(x, y) : 0 \leqslant x \leqslant 1 \quad \text{and} \quad 0 \leqslant y \leqslant f(x)\}.$$

We cannot draw an accurate sketch of $T$; however, a rough indication is contained in Figure 9.

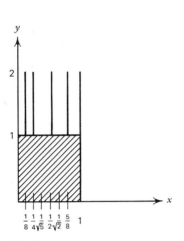

**Fig. 9**

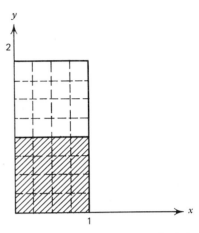

**Fig. 10** Inner grid $G_1$ contained in $\{(x, y) = 0 \leqslant x_1 y \leqslant 1\}$.

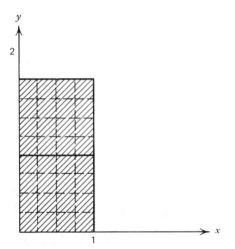

**Fig. 11** Outer grid $G_2$ is entire rectangle $R$.

One might call the set $T$ the "square with the fringe on top." The fringe is that portion of $T$ for which $1 < y \leqslant 2$. To show that we cannot define an area for the set $T$ we argue as follows. Let $R$ be the rectangle $\{(x, y) : 0 \leqslant x \leqslant 1, 0 \leqslant y \leqslant 2\}$. Then $T$ is contained in $R$. Partition $R$ into a grid $G$ and let $G_1$ and $G_2$ be the inner and outer grids, respectively (Figures 10 and 11).

Now if $R_i$ is any rectangle of the grid $G$ that contains points of the fringe of $T$, $R_i$ must also contain points of $R$ that are not in $T$. (Why?) Hence, such a rectangle $R_i$ cannot be a rectangle of the inner grid $G_1$. Therefore, the inner grid $G_1$ must be totally contained in the rectangle $\{(x, y) : 0 \leqslant x, y \leqslant 1\}$. Hence, $A(G_1) \leqslant 1$. On the other hand, the outer grid $G_2$ must be the entire rectangle $R$. Hence, $A(G_2) = 2$. Thus any number between 1 and 2 is on one hand greater than the area of an inner grid for $T$, and at the same time it is less than the area of an outer grid. Since there are infinitely many such numbers, the set $T$ has no area.

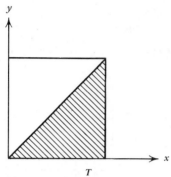

**Fig. 12**

***Example 2.***   Let $T = \{(x, y) : 0 \leqslant x \leqslant 1, 0 \leqslant y \leqslant x\}$. (Figure 12).

Let us show by the methods discussed in this section that the area of this triangle equals $\frac{1}{2}$. Partition $R$ into a grid the rectangles of which are 1/4 unit on each side. If $A_1$ and $A_2$ are the areas of the inner and outer grids, respectively (Figure 13), then $A_1 = 3/8$ and $A_2 = 5/8$.

We leave it as an exercise to verify that if the grid for $R$ is composed of squares $1/n$ unit on a side, then $A_1$, the area of the inner grid for $T$, equals $(n-1)/2n$. The area $A_2$ of the outer grid equals $(n+1)/2n$. Clearly for each $n$

$$\frac{n-1}{2n} < \frac{1}{2} < \frac{n+1}{2n}$$

and 1/2 is the only number with this property. Hence the area of the triangle $T$ equals 1/2.

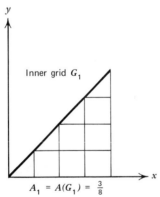

$$A_1 = A(G_1) = \tfrac{3}{8}$$

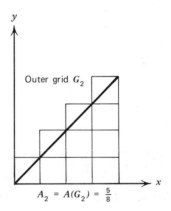

$$A_2 = A(G_2) = \tfrac{5}{8}$$

**Fig. 13**

## EXERCISES

Sketch each of the following regions $T$. Enclose $T$ in a suitable rectangle $R$ and partition $R$ into a grid made up of squares 1/4 unit on each side. Compute the resulting area of the inner grid $G_1$ and the outer grid $G_2$ for $T$. Do the same if the width of the square is 1/8, also 1/16. The calculation is simplified if coordinate paper is used for the sketch.

1.  $T = \{(x, y) : 0 \leqslant x \leqslant 2, 0 \leqslant y \leqslant 2x\}$.
2.  $T = \{(x, y) : 0 \leqslant |x| \leqslant 1, 0 \leqslant y \leqslant 1 - |x|\}$.
3.  $T = \{(x, y) : 0 \leqslant x \leqslant 1, 0 \leqslant y \leqslant x^2\}$.
4.  $T = \{(x, y) : 0 \leqslant |x| \leqslant 1, 0 \leqslant y \leqslant 1 - x^2\}$.
5.  $T = \{(x, y) : x^2 + y^2 \leqslant 4\}$.
6.  In Example 2 above suppose the grid is made up of squares having width $1/n$.

Show that the areas of the inner and outer grids respectively equal $(n-1)/2n$ and $(n+1)/2n$. Hence, conclude that the area of the triangle equals 1/2.

7. Let $T = \{(x, y) : 0 \leqslant x \leqslant 2,\ 0 \leqslant y \leqslant 2x\}$. Construct a grid for $T$ composed of squares of width $1/n$. Compute areas of the resulting inner and outer grids. Show as a result that the area of $T$ equals 4.

8. Let $T$ be a set contained in a rectangle $R$. Let $G_1$ be an inner grid for $T$ associated with one partition of $R$ and $G_2$ an outer grid for $T$ associated with another partition of $R$. Show that $A(G_1) \leqslant A(G_2)$. (Draw a sketch. What alteration must be made on the grids $G_1$ and $G_2$ to make this assertion clear?)

## 3.2 Step Functions and the Integral

Let $f$ be a nonnegative function defined on the interval $[a, b]$. The set of points

$$U_f = \{(x, y) : a \leqslant x \leqslant b \text{ and } 0 \leqslant y \leqslant f(x)\}$$

is called the *ordinate set* for the function $f$ on the interval $[a, b]$. This is the set of points $(x, y)$ in the plane lying between the lines $x = a$ and $x = b$ and between the graph of $f$ and the $x$ axis (Figure 14).

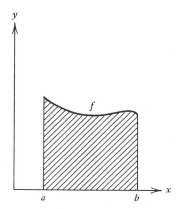

**Fig. 14**   Ordinate set $U_f$.

In this chapter we shall solve completely the area problem for ordinate sets. First, we shall define the area $A$ of the ordinate set of a function. Then if the area exists, we shall use the calculus to develop a method for computing it. We begin first with functions whose ordinate sets are grids. These are called step functions and are defined in the following way.

**Definition.**   *Let $[a, b]$ be a closed interval. By a partition of $[a, b]$ (Figure 15), we mean a division of $[a, b]$ into subintervals of the form $[x_k, x_{k+1}]$ where*

$$a = x_0 < x_1 < \cdots < x_n = b.$$

**Fig. 15**   Partition of $[a, b]$.

*The open intervals $(x_{k-1}, x_k)$ are called the open subintervals of the partition of $[a, b]$. A function $f$ with domain $[a. b]$, which is constant on each open subinterval of a partition of $[a, b]$ is called a step function (Figure 16).*

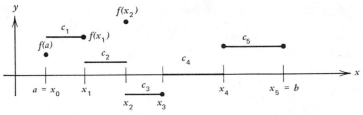

**Fig. 16**   Graph of a step function.

Thus, if $f$ is a step function, there are $n$ constants $c_1, \ldots, c_n$ such that

$$f(x) = c_k \text{ if } x_{k-1} < x < x_k.$$

We have said nothing about the value of $f$ at the points $x_0, x_1, \ldots, x_n$. Indeed, $f$ may have any value at these points, just so long as $f$ is defined at these points.

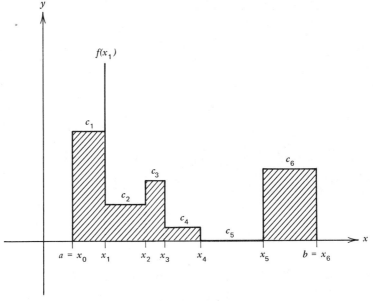

**Fig. 17**   Ordinate set of a nonnegative step function.

If $f$ is a nonnegative step function defined on an interval $[a, b]$, then the ordinate set $U_f$ for $f$ is a grid (Figure 17). The area $A(U_f)$ is the sum of the areas of the constituent rectangles. Hence, we have the formula

(1)        $$A(U_f) = c_1(x_1 - x_0) + c_2(x_2 - x_1) + \ldots + c_n(x_n - x_{n-1}).$$

We abbreviate the sum on the right-hand side of (1) by the symbol $\int_a^b f$, read "the integral of $f$ from $a$ to $b$." Thus for nonnegative step functions we have the following.

**Definition.**   *Let $f$ be a nonnegative step function, defined on the interval $[a, b]$, which is constant on the open subintervals $(x_{k-1}, x_k)$ of a partition $a = x_0 < x_1 < \cdots < x_k = b$ of the interval. If for each $k = 1, \ldots, n$,*

$$f(x) = c_k, \, x_{k-1} < x < x_k,$$

*then we define*

$$\int_a^b f = c_1(x_1 - x_0) + c_2(x_2 - x_1) + \cdots + c_n(x_n - x_{n-1}).$$

*If $A(U_f)$ is the area of the ordinate set for $f$ on $[a, b]$, then*

$$A(U_f) = \int_a^b f.$$

Let us compute $\int_a^b f$ for a few step functions $f$.

**Example 1.** Let $f$ be the postage stamp function introduced on p. 22. That is,

$$\begin{aligned} f(x) &= 6 & 0 \le x < 1 \\ &= 12 & 1 \le x < 2 \\ &= 18 & 2 \le x < 3 \text{ etc.} \end{aligned}$$

Compute $\int_1^5 f$ and $\int_{1/2}^{5/2} f$. (See Figure 18.)

First $\qquad \int_1^5 f = 12(2-1) + 18(3-2) + 24(4-3) + 30(5-4) = 84.$

Next $\qquad \int_{1/2}^{5/2} f = 6\left(1 - \frac{1}{2}\right) + 12(2-1) + 18\left(\frac{5}{2} - 2\right) = 24$

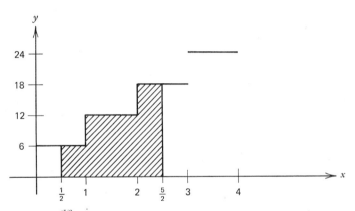

**Fig. 18** $\int_{1/2}^{5/2} f =$ area of ordinate set of $f$ on the interval $[1/2, 5/2]$.

**Example 2.** The greatest integer function is a convenient tool for constructing step functions. We define

$$[x] = \text{largest integer} \le x.$$

Thus $[2] = 2$, $[7/2] = 3$, $[\pi+1] = 4$, etc. It is apparent that $f(x) = [x]$ is a step function. The area of the ordinate set for $f$ on the interval $[1/3, 9/4]$ is

$$\int_{1/3}^{9/4} f = 0\left(1 - \frac{1}{3}\right) + 1(2-1) + 2 \cdot \left(\frac{9}{4} - 2\right) = \frac{3}{2}. \text{ (See Figure 19.)}$$

**Example 3.** Sketch the graph of $f(x) = [2x+1]$ for $x \ge 0$ and compute $\int_{1/2}^{11/2} 2f$. (See Figure 20). Now

$$\begin{aligned} 1 \le 2x+1 < 2 & \quad \text{if} \quad 0 \le x < \frac{1}{2} \\ 2 \le 2x+1 < 3 & \quad \text{if} \quad \frac{1}{2} \le x < 1 \\ 3 \le 2x+1 < 4 & \quad \text{if} \quad 1 \le x < \frac{3}{2} \text{ etc.} \end{aligned}$$

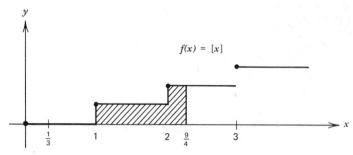

**Fig. 19**

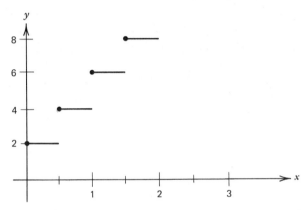

**Fig. 20**   Graph of function $2f(x) = 2[2x+1]$.

Therefore, $2f(x) = 2[2x+1] = 2$    if    $0 \leqslant x < \dfrac{1}{2}$

$$= 4 \quad \text{if} \quad \frac{1}{2} \leqslant x < 1$$

$$= 6 \quad \text{if} \quad 1 \leqslant x < \frac{3}{2}\text{etc.}$$

$$\int_{1/2}^{11/2} 2f = 4\left(1 - \frac{1}{2}\right) + 6\left(\frac{3}{2} - 1\right) + \cdots + 22\left(\frac{11}{2} - 5\right)$$

$$= \frac{1}{2}(4 + 6 + \cdots + 22)$$

$$= \frac{2}{2}(2 + 3 + \cdots + 11) = \frac{11 \cdot 12}{2} - 1$$

$$= 65.$$

For the latter calculation we are using the formula

$$1 + 2 + \cdots + n = \frac{n(n+1)}{2}$$

which was proved in Chapter 1.

### EXERCISES

1. Let $f$ be the postage stamp function. Write $f$, $2f$, $3f+1$ in terms of the greatest integer function. Compute $\int_{3/4}^{9/5} f$, $\int_{0}^{11/7} 2f$, $\int_{3/2}^{11/3} (3f+1)$.

2. Sketch the graph of $f(x) = [x^2]$ and $g(x) = [x]^2$. Compute $\int_{-2}^{3} f$ and $\int_{1}^{5/2} g$.

3. Sketch the graph of $f(x) = [4 - x^2]$. Compute the area of the ordinate set of $f$ on the interval $[-3/2, 7/4]$.

4. Sketch the graph of $f(x) = [|3-x|]$. Compute the area of that portion of the ordinate set of $f$ for which $f(x) \leq 5$. Express this area as an integral.

5. Sketch the graph of $f(x) = [|9-x^2|]$ and compute the area of that portion of the ordinate set of $f$ for which $f(x) \leq 2$. Express this area as an integral.

6. Let $f(x) = |[x]| + |[3x-1]|$. Sketch the ordinate set for $f$ on the interval $[-1, 2]$ and compute its area.

7. For a real number $x \geq 1$ the Euler $\phi$ function is defined by the rule

$$\phi(x) = \text{largest prime number} \leq x.$$

Thus, $\phi(5/2) = 2$; $\phi(8) = 7$; $\phi(\pi) = 3$, etc. Clearly $\phi$ is a step function. If

$$f(x) = \phi(x) + [x]$$

and $g(x) = \phi(2x)$, compute

$$\int_1^8 \phi, \int_2^{12} f, \int_1^4 g.$$

8. Let $g(x) = [x]$ be the greatest integers functions, and let $n$ and $m$ be integers with $n > m$. Derive a formula for $\int_m^n g$.

## 3.3 Fundamental Properties of the Integral

The two basic properties of the integral of step functions are contained in the next result.

**THEOREM 3.1**   *Let $f$ and $g$ be nonnegative step functions defined on an interval $[a, b]$ and let $c$ and $d$ be nonnegative real numbers. Then $cf + dg$ is a step function and*

(1)
$$\int_a^b (cf + dg) = c \int_a^b f + d \int_a^b g.$$

*Furthermore, if $f(x) \leq g(x)$ for each $x \in [a, b]$, then*

(2)
$$\int_a^b f \leq \int_a^b g.$$

*Proof.*   Suppose $f$ is constant on the open intervals $(x_{k-1}, x_k)$ of the partition $a = x_0 < x_1 < \cdots < x_n = b$ of the interval $[a, b]$. Then $cf$ is also constant on $(x_{k-1}, x_k)$ and assuming $f(x) = c_k, x_{k-1} < x < x_k$, we claim that

(3)
$$\int_a^b cf = c \int_a^b f.$$

The verification of (3) is immediate since

$$\int_a^b cf = cc_1(x_1 - x_0) + \cdots + cc_n(x_n - x_{n-1})$$
$$= c[c_1(x_1 - x_0) + \cdots + c_n(x_n - x_{n-1})]$$
$$= c \int_a^b f.$$

Next we show that if $f$ and $g$ are step functions, then $f + g$ is a step function and

(4)
$$\int_a^b (f+g) = \int_a^b f + \int_a^b g.$$

If $f$ and $g$ are constant on the same subintervals $(x_{k-1}, x_k)$, and $g(x) = d_k$,

$x_{k-1} < x < x_k$, then clearly $f + g$ is constant on these subintervals. Moreover

$$\int_a^b (f+g) = (c_1 + d_1)(x_1 - x_0) + \cdots + (c_n + d_n)(x_n - x_{n-1})$$
$$= c_1(x_1 - x_0) + \cdots + c_n(x_n - x_{n-1}) + d_1(x_1 - x_0) + \cdots + \\ + d_n(x_n - x_{n-1})$$
$$= \int_a^b f + \int_a^b g.$$

This is illustrated in Figure 21.

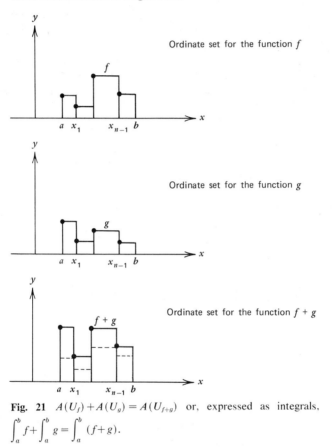

Ordinate set for the function $f$

Ordinate set for the function $g$

Ordinate set for the function $f + g$

**Fig. 21**   $A(U_f) + A(U_g) = A(U_{f+g})$   or, expressed as integrals,
$$\int_a^b f + \int_a^b g = \int_a^b (f+g).$$

If $g$ is not constant on the intervals $(x_{k-1}, x_k)$ of the partition for $f$, but is constant on the subintervals of some other partition $a = x_0' < x_1' < \cdots < x_m'$ $= b$, then we may construct a new partition on which both $f$ and $g$ are constant, by the simple device of taking the points $\{x_i\}$ and $\{x_i'\}$ together as division points. Now we may use the same argument as above to conclude that $f + g$ is a step function and to conclude that $\int_a^b (f+g) = \int_a^b f + \int_a^b g$. Formula (1) then follows from (3) and (4).

The partition of $[a, b]$ formed by taking the points $\{x_i\}$ and $\{x_i'\}$ together is called a *refinement* of the two given partitions (cf. Figure 22).

To show that $f \leq g$ implies that $\int_a^b f \leq \int_a^b g$ is equally easy. Since $f \leq g$, the ordinate set $U_f$ is a subset of $U_g$, hence intuitively $A(U_f) \leq A(U_g)$ or $\int_a^b f \leq \int_a^b g$ (see Figure 23).

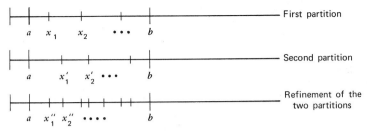

**Fig. 22**

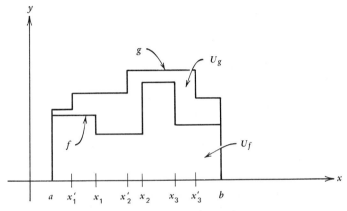

**Fig. 23**   $f \leqslant g$ hence $A(U_f) \leqslant A(U_g)$ or $\int_a^b f \leqslant \int_a^b g.$

To give an arithmetic argument, observe first that just as in the previous case we may assume that both $f$ and $g$ are constant on the subintervals of some partition $a = x_0 < x_1 < \cdots < x_n = b$ of $[a, b]$. If $f(x) = c_k$, $g(x) = d_k$, when $x_{k-1} < x < x_k$, then $c_k \leqslant d_k$ and

$$\int_a^b f = c_1(x_1 - x_0) + \cdots + c_n(x_n - x_{n-1})$$

$$\leqslant d_1(x_1 - x_0) + \cdots + d_n(x_n - x_{n-1}) = \int_a^b g.$$

This completes the proof.

Actually formula (2) may be derived as a consequence of (1). We leave that verification as an exercise.

As an immediate consequence of Theorem 3.1 we have the following useful result.

**THEOREM 3.2**   *Let $f$ be a nonnegative step function defined on an interval $[a, b]$. If $a \leqslant c \leqslant b$, then*

(3)
$$\int_a^b f = \int_a^c f + \int_c^b f.$$

**Proof.**   Let

$$\begin{aligned} f_1(x) &= f(x) &\quad a \leqslant x \leqslant c \\ f_1(x) &= 0 &\quad c < x \leqslant b \end{aligned}$$

and

$$\begin{aligned} f_2(x) &= 0 &\quad a \leqslant x \leqslant c \\ f_2(x) &= f(x) &\quad c < x \leqslant b. \end{aligned}$$

Then $f_1$ and $f_2$ are step functions. Furthermore, $f = f_1 + f_2$. By Theorem 3.1

$$\int_a^b f = \int_a^b f_1 + \int_a^b f_2.$$

But by definition of $f_1$ and $f_2$

$$\int_a^b f_1 = \int_a^c f \qquad \text{and} \qquad \int_a^b f_2 = \int_c^b f.$$

Hence

$$\int_a^b f = \int_a^c f + \int_c^b f$$

as required. Figure 24 illustrates the situation.

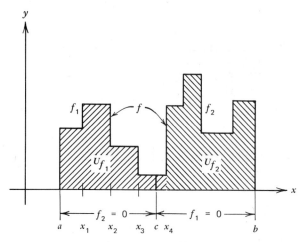

**Fig. 24** $f_1(x) = f(x), a \leqslant x \leqslant c \quad f_2 = 0, a \leqslant x \leqslant c$
$\qquad = 0, c < x \leqslant b; \qquad = f(x), c < x \leqslant b.$

$A(U_f) = A(U_{f_1}) + A(U_{f_2})$ or $\int_a^b f = \int_a^c f + \int_c^b f.$

Formula (3) states that if $a$, $b$, $c$ are three points in an interval on which a step function is defined and $a < c < b$, then

(3) $$\int_a^b f = \int_a^c f + \int_c^b f.$$

Formula (3) is very useful, and without any difficulty we can remove the restriction that $a < c < b$ by making the following conventions. First we set

$$\int_a^a f = 0.$$

Second, if $a < b$, we define

$$\int_b^a f = - \int_a^b f.$$

We leave the verification of the next result as an exercise.

**THEOREM 3.3** *If $a$, $b$, $c$ are three points in an interval $[d, e]$ on which a nonnegative step function is defined, then*

$$\int_a^b f = \int_a^c f + \int_c^b f.$$

**EXERCISES**

1. If $f$ is a nonnegative step function defined on the interval $[a, b]$, prove directly from the definition of the integral that $\int_a^b f \geqslant 0$.

2. If $f$ and $g$ are nonnegative step functions defined on $[a, b]$ and $f \geqslant g$, show that $f - g$ is a nonnegative step function, and hence $\int_a^b (f - g) \geqslant 0$.

3. Using the assumptions of Exercise 2 show that $\int_a^b f \geqslant \int_a^b g$ by writing $f = f + (g - f)$ and then applying Theorem 3.1 (1) and the results of Exercises 1 and 2.

4. Prove Theorem 3.3.

## 3.4    The Integral of More General Functions

We now wish to define the area of the ordinate set for nonnegative functions $f$ that are *more general* than just step functions. To do this we need to define $\int_a^b f$ for more general functions. This is done exactly as in the preliminary discussion of area in Section 3.1. First we assume that the nonnegative function $f$ is defined and bounded on the interval $[a, b]$. This means that for some constant $M$, $0 \leqslant f(x) \leqslant M$ for $a \leqslant x \leqslant b$. Now let $G$ and $g$ be nonnegative step functions and assume $g(x) \leqslant f(x) \leqslant G(x)$ for each $x \in [a, b]$. We shall call such a pair of step functions $G$ and $g$ *upper and lower approximating step functions* for $f$ (see Figure 25).

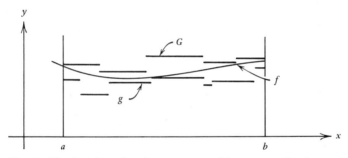

**Fig. 25**    The functions $G$ and $g$ are upper and lower approximating step functions for $f$.

By Theorem 3.1, since $g \leqslant G$, we have $\int_a^b g \leqslant \int_a^b G$. Hence the sets of numbers $S = \left\{ \int_a^b g : g \text{ a nonnegative step function and } g \leqslant f \right\}$, and $T = \left\{ \int_a^b G : G \text{ a nonnegative step function and } G \geqslant f \right\}$ are separated sets of real numbers. Thus, applying the completeness axiom for the real numbers (Section 1.5), it follows that there is at least one number $I$ that satisfies

(1) $$\int_a^b g \leqslant I \leqslant \int_a^b G$$

for *each pair* of upper and lower approximating step functions $G$ and $g$.

**Definition.**    *Let $f$ be a nonnegative bounded function defined on the interval $[a, b]$. If there is precisely one number $I$ that satisfies*

$$\int_a^b g \leqslant I \leqslant \int_a^b G$$

*for each pair of upper and lower approximating step functions $G$ and $g$, then $f$ is said to be integrable, and we define $I = \int_a^b f$.*

This solves the problem of the definition of the area of the ordinate set of $f$. (See Figure 26.)

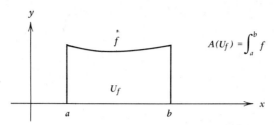

**Fig. 26**   The area of the ordinate set for $f$ can be defined if $f$ is an integrable function.

*Definition.   The ordinate set $U_f$ of a nonnegative function defined on the interval $[a, b]$ is said to have an area if $f$ is integrable  If $A(U_f)$ denotes that area, we define $A(U_f) = \int_a^b f$.*

In Chapter 1 (p. 14), we gave a condition on two separated sets of real numbers $S$ and $T$, which guarantees that the number $m$ separating $S$ and $T$ is unique. In the present context this condition specifies in terms of approximating step functions $g$, $G$ precisely when $f$ is integrable. This is called the Riemann condition in honor of the great 19th-century German mathematician, Bernhard Riemann, whose contributions to the theory of the integral were of great importance. Indeed, $\int_a^b f$ is commonly called the *Riemann integral* of $f$.

**THEOREM 3.4   (Riemann Condition).**   *A bounded nonnegative function $f$ defined on $[a, b]$ is integrable if and only if for each positive number $\epsilon$, nonnegative step functions $g$ and $G$ can be found such that*

(2) $$g \leqslant f \leqslant G$$

and

(3) $$\int_a^b G - \int_a^b g < \epsilon.$$

*In other words $f$ is integrable if and only if the difference of the integrals $\int_a^b G - \int_a^b g$, may be made arbitrarily small for suitably chosen upper and lower approximating step functions $G$ and $g$.*

The class of integrable functions is quite large and includes, in addition to all the step functions, every nonnegative function that is continuous on the interval $[a, b]$. Rather than prove that fact here, however, we shall postpone the argument to the appendix of Chapter 5.

*Example 1.*   Show that $f(x) = x^2$ is integrable on $[0, a]$ and compute $\int_0^a f$.

We first divide the interval $[0, a]$ into $n$ equal subintervals each of length $a/n$. To define upper and lower approximating step functions to $f$, set

(4) $$G(x) = \left[\frac{k}{n}a\right]^2, \text{ if } \left(\frac{k-1}{n}\right)a < x \leqslant \frac{k}{n}a.$$

and

(5) $$g(x) = \left[\frac{(k-1)}{n}a\right]^2, \text{ if } \frac{(k-1)}{n}a \leqslant x < \frac{k}{n}a. \text{ See Figure 27.}$$

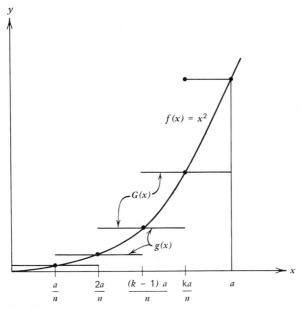

**Fig. 27**

Next we compute $\int_a^b G$ and $\int_a^b g$. By definition

(6) $$\int_0^a G = \left(\frac{a}{n}\right)^2\left(\frac{a}{n}-0\right) + \left(\frac{2a}{n}\right)^2\left(\frac{2a}{n}-\frac{a}{n}\right) + \cdots + \frac{a^2}{n^2}\left(a-\frac{(n-1)}{n}a\right)$$

$$= \frac{a^3}{n^3}(1^2 + 2^2 + \cdots + n^2).$$

and

(7) $$\int_0^a g = 0 \cdot \left(\frac{a}{n}\right) + \left(\frac{a}{n}\right)^2\left(\frac{2a}{n}-\frac{a}{n}\right) + \left(\frac{2a}{n}\right)^2\left(\frac{3a}{n}-\frac{2a}{n}\right) + \cdots + \left(\frac{n-1}{n}\right)^2 a^2\left(a-\frac{(n-1)}{n}a\right)$$

$$= \frac{a^3}{n^3}\left(1^2 + 2^2 + 3^2 + \cdots + \left(\frac{n-1}{n}\right)^2\right)$$

To show that $f(x) = x^2$ is integrable we need only verify the Riemann condition for the pair of step function approximations that we have chosen. Indeed, to show that $\int_a^b G - \int_a^b g$ can be made arbitrarily small, notice that from (6) and (7) it follows that

$$\int_0^a G - \int_0^a g = \frac{a^3}{n^3}(1^2 + \cdots + n^2) - \frac{a^3}{n^3}(1^2 + \cdots + (n-1)^2)$$

$$= \frac{a^3}{n^3} \cdot n^2 = \frac{a^3}{n}.$$

Clearly for $n$ large enough $a^3/n$ can be made smaller than any reassigned positive number $\epsilon$. Hence $f(x) = x^2$ is integrable on $[0, 2]$.

To compute $I = \int_0^a f$ we must find the number I that satisfies

$$\int_a^b g \leq I \leq \int_a^b G,$$

for our choice of upper and lower approximating step functions. To do this recall first the following fundamental inequality from Chapter 1 (p. 40). For each integer $n \geq 1$,

(8) $$1^2 + \cdots + (n-1)^2 < \frac{n^3}{3} < 1^2 + \cdots + n^2.$$

Combining this with (6) and (7) we have

$$\int_0^a g = \frac{a^3}{n^3}(1^2 + \cdots + (n-1)^2) < \frac{a^3}{n^3} \cdot \frac{n^3}{3} = \frac{a^3}{3}$$

and

$$\int_0^a G = \frac{a^3}{n^3}(1^2 + \cdots + n^2) > \frac{a^3}{n^3} \frac{n^3}{3} = \frac{a^3}{3}.$$

Hence, $a^3/3$ is the unique number with the property that

$$\int_0^a g \leq \frac{a^3}{3} \leq \int_0^a G$$

for each choice of upper and lower approximating step functions $G$ and $g$. Consequently it follows that if $f(x) = x^2$, then

(9) $$\int_0^a f = \frac{a^3}{3}.$$

Several remarks are in order. First, inequality (8) was not used to prove that $f(x) = x^2$ was integrable. It only entered in the actual computation of $\int_0^a f$. Thus, proving that a function is integrable appears to be easier than actually computing the integral.

Second, assuming that Theorem 3.3 holds for $f(x) = x^2$, we have

$$\int_0^b f = \int_0^a f + \int_a^b f.$$

Hence

$$\int_a^b f = \int_0^b f - \int_0^a f = \frac{b^3}{3} - \frac{a^3}{3}.$$

See Figure 28.

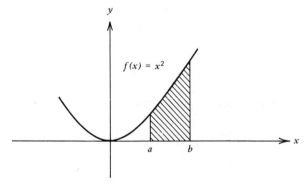

**Fig. 28** $\int_a^b f = \dfrac{b^3}{3} - \dfrac{a^3}{3}.$

Third, since $f(x) = x^2$ is integrable, let us define a new function $F$ by the rule

(10) $$F(x) = \int_0^x f, \text{ for } x \geq 0.$$

We know from our discussion that

$$F(x) = \frac{x^3}{3}.$$

But if we compute the derivative of $F$, we note that

$$F'(x) = x^2 = f(x).$$

Thus if we differentiate the integral of $f$, which is the function $F$, we get back the original function $f$. This is such a startling result that it is tempting to ask whether it holds in general. That is, if $f$ is integrable and we define

$$F(x) = \int_a^x f,$$

does it follow that

$$F'(x) = f(x)?$$

The answer is yes provide that $f$ is continuous at $x$. We shall derive these facts in the next section.

It is also interesting to note that the argument that $f(x) = x^2$ is integrable and the associated computation of $\int_0^a f = a^3/3$ were essentially given by Archimedes in the 3rd century B.C. Indeed, the approximation of $f$ by upper and lower step functions is the modern equivalent of the method of exhaustion developed by Eudoxus and his successors. However, the invention of the derivative and the discovery that if $F(x) = \int_0^x f$, then $F'(x) = f(x)$, had to wait for nearly 2000 years. These two discoveries, made independently by Newton and Leibnitz, were the crucial steps in the development of the calculus.

**EXERCISES**

1. Explain why the sets $S$ and $T$ (p. 103) are separated sets of real numbers, and explain why (1) holds.
2. Give a proof of Theorem 3.4, the Riemann condition.
3. If $f(x) = x^2$ compute $\int_1^5 f, \int_0^3 f$.
4. Compute the area of the region bounded by the $x$ axis, the lines $x = 2$, $x = 6$, and the function $f(x) = x^2$. Sketch the region.
5. Prove by induction or well ordering that

$$1^3 + 2^3 + \cdots + (n-1)^3 < \frac{n^4}{4} < 1^3 + \cdots + n^3.$$

6. In the same manner as was done for the example $f(x) = x^2$, show that $g(x) = x^3$ is integrable on the interval $[0, a]$. Using Exercise 5 show that

$$\int_0^a g = \frac{a^4}{4}.$$

7. Assuming Theorem 3.3 has been proved for $g(x) = x^3$ compute $\int_1^4 g$.
8. Sketch the region between $x = 0$, and $x = 1$, which lies between $f(x) = x$ and $g(x) = x^2$. Compute what you think should be the area of this region.
9. Sketch the region between $x = 0$ and $x = 2$, which lies between $f(x) = x^2$ and $g(x) = x^3$. Compute what you think should be the area of this region.
10. Suppose $k$ is a positive integer. Prove by induction that

$$1^k + 2^k + \cdots + (n-1)^k < \frac{n^{k+1}}{k+1} < 1^k + 2^k + \cdots + n^k$$

11. Prove that for each positive integer $k$, the function $f_k(x) = x^k$ is integrable on $[0, a]$. Furthermore, show that

$$\int_0^a f_k = \frac{a^{k+1}}{k+1}.$$

12. If $f_k(x) = x^k$, $k = 1, 2, 3, \ldots$ and $F_k(x) = \int_0^x f_k$ show that $F_k$ is a primitive for $f_k$.

13. If $f$ is a nonnegative function defined on the interval $[a, b]$ and if for each $\epsilon > 0$ there exist upper and lower approximating step functions $G$ and $g$ that satisfy

$$g(x) \leqslant f(x) \leqslant G(x), \qquad a \leqslant x \leqslant b$$

and

$$G(x) - g(x) \leqslant \epsilon, \qquad a \leqslant x \leqslant b,$$

can we conclude that $f$ is integrable? Explain.

### 3.5  Fundamental Theorems of the Calculus

Suppose that $f$ is a nonnegative integrable function defined on the interval $[a, b]$. It is the object of this section to prove the fundamental results of Newton and Leibnitz, which show that the problem of determining $\int_a^x f$ is in reality the same problem as determining a function $F$ whose derivative is $f$. It is convenient to separate the result into two parts. These are commonly called the first and second fundamental theorems of the calculus.

**THEOREM 3.5  (First Fundamental Theorem).** *Let $f$ be a nonnegative integrable function on the interval $[a, b]$. Set*

$$F(x) = \int_a^x f \qquad for \qquad a \leqslant x \leqslant b.$$

*If $f$ is continuous at the point $x_0$, $a \leqslant x_0 \leqslant b$, then*

$$F'(x_0) = f(x_0).$$

**THEOREM 3.6  (Second Fundamental Theorem).** *Let $f$ be a nonnegative integrable function on the interval $[a, b]$ and assume $f$ is continuous at each point of this interval. Then, if $G$ is a primitive for $f$, that is, if $G'(x) = f(x)$ for each $x$, $a \leqslant x \leqslant b$, it follows that*

$$\int_a^b f = G(b) - G(a).$$

The first of these results states that if $F(x)$ is the area of the ordinate set for $f$ on the interval $[a, x]$, then $F'(x_0) = f(x_0)$ at each point $x_0$ at which $f$ is continuous. (See Figure 29.)

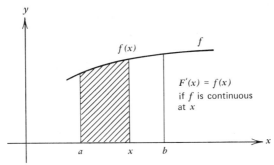

**Fig. 29**  $F(x) = \int_a^x f$ is the area of the ordinate set for $f$ between $a$ and $x$.

The second result reduces the problem of the computation of $\int_a^b f$ to the problem of finding a primitive for $f$. The result states that if $f$ is continuous and $G$ is any primitive for $f$, then

$$\int_a^b f = G(b) - G(a).$$

Before examining the proofs let us compute an example.

*Example 1.* Compute the area $A$ of the region above the $x$ axis and below the function $f(x) = 1 - x^4$. (See Figure 30.) This region is the ordinate set for $f$ on the interval $[-1, 1]$. A primitive for $f$ is $G(x) = x - (x^5/5)$. (Why?)

Therefore

$$A = \int_{-1}^{1} f = G(1) - G(-1)$$
$$= 1 - \frac{(+1)^5}{5} - \left(-1 - \frac{(-1)^5}{5}\right) = \frac{8}{5}.$$

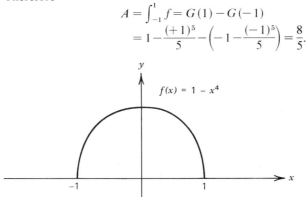

Fig. 30

To prove the fundamental theorems we need analogues for integrable functions of two results in Section 3.3. These are simple geometric facts but since the proofs are technical we omit them for now. The argument for Theorem 3.7 is sketched in the Exercise 27 at the end of this section. Theorem 3.8 will follow from what we do in the next section.

**THEOREM 3.7.** *If $f$ and $g$ are nonnegative integrable functions defined on the interval $[a, b]$ and $f \le g$, then*

(1)
$$\int_{a}^{b} f \le \int_{a}^{b} g.$$

Since $f \le g$, the ordinate set for $f$, $U_f$, is a subset of $U_g$, the ordinate set for $g$. If $A(U_f)$ and $A(U_g)$ are the areas of these ordinate sets, it is geometrically clear (Figure 31) that we must have $A(U_f) \le A(u_g)$. But since $A(U_f) = \int_{a}^{b} f$, and $A(U_g) = \int_{a}^{b} g$, this is just inequality (1)

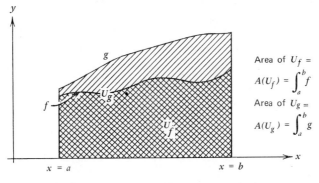

**Fig. 31** $f \le g$ implies $U_f \subset U_g$; hence $A(U_f) \le A(U_g)$.

**THEOREM 3.8** *Let $f$ be a nonnegative integrable function defined on the interval $[a, b]$. If $a \le x \le b$, then $f$ is integrable on the intervals*

$[a, x]$ and $[x, b]$. Furthermore,

(2) $$\int_a^b f = \int_a^x f + \int_x^b f.$$

Let $U_1$ and $U_2$ be the ordinate sets for $f$ on the intervals $[a, x]$ and $[x, b]$. Again it is clear that if $A(U_f) = \int_a^b f$ is the area of the ordinate set for $f$ on $[a, b]$, then $A(U_f)$ is the sum of the areas $A(U_1)$ and $A(U_2)$ (Figure 32).

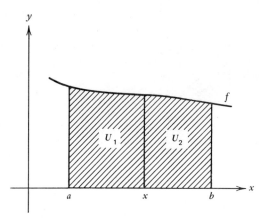

**Fig. 32**   $\int_a^b f = A(U_f) = A(U_1) + A(U_2) = \int_a^x f + \int_x^b f.$

We also extend the conventions on p. 102 to integrable functions. We define

(3) $$\int_a^a f = 0$$

and

(4) $$\int_b^a f = - \int_a^b f \text{ if } a < b.$$

Therefore, for any three numbers $c$, $d$, $e$ in an interval on which $f$ is integrable, we have

(5) $$\int_c^d f = \int_c^e f + \int_e^d f.$$

We now prove the first fundamental theorem. If $F(x) = \int_a^x f$ and $f$ is continuous at the point $x_0$, we must show that

$$F'(x_0) = \lim_{h \to 0} \frac{F(x_0 + h) - F(x_0)}{h} = f(x_0).$$

First, we give the geometric interpretation of the difference quotient $(F(x_0 + h) - F(x_0))/h$. Now $F(x_0 + h) = \int_a^{x_0 + h} f$ and $F(x_0) = \int_a^{x_0} f$. Therefore

$$\begin{aligned}
F(x_0 + h) - F(x) &= \int_a^{x_0 + h} f - \int_a^{x_0} f \\
&= \int_a^{x_0 + h} f + \int_{x_0}^a f \qquad \text{by (4)} \\
&= \int_{x_0}^{x_0 + h} f \qquad \text{by (5).}
\end{aligned}$$

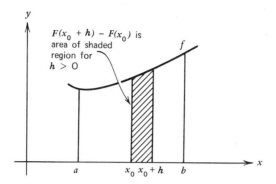

**Fig. 33** $F(x_0) =$ area of the ordinate set for $f$ on the interval $[a, x_0]$ and $F(x_0 + h) - F(x_0) =$ area of ordinate set for $f$ on the interval $[x_0, x_0 + h]$.

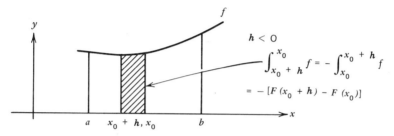

**Fig. 34** The area of the ordinate set for $f$ on the interval $[x_0 + h, x_0]$ equals $-(F(x_0 + h) - F(x_0))$.

Hence, if $h > 0$, then $F(x_0 + h) - F(x_0)$ is the area of the ordinate set for $f$ on the interval $[x_0, x_0 + h]$. (See Figure 33.) If $h < 0$, then $F(x_0 + h) - F(x_0)$ is the negative of the area of the ordinate set for $f$ on the interval $[x_0 + h, x_0]$. (See Figure 34.)

Thus, for both positive and negative $h$, $|F(x_0 + h) - F(x_0)|$ is the area of the ordinate set for $f$ between $x_0$ and $x_0 + h$. Since $|h|$ is the width of either interval $[x_0, x_0 + h]$ or $[x_0 + h, x_0]$, $(F(x_0 + h) - F(x_0))/h$ is the height of a rectangle with area $|F(x_0 + h) - F(x_0)|$ and width $|h|$. We are asserting that for $|h|$ small enough, this height $(F(x_0 + h) - F(x_0))/h$ approximates $f(x_0)$ as closely as desired.

To prove formally that $\lim_{h \to 0} (F(x_0 + h) - F(x_0))/h = f(x_0)$ it suffices to show that for a given $\epsilon > 0$, a corresponding $\delta$ may be chosen so that whenever $0 < |h| < \delta$, it follows that

$$\left| \frac{F(x_0 + h) - F(x_0)}{h} - f(x_0) \right| \leq \epsilon,$$

or equivalently that

(6) $$f(x_0) - \epsilon \leq \frac{F(x_0 + h) - F(x_0)}{h} \leq f(x_0) + \epsilon.$$

Now by assumption $f$ is continuous at $x_0$. Therefore, for a given $\epsilon > 0$ there is a positive number $\delta$ such that whenever $0 < |x - x_0| < \delta$ it follows that $|f(x) - f(x_0)| \leq \epsilon$. Equivalently

$$f(x_0) - \epsilon \leq f(x) \leq f(x_0) + \epsilon \qquad \text{if} \qquad |x - x_0| < \delta.$$

Therefore, if $0 < h < \delta$ and $f(x_0) - \epsilon \geq 0$, we may apply Theorem 3.7 and conclude that

$$\int_{x_0}^{x_0+h} (f(x_0) - \epsilon) \leq \int_{x_0}^{x_0+h} f \leq \int_{x_0}^{x_0+h} (f(x_0) + \epsilon).$$

But

$$\int_{x_0}^{x_0+h} (f(x_0) + \epsilon) = h \cdot (f(x_0) + \epsilon)$$

since this is the area of the rectangle of width $h$ and height $f(x_0) + \epsilon$. (See Figure 35.)

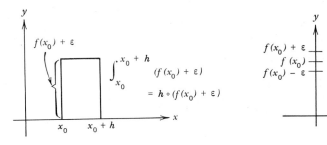

**Fig. 35**

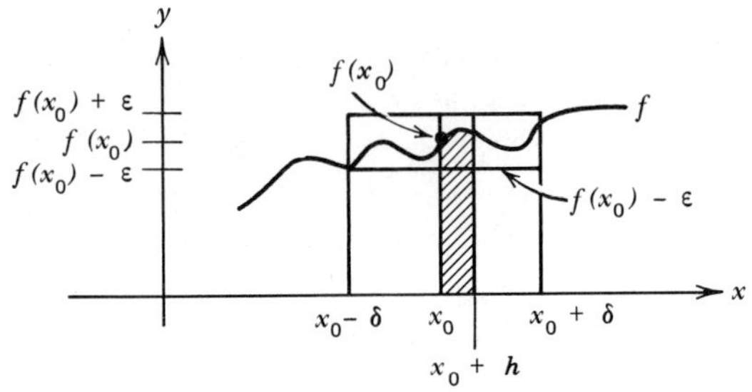

**Fig. 36**    Area of shaded strip $= F(x_0 + h) - F(x_0)$. This area $\leq (f(x_0) + \epsilon) \cdot h$ and $\geq (f(x_0) - \epsilon) \cdot h$.

Similarly

$$\int_{x_0}^{x_0+h} (f(x_0) - \epsilon) = h \cdot (f(x_0) - \epsilon).$$

Hence

(7)     $$h \cdot (f(x_0) - \epsilon) < \int_{x_0}^{x_0+h} f \leq h \cdot (f(x_0) + \epsilon).$$

If $f(x_0) - \epsilon < 0$, (7) still holds since $\int_{x_0}^{x_0+h} f \geq 0 > f(x_0) - \epsilon$.
Dividing (7) by $h$, we have

$$f(x_0) - \epsilon < \frac{1}{h} \int_{x_0}^{x_0+h} f \leq f(x_0) + \epsilon$$

which is the required statement (6) since

$$(F(x_0 + h) - F(x_0))/h = (1/h) \int_{x_0}^{x_0+h} f.$$

For $0 > h > -\delta$ the verification proceeds in a similar manner and is left as an exercise. Figure 36 illustrates the situation.

Next we prove Theorem 3.6. The result is, in reality, a corollary of the vanishing derivative theorem (Theorem 2.7), and the proof is very simple. We must show that if the nonnegative function $f$ is integrable and continuous and $G$ is any primitive for $f$, then

$$\int_a^b f = G(b) - G(a).$$

For $x \in [a, b]$, set $F(x) = \int_a^x f$. By Theorem 3.5, $F'(x) = f(x)$ since it is assumed that $f$ is continuous at each $x$. Thus $F$ is a primitive for $f$. But it follows from the vanishing derivative theorem that two primitives for $f$ differ only by a constant. So

$$F(x) = G(x) + k.$$

But $F(a) = 0$, hence $k = -G(a)$. Therefore

$$F(b) = \int_a^b f = G(b) + k = G(b) - G(a)$$

and we are finished.

There is a slight redundancy in Theorem 3.6. Namely, if $f$ is continuous on $[a, b]$, then necessarily $f$ is integrable. We give the proof of this fact in the appendix to Chapter 5. Thus, to assert that a nonnegative function is integrable on $[a, b]$, it suffices to know that it is continuous. We shall use this result freely in what follows.

For continuous functions $f$, the evaluation of $\int_a^b f$ can be broken down into two steps.

(i) Find a primitive or antiderivative $F$ for $f$.
(ii) Compute $F(b) - F(a)$.

To facilitate the computation of integrals we wish to change our notation for $\int_a^b f$ slightly. If $x$ denotes the independent variable of the function $f$, then in place of $\int_a^b f$ we write $\int_a^b f(x)\, dx$. The symbol $dx$ is introduced to specify which of the many letters present is to stand for the independent variable. Thus if $f(x) = \sqrt{t^2 + 4x^2}$ and we write $\int_0^1 \sqrt{t^2 + 4x^2}\, dx$, then with this notation it is clear that $x$ stands for a point in the interval $[0, 1]$ and $t$ is fixed. On the other hand, if we write $\int_0^1 \sqrt{t^2 + 4x^2}\, dt$, then $t$ is the independent variable that ranges over $[0, 1]$ and $x$ is fixed.

We also abbreviate $F(b) - F(a)$ by $F(x)\big|_{x=a}^{x=b}$ or just $F(x)\big|_a^b$. Thus, steps (i) and (ii) above become

(8)
$$\int_a^b f(x)\, dx = F(x)\Big|_a^b = F(b) - F(a).$$

The function $f$ appearing in the integral $\int_a^b f$ is called the *integrand*.

***Example 2.*** Compute $\int_1^4 (2x + 3x^2)\, dx$. Clearly, if $F(x) = x^2 + x^3$, then $F'(x) = 2x + 3x^2 = f(x)$. Hence

$$\int_1^4 (2x + 3x^2)\, dx = (x^2 + x^3)\Big|_1^4 = 80 - 2 = 78.$$

Formula (8) simplifies slightly the computation of integrals because it allows us to separate the determination of the primitive $F$ from the actual evaluation of $F(b) - F(a)$.

***Example 3.*** For $n \neq -1$ and $0 < a < b$, compute $\int_a^b x^n dx$. The function $f(x) = x^{n+1}/(n+1)$ is a primitive for $f(x) = x^n$. Hence

$$\int_a^b x^n dx = \frac{x^{n+1}}{n+1}\Big|_a^b = \frac{b^{n+1}}{n+1} - \frac{a^{n+1}}{n+1}.$$

***Example 4.*** Compute $\int_0^2 (t + 3x^2)\, dt$ and $\int_0^2 (t + 3x^2)\, dx$. For the first integral the symbol $t$ is the independent variable and $x$ is fixed. Hence $F(t) = t^2/2 + 3x^2 t$ is a primitive for $f(t) = t + 3x^2$. Therefore

$$\int_0^2 (t + 3x^2)\, dt = \frac{t^2}{2} + 3x^2 t\Big|_{t=0}^{t=2} = 2 + 6x^2.$$

On the other hand, if $x$ is the independent variable, $G(x) = tx + x^3$ is a primitive for $g(x) = t + 3x^2$. Hence

$$\int_0^2 (t + 3x^2) \, dx = tx + x^3 \Big|_{x=0}^{x=2} = 2t + 8.$$

## EXERCISES

Compute the following integrals $\int_a^b f$. Determine primitives for the integrands $f$ by inspection.

1. $\int_1^2 (3x - 1) \, dx.$

2. $\int_4^5 (x^2 - 4x) \, dx.$

3. $\int_1^3 \left(4x^3 + \dfrac{1}{x^2}\right) dx.$

4. $\int_2^3 \left(\dfrac{3}{x^2} + \dfrac{4}{x^3}\right) dx.$

5. $\int_1^2 (2x^{3/2} + x^{1/2}) \, dx.$

6. $\int_1^4 \left(2x + 3x^2 + \dfrac{4}{x^2}\right) dx.$

7. $\int_1^3 (2t^3 + at) \, dt.$

8. $\int_2^5 \left(sb + \dfrac{1}{(sa)^2}\right) ds.$

9. $\int_{-1}^1 (x - t)^2 dx.$

10. $\int_0^1 (x - t)^2 dt.$

11. $\int_1^3 (x + 4)^{10} dx.$

12. $\int_0^2 (2x - 1)^6 dx.$

13. $\int_1^3 \dfrac{1}{(x+2)^{1/2}} dx.$

14. $\int_{-1}^1 (2t - a)^4 dt.$

15. $\int_0^a y(a - y^2) \, dy.$

16. Compute the area of the ordinate set for $f(x) = x^3 - 2$ on the interval $[2, 3]$.
17. Compute the area of the ordinate set of $f(x) - 2x^4 + 1$ on the interval $[-1, 2]$.
18. Compute the area of the ordinate set of $f(x) = 2\sqrt{x}$ which lies to the left of the line $x = 5$.
19. Compute the area of the region under the parabola $y = 8 - 2x^2$ and above the $x$ axis.
20. Compute the area of the region under the parabola $y = -x^2 + 4x + 5$ and above the $x$ axis.
21. Compute the area of the ordinate set for $f(x) = (1 + x)^{1/2}$ lying to the left of the line $x = 8$.
22. If $F(x) = \int_1^x \sqrt{1 + t^2} \, dt$, compute $F'(x), F'(2)$.

23. Compute $\dfrac{d}{dx} \int_1^x \sqrt{1 + t^3} \, dt.$

24. Compute $\dfrac{d}{ds} \int_1^s (1 + t)^3 dt.$

25. If $F(x) = \int_1^x \sqrt{4 - t^2} \, dt$, compute $F''(1)$.

26. If $F(x) = \int_1^x \dfrac{1}{t} dt, x > 0$, for what value of $x$ does $F''(x) = -2$?

*27. If $f$ and $g$ are nonnegative functions that are integrable on $[a, b]$, show that $f \le g$ implies $\int_a^b f \le \int_a^b g$. (Suppose not. Then $\int_a^b g < \int_a^b f$; and if $\int_a^b f - \int_a^b g = \epsilon$, then $\epsilon > 0$. Since $f$ is integrable, the Riemann condition tells us that there must exist a step function $f_1$ such that $f_1 \le f$ and

$$\int_a^b f_1 > \int_a^b f - \dfrac{\epsilon}{2}.$$

But then

$$\int_a^b f_1 > \int_a^b g.$$

Show, however, that this is impossible. Hence we have a contradiction and the proof is complete.)

## 3.6   Linearity of the Integral

If $f$ is integrable on the interval $[a, b]$ and we set $F(x) = \int_a^x f$, then the fundamental theorem of the calculus asserts that $F'(x) = f(x)$ if $f$ is continuous. From our work on the derivative in Chapter 2 we know that differentiation is a linear operation. That is, if $f$ and $g$ are differentiable and $c$ and $d$ are real numbers, then $cf + dg$ is differentiable. Furthermore

(1) $$(cf + dg)' = cf' + dg'.$$

We ask now if this linearity property holds for integrals. If $f$ and $g$ are integrable, is $cf + dg$ integrable? If so, is it true that

(2) $$\int_a^b (cf + dg) = c \int_a^b f + d \int_a^b g?$$

The answer is yes, but in order to establish this result, we must first define the notion of the integral for functions that are not necessarily positive. If we examine our definition of the integral, we see immediately that we have not used the fact that the function is nonnegative except to establish the connection between the area of the ordinate set of the function $f$ and the integral $\int_a^b f$.

To see how to define the integral for bounded functions that are not necessarily positive, we first consider step functions. If $f$ is a step function on the interval $[a, b]$, and $f(x) = c_k$ for $x_{k-1} < x < x_k$, where $a = x_0 < x_1 \cdots < x_n = b$, then the formula

$$\int_a^b f = c_0(x_1 - x_0) + c_2(x_2 - x_1) + \cdots + c_n(x_n - x_{n-1})$$

makes perfectly good sense regardless of whether the numbers $c_k$ are nonnegative or not. The only change is that now $\int_a^b f$ is *not* the area of the region between the graph of $f$ and the $x$ axis. Clearly, if $c < 0$, and $f(x) = c$ for $a < x < b$, then the area $A$ of the region between the graph of $f$ and the $x$ axis is $|c|(b - a)$, that is, $A = \int_a^b |f|$. See Figure 37.

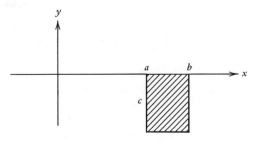

**Fig. 37**

Indeed, for an arbitrary step function that is not necessarily positive, it is easily seen that the same formula is valid. The area $A$ of the region between the graph of $f$ and the $x$ axis is given by

$$A = \int_a^b |f|.$$

See Figure 38.

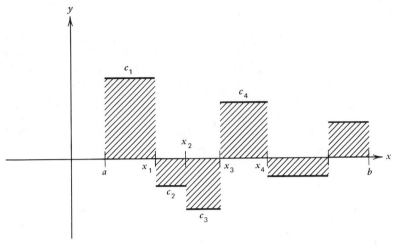

**Fig. 38**

Let us formalize our discussion with the following.

**Definition.** *If $f$ is a step function on the interval $[a, b]$ and $f(x) = c_k$, $x_{k-1} < x < x_k$, where $a = x_0 < x_1 < \cdots < x_n = b$, then we define*

$$\int_a^b f = c_1(x_1 - x_0) + \cdots + c_n(x_n - x_{n-1}).$$

**Example 1.** Compute $\int_{-1}^2 [2x-1]\,dx$ and compute the area $A$ of the region between the graph of $f(x) = [2x-1]$ and the $x$ axis, which lies between the lines $x = -1$ and $x = 2$. The function $g(x) = [2x-1]$ is indeed a step function. By the definition of the greatest integer function (p. 97).

$$[2x-1] = -3 \text{ if } -1 \leq x < -\frac{1}{2}$$

$$= -2 \text{ if } -\frac{1}{2} \leq x < 0 \cdots \text{ etc.}$$

$$= 2 \text{ if } \frac{3}{2} \leq x < 2.$$

See Figure 39. Therefore

$$\int_{-1}^2 [2x-1]\,dx = -3\left(\frac{1}{2}\right) + -2\left(\frac{1}{2}\right) + \cdots + 2\left(\frac{1}{2}\right)$$

$$= -\frac{3}{2}.$$

The desired area $A = \int_{-1}^2 |[2x-1]|\,dx$. Since

$$|[2x-1]| = 3 \qquad \text{if} \quad -1 \leq x < -\frac{1}{2}$$

$$= 2 \qquad \text{if} \quad -\frac{1}{2} \leq x < 0 \cdots \text{ etc.}$$

$$= 2 \qquad \text{if} \quad \frac{3}{2} \leq x < 2,$$

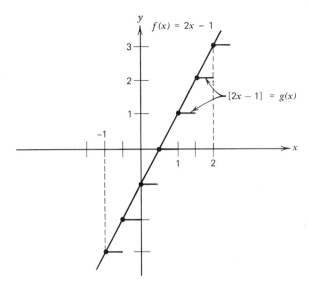

**Fig. 39**

we have

$$\int_{-1}^{2} |[2x-1]|\,dx = 3\left(\frac{1}{2}\right)+2\left(\frac{1}{2}\right)+1\left(\frac{1}{2}\right)+1\left(\frac{1}{2}\right)+2\left(\frac{1}{2}\right)=\frac{9}{2}$$

The next result is the analogue of Theorem 3.1. The proof is exactly the same as in the previous case, and hence we leave it for an exercise.

**THEOREM 3.9.** *Let f and g be step functions defined on the interval* $[a,b]$, *and let c and d be real numbers. Then cf + dg is a step function, and*

$$\int_{a}^{b} (cf+dg) = c \int_{a}^{b} f + d \int_{a}^{b} g.$$

*Furthermore, if* $f(x) \le g(x)$ *for each* $x \in [a,b]$, *then*

$$\int_{a}^{b} f \le \int_{a}^{b} g.$$

A function $f$ defined on the interval $[a,b]$ is said to be *bounded* (Figure 40) if there exists a constant $M$, satisfying $-M \le f(x) \le M$ for all $x$ in $[a,b]$. For bounded functions, the definition of the integral is exactly the same as for nonnegative bounded functions.

**Definition.** *A bounded function f is said to be integrable if there is precisely one number I with the property that*

$$\int_{a}^{b} g \le I \le \int_{a}^{b} G$$

*for each pair of step functions g, G satisfying* $g \le f \le G$. *If this condition holds we define*

$$\int_{a}^{b} f = I.$$

The approximation of a bounded function $f$ by upper and lower approximating step functions $G$ and $g$ is illustrated in Figure 41. The symbol

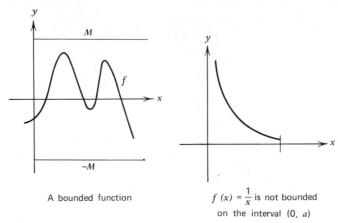

A bounded function

$f(x) = \dfrac{1}{x}$ is not bounded on the interval $(0, a)$

**Fig. 40**

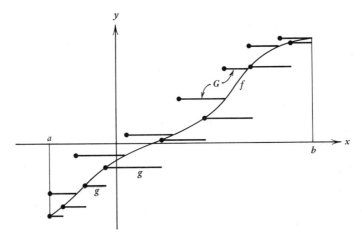

**Fig. 41** Approximation to $f$ by step functions $g$ and $G$.

$\int_a^b f$ is called the integral of $f$ from $a$ to $b$. Just as for non-negative functions, $f$ is integrable if and only if the following Riemann condition holds.

**THEOREM 3.10 (Riemann Condition).** *A bounded function $f$ defined on the interval $[a, b]$ is integrable if and only if for each positive number $\epsilon$, there exist step functions $g$, $G$ satisfying*

(3)
$$g \leqslant f \leqslant G$$

*and*

(4)
$$\int_a^b G - \int_a^b g < \epsilon.$$

It is an immediate consequence of Theorem 3.10 that $\int_a^b f$ may be approximated by finite sums in the following way. Let $a = x_0 < x_1 < \cdots < x_n = b$ be a partition of $[a, b]$. Set $\Delta x_i = x_i - x_{i-1}$, and let $x_i'$ be a point in the interval $[x_{i-1}, x_i]$. Form the sum $f(x_1')\Delta x_1 + \cdots + f(x_n')\Delta x_n$. This sum is called a *Riemann sum approximation* to the integral $\int_a^b f$. Indeed we may infer from 3.10 that for each $\epsilon > 0$, there is a partition of $[a, b]$ such that

(5) $$\left| \int_a^b f - (f(x_1')\Delta x_1 + \cdots + f(x_n')\Delta x_n) \right| < \epsilon.$$

no matter how the points $x_i'$ are chosen in the interval $[x_{i-1}, x_i]$. To see that this follows from 3.10 let $g$ and $G$ be step functions defined on a common partition of $[a, b]$ and satisfying $g \leq f \leq G$ and $\int_a^b G - \int_0^b g < \epsilon$. Then no matter how the points $x_i'$ are chosen from the intervals $[x_{i-1}, x_i]$

(6) $$\int_a^b g \leq f(x_1')\Delta x_1 + \cdots + f(x_n')\Delta x_n \leq \int_a^b G$$

Since $\int_a^b g \leq \int_a^b f \leq \int_a^b G$, (6) implies that

$$\left| \int_a^b f - (f(x_1')\Delta x_1 + \cdots + f(x_n')\Delta x_n) \right| < \int_a^b G - \int_a^b g < \epsilon.$$

This proves (5).

The fundamental theorems of the calculus, Theorems 3.5 and 3.6, go through now exactly as for nonnegative functions. We state the result as one theorem.

**THEOREM 3.11 (Fundamental Theorem of the Calculus).** *Let $f$ be integrable on the interval $[a, b]$. For $a \leq x \leq b$ set*

$$F(x) = \int_a^x f.$$

(i) *if $f$ is continuous at the point $x$, then $F'(x) = f(x)$.*

(ii) *if $f$ is continuous at all points $x$ of the interval $[a, b]$, and $G$ is any primitive for $f$, then*

$$\int_a^b f = G(b) - G(a).$$

We now state our desired linearity theorem for integrable functions.

**THEOREM 3.12** *Let $f$ and $g$ be integrable on the interval $[a, b]$ and let $c, d$ be real numbers. Then $cf + dg$ is integrable and*

(7) $$\int_a^b (cf + dg) = c \int_a^b f + d \int_a^b g.$$

*Furthermore, if $f(x) \leq g(x)$ for each $x$ in $[a, b]$, then*

(8) $$\int_a^b f \leq \int_a^b g.$$

The proof of this theorem, although straightforward, is rather long, so it is presented as a series of exercises. We shall use the result freely, however. A function of the form $cf + dg$ where $c$ and $d$ are real numbers is called a *linear combination* of $f$ and $g$. One immediate consequence of Theorem 3.9

is that a polynomial $f(x) = a_n x^n + \cdots + a_0$ is integrable. Indeed

$$\int_a^b f(x)dx = a_n \int_a^b x^n dx + \cdots + a_0 \int_a^b 1\,dx$$

$$= a_n \left(\frac{x^{n+1}}{n+1}\right)\Big|_a^b + \cdots + a_0 x\Big|_a^b$$

$$= a_n \left(\frac{b^{n+1} - a^{n+1}}{n+1}\right) + \cdots + a_0(b - a).$$

If the functions $f$ and $g$ in Theorem 3.12 are known to be continuous on $[a, b]$, then (7) follows immediately from the fundamental theorem, since the functions $\int_a^x (cf + dg)$ and $c\int_a^x f + d\int_a^x g$ then will have the same derivative, namely $cf + dg$. The verification of this is presented as Exercise 34.

**Example 2.** Show that $f(x) = 2x^2 - 3x^5$ is integrable and compute $\int_1^2 (2x^2 - 3x^5)\,dx$. Since $f$ is a linear combination of $x^2$ and $x^5$, $f$ is integrable by Theorem 3.12. Moreover,

$$\int_1^2 (2x^2 - 3x^5)dx = 2\int_1^2 x^2 dx - 3\int_1^2 x^5 dx$$

$$= 2(x^3/3)\Big|_1^2 - 3(x^6/6)\Big|_1^2$$

$$= 2\left(\frac{8}{3} - \frac{1}{3}\right) - 3\left(\frac{64}{6} - \frac{1}{6}\right)$$

$$= \frac{14}{3} - \frac{63}{2} = -\frac{35}{6}.$$

Theorem 3.8 states that if $f$ is nonnegative and integrable on the interval $[a, b]$, then

$$\int_a^b f = \int_a^c f + \int_c^b f, \quad a \leqslant c \leqslant b.$$

This fact is valid for arbitrary integrable functions as well. We state this result next.

**THEOREM 3.13**  Let $f$ be integrable on the interval $[a, b]$ and let $a \leqslant c \leqslant b$. Then $f$ is integrable on the intervals $[a,c]$ and $[c,b]$, and furthermore

$$\int_a^b f = \int_a^c f + \int_c^b f.$$

The proof involves an approximation by upper and lower step functions and is sketched in Exercise 33, p. 125. We can extend the above formula by induction to any partition of $[a,b]$. Indeed, if $a = x_0 < x_1 < \cdots < x_n = b$, then

$$\int_a^b f = \int_a^{x_1} f + \int_{x_1}^{x_2} f + \cdots + \int_{x_{n-1}}^b f$$

Figure 42 illustrates this additivity property of the integral. Just as for nonnegative functions we want the formula

(9) $$\int_a^b f = \int_a^c f + \int_c^b f$$

to hold regardless of the order of the points $a$, $b$, and $c$. Hence, we define

$$\int_a^a f = 0 \quad \text{and} \quad \int_b^a f = -\int_a^b f \quad \text{if} \quad a < b.$$

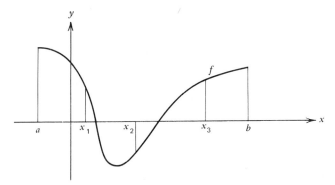

**Fig. 42** $\displaystyle\int_a^b f = \int_a^{x_1} f + \int_{x_1}^{x_2} f + \cdots + \int_{x_{n-1}}^b f.$

Thus, if $c$, $d$, $e$ are any three numbers in an interval on which $f$ is integrable, we have

$$\int_c^d f = \int_c^e f + \int_e^d f.$$

It is often important to estimate the size of an integral. One easily obtained estimate is the following. It $m$ and $M$ are constants, and $m \leqslant f(x) \leqslant M$ on $[a, b]$, then by (8)

$$\int_a^b m \cdot dx \leqslant \int_a^b f(x)dx \leqslant \int_a^b Mdx.$$

But

$$\int_a^b mdx = m(b-a) \qquad \text{and} \qquad \int_a^b Mdx = M(b-a).$$

Hence

(10) $$m(b-a) \leqslant \int_a^b f(x)dx \leqslant M(b-a).$$

In particular if $-M \leqslant f(x) \leqslant M$ then

$$-M(b-a) \leqslant \int_a^b f(x)dx \leqslant M(b-a)$$

or

$$\left| \int_a^b f(x)dx \right| \leqslant M(b-a).$$

A sharper estimate can be obtained by using the fact that

$$-|f|(x) \leqslant f(x) \leqslant |f|(x)$$

where the "absolute value of $f$" is defined by $|f|(x) = |f(x)|$. If $f$ is integrable, it follows that $|f|$ is integrable. (See Exercise 31.) Hence by (8)

$$-\int_a^b |f|(x)dx \leqslant \int f(x)dx \leqslant \int_a^b |f|(x)dx$$

or

(11) $$\left| \int_a^b f(x)dx \right| \leqslant \int_a^b |f|(x)dx.$$

This estimate on the size of $\left| \int_a^b f \right|$ is often useful in theoretical investigations.

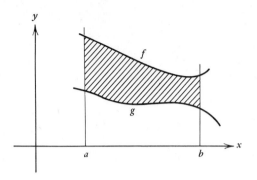

**Fig. 43**                                **Fig. 44**

Let us return to the area problem and give a definition of the area $A$ of the region between two integrable functions. Clearly if $f \geqslant g$, then we should define $A = \int_a^b (f(x) - g(x))dx$. (See Figure 43.) However, if the graphs of $f$ and $g$ cross as in Figure 44, we should define $A = \int_a^c (f-g) + \int_c^b (g-f)$. The right-hand side may be conveniently abbreviated $\int_a^b |f-g|$ since if $f \geqslant g$, $|f-g| = f-g$; and if $g \geqslant f$, $|f-g| = g-f$. Summarizing, we define the area $A$ of the region between two integrable functions which lie between the lines $x = a$ and $x = b$ by the formula

$$A = \int_a^b |f-g|.$$

To actually compute this area one must locate the points at which $f-g$ changes sign.

**Example 3.**  Compute the area $A$ of the region bounded by the graph of $y = x(x-1)(x-2)$, the $x$ axis, and the lines $x = 0$ and $x = 2$ (Figure 45).

$$A = \int_0^2 |x(x-1)(x-2)|dx.$$

Now for $0 \leqslant x \leqslant 1$

$$|x(x-1)(x-2)| = x(x-1)(x-2) = x^3 - 3x^2 + 2x$$

and if $1 \leqslant x \leqslant 2$

$$|x(x-1)(x-2)| = -x(x-1)(x-2) = -(x^3 - 3x^2 + 2x).$$

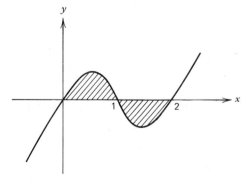

**Fig. 45**   $A = \int_0^2 |x(x-1)(x-2)|\, dx.$

Hence

$$A = \int_0^1 (x^3 - 3x^2 + 2x)\,dx - \int_1^2 (x^3 - 3x^2 + 2x)\,dx$$

$$= \left(\frac{x^4}{4} - x^3 + x^2\right)\Big|_0^1 - \left(\frac{x^4}{4} - x^3 + x^2\right)\Big|_1^2$$

$$= \frac{1}{4} - \left(-\frac{1}{4}\right) = \frac{1}{2}$$

## Exercises
Compute the following integrals

1. $\int_1^2 (x - 3x^2)\,dx.$

2. $\int_{-1}^0 (x + x^5)\,dx.$

3. $\int_{-10}^{10} (4x + x^3)\,dx.$

4. $\int_{-2}^3 [4x - 1]\,dx$ where $[x]$ is the greatest integer function.

5. $\int_1^3 \left(x^4 - 3x^3 + \frac{x^2}{2} - x + 1\right)dx.$

6. $\int_{-3}^2 x^{2n}\,dx.$

7. $\int_{-1}^2 (x^2 - x)\,dx.$

8. $\int_{-1}^1 |2x + 1|\,dx.$

9. $\int_{-1}^2 |x^2 - x|\,dx$

10. $\int_{-2}^3 |x^3 + x^2|\,dx.$

11. $\int_{-2}^2 |[1 - 2x]|\,dx$ where $[x]$ is the greatest integer function.

12. $\int_{-1}^2 |[x^2 - 1]|\,dx.$

Write the areas of the following regions as integrals and compute these in-tegrals. Draw a sketch of the regions involved.
13. Region between $y = 3x - 2$, the $x$ axis, and the lines $x = -1, x = 4$.
14. Region between $y = 2x - 2$, the parabola $y = 2x^2 - x$, and the lines $x = -1$, $x = 2$.
15. Region above the parabola $y = 2x^2 + 1$ and below the parabola $y = x^2 + 5$.
16. Region between the curve $y = x^3$ and the line $y = 4x$.
17. Region between $f(x) = [x - 2]$, the $x$ axis, and the lines $x = -2$ and $x = 4$.
18. Region between $g(x) = x^2, f(x) = [x + 5], x = -2, x = 3$.
19. Region between $y = x^3 + x^2 + x - 3$ and the line $y = 3x - 3$.
20. Show that the area of the region above the parabola $y = x^2$ and below the line $y = a$ is $\frac{4}{3}$ of the area of the triangle formed by the points $(-\sqrt{a}, a)$, $(0, 0)$, $(\sqrt{a}, a)$.
21. Let $P$ and $Q$ be points on the parabola $y = x^2$. Let $R$ be the point on the parabola such that the tangent to the parabola at $R$ is parallel to the line $PQ$. (See Figure 46.)
Show that the area of the parabolic segment $PRQ$ is $\frac{4}{3}$ the area of the triangle $PRQ$. (This result was first proved by Archimedes in the 3rd century B.C. by the method of exhaustion.)
22. Assuming that the following functions $f$ are integrable, use (10) to determine constants $A$ and $B$ satisfying

$$A \le \int_a^b f(x)\,dx \le B$$

if

(a) $f(x) = \sqrt{1 + x^2}, a = 1, b = 3$
(b) $f(x) = \sqrt{1 + x^3}, a = 0, b = 2$
(c) $f(x) = 1/x, \qquad a = 1, b = 3$

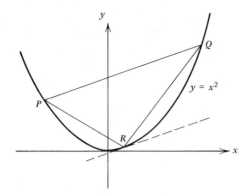

**Fig. 46**

23. Using the fundamental theorem of the calculus determine $F'(x)$ if

(a) $F(x) = \int_1^x \dfrac{dt}{t}$

(b) $F(x) = \int_1^x \sqrt{1+t^2}\, dt$

(c) $F(x) = \int_x^2 \dfrac{dt}{t}, x > 0$

(d) $F(x) = \int_1^{3x} \sqrt{1+t^2}\, dt$ (The chain rule must be used here.)

(e) $F(x) = \int_{2x}^{3x} \dfrac{dt}{1+t}$, $x > 0$. (Pick a point $a$ between $2x$ and $3x$, then apply (5) and the chain rule.)

(f) $F(x) = \int_a^{x^2} \sqrt{1+t^2}\, dt$.

The following group of exercises shows that if $f$ and $g$ are integrable on the interval $[a, b]$ and $c$ and $d$ are real numbers, then $cf + dg$ is integrable and

$$\int_a^b (cf + dg) = c \int_a^b f + d \int_a^b g.$$

*24. If $f$ is a step function and $c$ is a real number, show that $cf$ is a step function and also that $\int_a^b cf = c \int_a^b f$.

*25. If $f$ and $g$ are step functions, prove that $f + g$ is a step function and that

$$\int_a^b (f + g) = \int_a^b f + \int_a^b g.$$

(This result is easy if both $f$ and $g$ are constant on the subintervals of one partition of $[a,b]$. In fact, just as in the proof of Theorem 3.1, we may assume this for $f$ and $g$ by constructing a new partition of $[a, b]$ consisting of all the points of the partition for $f$ together with all the points of the partition for $g$.)

*26. If $f$ and $g$ are step functions defined on the interval $[a, b]$ and $c$, $d$ are real numbers, show that $cf + dg$ is a step function and show that

$$\int_a^b (cf + dg) = c \int_a^b f + d \int_a^b g.$$

*27. If $f$ and $g$ are step functions and $f \leqslant g$, show that $\int_a^b f \leqslant \int_a^b g$. (This completes the proof of Theorem 3.9.)

*28. If $f$ is integrable on $[a, b]$ and $c$ is a real number, show that $cf$ is integrable and, furthermore,

$$\int_a^b cf = c \int_a^b f.$$

(To show $cf$ integrable we must verify that for each $\epsilon > 0$ there exist step functions $g$, $G$ satisfying $g \leqslant cf \leqslant G$ and

$$\int_a^b G - \int_a^b g < \epsilon.$$

But if $g$ and $G$ are step function approximations to $f$ satisfying $g \leqslant f \leqslant G$ and

$$\int_a^b G - \int_a^b g < \epsilon,$$

then $cg \leqslant cf \leqslant cG$ if $c \geqslant 0$. If $c < 0$, $cG \leqslant cf \leqslant cg$. In either case

$$\left| \int_a^b cG - \int_a^b cg \right| = |c| \left\{ \int_a^b G - \int_a^b g \right\} < |c|\epsilon.$$

Explain why this shows that $cf$ is integrable. Why is it true that

$$\int_a^b cf = c \int_a^b f?)$$

*29. If $f$ and $g$ are integrable on $[a, b]$, show that $f + g$ is integrable and, furthermore,

(12)
$$\int_a^b (f + g) = \int_a^b f + \int_a^b g.$$

(The argument is similar to Problem 28.
Let $f_1, F_1, g_1, G_1$ be step functions satisfying $f_1 \leqslant f \leqslant F_1, g_1 \leqslant g \leqslant G_1$, and the chain rule.)

$$\int_a^b F_1 - \int_a^b f_1 < \epsilon \qquad \text{and} \qquad \int_a^b G_1 - \int_a^b g_1 < \epsilon.$$

Then $f_1 + g_1 \leqslant f + g \leqslant F_1 + g_1$ and

$$\int_a^b (F_1 + G_1) - \int_a^b (f_1 + g_1) < 2\epsilon.$$

Explain why this shows that $f + g$ is integrable. Why must (12) hold?)

*30. If $f$ and $g$ are integrable and $f \leqslant g$, show that $\int_a^b f \leqslant \int_a^b g$. This completes the proof of Theorem 3.12.

*31. If $f$ is integrable, show that $|f|$ is integrable. We divide this exercise into several parts.

(a) If $f$ is a function define the "positive part of $f$," written $f^+$, by the formula $f^+(x) =$ maximum of the two numbers $f(x)$ and 0. Thus if $f(x) \geqslant 0$, $f^+(x) = f(x)$. If $f(x) < 0$, $f^+(x) = 0$. Show that $|f| = f^+ + (-f)^+$ and $f = f^+ - (-f)^+$.

(b) If $f$ is a step function, show that $f^+$ is a step function. If $g$ is a step function, and $f \leqslant g$, show that $f^+ \leqslant g^+$ and that $g - f \geqslant g^+ - f^+$.

(c) If $f$ is integrable, show that $f^+$ is integrable.

(d) If $f$ is integrable, show that $|f|$ is integrable.

32. If $|f|$ is integrable, does it follow that $f$ is integrable? Explain.

*33. Prove Theorem 3.13. (Let $G$ and $g$ be step functions such that $g \leqslant f \leqslant G$ and assume $\int_a^b G - \int_a^b g < \epsilon$. Show first that, if $a \leqslant c \leqslant b$ then $\int_a^b G = \int_a^c G + \int_c^b G$ and $\int_a^b g = \int_a^c g + \int_c^b g$. Next verify that $\int_a^c G - \int_a^c g < \epsilon$ and $\int_c^b G - \int_c^b g < \epsilon$. Conclude as a result that $f$ is integrable on the intervals $[a, c]$ and $[c, b]$. To show $\int_a^c f + \int_c^b f = \int_a^b f$, verify that $\int_a^c f + \int_c^b f \leqslant \int_a^c G + \int_c^b G < \int_a^b f + \epsilon$ and $\int_a^c f + \int_c^b f \geqslant \int_a^c g + \int_c^b g > \int_a^b f - \epsilon$. Since this holds for each choice of $\epsilon$, it follows that $\int_a^c f + \int_c^b f = \int_a^b f$.)

34. If $f$ and $g$ are continuous on $[a, b]$, use the fundamental theorem of the calculus to show that $\int_a^b (cf + dg) = c \int_a^b f + d \int_a^b g$.

## 3.7 Conditions for Integrability of a Bounded Function

We now need some simple tests that will tell us if a given function is integrable. The tests should be easy to use and must be applicable in a wide variety of situations.

The first of these tests covers most of the functions that we shall discuss. First, let us define some terms.

*Definition* Let $f$ be defined on the interval $[a, b]$. If for each pair of points $x_1, x_2$; $a \le x_1 < x_2 \le b$, it follows that $f(x_1) < f(x_2)$, then $f$ is said to be increasing *on* $[a, b]$. *If for each such pair* $x_1, x_2, f(x_1) > f(x_2)$, *then* $f$ *is said to be* decreasing. *If* $f(x_1) \le f(x_2)$, *then* $f$ *is said to be* nondecreasing. *If* $f(x_1) \le f(x_2)$, *then* $f$ *is said to be* nonincreasing. *It is important to note that the appropriate condition of* $f$ *is assumed to hold for* each pair *of points* $x_1, x_2$ *satisfying* $a \le x_1 < x_2 \le b$. *If* $f$ *satisfies one of these four conditions on* $[a, b]$, *then* $f$ *is said to be* monotone *on the interval (see Figure 47).*

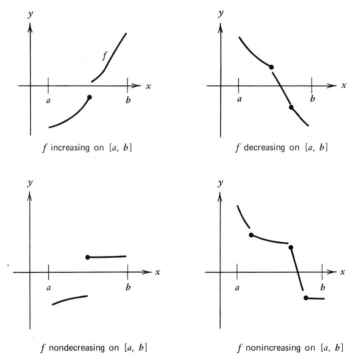

$f$ increasing on $[a, b]$

$f$ decreasing on $[a, b]$

$f$ nondecreasing on $[a, b]$

$f$ nonincreasing on $[a, b]$

**Fig. 47** Examples of monotone functions.

*Definition.* *The function* $f$ *is said to be* piecewise *monotone on the interval* $[a, b]$ *if there is a partition of* $[a, b]$, $a = x_0 < x_1 < \cdots < x_n = b$, *such that on each of the open* subintervals, $x_{k-1} < x < x_k$, $f$ *is monotone.*

Thus, $f$ is piecewise monotone if the graph of $f$ can be broken up into finitely many pieces, each of which is either rising, falling, or constant (Figure 48).

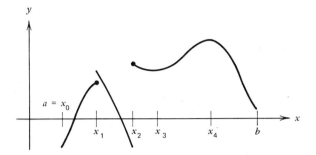

**Fig. 48**   A piecewise monotone function.

Polynomials are piecewise monotone on every interval $[a, b]$. A rational function, the quotient $p/q$ of two polynomials, is also piecewise monotone. However, not every continuous function on $[a, b]$ is piecewise monotone. There are easy examples of continuous functions with an infinite number of oscillations. One such example is given in Exercise 24, p. 34. However, bounded piecewise monotone functions are necessarily integrable. This will follow from our next result.

**THEOREM 3.14**   *Let $f$ be a bounded function defined on $[a, b]$. If $f$ is monotone on the open interval $(a, b)$, then $f$ is integrable on $[a, b]$.*

**Proof.**   Assume $f$ is nondecreasing. The argument in the case that $f$ is nonincreasing is exactly the same and is left as an exercise. We verify that the Riemann condition holds for $f$ in virtually the same manner as was done for $f(x) = x^2$ in the example on page 104. Since the function $f$ is bounded, there are constants $m, M$ such that

$$m \leqslant f(x) \leqslant M \text{ for each } x, a \leqslant x \leqslant b.$$

Divide the interval $[a, b]$ into $n$ equal subintervals and define a pair of approximating step functions in the following way:

$$
\begin{aligned}
g(x) &= m & a \leqslant x < x_1 \\
&= f(x_1) & x_1 \leqslant x < x_2 \\
&= f(x_2) & x_2 \leqslant x < x_3, \text{ etc.} \\
&= f(x_{n-1}) & x_{n-1} \leqslant x < b \\
&= m & x = b.
\end{aligned}
$$

$$
\begin{aligned}
G(x) &= M & x = a \\
&= f(x_1) & a < x \leqslant x_1 \\
&= f(x_2) & x_1 < x \leqslant x_2, \text{ etc.} \\
&= f(x_{n-1}) & x_{n-2} < x \leqslant x_{n-1} \\
&= M & x_{n-1} < x \leqslant b.
\end{aligned}
$$

Then

(1)   $$\int_a^b g = m(x_1 - a) + f(x_1)(x_2 - x_1) + \cdots + f(x_{n-1})(b - x_{n-1})$$

$$= (m + f(x_1) + \cdots + f(x_{n-1})) \frac{b - a}{n}.$$

(2)   $$\int_a^b G = f(x_1)(x_1 - a) + \cdots + f(x_{n-1})(x_{n-1} - x_{n-2}) + M(b - x_{n-1})$$

$$= (f(x_1) + \cdots + f(x_{n-1}) + M) \frac{b - a}{n}.$$

But from (1) and (2)

(3)
$$\int_a^b G - \int_a^b g = (M-m)\frac{(b-a)}{n},$$

since all terms involving $f(x_1),\dots,f(x_2)$ calcel out. Clearly, if $n$ is taken large enough, $(M-m)(b-a)/n$ is smaller than a given positive number $\epsilon$. Hence the Riemann condition is satisfied for $f$, and the function is integrable.

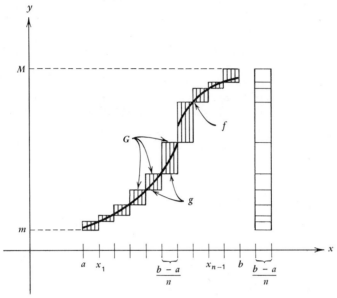

**Fig. 49** $\int_a^b G - \int_a^b g = (M-m)\frac{(b-a)}{n}$, which is the area of the rectangle to the right of the graph of $f$.

Note that $\int_a^b G - \int_a^b g$ is the area of the shaded portion in Figure 49. Also note that this proof gives no indication as to the value of $\int_a^b f$. Just as in the case $f(x) = x^2$, we must use some other technique, e.g., the fundamental theorem of the calculus, to calculate $\int_a^b f$.

**COROLLARY.** *If the bounded function $f$ is piecewise monotone on the interval $[a, b]$, then $f$ is integrable.*

The proof, which consists of applying the argument of Theorem 3.14 to finitely many intervals, is left as an exercise.

In short, a bounded function is integrable if the graph can be broken up into finitely many pieces, each of which is either rising or falling or constant.

***Example 1.*** If $x > 0$, then for any rational number $r$

$$f(x) = x^r$$

is a monotone function. If $r > 0$, $f$ is increasing; if $r = 0$, $f$ is constant; if $r < 0$, $f$ is decreasing. Therefore, by Theorem 3.14, $f$ is integrable on any interval $[a, b]$, $0 < a < b$. (If $r < 0$, we must have $a > 0$ for otherwise $x^r$ is not bounded.) Now if $r \neq -1$, then

$$F(x) = \frac{x^{r+1}}{r+1}$$

is a primitive for $x^r$. Hence

$$(4) \qquad \int_a^b x^r \, dr = \frac{x^{r+1}}{r+1} \Big|_a^b = \frac{1}{r+1}(b^{r+1} - a^{r+1}).$$

If $r = -1$, then we cannot determine a primitive for $f(x) = 1/x$ from formula (4). However, $f(x) = 1/x$ is decreasing on any interval $[a, b]$, $0 < a < b$. Hence, by Theorem 3.14, $\int_a^b (1/t) \, dt$ exists. Moreover, by the fundamental theorem of the calculus,

$$F(x) = \int_a^x \frac{1}{t} \, dt$$

is a primitive for $f(x) = 1/x$. In the next chapter we shall investigate this integral further.

We close this section with a second condition for integrability. We shall show in the appendix to Chapter 5 that each continuous function on $[a, b]$ satisfies this condition. Hence, each continuous function is integrable.

**THEOREM 3.15** *If $f$ is a function defined on $[a, b]$ with the property that for each $\epsilon > 0$, there exists a step function $h$ such that*

$$|f(x) - h(x)| < \epsilon$$

*for each $x$ in the interval $[a, b]$, then $f$ is integrable on $[a, b]$.*

**Proof.** The hypothesis of the theorem implies that the function $f$ can be approximated by a step function $h$ with an error that does not exceed $\epsilon$ for all $x$ in the interval $[a, b]$. (See Figure 50.)

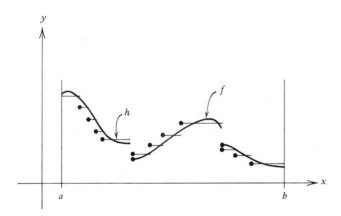

**Fig. 50** $h$ is a step function approximation to $f$.

To prove the theorem we verify the Riemann condition, Theorem 3.10. We must show that for each $\epsilon > 0$, there are step functions $g$ and $G$ satisfying $g \le f \le G$ and $\int_a^b G - \int_a^b g \le \epsilon$. Set $\delta = \epsilon/2(b-a)$. By hypothesis, there is a step function $h$ satisfying

$$|f(x) - h(x)| < \delta$$

for each $x$ in $[a, b]$. This means that

$$h(x) - \delta < f(x) < h(x) + \delta$$

for each $x$ in $[a, b]$. Set $G(x) = h(x) + \delta$ and $g(x = h(x) - \delta$. Then $G$, $g$ are step functions, and $g(x) \leqslant f \leqslant G(x)$. Moreover, $G(x) - g(x) = 2\delta = \epsilon/(b-a)$. Hence,

$$\int_a^b G - \int_a^b g = \int_a^b (G - g) = \int_a^b \frac{\epsilon}{b-a} = \epsilon.$$

Thus the Riemann condition is satisfied and $f$ is integrable.

There are integrable functions that do not satisfy the approximation condition of Theorem 3.15. However, all integrable functions that we shall consider do satisfy this condition. Necessary and sufficient conditions for a function to be integrable were formulated first by Riemann in terms of the oscillation of the function. These conditions are essentially those of Theorem 3.10. It was later shown by the French mathematician Henri Lebesgue around 1900 that a bounded function is integrable if and only if its set of discontinuities is "sufficiently small." We cannot go into these matters here, however.

## EXERCISES

1. For each of the following integrals, the integrand is piecewise monotone. Evaluate the integral by using the fundamental theorem of the calculus.

(a) $\int_1^2 3\sqrt{x}\, dx$

(e) $\int_2^4 \left(4 - s^2 + \frac{2}{s^{1/3}}\right) ds$

(b) $\int_2^3 \left(\frac{1}{\sqrt{x}} - \frac{3}{\sqrt[3]{x}}\right) dx$

(f) $\int_1^4 \left(2x - \frac{5}{\sqrt[3]{x^2}}\right) dx$

(c) $\int_0^4 (1 + 3x^2 - 4x^{5/2})\, dx$

(g) $\int_1^3 \left(y^3 - \left(\frac{2}{y}\right)^2 + \sqrt{1/y}\right) dy$

(d) $\int_1^3 (3/t^{1/3} - 5/t^{7/2})\, dt$

(h) $\int_2^5 \left(ct^2 + \frac{d}{t^{2/3}}\right) dt$, $c$ and $d$ constants

2. Compute the area of the region in the upper half plane which lies below the curve $x = y^2$ and to the left of the line $x = 4$. Sketch the region.
3. Compute the area of the region in the first quadrant which lies below the curve $2x^2 = y^3$ and above the line $y = 3x - 4$. Sketch the region.
4. Compute the area of the region in the upper half plane which lies below the curve $x = y^2$ and above the line $y = x - 2$. Sketch the region.
5. Compute the area of the region below the curve $x = y^4$ and above the parabola $y = x^2$. Sketch the region.
6. For each of the following integrals the integrands are piecewise monotone. Try by inspection to determine primitives for the integrands. Then use the fundamental theorem of the calculus to evaluate the integral.

(a) $\int_1^2 \sqrt{x+1}\, dx$

(f) $\int_0^1 (2x-1)^{25}\, dx$

(b) $\int_1^2 (x+1)^{10} dx$

(g) $\int_1^2 (4x-2)^{1/5}\, dx$

(c) $\int_3^5 \sqrt[3]{x-2}\, dx$

(h) $\int_{-1}^2 \frac{3}{(2x-5)^3}\, dx$

(d) $\int_1^2 \frac{1}{(x-4)^3}\, dx$

(i) $\int_1^3 (2x^3 + \sqrt{4x-1})\, dx$

(e) $\int_0^1 \frac{dx}{(3x+1)^2}$

(j) $\int_1^2 \left(\sqrt{2x-1} + \frac{1}{\sqrt[3]{3x+4}}\right) dx$

(k) $\int_{-2}^{1} \left( (2x-1)^5 + \dfrac{1}{(x+4)^5} \right) dx$

(n) $\int_{1}^{2} \dfrac{x}{\sqrt[3]{x^2+2}} dx$

(l) $\int_{2}^{3} \dfrac{dx}{\sqrt{1+x}-\sqrt{x}}$

(o) $\int_{0}^{1} x^2(x^3+1)^{10} dx$

(m) $\int_{-1}^{2} 2x\sqrt{x^2+1}\, dx$

7. Compute the area of the region lying below the curve $y = 4 - x^2$ and above $y = [x]$. Sketch the region.

8. For $t > 0$, the function $f(t) = 1/t$ is monotone decreasing, hence integrable on any interval $[a, b]$, $0 < a < b$. Sketch the graph of $f$. By using appropriate estimates of the magnitude of $f(t)$ show that

$$\frac{1}{2} \leqslant \int_{1}^{2} \frac{dt}{t} \leqslant 1.$$

9. If $x > 1$, show that

$$1 - \frac{1}{x} \leqslant \int_{1}^{x} \frac{dt}{t} \leqslant x - 1.$$

10. Show that the inequality in Exercise 9 is valid if $0 < x < 1$.

11. Show that for each positive integer $n$

$$\frac{1}{2} + \cdots + \frac{1}{n} \leqslant \int_{1}^{n} \frac{dt}{t} \leqslant 1 + \frac{1}{2} + \cdots + \frac{1}{n-1}.$$

12. Is a step function piecewise monotone?

13. Give examples of each of the four types of monotone functions.

14. Give examples of functions that are not monotone.

15. Give an example of a function that is *not* piecewise monotone.

16. Prove Theorem 3.13 for nonincreasing functions.

17. Prove the corollary to Theorem 3.13.

## 3.8   Further Applications of the Integral

Suppose a physical problem of some sort is given. How does one determine that the solution to the problem is an integral? Assuming that the solution to the problem is known to be an integral, how does one go about determining the interval $[a, b]$ and the function $f$? There are two fundamental properties of the integral $\int_{a}^{b} f$ that provide the solution to these questions. We have met both properties before, but for convenience we state them again. For the moment let us abbreviate $\int_{a}^{b} f$ by $I(a, b)$. The first fundamental property is that the integral $I(a, b)$ is additive on subintervals of $[a, b]$. By this we mean that if $a < t_1 < \cdots < t_n < b$ is a partition of the interval $[a, b]$, then

(1) $$I(a, b) = I(a, t_1) + I(t_1, t_2) + \cdots + I(t_{n-1}, G).$$

This is clear since (1) just says that

$$\int_{a}^{b} f = \int_{a}^{t_1} f + \int_{t_1}^{t_2} f + \cdots + \int_{t_{n-1}}^{b} f.$$

This property of the integral was discussed on p. 120. The second principle is the familiar estimate on the size of an integral, which was given on p. 121. That is, on any interval $[c, d]$ if $m \leqslant f(x) \leqslant M$ for $c \leqslant x \leqslant d$, then

(2) $$m(d-c) \leqslant I(c, d) \leqslant M(d-c).$$

Again, since $I(c, d) = \int_c^d f$, this just expresses the inequality

$$m(d-c) \leqslant \int_c^d f \leqslant M(d-c) \qquad \text{if} \qquad m \leqslant f(x) \leqslant M \text{ for } c \leqslant x \leqslant d.$$

Suppose that we have a problem associated with an interval $[a, b]$. Let $A(a, b)$ be the desired answer to the problem. We ask two questions. Is the problem additive on subintervals of $[a, b]$? That is, does

$$(3) \qquad A(a, b) = A(t, t_1) + A(t_1, t_2) + \cdots + A(t_n, b)$$

where $A(t_k, t_{k-1})$ is the answer to the problem on the subinterval $[t_k, t_{k+1}]$. Moreover, does there exist an integrable function $f$ defined on all of $[a, b]$ such that on each subinterval $[c, d]$

$$(4) \qquad m(d-c) \leqslant A(c, d) \leqslant M(d-c)$$

whenever $m \leqslant f(x) \leqslant M$ for $x \in [c, d]$? If the answer to both questions is yes, then the solution $A(a, b)$ to the problem, must be given by

$$(5) \qquad A(a, b) = \int_a^b f.$$

Before discussing some examples let us justify (5). The reason is very simple. Let $g$ and $G$ be step functions defined on $[a, b]$ and assume that $g \leqslant f \leqslant G$. Then $\int_a^b g \leqslant \int_a^b G$ and statements (3) and (4) imply that the number $A(a, b)$ satisfies

$$(6) \qquad \int_a^b g \leqslant A(a, b) \leqslant \int_a^b G. \qquad \text{(Why?)}$$

But since $f$ is integrable there is only one number, namely $\int_a^b f$, with this property. Hence

$$A(a, b) = \int_a^b f.$$

If the solution of a given problem is an integral, $\int_a^b f$, then neither the interval $[a, b]$ nor the integrand $f$ is necessarily unique. This is even true with area problems.

*Example 1.* Determine the area $A$ of the region to the right of the curve $x = y^2$ and to the left of the line $y = x - 2$. If we choose the interval $[a, b]$ on the $x$ axis, then $a = 0, b = 2$.

Area is certainly additive on subintervals of $[0, 2]$. If $[c, d]$ is a subinterval of $[0, 2]$ as in Figure 51, then the area $A(c, d)$ of the strip between $x = c$ and $x = d$ satisfies

$$m(d-c) \leqslant A(c, d) \leqslant M(d-c)$$

where $m$ is the height of a rectangle contained in the strip and $M$ is the height of a rectangle containing the strip. If $f(x) = \sqrt{x}, 0 \leqslant x \leqslant 4$ and $g(x) = -\sqrt{x}, 0 \leqslant x < 1$, $g(x) = x - 2, 1 \leqslant x \leqslant 4$, then $m \leqslant f(x) - g(x) \leqslant M$. Hence the desired area $A = \int_0^4 [f(x) - g(x)] dx$. If $0 \leqslant x \leqslant 1, f(x) - g(x) = 2\sqrt{x}$, and if $1 \leqslant x \leqslant 4, f(x) - g(x) = \sqrt{x} - x + 2$. Hence

$$A = \int_0^1 2\sqrt{x}\, dx + \int_1^4 (\sqrt{x} - x + 2) dx$$

$$= 2 \cdot \frac{2}{3} x^{3/2} \Big|_0^1 + \frac{2}{3} x^{3/2} - (x^2/2) + 2x \Big|_1^4$$

$$= \frac{4}{3} + \left(\frac{2}{3} \cdot 8 - 8 + 8\right) - \left(\frac{2}{3} - \frac{1}{2} + 2\right) = \frac{9}{2}.$$

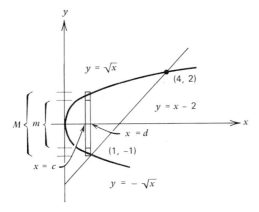

**Fig. 51**

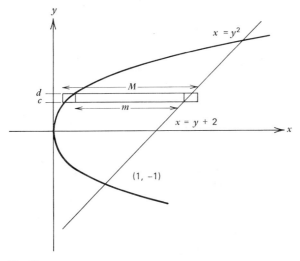

**Fig. 52**

On the other hand, we may choose our interval $[a, b]$ along the $y$ axis as in Figure 52. Then $a = -1$ and $b = 2$. Let $[c, d]$ be a subinterval of $[-1, 2]$, and let $m$ be the length of a rectangle contained in the strip through $c \leqslant y \leqslant d$. Let $M$ be the length of a rectangle containing this strip (Figure 52). Then

$$m \leqslant y + 2 - y^2 \leqslant M.$$

Setting $f(y) = y + 2 - y^2$, we have

$$A = \int_{-1}^{2} f(y)\,dy = \int_{-1}^{2} (y + 2 - y^2)\,dy$$

$$= (y^2/2) + 2y - (y^3/3)\Big|_{-1}^{2} = 2 + 4 - \frac{8}{3} - \left(\frac{1}{2} - 2 + \frac{1}{3}\right) = \frac{9}{2}.$$

Area is an example of a function that is additive on subintervals. This is true of volume as well. Next we derive a formula for the volume of a solid of revolution.

*Example 2.* Compute the volume $V$ of the solid formed by rotating about the $x$ axis the region bounded by $y = x^2$, the $x$ axis, and the line $x = 2$. See Figure 53.

In this case, the desired interval is $[0, 2]$. Let us denote the volume by $V(0, 2)$. If we partition $[0, 2]$ by points $t_1, \dots, t_n$ such that $0 < t_1 < t_2 < \cdots < t_n < 2$, then

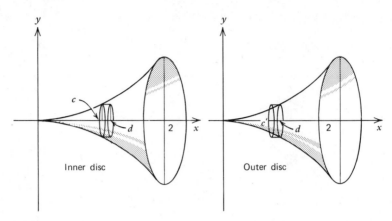

**Fig. 53**

$V(0, 2)$ is certainly the sum of the volumes of the respective strips. That is,

$$V(0, 2) = V(0, t_1) + V(t_1, t_2) + \cdots + V(t_n, 2).$$

Thus volume is additive on subintervals. Next we estimate the volume of a slice between $x = c$ and $x = d$. Let $v$ be the volume of a disc with thickness $d - c$, and radius $r$ which lies completely within the region. Then $r \leqslant x^2$, $c \leqslant x \leqslant d$ and $v = \pi r^2(d - c)$. Similarly, let $V$ be the volume of a disc of radius $R$ which contains that portion of the solid between $x = c$ and $x = d$, then $R \geqslant x^2$, $c \leqslant x \leqslant d$ and $V = \pi R^2(d - c)$. Thus if $V(c, d)$ is the volume of the strip between $x = c$ and $x = d$,

(7) $$\pi r^2(d - c) \leqslant V(c, d) \leqslant \pi R^2(d - c).$$

Moreover (7) holds for any choice of constants $r$, and $R$ just so long as $\pi r^2 \leqslant \pi x^4 \leqslant \pi R^2$ for $c \leqslant x \leqslant d$. Hence $V = \int_0^2 f$ where $f(x) = \pi x^4$ or

$$V = \pi \int_0^2 x^4 dx = \pi \frac{x^5}{5}\Big|_0^2 = \frac{32\pi}{5}.$$

This argument shows that if the ordinate set of a function $f$ defined on the interval $[a, b]$ is rotated about the $x$ axis, then the volume $V$ of the resulting solid is given by

(8) $$V = \pi \int_a^b [f(x)]^2 dx.$$

See Figure 54.

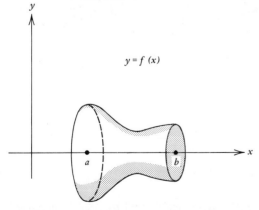

**Fig. 54** The volume $V$ of a solid of revolution formed by rotating the ordinate set for $f$ about the $x$ axis is given by $V = \pi \int_a^b [f(x)]^2 \, dx$.

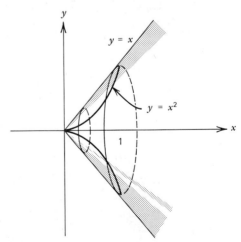

**Fig. 55**

*Example 3.* Determine the volume of the solid obtained by rotating the region below the graph of $y = x$ and above the graph of $y = x^2$ about the $x$ axis. (See Figure 55.)

The desired volume $V$ is clearly the difference $V_1 - V_2$ where

$$V_1 = \pi \int_0^1 x^2 \, dx \qquad \text{and} \qquad V_2 = \pi \int_0^1 x^4 \, dx.$$

Hence

$$V_1 - V_2 = \pi \int_0^1 x^2 \, dx - \pi \int_0^1 x^4 \, dx = \pi \left[ \frac{x^3}{3} - \frac{x^4}{4} \right] = \frac{\pi}{12}.$$

In general, then, if $0 \leq g \leq f$, and the region between the graphs of $f$ and $g$ is rotated about the $x$ axis, the volume $V$ of the resulting solid is given by

(9)
$$V = \pi \int_a^b [f^2(x) - g^2(x)] dx.$$

It is instructive to deduce this formula directly from the general principles we have been applying. The volume $V$ is certainly additive on subintervals, so to deduce formula (9) we need only to estimate the volume $V(c, d)$ of a strip cut off by a subinterval $[c, d]$. We estimate $V(c, d)$ from below and above by computing the volumes of washers of thickness $d - c$ which, on one hand, lie within the strip and, on the other hand, contain the strip. Now the volume of a washer is given by $\pi(r_1^2 - r_2^2)h$ where $h$ is the thickness and $r_1, r_2$ are the outer and inner radii. Referring to Figure 56, we see that

$$\pi(r_1^2 - r_2^2)(d - c) \leq V(c, d) \leq \pi(R_1^2 - R_2^2)(d - c).$$

But a washer with radii $r_1, r_2$ lies within the strip if

$$\pi(r_1^2 - r_2^2) \leq \pi[f^2(x) - g^2(x)], \text{ for } c \leq x \leq d.$$

On the other hand it contains the strip if $\pi(R_1^2 - R_2^2) \geq \pi[f^2(x) - g^2(x)]$ for $c \leq x \leq d$.

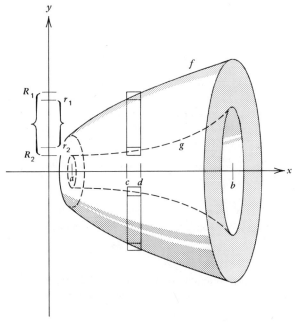

**Fig. 56**

Hence, letting $m = \pi(r_1{}^2 - r_2{}^2)$ and $M = \pi(R_1{}^2 - R_2{}^2)$, we conclude that

$$m(d-c) \leqslant V(c,d) \leqslant M(d-c)$$

whenever

$$m \leqslant \pi[f^2(x) - g^2(x)] \leqslant M$$

for each $x \in [c,d]$. This implies, by formula (5), that

$$V(a,b) = \int_a^b \pi[f^2(x) - g^2(x)]\, dx$$

which is the desired result.

As another illustration of this principle we consider the problem of work. If an object is moved through an interval $[a, b]$ as a result of the application of a constant force $F$, then the *work* done is defined by the formula

$$W = F \cdot (b-a).$$

Suppose the force $F$ has a constant value $F_k$ on each subinterval $(x_k, x_{k-1})$ of a partition $a < x_1 < \cdots < x_n < b$. Then the work done as the object is moved through the $k$th interval is

$$W_k = F_k(x_k - x_{k-1})$$

and the total work is defined as the sum

$$W = F_1 \cdot (x_1 - a) + \cdots + F_n(b - x_n).$$

Suppose now we have a variable force $F(x)$. What now should be the definition of the work done by this force function in moving an object from $a$ to $b$? It is reasonably clear that the work $W$ should be additive on subintervals. That is, if $a < x_1 < \cdots < x_n < b$ then

$$W(a,b) = W(x_1, a) + \cdots + W(b, x_n).$$

Moreover, if the force is increased, then the work done should increase. Thus on any subinterval $[c, d]$ if $m \leqslant F(x) \leqslant M$, then

$$m(d-c) \leqslant W(c,d) \leqslant M(d-c).$$

From this it follows that we must have

(10) $$W = \int_a^b F(x)dx.$$

*Example 4.*　Hooke's law states that an elastic body, e.g., a spring, rubber band, etc., will stretch an amount proportional to the force applied if the applied force is not too great. If a spring is stretched $\frac{1}{2}$ inch when a 2-pound force is applied to it, how much work is done in stretching the spring an additional $\frac{1}{2}$ inch? If the work $W = \int_a^b F(x)dx$, the problem is to determine the force function $F$ and the interval $[a, b]$. By Hooke's law if $x$ denotes the distance the spring has been stretched, then the force function

$$F(x) = kx.$$

But when $x = \frac{1}{2}$, $F(x) = 2$. Hence $k/2 = 2$ or $k = 4$, and it follows that $F(x) = 4x$. The work done in stretching the spring an additional $\frac{1}{2}$ inch would be

$$W = \int_{1/2}^1 4x\,dx = 4\frac{x^2}{2}\Big|_{1/2}^1 = \frac{3}{2}.$$

If the force is measured in pounds and the distance in feet (inches), then the unit of measurement for work is the foot-pound (inch-pound). Hence for this problem $W = \frac{3}{2}$ inch-pounds.

**EXERCISES**

Determine the areas of the following regions in two ways, first as an integral over an interval on the $x$ axis, then as an integral over an interval on the $y$ axis. Sketch the regions involved.

1. The region lying to the right of the line $y = 2x$ and above the parabola $y = x^2$.
2. The region lying to the right of the parabola $y^2 = 1 - x$ and to the left of the parabola $4y^2 = 4 - x$.
3. The region lying between the curve $x^2 y = 1$, the $x$ axis, and the lines $x = 1$ and $x = 4$.
4. The region in the right half plane lying above the curve $y = 1/x^2$, below the curve $y = x^2$, and to the left of the line $x = 4$.
5. The region lying below the parabola $y = 4 - (x - 2)^2$ and above the step function $y = [x]$.
6. Write an integral that represents the area of the region in the right half plane lying above the hyperbola $xy = 4$ and below the line $y = 5 - x$.
7. Find by integration the volume of a right circular cone if the radius of the base is 4 and the height is 8.
8. Find by integration the volume of the section of a sphere of radius $r$ and height $h$.
9. A football is formed by rotating the ellipse $3x^2 + 4y^2 = 48$ about the $x$ axis. What is its volume?
10. The region in the first quadrant between the hyperbola $xy = 1$ and the line $y = 4 - x$ is revolved about the $y$ axis. What is the volume of the resulting solid?
11. The region to the right of the parabola $x = y^2$ and to the left of the line $x = 4$ is revolved about the $y$ axis. Determine the volume of the resulting solid.
12. The region below the parabola $y = 4 - x^2$ and above the line $y = 2 - x$ is revolved about the line $x = 3$. What is the volume of the resulting solid?
13. A hole 2 inches in diameter is bored through the center of a solid sphere 4 inches in diameter. What is the volume of the resulting solid?

Formulas (8) and (9) are special cases of the following formula for the volume and of a solid region $S$. Fix a line $l$, and let $P_x$ be the plane through $x$ perpendicular to $l$. Let the area of the intersection of the plane $P_x$ with $S$ be denoted by $A(x)$. (See Figure 57.)

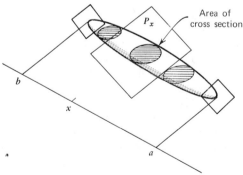

Fig. 57

Then

(11) $$V = \int_a^b A(x)\,dx.$$

Formula (8) is the special case where the plane cross sections are discs of radius $f(x)$. Hence $A(x) = \pi f^2(x)$. In formula (9) the cross sections are washers with inner radius $g(x)$ and outer radius $f(x)$. Hence $A(x) =$

$\pi[f^2(x) - g^2(x)]$. In the case that plane cross sections of $S$ are rectangles, triangles, and other simple figures, formula (11) is easy to establish. In the following exercises the area $A$ of each plane section must be determined from the geometry of the figure. It helps to make a sketch of the region.

14. A solid has a circular base of radius 2. What is the volume of the solid if each cross section perpendicular to a diameter of the circle is a square. (See Figure 58.) If $(x, y)$ is a point on the circle $x^2 + y^2 = 4$, then the area of the plane cross section through this point is $4y^2$. Hence $A(x) = 4y^2 = 16 - 4x^2$.)

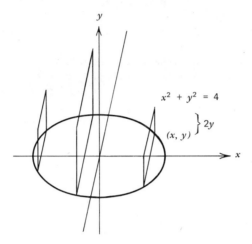

Fig. **58**

15. A solid has a circular base of radius 2. What is the volume of the solid if each cross section perpendicular to a diameter of the circle is an equilateral triangle?

16. A solid has as its base the region in the upper half plane lying below the parabola $y = 8 - 4x^2$. What is the volume of the solid if the cross sections perpendicular to the $y$ axis are squares? What is the volume if the cross sections perpendicular to the $x$ axis are squares?

17. The base of a solid is the region in the first quadrant bounded by the coordinate axes and the line $3x + 4y = 5$. What is the volume if cross sections perpendicular to the $x$ axis are semicircles?

18. Find the volume of an elliptical cone with height $h$ if the base is the ellipse

$$(x^2/a^2) + (y^2/b^2) = 1.$$

(Use similar triangles to determine the area of each elliptical cross section. The area of the elliptical base is $\pi ab$.)

19. A 60-pound force stretches an 8-foot spring by 1 foot. How much work has been done when the spring has been stretched to 12 feet?

20. How much work is done in stretching the spring of Exercise 19 from 12 to 14 feet?

21. A 100-foot cable weighing 2 lb/ft is hanging from a drum. What is the work done in winding all of the cable onto the drum? What is the work done in winding up the first 50 feet?

22. Answer Problem 21 if in addition there is a 50-lb weight hanging on the end.

23. The gravitational force $F$ exerted by the earth on a body weighing $w$ pounds which is $x$ miles from the center of the earth is, according to the inverse square law, given by

$$F = F(x) = \frac{kw}{x^2} \text{ for a suitable constant } k.$$

How much work is done in boosting a 10-ton rocket to a height 1000 miles above the earth's surface? Take the radius of the earth to be 4000 miles.

24. In Problem 23, as the rocket rises its weight is reduced because of the propellant burned. Assume the rocket contains 8 tons of fuel and the fuel is consumed at a rate proportional to the distance traveled. If burnout occurs 10 miles above the earth's surface, write an integral that expresses the work done in lifting the rocket to a height of 100 miles. Can you evaluate the integral?

25. A cylindrical cistern 10 feet deep and 4 feet in diameter is full of water weighing 62.4 lb/ft$^3$. How much work is done in pumping all of the water out through a hole in the top?

26. A conical tank 10 feet deep and 10 feet in diameter, which is full of water, has its vertex pointed down. How much work is done in pumping the tank out through a hole in the top? How much work is done by the water if it drains out through a hole in the bottom?

27. Justify inequality (6).

28. Derive formula (11) by the methods of this section if the plane cross sections of the solid $S$ are squares.

29. Derive formula (11) by the methods of this section if the plane cross sections of the solid $S$ are equilateral triangles.

# CHAPTER FOUR

# TRANSCENDENTAL FUNCTIONS

## 4.1 The Logarithm

In elementary textbooks the following definition is often encountered. For a positive number $x$, $\log_{10} x$ is defined as the number $y$ such that $10^y = x$. For irrational numbers $y$, for example, $y = \sqrt{2}$, this presents certain difficulties. What is meant by $10^{\sqrt{2}}$? Assuming we know what the symbol $10^y$ means, how can we show that for a given number $x$ there is a number $y$ that satisfies $x = 10^y$?

To avoid these difficulties we shall reverse the process of definition. First, we shall define a logarithm function and then define the exponential function in terms of it. It is rather surprising that we can use the tools of the calculus to accomplish this. To begin, let us note that the importance of logarithms as a computational device derives from the following fact. If for $x > 0$ we write $f(x) = \log_{10} x$, then

$$(1) \qquad f(xy) = f(x) + f(y) \text{ for all } x, y > 0.$$

Let us start with this logarithm identity and attempt to determine a *differentiable* function that satisfies it. If there is a differentiable function $f$ satisfying (1), then we may determine $f'$ by differentiating (1) with respect to $x$. If we hold $y$ fixed and apply the chain rule to (1), we obtain

$$(2) \qquad f'(xy)y = f'(x)$$

Since $f(y)$ is constant, its derivative with respect to $x$ is zero. Now set $x = 1$. Then it follows from (2) that

$$f'(y)y = f'(1) \text{ for each } y > 0.$$

Since $f'(1)$ is some constant, say $c$, we have shown that for all positive numbers $y$,

$$(3) \qquad f'(y) = \frac{c}{y}.$$

It is easy to show, and we leave the verification as an exercise, that if $c = 0$, then $f$ must be identically zero. The first fundamental theorem of the calculus (Theorem 3.5) immediately gives us one function $f$ that satisfies (3). Indeed if we define for $x > 0$,

$$L(x) = \int_1^x \frac{dt}{t},$$

then $L'(x) = 1/x$. Moreover, if we set

$$f(x) = cL(x),$$

then

$$f'(x) = cL'(x) = \frac{c}{x}.$$

But we also know that if two functions have the same derivative they differ by a constant (Corollary 2.8). Hence this implies that if $f$ satisfies (1) and is differentiable, then

$$f(x) = cL(x) + k.$$

We next show that $k = 0$. Certainly $L(1) = \int_1^1 dt/t = 0$. Since $f$ satisfies (1) by assumption,

$$f(1) = f(1 \cdot 1) = f(1) + f(1) = 2f(1)$$

or

$$f(1) = 0.$$

Therefore

$$0 = f(1) = cL(1) + k = k.$$

So if $f$ satisfies (1), then

$$f(x) = cL(x).$$

Next we must show that $L(x)$ actually satisfies the logarithm identity. (Why doesn't that follow from what we have done?) To do this, fix $y$ again and set

(4) $$G(x) = L(xy) - L(x) = \int_1^{xy} \frac{dt}{t} - \int_1^x \frac{dt}{t}.$$

We assert $G'(x) = 0$. To see this, differentiate (4) and apply the chain rule. This yields

$$G'(x) = L'(xy) \cdot y - L'(x)$$
$$= \frac{1}{xy} y - \frac{1}{x} = 0.$$

Therefore, $G(x) = $ constant. But $G(1) = L(y) - L(1) = L(y)$ since $L(1) = \int_1^1 (1/t)\,dt = 0$. Therefore, for each $x$,

$$G(x) = G(1) = L(y).$$

But by definition of $G$

$$G(x) = L(xy) - L(x).$$

Therefore, $L(xy) = L(x) + L(y)$, and the function $L$ satisfies the logarithm identity (1).

We have now completely proved the following result.

**THEOREM 4.1**   *A differentiable function $f$ satisfies the logarithm identity*

$$f(xy) = f(x) + f(y), \qquad \text{for} \qquad x, y > 0$$

*if and only if*

$$f(x) = c \int_1^x \frac{dt}{t} \qquad \text{for some constant } c.$$

Equation (1) is an example of *a functional equation*. The unknown is not a real number as in an algebraic equation, but is instead a function. Another such equation, which we shall study shortly, is the exponential equation

$$f(x + y) = f(x) \cdot f(y).$$

## 4.2   The Graph of the Function $L(x)$

If we draw the graph of the function $f(x) = 1/x$, then for $x > 1$, $L(x)$ represents the area of the ordinate set for $f$ between 1 and $x$. (See Figure 1.) If $0 < x < 1$, then $L(x)$ is the negative of the area of the ordinate set for $f$.

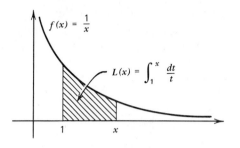

**Fig. 1**                                          **Fig. 2**

Let us note some other qualitative facts about the function $L(x)$. First $L(1) = 0$, and if $x < y$, then $L(x) < L(y)$. That is, $L$ is an increasing function (Figure 2). To prove this just write

$$L(y) - L(x) = \int_1^y \frac{dt}{t} - \int_1^x \frac{dt}{t}$$

$$= \int_1^y \frac{dt}{t} + \int_x^1 \frac{dt}{t} = \int_x^y \frac{dt}{t} > 0$$

Also, since $L$ has a derivative, the function $L$ must be continuous. That is, for each $x > 0$ $\lim_{h \to 0} L(x+h) = L(x)$.

Next we wish to show that for each real number $y$ there is exactly one positive real number $x$ satisfying $L(x) = y$. Since the function $L$ is increasing there certainly can be no more than one value of $x$ with this property. The problem is to show that there is such a number $x$. To accomplish this we must appeal to the completeness axiom for the real numbers in exactly the same way as was done in Chapter 1 to prove that there is a number $x$ satisfying $x^2 = 2$. Let us suppose for definiteness that $y > 0$. Let $S = \{s : L(s) < y\}$ and $T = \{t : L(t) > y\}$. We assert $S$ and $T$ are separated. To do this we must show first that $S$ and $T$ are both nonempty. $S$ is clearly nonempty since $1 \in S$. To show that $T \neq \phi$ we argue as follows. Certainly $L(2) > 0$. Then by (1), $L(4) = L(2 \cdot 2) = L(2) + L(2) = 2L(2)$. By induction we may infer that for each positive integer $n$,

$$L(2^n) = nL(2).$$

Pick an integer $n$ for which $nL(2) > y$. Then $L(2^n) > y$, or $2^n \in T$.

To finish the proof that $S$ and $T$ are separated we must show that if $s \in S$ and $t \in T$, then

$$s \leq t.$$

This is clear, since if $s > t$, then $L(s) > L(t)$. But $L(t) > y$ and $y > L(s)$ by assumption. Since this is a contradiction, it is impossible that $s > t$.

Now let $x$ be a number that separates $S$ and $T$. That is, $x \geq s$ for each $s \in S$ and $x \leq t$ for each $t \in T$. We now assert $L(x) = y$. To do this we show that the other possibilities $L(x) < y$ and $L(x) > y$ are impossible. Suppose $L(x) < y$. Since $L$ is a continuous function, $\lim_{h \to 0} L(x+h) = L(x)$. This means that for $h$ small enough and greater than zero

$$L(x) < L(x+h) < y.$$

(Why?) This is clearly impossible since if $L(x+h) < y$, then $x+h \in S$. This contradicts the fact that $x \geqslant s$ for each $s \in S$. In a similar way we show that it cannot happen that $L(x) > y$. Hence $L(x) = y$, and we have proved our assertion. The graph of $L(x)$ is sketched in Figure 3.

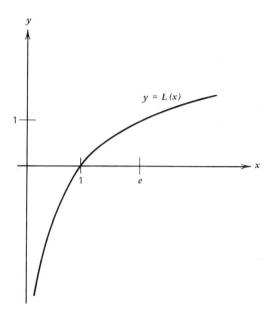

**Fig. 3**

The number $x$ such that $L(x) = 1$ is denoted by $e$. Approximately $e = 2.71828\ldots$. In place of $L(x)$ it is customary to write $\log x$. However, in physics and engineering books the function $L(x)$ is often written $\ln(x)$ or $\log_e x$, but we shall not use this notation. The logarithm function $\log x$ is also called the natural logarithm function. Its values are sometimes called natural or Naperian logarithms in honor of their inventor, Sir John Napier (1550–1617).

Next we wish to define the function $\log_{10} x$. Of course, this function must be differentiable and must satisfy the logarithm identity

(1) $$f(xy) = f(x) + f(y).$$

Also $\log_{10} 10$ should equal 1. The first two conditions imply that

$$\log_{10} x = c \log x$$

for some constant $c$. We choose the constant $c$ in order to achieve

$$\log_{10} 10 = 1.$$

This implies $c = (1/\log 10)$. Consequently, we define

$$\log_{10} x = \frac{\log x}{\log 10}.$$

Indeed, for any positive number $b \neq 1$, we define

$$\log_b x = \frac{\log x}{\log b}.$$

The function $\log_b x$ satisfies (1), and $\log_b b = 1$. The number $b$ that satisfies $\log_b b = 1$ is called the *base* of the logarithm function $\log_b x$. The derivative of $\log_b x$ is easily computed:

$$\frac{d}{dx}(\log_b x) = \frac{d}{dx}\left(\frac{\log x}{\log b}\right) = \frac{1}{x \log b}.$$

Graphs of the function $\log_b x$ for various values of $b$ are sketched in Figure 4.

The computational aspects of logarithms have become outmoded with the development of high-speed computing machines. However, the logarithm function has great theoretical importance in mathematics and the physical sciences.

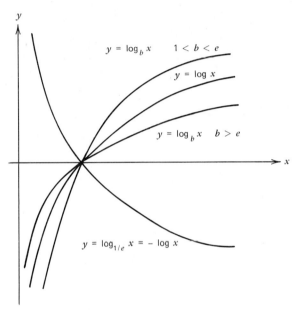

**Fig. 4** Graphs of the functions $f(x) = \log_b x$ for various values of $b$.

If $f$ is a positive differentiable function, then we may compose the function $\log x$ with $f$ to obtain a new function

(2)
$$h(x) = \log f(x).$$

We may compute the derivative of $h$ by applying the chain rule to equation (2). Recall that if $p(x) = g(f(x))$, then

$$p'(x) = g'(f(x)) \cdot f'(x).$$

Applying this to (2) we obtain

$$h'(x) = \frac{f'(x)}{f(x)}$$

or

$$\frac{d}{dx}(\log f) = \frac{1}{f}f' = \frac{1}{f}\frac{d}{dx}(f).$$

In terms of primitives, this equation states that if $f$ is positive, then $\log f$ is a primitive for $f'/f$.

On the other hand, if $f$ is negative then $\log(-f)$ is a primitive for $f'/f$. This is clear since

$$\frac{d}{dx}(\log(-f)) = \frac{-f'}{-f} = \frac{f'}{f}.$$

Taking these two statements together we see that if $f$ does not vanish on the interval $[a, b]$, a primitive for $f'/f$ is given by $\log|f|$.

It is convenient at this point to introduce a notation for primitives or antiderivatives. Indeed if $F'(x) = f(x)$, then we write

$$F(x) = \int f(x)dx.$$

This symbol $\int f(x)dx$ is called the *indefinite integral* of the function $f$. Thus the symbol $\int f(x)dx$ stands for any function $F$ satisfying $F'(x) = f(x)$. Since all such functions differ by a constant, it is customary to write

$$\int f(x)dx = F(x) + C$$

where $F$ is one function with the property that $F'(x) = f(x)$. Thus if $n \neq -1$,

$$\int x^n dx = \frac{x^{n+1}}{n+1} + C$$

and $\int f(x)^n f'(x)dx = f(x)^{n+1}/(n+1) + C$. For $n = -1$ we have

$$\int \frac{dx}{x} = \log|x| + C$$

and

$$\int \frac{f'(x)}{f(x)}dx = \log|f(x)| + C.$$

**Example 1.** If $f(x) = \log(x^2 + \log x)$, compute $f'(x)$. Applying the chain rule we have

$$f'(x) = \frac{1}{x^2 + \log x} \cdot \frac{d}{dx}(x^2 + \log x)$$

$$= \frac{2x + (1/x)}{x^2 + \log x} = \frac{2x^2 + 1}{x^3 + x \log x}.$$

**Example 2.** Compute $\int 2x/(x^2 + 1)dx$. If $f(x) = x^2 + 1$, $f'(x) = 2x$. Hence

$$\int \frac{2xdx}{x^2 + 1} = \log(x^2 + 1) + C.$$

We may omit the absolute value sign here since $x^2 + 1$ is always positive.

If $dF/dx = f$ and $c$ is a constant, then $d(cF)/dx = cf$. Using indefinite integrals this becomes

$$\int cf(x)dx = c \int f(x)dx.$$

We may use this fact to adjust indefinite integrals by a multiplicative constant if needed.

**Example 3.** Compute $\int x/(4 - x^2)\, dx$. If $f(x) = 4 - x^2$, then $f'(x) = -2x$. Therefore

$$\int \frac{(-2x)}{4 - x^2}dx = \log|4 - x^2|.$$

Hence, for our original problem,

$$\int \frac{x}{4-x^2}\,dx = \frac{(-2)}{(-2)}\int \frac{x}{4-x^2}\,dx = -\frac{1}{2}\int \frac{(-2x)}{4-x^2}\,dx$$

$$= -\frac{1}{2}\log|4-x^2|.$$

This device of writing an indefinite integral $\int f(x)\,dx$ as $1/c \int cf(x)\,dx$ where $c$ is some nonzero constant will be found to be quite useful.

### EXERCISES

1. Suppose the function $f$ satisfies the logarithm identity $f(xy) = f(x) + f(y)$.
   (a) If 0 belongs to the domain of $f$, show that $f$ must be identically zero.
   (b) Show that if the domain of $f$ consists of all real numbers $\neq 0$ then $f(-x) = f(x)$. (Show first that $f(-1) = f(1) = 0$ and exploit this fact.)
2. Suppose that $f$ satisfies the logarithm identity, and $f'(x) = c/x$. Show that if $c = 0$, then $f$ is identically zero.
3. Let $a, b > 0$, and $a, b \neq 1$. Verify the following identities.
   (a) $\log_b x = \log_b a \log_a x$
   (b) $\log_b x = \dfrac{\log_a x}{\log_a b}$.
4. Compute the derivatives of the following functions.
   (a) $f(x) = x \log x$
   (b) $f(x) = x \log_{10}(x^2 + 1)$
   (c) $f(x) = (x+1)\,[\log x]^2$
   (d) $f(x) = \log_2\left(\dfrac{x+1}{x-1}\right)$
   (e) $f(x) = \log(\sqrt{1+x^2})$
   (f) $f(x) = [\log(x^2+4)]^4$
   (g) $f(x) = \log \log x$
   (h) $f(x) = \log \int_1^x \sqrt{1+t^2}\,dt$
   (i) $f(x) = x \log(x^2 \log x)$
   (j) $f(x) = \log \int_x^2 \sqrt{1+t^3}\,dt$
   (k) $f(x) = \log\left(\dfrac{x^2-1}{x^2+1}\right)$
5. Compute the following indefinite integrals.
   (a) $\displaystyle\int \frac{dx}{x+4}$
   (b) $\displaystyle\int \frac{2x\,dx}{x^2-4}$
   (c) $\displaystyle\int \frac{3x^3\,dx}{x^3+a^3}$
   d) $\displaystyle\int 2x(x^2+1)^2\,dx$
   (e) $\displaystyle\int \frac{x}{\sqrt{x^2+1}}\,dx$
   (f) $\displaystyle\int \frac{dx}{x \log x}$
6. Compute the following indefinite integrals. It may be necessary to write $\int f(x)\,dx = \left(\dfrac{1}{c}\right)\int cf(x)\,dx$ for some appropriate constant $c \neq 0$.
   (a) $\displaystyle\int x\sqrt{x^2+1}\,dx$
   (b) $\displaystyle\int \frac{x\,dx}{\sqrt{1-x^2}}$
   (c) $\displaystyle\int \frac{dx}{2x+4}$
   (d) $\displaystyle\int (3x-2)^2\,dx$
   (e) $\displaystyle\int x^2(x^3+a^3)^4\,dx$
   (f) $\displaystyle\int \frac{x}{\sqrt{4x^2-1}}\,dx$
   (g) $\displaystyle\int \frac{x^3}{x^4+1}\,dx$
   (h) $\displaystyle\int \frac{dx}{(2x+1)\log(2x+1)}$
   (i) $\displaystyle\int \frac{x-1}{x+1}\,dx$
   (j) $\displaystyle\int \frac{x}{(x^2+1)\log(x^2+1)}\,dx$
   (k) $\displaystyle\int \frac{2dx}{x \log_{10} x}$

7. Let $f(x) = \log\ x - (x-1) - \frac{1}{2}(x-1)^2$. Show that $f(1) = f'(1) = f''(1) = 0$. If $g(x) = f(x) + a(x-1)^3$, determine $a$ so that $g'''(1) = 0$.

8. Determine the area of the region in the first quadrant lying below the line $2x + y = 9$ and above the hyperbola $xy = 4$.

9. Determine the area of the region in the first quadrant lying below the graph of the function $y = 2 - |x-2|$ and above the graph of $xy = 3$.

10. The curve $xy^2 = 1$ is rotated about the $x$ axis. What is the volume of the resulting solid which lies between the lines $x = 1$ and $x = 4$?

11. That portion of the graph of $y = 1/\sqrt{x}$ between $x = \frac{1}{2}$ and $x = 2$ is revolved about the line $x = 1$. Determine the volume of the resulting solid.

12. Approximate $f(t) = 1/t$ by suitable constant functions on the interval $[1, x]$ and derive the inequalities

$$1 - \frac{1}{x} < \log x < x - 1.$$

Show that these inequalities remain valid if $0 < x < 1$. What happens if $x = 1$? Deduce the inequality $e > 2$.

13. Approximate $f(x) = 1/x$ by suitable step functions on the interval $[1, n]$, $n$ a positive integer and derive the inequalities

$$\frac{1}{2} + \frac{1}{3} + \cdots + \frac{1}{n} < \log n < 1 + \frac{1}{2} + \cdots + \frac{1}{n-1}.$$

14. By comparing the integrals

$$\int_1^x \frac{dt}{t} \quad \text{and} \quad \int_1^x \frac{dt}{\sqrt{t}},$$

show that for $x > 1$, $\log x \leq 2(\sqrt{x} - 1)$.

15. By an argument similar to the one given in this section, show that for each $y < 0$ there exists exactly one $x$ satisfying

$$\log x = y.$$

16. Prove by induction that for an integer $n$, $\log x^n = n \log x$. Extend this result to rational exponents and prove that for each positive number $x$ and rational number $p/q$,

$$\log x^{p/q} = \frac{p}{q} \log x.$$

In particular, show that

$$\log e^{p/q} = \frac{p}{q}.$$

If a function $f$ is given as the product of a large number of factors, the logarithm can be used to facilitate the computation of $f'$. This process is called *logarithmic differentiation*.

*Example 4.*   If

$$f(x) = \frac{(x-1)^2 \sqrt{2x+3}}{(x+4)^3(2x-1)^4}$$

compute $f'(x)$. We first compute $\log |f(x)|$. Indeed

$$\log |f(x)| = 2 \log |x-1| + \frac{1}{2} \log |2x+3| - 3 \log |x+4| - 4 \log |2x-1|.$$

Differentiating and using the fact that $(d/dx) \log |f(x)| = (f'(x)/f(x))$, we obtain

$$\frac{f'(x)}{f(x)} = \frac{2}{(x-1)} + \frac{1}{2x+3} - \frac{3}{x+4} - \frac{8}{2x-1}.$$

Therefore

$$f'(x) = \left( \frac{2}{x-1} + \frac{1}{2x+3} - \frac{3}{x+4} - \frac{8}{2x-1} \right) \left( \frac{(x-1)^2 \sqrt{2x+3}}{(x+4)^3(2x-1)^4} \right).$$

17. Using this procedure of logarithmic differentiation compute $f'$ if

(a) $f(x) = \dfrac{(x-1)^2(2x+3)^3}{(x-4)}$

(b) $f(x) = \dfrac{\sqrt{2x-3}}{(x+4)^3}$

(c) $f(x) = \dfrac{(2x-1)^3\sqrt{4x-3}}{(7x+3)^{10}(x-1)^{2/3}}$

(d) $f(x) = \dfrac{(x-1)^3(x-2)^2(x-3)}{x^3(x+1)^2(x+2)}$

(e) $f(x) = \sqrt{(x-a_1)(x-a_2)\cdots(x-a_n)}$

## 4.3   THE EXPONENTIAL FUNCTION

In the previous section we verified that to each real number $x$ there was one and only one positive number $y$ satisfying $\log y = x$. Thus the set $E = \{(x,y) : \log y = x\}$ has the property that each vertical line intersects $E$ exactly once. Hence $E$ defines a function, and using the usual terminology if $(x,y) \in E$, we write $y = E(x)$. Thus $y = E(x)$ if $\log y = x$. (See Figure 5.)

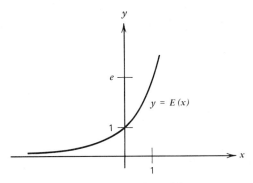

**Fig. 5**   The exponential function $E(x)$.

Knowing the graph of the logarithm function enables us to sketch the graph of the function $E$. Indeed, since we have just interchanged $x$ and $y$, this can be accomplished by reflecting the graph of $y = \log x$ about the diagonal $y = x$. Since $\log 1 = 0$, $E(0) = 1$ and $\log e = 1$ implies $E(1) = e$. Now $E(2)$ is the number $y$ such that $\log y = 2$. But from the logarithm identity it follows that $\log e^2 = 2$. Indeed (Exercise 16, Section 4.2) for each rational number $p/q$, $\log e^{p/q} = p/q$.

Thus

(1)
$$E\left(\frac{p}{q}\right) = e^{p/q}.$$

The function $E(x)$ is called the exponential function and equation (1) suggests that we write $e^x$ in place of $E(x)$. This is just what we do, but we emphasize again that for rational numbers $p/q$ we *prove* that

$$E\left(\frac{p}{q}\right) = e^{p/q},$$

but for irrational numbers $x$ we *define* $e^x$ to be equal to $E(x)$. Notice that the domain of the function $E(x)$ is the range of the logarithm function, and this is the set of all real numbers. Furthermore, the range of the exponential function $E$ is the domain of the logarithm function and this is the set of positive real numbers. In addition, $E$ is an increasing function. That is, $E(x) < E(x_1)$ whenever $x < x_1$.

We wish now to prove the "law of exponents" for the function $E$. This is the functional equation

(2) $$E(x+y) = E(x) \cdot E(y),$$

which we assert holds for all real numbers $x$ and $y$. To establish (2) we need only show that the numbers $E(x+y)$ and $E(x) \cdot E(y)$ have the same logarithm. For if $\log z = \log w$, then $z = w$ since the logarithm is an increasing function. By definition of $E$,

$$\log(E(x+y)) = x+y.$$

However, from the logarithm identity of Section 4.1, we infer that

$$\log(E(x) \cdot E(y)) = \log(E(x)) + \log(E(y)) = x+y.$$

Thus, $\log(E(x+y)) = \log(E(x) \cdot E(y))$ and hence $E(x+y) = E(x) \cdot E(y)$.

Next we verify that the exponential function $E$ is differentiable. Geometrically this means that for each $x$, the function $E$ will have a nonvertical tangent line passing through the point $(x, E(x))$. Since the graph of $y = E(x)$ has been obtained by reflecting the graph of $y = \log x$ about the line $y = x$, this fact is clear on geometrical grounds. (See Figure 6.) It is instructive, however, to give a purely analytical proof of the differentiability of the function $E$.

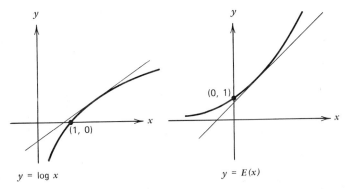

**Fig. 6** If the graph of $y = \log x$ is reflected about the line $y = x$, then the tangent line to $t = \log x$ is reflected into the tangent line for $y = E(x)$.

First, however, we must show that $E$ is continuous at each real number $x$. That is, we must show $\lim_{h \to 0} E(x+h) = E(x)$. Equivalently, for each $\epsilon > 0$ we must show that there is a corresponding $\delta$ so that whenever $0 < |h| < \delta$ it follows that

(3) $$\left| E(x+h) - E(x) \right| < \epsilon.$$

Fix $x$ and let $y = E(x)$. Since we must verify (3) only for small values of $\epsilon$, we may assume $\epsilon < E(x)$. Now choose $x_1$ and $x_2$ so that $E(x_1) = y - \epsilon$, and $E(x_2) = y + \epsilon$ as in Figure 7.
Next, choose $\delta$ so that

$$x_1 < x - \delta < x < x + \delta < x_2.$$

Now since $E$ is increasing, it follows that $x < x'$ implies $E(x) < E(x')$.

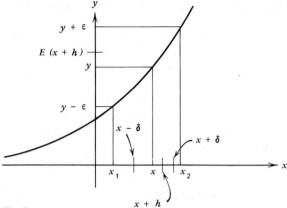

**Fig. 7**

Hence, if $0 < |h| < \delta$, then $x - \delta < x + h < x + \delta$ and, consequently,

$$y - \epsilon < E(x - \delta) < E(x + h) < E(x + \delta) < y + \epsilon.$$

Since $y - \epsilon < E(x) < y + \epsilon$ as well, it follows that

$$\left| E(x) - E(x + h) \right| < \epsilon.$$

This proves the function $E$ is continuous at the point $x$.

Next we wish to show that the function $E$ is differentiable. To do that we must show that

$$\lim_{h \to 0} \frac{E(x + h) - E(x)}{h}$$

exists for each real number $x$. To this end, fix $x$ and let

$$y = E(x)$$

and

$$y + k = E(x + h).$$

Then $x = \log y$ and $x + h = \log(y + k)$. Consequently,

$$\frac{E(x + h) - E(x)}{h} = \frac{y + k - y}{h} = \frac{k}{h}$$

(4)
$$= \frac{k}{x + h - x} = \frac{k}{\log(y + k) - \log y}$$

$$= \frac{1}{\dfrac{\log(y + k) - \log y}{k}}.$$

Figure 8 illustrates this quotient.

Now if $h \to 0$ it follows that $k \to 0$. (This is just the statement that $E$ is continuous at $x$!). Hence, from equation (4), we get

$$\lim_{h \to 0} \frac{E(x + h) - E(x)}{h} = \lim_{k \to 0} \frac{1}{\dfrac{\log(y + k) - \log y}{k}} = \frac{1}{\log'(y)}.$$

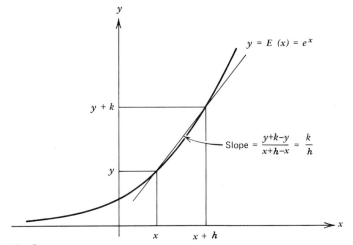

Fig. 8

But the derivative of $\log y$ is $1/y$. Hence

$$\lim_{h \to 0} \frac{E(x+h) - E(x)}{h} = \frac{1}{1/y} = y = E(x).$$

We summarize these results in the following.

**THEOREM 4.2**   *The function $E$ is differentiable, and for all values of $x$*

$$E'(x) = E(x).$$

At this point we drop the $E(x)$ notation and write $e^x = E(x)$.

***Example 1.***   Compute $f'$ if $f(x) = e^{\sqrt{x^2+1}}$. Now $f(x) = e^{g(x)}$ where $g(x) = \sqrt{x^2+1}$. Hence, by the chain rule,

$$f'(x) = e^{g(x)} g'(x)$$

or

$$f'(x) = e^{\sqrt{x^2+1}} \cdot \frac{1}{2} \frac{1}{\sqrt{x^2+1}} \cdot 2x = \frac{x}{\sqrt{x^2+1}} e^{\sqrt{x^2+1}}.$$

Now by definition of the exponential function, $y = e^x$ if $\log y = x$. This says that for real numbers $x$

$$(5) \qquad\qquad x = \log(e^x)$$

and for positive numbers $y$

$$(6) \qquad\qquad y = e^{\log y}.$$

Equations (5) and (6) state that the exponential and logarithmic functions are *inverses* of one another. That is, if we first apply the exponential function to a number $x$ obtaining $e^x$ and then apply the logarithm function to this number, the result is the number $x$ with which we started. Furthermore, if we compute $\log y$ for a positive number $y$, and then apply the exponential function to the number $\log y$, the result is the given number $y$. We shall explore this relation in detail in the next section.

Equation (6) allows us to define $a^x$ for all positive numbers $a$ and real numbers $x$. By (6)

$$a = e^{\log a}$$

and

$$a^2 = e^{\log a^2} = e^{2 \log a}.$$

Indeed, if $p/q$ is a rational number

$$a^{p/q} = e^{\log a^{p/q}} = e^{p/q \log a}.$$

Therefore, just as in the case of $e^x$ we *define*

(7)
$$a^x = e^{x \log a}.$$

We leave it as an exercise to verify the law of exponents for $a^x$, i.e.,

$$a^{x+y} = a^x \cdot a^y.$$

Figure 9 is a sketch of the graph of $a^x$ for various values of $a$. The chain rule immediately yields the differentiation formula for $a^x$. Indeed, if

$$f(x) = a^x = e^{x \log a}$$

then

$$f'(x) = e^{x \log a} \log a = a^x \log a.$$

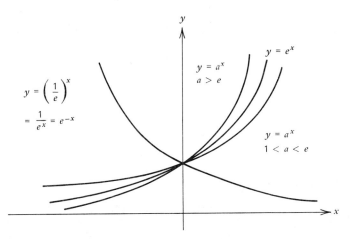

Fig. 9

Since the derivative of $e^x$ is again $e^x$, $e^x$ is a primitive for itself and the fundamental theorem of the calculus immediately tells us that

(8)
$$\int_a^b e^x dx = e^x \Big|_a^b = e^b - e^a.$$

For $a \ne e$ we can easily derive a similar formula for $a^x$. If

$$f(x) = \frac{a^x}{\log a},$$

then by the chain rule

$$f'(x) = \frac{a^x}{\log a} \log a = a^x.$$

Hence $a^x/\log a$ is a primitive for $a^x$ and, consequently, $\int a^x dx = (a^x/\log a) + C$ for some constant $C$. The definite integral

$$\int_b^c a^x dx = \frac{1}{\log a}\, a^x \Big|_b^c = \frac{1}{\log a}(a^c - a^b).$$

If we compose $e^x$ with a differentiable function $f(x)$ and write $g(x) = e^{f(x)}$, then by the chain rule $g'(x) = e^{f(x)}f'(x)$. The analogous formula for indefinite integrals is

(9) $$\int e^{f(x)}f'(x)dx = e^{f(x)} + C.$$

**Example 2.**   Differentiate $f(x) = 10^{x^2}$. By (7)

$$f(x) = e^{x^2 \log 10}.$$

Therefore, applying the chain rule, we obtain

$$f'(x) = e^{x^2 \log 10} \cdot 2x \log 10 = (2 \log 10)x10^{x^2}.$$

**Example 3.**   Differentiate $f(x) = x^x$. Again by (7)

$$f(x) = e^{x \log x}.$$

Hence

$$f'(x) = e^{x \log x} \cdot (\log x + 1) = x^x(\log x + 1).$$

**Example 4.**   Compute $\int xe^{x^2}dx$. Now $d(x^2)/dx = 2x$. Hence by (9)

$$\int 2xe^{x^2}dx = e^{x^2} + C.$$

Therefore

$$\int xe^{x^2}dx = \frac{1}{2}\int 2xe^{x^2}dx = \frac{1}{2}e^{x^2} + C.$$

**EXERCISES**

1. Sketch the graph of $f(x) = a^x$ if $a = 2, 10, \frac{1}{2}$.
2. Verify that the function $a^x$ satisfies the functional equation

$$a^{x+y} = a^x \cdot a^y.$$

3. Determine the first and second derivatives of the following functions.

   (a) $e^{x^2+1}$

   (b) $xe^x$

   (c) $xe^{\log(x+1)}$

   (d) $\dfrac{e^{-x}}{x}$

   (e) $\log(e^x + x)$

4. Differentiate the following functions.

   (a) $10^{(\log_{10} x + x)}$

   (b) $\dfrac{\log_2 x}{2^x}$

   (c) $\dfrac{e^x - e^{-x}}{e^x + e^{-x}}$

   (d) $10^{\log x}$

   (e) $\log(\log(\log x))$

   (f) $x^{1/x}$

   (g) $x^{\log x}$

   (h) $\log(x^x)$

5. Determine $y'$ as a function of $x$ and $y$ if the differentiable function $y = f(x)$ satisfies the following equations.

   (a) $x + y = e^{x+y}$

   (b) $e^x + e^y = e^{x+y}$

   (c) $x \log y = y^y$

6. If $y = e^{2x} - e^{-2x}$, show that $y'' = 4y$.
7. If $a$ and $b$ are constants and $\omega$ is a real number, show that $y = ae^{\omega t} + be^{-\omega t}$ satisfies

$$\frac{d^2y}{dt^2} - \omega^2 y = 0.$$

8. Let $f(x) = e^x$, determine a polynomial $p(x) = a_0 + a_1 x + a_2 x^2$ such that for $x = 0$, $p(0) = f(0)$, $p'(0) = f'(0)$, and $p''(0) = f''(0)$. Sketch the graph of $p$.

9. Generalize Exercise 8 to a polynomial of order $n$. That is, determine the coefficients $a_k$ for $p(x) = a_0 + a_1 x + \cdots + a_n x^n$ so that if $f(x) = e^x$ then $p(0) = f(0)$ and $p^{(k)}(0) = f^{(k)}(0)$, for $k = 1, \ldots, n$.

10. Compute

    (a) $\int_1^2 2^x \, dx$

    (b) $\int_{-2}^3 10^x \, dx$

11. Determine a formula for the area of the ordinate set for the function $f(t) = e^{-t}$ on the interval $[1, x]$. What happens when $x$ increases without bound?

12. Compute the area of the region between the graphs of $f(x) = e^x$ and the polynomial $p(x)$ of Exercise 8 on the interval $[-1, 1]$. (It may be shown that $p(x) > e^x$ if $x < 0$ and $p(x) < e^x$ if $x > 0$.)

13. Sketch the graph of the function $f(x) = (e^x + e^{-x})/2$. (Sketch the graphs of $e^x$, $e^{-x}$, and add.) The ordinate set for $f$ on the interval $[-1, 1]$ is revolved about the $x$ axis. Determine the volume of the resulting solid.

14. Compute the following indefinite integrals.

    (a) $\int x^2 e^{x^3} \, dx$

    (f) $\int \frac{x e^{x^2}}{e^{x^2} + 1} \, dx$

    (b) $\int x \, 10^{x^2} \, dx$

    (g) $\int x^x (\log x + 1) \, dx$

    (c) $\int \left(\frac{x+1}{x}\right) e^{x + \log x} \, dx$

    (h) $\int \frac{2^x}{10^x} \, dx$

    (d) $\int \frac{e^x}{e^x + 1} \, dx$

    (e) $\int \frac{x}{2^{x^2}} \, dx$

15. Compute $F''(x)$ if

    (a) $F(x) = \int_1^{x^2} e^{t^2} \, dt$

    (c) $F(x) = \int_x^{x^2} e^{t^2} \, dt$

    (b) $F(x) = \int_x^2 \frac{e^t}{\log t} \, dt$

16. If $x$ is positive and $r$ is a rational number, verify that $x^r = e^{r \log x}$. For arbitrary real numbers $r$, we *define* $x^r$ by the formula

$$x^r = e^{r \log x}.$$

    Verify that for arbitrary real numbers $r$, if $y = x^r$ then $y' = r x^{r-1}$.

17. Using the definition $a^x = e^{x \log a}$, establish the following properties of exponents.

    (a) $(ab)^x = a^x b^x$

    (c) $\log a^x = x \log a$

    (b) $(a^x)^y = a^{xy}$

    (d) For $a \neq 1$, $y = a^x$ if and only if $x = \log_a y$

18. Using mathematical induction verify the following differentiation formulas.

    (a) If $y = x e^x$, then $y^{(n)} = (x + n) e^x$.

    (b) If $y = x^n \log x$, then $y^{(n+1)} = n!/x$.

19. Explain why $\lim_{h \to 0} e^{f(h)} = e^L$ if $f$ is a real valued function and $\lim_{h \to 0} f(h) = L$. (This is a previous theorem.)

20. For $x > 0$ verify that

$$\frac{\log(x+h) - \log x}{h} = \log\left(1 + \frac{h}{x}\right)^{1/h}.$$

Using this and Exercise 19, deduce that

$$e^{1/x} = \lim_{h \to 0} \left(1 + \frac{h}{x}\right)^{1/h}.$$

Letting $h = 1/n$ it then follows that for $x > 0$,

$$\lim_{n \to \infty} \left(1 - \frac{x}{n}\right)^{-n} = e^x = \lim_{n \to \infty} \left(1 + \frac{x}{n}\right)^n.$$

21.  Prove that for $x > 0$

$$\frac{\log(x+h) - \log x}{h} < \frac{1}{x} \text{ if } h > 0$$

and

$$\frac{\log(x+h) - \log x}{h} > \frac{1}{x} \quad \text{if} \quad h < 0 \text{ and } |h| < x.$$

(Write the difference quotient as an integral, and then make a suitable step function approximation to the integral.)

22.  Show on the basis of Exercise 21 that

$$\left(1 + \frac{x}{n}\right)^n < e^x \text{ and } e^x < \left(1 - \frac{x}{n}\right)^{-n} \text{ if } x < n.$$

By taking an appropriate value of $n$ show that

$$2.5 < e < 2.99.$$

23.  If a sum of money $P$ is deposited at a yearly interest rate $r$, then the amount $S$ on deposit after $t$ years is given by

$$S = P(1 + r)^t.$$

If this interest is compounded $n$ times a year, then the amount on deposit is

$$S = P\left(1 + \frac{r}{n}\right)^{tn}.$$

Some California savings and loan institutions compound their interest daily. Give an upper estimate of the amount on deposit after one year if $1000 is deposited at 6% compounded daily. If the bank compounded the interest "continuously" how much would be on deposit at the end of a year? What rate of simple interest (compounded at the end of the year) would yield the same return?

## 4.4   Inverse Functions

We have shown that the functions $f(x) = \log x$ and $g(x) = e^x$ discussed in the previous section satisfy the following two relations. For each real number $x$

$$\log(e^x) = x,$$

and for each positive number $x$,

$$e^{\log x} = x.$$

Another pair of functions with similar properties are $f(x) = x^n$ where $n$ is an odd positive integer and $g(x) = x^{1/n}$. Indeed, for each real number $x$

$$g(f(x)) = (x^n)^{1/n} = x$$

and for each positive number $x$,

$$f(g(x)) = (x^{1/n})^n = x.$$

If $n$ is an even positive integer, then these relations remain true if we consider only positive values of $x$.

Suppose that $f$ is an arbitrary function. We ask the following question. When does there exist a function $g$ with the following two properties?

For each $x$ in the *domain* of $f$,

(1) $$g(f(x)) = x,$$

and for each $y$ in the *range* of $f$

(2) $$f(g(y)) = y.$$

In other words when does there exist a function $g$ that reverses, or undoes, the action of $f$? If such a function $g$ exists, we call $g$ an *inverse* for $f$. Moreover, it is easy to show that there is only one such function, and hence we write $g = f^{-1}$. Using the circle notation $\circ$ for function composition, (1) and (2) become

(3) $$(f^{-1} \circ f)(x) = f^{-1}(f(x)) = x$$

for each $x$ in the domain of $f$ and

(4) $$(f \circ f^{-1})(y) = f(f^{-1}(y)) = y$$

for each $y$ in the range of $f$. When $f^{-1}$ exists, the domain of $f$ is the range of $f^{-1}$, and the range of $f$ is the domain of $f^{-1}$. The relationship between $f$ and $f^{-1}$ is illustrated in Figure 10.

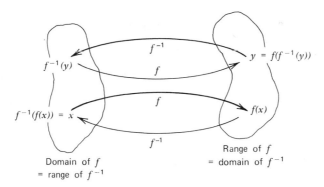

Fig. 10

The condition on $f$ that guarantees that $f^{-1}$ exists is very simple. It is just the following:

*To each $y$ in the range of $f$ there exists precisely one $x$ in the domain of $f$ such that $f(x) = y$.*

A function that satisfies this condition is called *one-to-one* (written $1:1$). Thus if $x_1$ and $x_2$ are points in the domain of a $1:1$ function and $x_1 \neq x_2$, then it follows that $f(x_1) \neq f(x_2)$. For a function $f$ whose domain and range are sets of real numbers, this $1:1$ condition may be interpreted very simply. Such a function is $1:1$ if and only if each *horizontal* line intersects the graph of $f$ at most once. See Figure 11.

The fundamental result on inverse functions is the following.

**THEOREM 4.3**   *An inverse exists for the function $f$ if and only if $f$ is $1:1$. Furthermore, this inverse function $g$ is unique.*

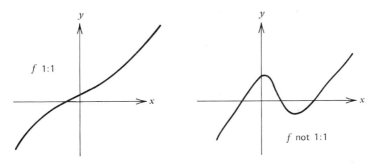

**Fig. 11**

***Proof.*** Let $D$ be the domain of the function $f$ and let $E$ be the range. Suppose $f$ is $1:1$. Then to each $y \in E$ there corresponds precisely one $x \in D$ such that $f(x) = y$. But this means we can define a function $g$ with domain $E$ and range $D$ by the formula $g(y) = x$ if $y = f(x)$. By definition of $g$,

(5) $$g(y) = g(f(x)) = x$$

and

(6) $$f(x) = f(g(y)) = y.$$

Conversely, if there exists a function $g$ satisfying (5) and (6) for each $x \in D$ and $y \in E$, then we assert that $f$ is $1:1$. To see this, fix $y \in E$ and observe first that (6) implies that $g(y)$ equals some point $x$ with the property that $f(x) = y$. If $x'$ is another point with this property then (5) implies that

$$g(y) = g(f(x')) = x'.$$

But since $g$ is a function, we must have $x = x'$. Thus there can only be one point $x$ with the property that $f(x) = y$. Thus $f$ must be $1:1$.

We leave it as an exercise to verify that the inverse function $g$ is unique.

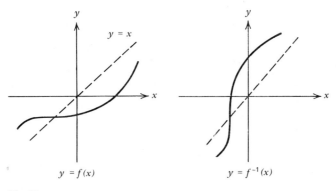

**Fig. 12**

Just as for the logarithm and exponential functions, a sketch of the graph of $f^{-1}$ may be obtained from the graph of $f$ by reflecting the graph of $f$ about the diagonal $y = x$. This is illustrated in Figure 12. Let us consider some examples.

***Example 1.***   If $f$ is a nonconstant linear function, then $f(x) = mx + b$ and $m \neq 0$. Clearly, the formula for $f^{-1}$ can be obtained by solving the equation

$$y = mx + b$$

for $x$. Indeed

$$x = \frac{1}{m}y - \frac{b}{m}.$$

Hence

$$x = f^{-1}(y) = \frac{1}{m}y - \frac{b}{m}.$$

For a real number $x$ then, $f^{-1}(x) = x/m - b/m$ is the inverse function for $f(x) = mx + b$. (See Figure 13.)

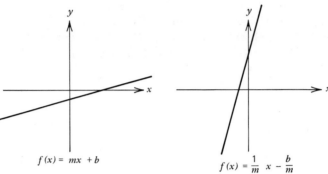

$$f(x) = mx + b \qquad\qquad f(x) = \frac{1}{m}x - \frac{b}{m}$$

**Fig. 13**

***Example 2.***   If $n$ is an odd positive integer then $f(x) = x^n$ is $1:1$, and the inverse function $f^{-1}$ is given by

$$f^{-1}(x) = x^{1/n}$$

See Figure 14.

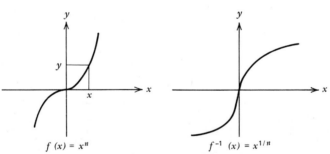

$$f(x) = x^n \qquad\qquad f^{-1}(x) = x^{1/n}$$

**Fig. 14**

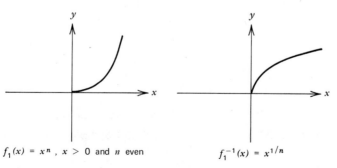

$$f_1(x) = x^n, \; x > 0 \text{ and } n \text{ even} \qquad\qquad f_1^{-1}(x) = x^{1/n}$$

**Fig. 15**

*Example 3.* If $n$ is an even positive integer, and $f(x) = x^n$ is defined for all real numbers $x$, then clearly $f$ is not $1:1$. Hence $f$ has no inverse function. However, we may restrict the domain of $f$ so that the restricted function $f$ will be $1:1$. This new function will, of course, have an inverse. Indeed, if $f_1(x) = x^n$, $x \geq 0$ and $f_2(x) = x^n$, $x \leq 0$, then both functions $f_1$ and $f_2$ are $1:1$. The inverses for $f_1$ and $f_2$ are given by $f_1^{-1}(x) = x^{1/n}$ and $f_2^{-1}(x) = -x^{1/n}$. (See Figures 15 and 16.)

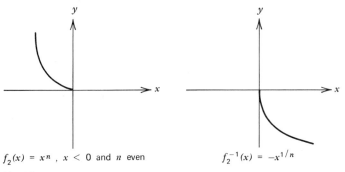

$f_2(x) = x^n$ , $x < 0$ and $n$ even       $f_2^{-1}(x) = -x^{1/n}$

**Fig. 16**

### EXERCISES
Determine which of the following sets define $1:1$ functions. Sketch the graph of the function and determine the inverse function if it exists. Specify the domain and range of the inverse function.

1. $G = \{(x, y) : 3x - 2y = 4\}$.

2. $G = \{(x, y) : y = 3x^2 - 4, x \geq 0\}$.

3. $G = \{(x, y) : y = 1 + |x|\}$.

4. $G = \left\{(x, y) : y = \dfrac{1}{x}\right\}$.

5. $G = \left\{(x, y) : y = \dfrac{1}{x^n}, n \text{ an odd positive integer}\right\}$.

6. $G = \left\{(x, y) : y = \dfrac{1}{x^n}, n \text{ an even positive integer}\right\}$.

7. $G = \left\{(x, y) : y + \dfrac{1}{2 + x} = 0\right\}$.

8. $G = \left\{(x, y) : y = \dfrac{x^2 - 4}{x + 2}\right\}$.

9. $G = \left\{(x, y) : y = \dfrac{x + 2}{x - 2}\right\}$.

10. $G = \{(x, y) : y = 2x^2 + x - 1\}$.

11. $G = \{(x, y) : y = \sqrt{1 - x^2}\}$.

The following functions $f$ are not $1:1$. Determine a number $a$ such that if $f_1(x) = f(x)$ for $x > a$ and $f_2(x) = f(x)$ for $x < a$ then both functions $f_1$ and $f_2$ are $1:1$. Determine the inverses $f_1^{-1}$ and $f_2^{-1}$.

12. $f(x) = x^4$.

13. $f(x) = 1/x^2$.

14. $f(x) = 1 - |1 - x|$.

15. $f(x) = x^2 - 2x + 1$.

16. $f(x) = x^2 - 3x + 2$.

17. $f(x) = x^2 - 2x + 2$.

18. Finish the proof of Theorem 4.3. That is, show that if the function $f^{-1}$ exists, it is necessarily unique.

## 4.5   Inverses of Monotone Functions

We assume now that $f$ is a continuous $1:1$ function defined on an interval $[a, b]$. What does the graph of $f$ look like? For real valued functions $f$ we have seen that $f$ is $1:1$ if and only if each horizontal line intersects the graph of $f$ at most once. Now imagine a point $P$ moving along the graph of $f$ as the $x$ coordinate of $P$ moves to the right. Since the graph of $f$ is unbroken,

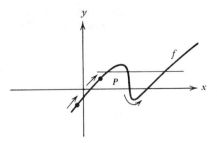

**Fig. 17**   If the graph of a continuous function rises and falls on the interval $[a, b]$, then $f$ is not $1:1$.

there appear to be only two possibilities. Either the point $P$ moves steadily up as $x$ moves to the right or $P$ moves steadily down. If at some point $x$, $P$ is going up and at some point $x'$, $P$ is coming down, then it appears from Figure 17 that there should exist a horizontal line intersecting the graph of $f$ at least twice. Hence $f$ would not be $1:1$.

Thus for a continuous $1:1$ function defined on an interval $[a, b]$ there appear to be only two possibilities. Either $f$ is increasing on $[a, b]$, in which case the range of $f$ is the interval $[f(a), f(b)]$, or $f$ is decreasing on $[a, b]$ in which case the range of $f$ is the interval $[f(b), f(a)]$. (See Figure 18.) However, the above discussion is not a proof of these facts. For the moment we take them for granted; we shall prove them in the next chapter. The following result summarizes the facts about $1:1$ continuous functions.

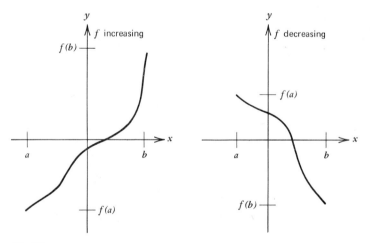

**Fig. 18**

**THEOREM 4.4**   *Let the function $f$ be continuous and $1:1$ on the interval $[a, b]$. Then $f$ is either increasing on $[a, b]$ or decreasing on $[a, b]$. The range of $f$ is the interval $[f(a), f(b)]$ if $f$ is increasing and $[f(b), f(a)]$ if $f$ is decreasing. Moreover the inverse function $f^{-1}$ is continuous. If $f$ is differentiable at the point $x$ and if $f'(x) \neq 0$, then $f^{-1}$ s differentiable at the corresponding point $y = f(x)$. Indeed,*

(1)
$$f^{-1\prime}(y) = \frac{1}{f'(x)}.$$

Accepting the fact that a $1:1$ continuous function defined on an interval $[a, b]$ is either increasing or decreasing on this interval, we have already given a complete proof of the balance of the theorem for the special case of $f(x) = \log x$ and $f^{-1}(x) = e^x$. The arguments given in sections 4.1–4.3 hold verbatim in the general case. The verification of these facts we leave as exercises. However, a comment on formula (1) is in order. Let us assume $f'(x) \neq 0$. If it is known that $f^{-1}$ is differentiable at the point $y = f(x)$, then we may derive formula (1) as a consequence of the chain rule. To do this we just differentiate the equation

$$f^{-1}(f(x)) = x.$$

Applying the chain rule we get

$$f^{-1\prime}(f(x)) \cdot f'(x) = 1$$

or

$$f^{-1\prime}(y) = \frac{1}{f'(x)} \text{ when } y = f(x).$$

Let us use this technique to derive the formula for the derivative of $E(x) = e^x$. Since $\log x$ is the inverse of $E(x)$,

$$\log E(x) = x.$$

Differentiating with respect to $x$ we obtain

$$\frac{1}{E(x)} E'(x) = 1$$

or

$$E'(x) = E(x).$$

We shall use this procedure several times in the future to derive the formulas for the derivatives of various inverse functions.

## EXERCISES

1. Suppose $f$ is $1:1$ and defined on the interval $[a, b]$. If $f$ is not assumed to be continuous, must $f$ be either increasing on $[a, b]$ or decreasing on $[a, b]$? If so, prove it; if not, give an example.
2. Assuming $d(e^x)/dx = e^x$ verify by the chain rule that

$$\frac{d}{dx} \log x = \frac{1}{x}.$$

3. If $x \geq 0$ and $f(x) = x^n$, Then $f^{-1}(x) = x^{1/n}$. Knowing the derivative for $f$, derive the formula for the derivative of $f^{-1}$ by the methods of this section.
4. Suppose $f$ is continuous and increasing on the interval $[a, b]$. Show that for each real number $y$ satisfying $f(a) < y < f(b)$ there is a real number $x$ satisfying $a < x < b$ and $f(x) = y$. (Adapt the corresponding argument for $f(x) = \log x$ given in Section 4.2.)
5. Prove that the inverse function of a continuous increasing function is continuous. (Adapt the corresponding argument for the inverse of $\log x$ given in Section 4.3.)
6. Prove that the inverse function $f^{-1}$ of a differentiable increasing function $f$ has a derivative at all points $x$ such that $f'(x) \neq 0$. For such points show that

$$f^{-1\prime}(y) = \frac{1}{f'(x)} \text{ if } y = f(x).$$

*7. For $-1 < x < 1$ define the function $f$ by the formula

$$f(x) = \int_0^x \frac{dt}{\sqrt{1-t^2}}.$$

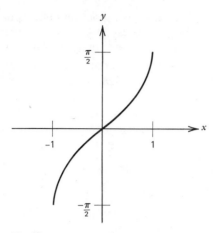

Fig. 19

(Do not attempt to evaluate this integral.) Figure 19 is a sketch of the graph of $f$.
(a) Compute $f'(x)$.
(b) Let $F$ be the inverse of $f$. (Then $F(y) = x$ when $y = f(x)$.) Define $G(y) = F'(y)$. Using formula (1), verify that

$$[F(y)]^2 + [G(y)]^2 = 1.$$

What is your guess for the functions $F$ and $G$. Which is which?

8. The difficulty in proving that a continuous $1:1$ function defined on $[a, b]$ is either increasing on $[a, b]$ or decreasing on $[a, b]$ is proving the following fact:

   [P] *If $f(a) < f(b)$, then for each $y$ satisfying $f(a) < y < f(b)$ there is an $x$ in $[a, b]$ for which $f(x) = y$.*

   If $f$ is assumed to be increasing, then this follows from the completeness property of the real numbers in a straightforward fashion (Exercise 4). If $f$ is not assumed to be $1:1$, the statement [P], called the *intermediate value property*, is still true, but the argument is more involved. Geometrically the statement seems obvious because we interpret the graph of a continuous function as unbroken. Assuming the intermediate value property for a continuous function $f$ defined on $[a, b]$, prove that if $f$ is in addition $1:1$ and $f(a) < f(b)$, then $f$ is increasing on $[a, b]$.

## 4.6   The Trigonometric Functions

We define the sine, cosine, and tangent functions in the following way. Let $C = \{(x, y) : x^2 + y^2 = 1\}$ be the unit circle in the $(x, y)$ plane. Denote the origin by $0$ and the point $(1, 0)$ by $P$. For each $s$, $0 \leqslant s \leqslant 2\pi$ let $Q$ be the point on $C$ such that the circular arc $\overline{PQ}$, measured in a counterclockwise direction, has length equal to $s$. If $(x, y)$ are the coordinates of $Q$, then we define

$$\sin s = y$$
(1)
$$\cos s = x$$
$$\tan s = (y/x) \qquad \text{if} \qquad x \neq 0.$$

See Figure 20.

We extend the domains of these functions to the set of all real numbers by requiring in addition that each of the above functions satisfy the identity

(2) $$f(s) = f(s + 2\pi).$$

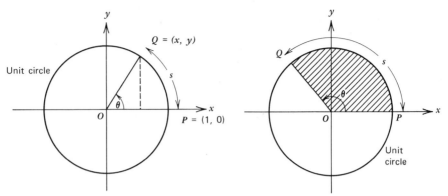

**Fig. 20**   Sin $s = y$,   cos $s = x$,   tan $s = y/x$   if $x \neq 0$.

**Fig. 21**   $s =$ length $PQ = 2$ area $POQ$.

Thus if the values of $f$ are known on the interval $[0, 2\pi]$, equation (2) gives the values of $f$ on the intervals $[2\pi, 4\pi]$, $[4\pi, 6\pi]$, etc., and on the intervals $[-2\pi, 0]$, $[-4\pi, -2\pi]$, ... etc. Of course, in order for this process to make sense we must check that the trigonometric functions defined by (1) have the same values at $s = 0$ and $s = 2\pi$. For the functions sine, cosine, and tangent this is clear since, by definition,

and
$$\sin 0 = \sin 2\pi = 0$$
$$\cos 0 = \cos 2\pi = 1$$
$$\tan 0 = \tan 2\pi = 0.$$

Before discussing elementary properties of the trigonometric functions let us comment on one subtlety in this definition. We have assumed that it is "obvious" that to each real number $s$, $0 \le s \le 2\pi$, there exists a point $Q$ on the circumference of the unit circle such that the arc $\overset{\frown}{PQ}$ has length $s$. If we try to prove that statement analytically we are immediately in trouble. The first problem is that we have not even given a definition of the length of a circular arc. Assuming we can do this, then there is still the problem of proving that for a given number $s$ there is a point $Q$ such that the arc $\overset{\frown}{PQ}$ has length $s$. It is an interesting exercise to try to surmount these difficulties. We can avoid the necessity of defining arc length by replacing the length of the circular arc $\overset{\frown}{PQ}$ by twice the area of the circular segment $POQ$. Once the arc length has been suitably defined it can be shown that for each point $Q$ on the unit circle

$$\text{length } \overset{\frown}{PQ} = 2\text{area } POQ.$$

See Figure 21.

If we replace the length $\overset{\frown}{PQ}$ by twice the area of the segment $POQ$, then we still must show that for each $s$ in the interval $[0, 2\pi]$ there is a point $Q$ on the unit circle satisfying

$$2 \text{ area } POQ = s.$$

A sketch of how to do this is included with the exercises at the end of the section. However, for the balance of our discussion we shall ignore this subtlety, and shall assume the existence of the point $Q$ such that the length of the arc $\overset{\frown}{PQ}$ equals the given number $s$.

If $\theta$ is the central angle $POQ$, then the number $s$ is called the *radian measure* of the angle $\theta$. Moreover, we may define

$$\sin \theta = \sin s$$

and
$$\cos \theta = \cos s$$
$$\tan \theta = \tan s.$$

It causes no difficulty if we identify the angle $\theta$ with its radian measure. Hence, we shall often use phrases like "let $\theta$ be an angle between 0 and $\pi$" where to be correct we should say "let $\theta$ be an angle whose radian measure lies between 0 and $\pi$."

Referring to Figure 20 we may now define the remaining three trigonometric functions, cotangent cosecant, and secant. Indeed for $0 \leqslant s \leqslant 2\pi$ if $s = $ length $\overset{\frown}{PQ}$ where $Q = (x, y)$, then

$$
\begin{array}{lll}
\cot s = x/y & \text{if} & y \neq 0 \\
\csc s = 1/y & \text{if} & y \neq 0 \\
\sec s = 1/x & \text{if} & x \neq 0.
\end{array}
$$

Clearly

$$\cot s = \frac{1}{\tan s} \qquad \text{if} \qquad \tan s \neq 0$$

and

$$\csc s = \frac{1}{\sin s} \qquad \text{if} \qquad \sin s \neq 0$$

$$\sec s = \frac{1}{\cos s} \qquad \text{if} \qquad \cos s \neq 0.$$

These functions are extended to all values of $s$ in the same way as was done for the functions sine, cosine, and tangent by requiring that each of them satisfy the identity

$$f(s) = f(s + 2\pi)$$

for all values of $s$ for which they are defined.

Whenever a function $g$ satisfies an identity of the form

$$g(x + a) = g(x)$$

for all values of $x$, then we say $g$ is *periodic*.

If $a$ is the smallest positive number for which this is true, then the number $a$ is called the *period* of the function $g$. Each of the trigonometric functions is periodic. The periods of the sine, cosine, secant, and cosecant functions is $2\pi$, whereas the period of the tangent and cotangent is $\pi$. If a given function $g$ is defined on an interval $[a, b]$ of length $l = b - a$ and $g(a) = g(b)$, then we may extend the domain of $g$ to the set of all real numbers by requiring that for all real numbers $x$

$$g(x) = g(x + l).$$

Just as for the trigonometric functions this amounts to successively shifting the graph of $g$ by an interval of width $l$. The extended function is called the *periodic extension* of the origional function $g$ (Figure 22).

The graphs of the six trigonometric functions on the interval $[0, 2\pi]$ are shown in Figure 23. The independent variable is denoted as usual by $x$ rather than $s$.

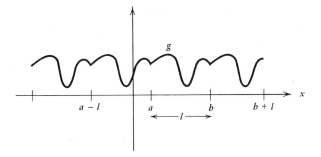

**Fig. 22** Periodic extension of a function $g$.

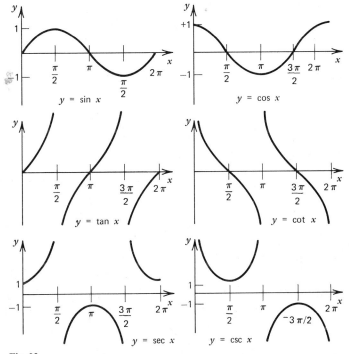

**Fig. 23**

The trigonometric functions satisfy numerous identities that are assumed to be familiar to the student. These all can be deduced from the following four formulas that are valid for all real numbers $x$ and $y$.

(3) $$\sin(-x) = -\sin x$$

(4) $$\cos(-x) = \cos x$$

(5) $$\sin(x+y) = \sin x \cos y + \cos x \sin y$$

(6) $$\cos(x+y) = \cos x \cos y - \sin x \sin y.$$

The symmetry properties (3) and (4) hold, of course, for many functions. Indeed, a function $f$ that satisfies the identity

(7) $$f(x) = f(-x)$$

is said to be *even*; and if $f$ satisfies

(8) $$f(-x) = -f(x),$$

*f* is said to be *odd*. Formulas (5) and (6) are called the addition formulas for the trigonometric functions. Many other useful identities can be deduced from these four formulas.

**Example 1.**    Deduce that for all real numbers *x* the following double angle formulas are valid

$$\sin 2x = 2 \sin x \cos x$$

$$\cos 2x = \cos^2 x - \sin^2 x.$$

These follow immediately from (5) and (6) by setting *x* = *y*. The derivation of other identities is left as exercises.

**EXERCISES**

1. Establish the following trigonometric identities.
    (a) $\sin^2 x + \cos^2 x = 1$
    (b) $\sin (\pi/2 + x) = \cos x$
    (c) $\cos (\pi/2 + x) = -\sin x$
    (d) $\sin (\pi + x) = -\sin x$
2. Establish the following trigonometric identities and discuss when $\pm$ is to be used.
    (a) $\sin \dfrac{x}{2} = \pm \sqrt{\dfrac{1 - \cos x}{2}}$
    (b) $\cos \dfrac{x}{2} = \pm \sqrt{\dfrac{1 + \cos x}{2}}$
3. Show that for all real numbers *x* and *y*

    (a) $\sin x - \sin y = 2 \cos \left(\dfrac{x+y}{2}\right) \sin \left(\dfrac{x-y}{2}\right)$

    (b) $\cos x - \cos y = -2 \sin \left(\dfrac{x+y}{2}\right) \sin \left(\dfrac{x-y}{2}\right)$

    (c) $\tan x - \tan y = (1 + \tan x \tan y) \tan (x - y)$
4. Sketch the graph of

    $$f(x) = 2 \cos (x + \pi/4),$$

    and determine constants *a*, *b* such that $f(x) = a \cos x + b \sin x$.
5. If constants $\alpha$ and $\beta$ are given, show that numbers *a* and *b* can be found satisfying

    $$\alpha \cos (x + \beta) = a \cos x + b \sin x.$$

6. Show that $a \cos x + b \sin x = 0$ for all *x* implies $a = b = 0$.
7. Determine all real numbers *x* such that
    (a) $\sin x = 0$          (d) $\cos x = 1$
    (b) $\cos x = 0$          (e) $\sin x = -1$
    (c) $\sin x = 1$          (f) $\cos x = -1$
8. Solve the following trigonometric equations, i.e., determine all numbers *x* for which the identities hold.
    (a) $\sin x = \cos x$
    (b) $\sin x = \tan x$
    (c) $\sin x - \cos x = 1$
9. Show that $f(x) = \tan x$ is periodic of period $\pi$. Show that $|\sin x|$ and $|\cos x|$ are periodic of period $\pi$. What is the period of $|\tan x|$? Sketch the graphs of these functions.
10. Let $f(x) = x$ for $-1 \leqslant x \leqslant 1$. Extend *f* periodically and sketch the graph. Do the same for $2(x) = |x|$.
11. For an arbitrary function *f* show that

    $$g(x) = f(x) + f(-x) \text{ is even}$$

and
$$h(x) = f(x) - f(-x) \text{ is odd.}$$

From this deduce that any function can be written as the sum of even and odd functions.

12. Write $f(x) = x^2 - x + 1$ as the sum of even and odd functions. Do the same for $g(x) = e^x$. Sketch the graphs of the respective even and odd functions.

*13. Show that for each $s$ in the interval $[0, 2\pi]$ there exists a point $Q$ on the unit circle such that if $A$ denotes the area of the sector $POQ$ (Figure 24) then $2A = s$.

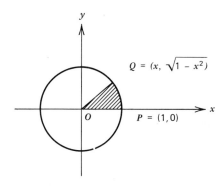

**Fig. 24**

(Consider first those values of $s$ in the interval $[0, \pi]$.
For $Q = (x, \sqrt{1-x^2})$ on the upper half of the semicircle, let $A(x)$ denote the area of the sector $POQ$.
   Show that

$$A(x) = \frac{x\sqrt{1-x^2}}{2} + \int_x^1 \sqrt{1-t^2}\, dt.$$

Clearly $A(1) = 0$ and $A(-1) = \pi/2$, since $\pi$ is defined to be the area of the unit circle. Convince yourself that $A(x)$ is a continuous decreasing function of $x$ on the interval $[-1, 1]$. Now use Exercise 4, p. 161 to complete the proof. The case $\pi \leqslant s \leqslant 2\pi$ is Exercise 16.)

*14. Let $A(x)$ be the function of Exercise 13. Show that

$$A(x) = \frac{\pi}{4} - \frac{1}{2} \int_0^x \frac{dt}{\sqrt{1-t^2}}.$$

As a result evaluate

$$\int_0^1 \frac{dt}{\sqrt{1-t^2}}.$$

$\left( \text{Let } f(x) = A(x) - (\pi/4) + \frac{1}{2} \int_0^x 1/\sqrt{1-t^2}\, dt \right).$ Show that $f'(x) = 0$ and apply vanishing derivative theorem.

*15. If $-1 \leqslant x \leqslant 1$, show that

$$x = \sin \int_0^x \frac{dt}{\sqrt{1-t^2}}.$$

(See Exercise 14.)

*16 Show that to each number $s$ in the interval $[\pi, 2\pi]$ there is a point $Q$ on the lower half of the unit circle such that if $A$ denotes the area of the sector $POQ$ then $2A = s$. (If $Q$ is on the lower half of the unit circle, $Q = (x, -\sqrt{1-x^2})$. Derive a formula for the area $A(x)$ of the sector as a function of $x$. Now apply the same reasoning as in the case when $0 \leqslant s \leqslant \pi$. See Figure 25.)

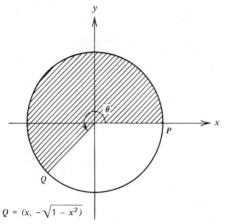

$Q = (x, -\sqrt{1-x^2})$

**Fig. 25**

### 4.7 Differentiation Formulas for the Trigonometric Functions

If $f(x) = \sin x$, then to derive a formula for $f'(x)$ we proceed as follows. By definition

$$f'(x) = \lim_{h \to 0} \frac{\sin(x+h) - \sin x}{h}.$$

But applying the addition formula (5) p. 165 to the difference quotient, we obtain

$$\frac{\sin(x+h) - \sin x}{h} = \frac{\sin x \cos h + \cos x \sin h - \sin x}{h}$$

$$= \sin x \left(\frac{\cos h - 1}{h}\right) + \cos x \frac{\sin h}{h}.$$

Thus, to compute $d(\sin x)/dx$ we must first determine $\lim_{h \to 0}(\cos h - 1)/h$ and $\lim_{h \to 0}(\sin h/h)$. As we shall see, these have value 0 and 1 respectively. Hence

$$\frac{d}{dx}(\sin x) = \cos x.$$

Now to derive these limits we must first establish some inequalities involving the trigonometric functions. These are evident from Figure 26, and can be easily established analytically.

Let $Q = (p, q)$ be a point on the unit circle in the first quadrant. Denote the central angle by $\theta$. We compare the areas of the triangles $ORQ$, $OPQ$, and $OPS$ with the area of the sector $OPQ$. By inspection of Figure 26 it is clear that

(1)     Area $\triangle ORQ \leqslant$ Area $\triangle OPQ \leqslant$ Area sector $OPQ \leqslant$ Area $\triangle OPS$.

Now

$$\text{Area } \triangle ORQ = \frac{1}{2} pq = \frac{1}{2} \sin\theta \cos\theta,$$

$$\text{Area } \triangle OPQ = \frac{1}{2} pq + \frac{1}{2}(1-p)q = \frac{1}{2} q = \frac{1}{2} \sin\theta,$$

$$\text{Area sector } OPQ = \theta/2,$$

and

$$\text{Area } \triangle OPS = \frac{1}{2} \tan\theta.$$

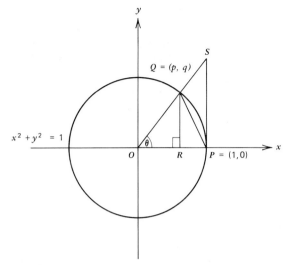

**Fig. 26**

Therefore, for $0 \leqslant \theta \leqslant \pi/2$, it follows from (1) that

(2) $$\sin \theta \cos \theta \leqslant \sin \theta \leqslant \theta \leqslant \tan \theta.$$

Let us use (2) to establish first that $\lim\limits_{x \to 0} \sin x = 0$ and $\lim\limits_{x \to 0} \cos x = 1$. Setting $\theta = x$ we infer from (2) that for $0 \leqslant x \leqslant \pi/2$

$$0 \leqslant \sin x \leqslant x.$$

Therefore, certainly

$$-x \leqslant 0 \leqslant \sin x \leqslant x,$$

or

(3) $$|\sin x| \leqslant \|x\|.$$

Now (3) holds for $-\pi/2 \leqslant x \leqslant 0$ as well. For if $-\pi/2 \leqslant x \leqslant 0$, then $0 \leqslant -x \leqslant \pi/2$. But $|\sin x| = |-\sin x| = |\sin(-x)|$. Applying (3) we infer that if $0 \leqslant -x \leqslant \pi/2$,

$$|\sin(-x)| \leqslant |-x| = |x|.$$

Therefore, if $-\pi/2 \leqslant x \leqslant \pi/2$,

$$|\sin x| \leqslant |x|.$$

Applying the definition of a limit to this inequality we infer that

$$\lim_{x \to 0} \sin x = 0.$$

To prove that $\lim\limits_{x \to 0} \cos x = 1$ observe that for $-\pi/2 \leqslant x \leqslant \pi/2$

$$\cos x = \sqrt{1 - \sin^2 x}.$$

Since $\lim\limits_{x \to 0} \sin x = 0$, it follows by the continuity of the function $\sqrt{1 - y^2}$ that $\lim\limits_{x \to 0} \cos x = 1$.

Next we verify that $\sin x$ and $\cos x$ are continuous functions for all values of $x$. For $\sin x$ we must show

(4) $$\lim_{h \to 0} \sin(x + h) = \sin x.$$

But $\sin(x+h) = \sin x \cos h + \cos x \sin h$. Since as $h \to 0$, $\cos h \to 1$, and $\sin h \to 0$, (4) follows by an application of the limit theorems. The continuity of $\cos x$ can be derived in a similar fashion and is left as an exercise.

Now we must establish the following fundamental limit relation

(5) $$\lim_{x \to 0} \frac{\sin x}{x} = 1.$$

Again from (2) if $0 < x \leqslant \pi/2$, then $\sin x \cos x \leqslant x \leqslant \tan x$. Dividing by $\sin x$ we get

(6) $$\cos x \leqslant \frac{x}{\sin x} \leqslant \frac{1}{\cos x}.$$

This inequality holds for $-\pi/2 \leqslant x < 0$ as well since (6) is unchanged if $-x$ is substituted for $x$. Both $\cos x$ and $1/\cos x$ tend to 1 as $x$ tends to zero. Hence $\lim_{x \to 0} (x/\sin x) = 1$ and therefore (5) holds.

As a corollary of (5) we have that

(7) $$\lim_{x \to 0} \frac{1 - \cos x}{x} = 0,$$

since for $x \neq 0$

$$\frac{1 - \cos x}{x} = \frac{1 - \cos x}{x} \cdot \frac{1 + \cos x}{1 + \cos x} = \frac{\sin^2 x}{x} \frac{1}{1 + \cos x}.$$

Therefore

$$\lim_{x \to 0} \frac{1 - \cos x}{x} = \lim_{x \to 0} \frac{\sin x}{x} \frac{\sin x}{1 + \cos x}$$

$$= \lim_{x \to 0} \frac{\sin x}{x} \cdot \lim_{x \to 0} \frac{\sin x}{1 + \cos x}$$

$$= 1 \cdot 0 = 0.$$

Before computing the derivatives of $\sin x$ and $\cos x$ let us examine the argument used to prove (5). We asserted that $(x/\sin x)$ tended to 1 as $x$ tended to 0 because $(x/\sin x)$ was always between the two numbers $\cos x$ and $1/\cos x$, both of which tended to 1. This procedure occurs repeatedly so we state the principle as a theorem.

**THEOREM 4.5**   *If for each $x \neq a$ in some interval $b < a < c$, it follows that*

$$g(x) < f(x) < h(x)$$

*and if it is known that*

$$\lim_{x \to a} g(x) = L = \lim_{x \to a} h(x), \qquad then \qquad \lim_{x \to a} f(x) = L.$$

The proof of this result follows directly from the definition of a limit and is left as an exercise.

**THEOREM 4.6**   *The functions $\sin x$ and $\cos x$ are differentiable for all values of $x$. Furthermore,*

(8) $$\frac{d}{dx}(\sin x) = \cos x$$

*and*

(9) $$\frac{d}{dx}(\cos x) = -\sin x.$$

*Proof.*  It suffices to verify formulas (8) and (9) for all values of $x$. By definition

$$\frac{d}{dx}(\sin x) = \lim_{h \to 0} \frac{\sin(x+h) - \sin x}{h}.$$

But applying the addition formula (5), p. 165 we see that

$$\frac{\sin(x+h) - \sin x}{h} = \frac{\sin x \cos h + \cos x \sin h - \sin x}{h}$$

$$= \sin x \frac{(\cos h - 1)}{h} + \cos x \cdot \frac{\sin h}{h}.$$

Therefore

$$\lim_{h \to 0} \frac{\sin(x+h) - \sin x}{h} = \sin x \cdot \lim_{h \to 0} \frac{(\cos h - 1)}{h} + \cos x \lim_{h \to 0} \frac{\sin h}{h}$$

$$= \sin x \cdot 0 + \cos x \cdot 1 = \cos x$$

since $\lim_{h \to 0} (\cos h - 1)/h = 0$ and $\lim_{h \to 0} \sin h/h = 1$. Similarly

$$\frac{d}{dx}(\cos x) = \lim_{h \to 0} \frac{\cos(x+h) - \cos x}{h}$$

$$= \lim_{h \to 0} \left[ \cos x \left( \frac{\cos h - 1}{h} \right) - \sin x \frac{\sin h}{h} \right]$$

$$= -\sin x.$$

The formulas for the derivatives of the four remaining trigonometric functions are

(10) $$\frac{d}{dx}(\tan x) = \sec^2 x$$

(11) $$\frac{d}{dx}(\cot x) = -\csc^2 x$$

(12) $$\frac{d}{dx}(\sec x) = \sec x \tan x$$

(13) $$\frac{d}{dx}(\csc x) = -\csc x \cot x.$$

These can be immediately derived from Theorem 4.6 using formulas (8) and (9) and standard differentiation techniques.

*Example 1.*  To derive (10) note that

$$\frac{d}{dx}(\tan x) = \frac{d}{dx}\left(\frac{\sin x}{\cos x}\right) = \frac{\cos x \frac{d}{dx}(\sin x) - \sin x \frac{d}{dx}(\cos x)}{\cos^2 x}$$

$$= \frac{\cos^2 x + \sin^2 x}{\cos^2 x} = \frac{1}{\cos^2 x} = \sec^2 x.$$

*Example 2.*  Compute

$$\frac{d}{dx}\cos^2\left(\frac{1+e^x}{1-e^x}\right).$$

Applying the chain rule we get

$$\frac{d}{dx}\cos^2\left(\frac{1+e^x}{1-e^x}\right) = 2\cos\left(\frac{1+e^x}{1-e^x}\right)\cdot -\sin\left(\frac{1+e^x}{1-e^x}\right)\cdot\frac{d}{dx}\left(\frac{1+e^x}{1-e^x}\right)$$

$$= -\sin 2\left(\frac{1+e^x}{1-e^x}\right)\cdot\frac{(1-e^x)e^x-(1+e^x)(-e^x)}{(1-e^x)^2}$$

$$= \frac{-2e^x}{(1-e^x)^2}\sin 2\left(\frac{1+e^x}{1-e^x}\right).$$

Formulas (8) to (13) immediately translate into appropriate formulas for indefinite integrals.

$$(14) \qquad\qquad \int \sin x\, dx = -\cos x + C$$

$$(15) \qquad\qquad \int \cos x\, dx = \sin x + C$$

$$(16) \qquad\qquad \int \sec^2 x\, dx = \tan x + C$$

$$(17) \qquad\qquad \int \csc^2 x\, dx = -\cot x + C$$

$$(18) \qquad\qquad \int \sec x \tan x\, dx = \sec x + C$$

$$(19) \qquad\qquad \int \csc x \cot x\, dx = -\csc x + C.$$

If we compose the trigonometric functions sine, cosine, and tangent with another function $f$, then the chain rule yields the following formulas for the derivatives

$$(20) \qquad\qquad (\sin f)' = (\cos f)f'$$

$$(21) \qquad\qquad (\cos f)' = (-\sin f)f'$$

$$(22) \qquad\qquad (\tan f)' = (\sec^2 f)f'.$$

The analagous formulas for indefinite integrals are

$$(23) \qquad\qquad \int \cos f(x)f'(x)dx = \sin f(x) + C$$

$$(24) \qquad\qquad \int \sin f(x)f'(x)dx = -\cos f(x) + C$$

$$(25) \qquad\qquad \int \sec^2 f(x)f'(x)dx = \tan f(x) + C.$$

**Example 3.**   Compute $\int x\sin x^2\, dx$. Let $f(x) = x^2$. Then from (24)

$$\int (2x)\sin x^2\, dx = -\cos x^2 + C.$$

Hence

$$\int x\sin x^2\, dx = \frac{1}{2}\int (2x)\sin x^2\, dx$$

$$= -\frac{1}{2}\cos x^2 + C.$$

**EXERCISES**
1. Differentiate the following functions
   (a) $f(x) = \sin(3x+2)$        (c) $f(x) = \cos^5 x$
   (b) $f(x) = \tan(2-x^2)$        (d) $f(x) = \sin^3(x^2-1)$

(e) $f(x) = \cos \left( \dfrac{1+x}{1-x} \right)$

(i) $f(x) = \log \left( \dfrac{1+\sin x}{1-\cos x} \right)$

(f) $f(x) = e^x \sin 2x$

(g) $f(x) = (\tan^2 x + 1)^{1/2}$

(j) $f(x) = \sin^2 x \int_0^x \sqrt{1-t^2}\, dt$

(h) $f(x) = \dfrac{e^x}{1+\cos x}$

(k) $f(x) = \int_0^{\cos x} \sqrt{1-t^2}\, dt$

2. Derive the formula for $D(\cos x)$ by using the fact that

(a) $\cos x = \sin \left( \dfrac{\pi}{2} - x \right)$

(b) $\cos x = \sqrt{1 - \sin^2 x}$

3. Derive formulas 10 to 12.

4. Compute $y'$ if

(a) $y = \dfrac{\tan x + \sec x}{x+1}$

(c) $y = \sqrt{\dfrac{1+\sec x}{1+\tan x}}$

(b) $y = \csc^2 (1 + e^{2x})$

(d) $y = \dfrac{\csc 2x}{\cot (1+x^2)}$

5. Find $y'$ if

(a) $x \sin y = \cos (x+y)$

(b) $\cos x + \tan y = 1$

(c) $\cot (x+y) = \sec x + \csc y$

6. Verify that $y'' + a^2 y = 0$ if $y = A \sin ax + B \cos ax$ where $A$ and $B$ are constants.

7. Compute

(a) $\displaystyle\int x \cos x^2 \, dx$

(f) $\displaystyle\int \cos x e^{\sin x} \, dx$

(b) $\displaystyle\int e^x \sin e^x \, dx$

(g) $\displaystyle\int \cot x \, dx$

(c) $\displaystyle\int \dfrac{x}{\cos^2 (x^2)} \, dx$

(h) $\displaystyle\int x \sin x^2 e^{\cos x^2} \, dx$

(d) $\displaystyle\int x^2 \sin (x^3 + 1) \, dx$

(i) $\displaystyle\int x \tan x^2 \sec x^2 \, dx$

(e) $\displaystyle\int \dfrac{\sin x}{\cos x} \, dx$

(j) $\displaystyle\int e^x \sec^2 e^x \, dx$

8. Evaluate the following definite integrals

(a) $\displaystyle\int_0^{\pi/4} \sin 2x \, dx$

(c) $\displaystyle\int_0^{2\pi} \cos 4x \, dx$

(b) $\displaystyle\int_{-\pi/4}^{\pi/4} \sec^2 x \, dx$

(d) $\displaystyle\int_{\pi/4}^{\pi/2} \cos (2x + \pi) \, dx$

9. Using the double angle formula $\cos^2 x = (1 + \cos 2x)/2$ derive a formula for $\displaystyle\int \cos^2 x \, dx$. Show that $\displaystyle\int_0^{\pi/2} \cos^2 x \, dx = \pi/4$.

10. Using the formula $\cos^2 x - \sin^2 x = \cos 2x$ show that

$$\int_0^{\pi/2} \cos^2 x \, dx = \int_0^{\pi/2} \sin^2 x \, dx.$$

11. Determine the area of the ordinate set for the function $f(x) = \sec^2 x$ on the interval $[-\pi/4, \pi/4]$. Sketch the graph of $f$.

12. Determine the area of the ordinate set for $f(x) = (1 + \sin 2x)/2$ on the interval $[0, \pi/2]$. Sketch the graph of $f$.

13. (a) Show that if $|x| \leqslant \pi/2$, then $|\cos x - 1| \leqslant |x|$. From this show that $\displaystyle\lim_{x \to 0} \cos x = 1$.

(b) Prove that for each $x$, $\displaystyle\lim_{h \to 0} \cos (x+h) = \cos x$.

14. Show that

(a) $\lim\limits_{\theta\to 0}\dfrac{\sin 3\theta}{\theta}=3$        (c) $\lim\limits_{\theta\to 0}\dfrac{\tan\theta}{\theta}=1$

(b) $\lim\limits_{\theta\to 0}\dfrac{\cos\theta-1}{\sin\theta}=0$        (d) $\lim\limits_{\theta\to 0}\dfrac{1-\cos\theta}{\theta^2}=\dfrac{1}{2}$

15. A lighthouse is 3000 feet from a straight shore and the beacon revolves at a rate of two revolutions per minute. How fast is the lightbeam moving up the coast at the nearest point to the lighthouse? How fast is it moving at a point on the shore 5000 feet from the lighthouse?

16. How fast in feet per hour is the shadow of a 50-foot flag pole lengthening when the sun is at a 45° angle? Assume that the sun and the pole are always in the same plane.

17. The ordinate set for $f(x)=\sin x$ on the interval $[0,\pi]$ is revolved about the $x$ axis. Determine the volume of the resulting solid. (See Exercises 9 and 10.)

## 4.8  Inverse Trigonometric Functions

The function $y=\sin x$ is not $1:1$ on the interval $[0,2\pi]$. Hence we cannot talk of an inverse function $F$ for the sine function that satisfies

$$y=\sin F(y) \qquad -1\leqslant y\leqslant 1$$

and

$$x=F(\sin x) \qquad 0\leqslant x\leqslant 2\pi.$$

However, if we restrict the sine function to an interval on which it is $1:1$, then an inverse function $F$ may be defined. The customary choice is to take $-\pi/2\leqslant x\leqslant\pi/2$. The resulting inverse function $F$ is usually written $F(y)=\sin^{-1}y$ or $F(y)=\arc\sin y$. We shall use the latter terminology. Thus $x=\arc\sin y$ if $-\pi/2\leqslant x\leqslant\pi/2$ and $y=\sin x$. To sketch the graph of the function arc sin $x$ we need only rotate the graph of $y=\sin x$ about the line $y=x$ interchanging the $x$ and $y$ axes. (See Figure 27.)

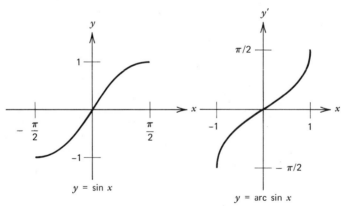

$y=\sin x$            $y=\arc\sin x$

**Fig. 27**

To compute $d(\arc\sin x)/dx$ we use the procedure discussed in Section 4.5, p. 161. Write $F(x)=\arc\sin x$ and apply the chain rule to the identity

(1) $$x=\sin F(x).$$

This yields

$$1=\cos F(x)\cdot F'(x).$$

or

(2) $$F'(x) = \frac{1}{\cos F(x)}.$$

Now

$$\cos^2 F(x) + \sin^2 F(x) = 1$$

and since

$$\frac{-\pi}{2} \leqslant F(x) \leqslant \frac{\pi}{2},$$

we have $\cos F(x) \geqslant 0.$

Hence

(3) $$\cos F(x) = \sqrt{1 - \sin^2 F(x)} = \sqrt{1 - x^2}.$$

Substituting (3) in (2) we obtain

(4) $$F'(x) = \frac{1}{\sqrt{1 - x^2}}.$$

Thus, $d(\arc \sin x)/dx = 1/\sqrt{1 - x^2}$. Knowing the derivative of $\arc \sin x$ enables us to write $\arc \sin x$ as an integral. Indeed, by the fundamental theorem of the calculus, if

$$\frac{d}{dx}(\arc \sin x) = \frac{1}{\sqrt{1 - x^2}},$$

then

$$\arc \sin x = \int_0^x \frac{dt}{\sqrt{1 - t^2}} + C.$$

But since $\arc \sin 0 = 0$, the constant $C = 0$ in the above equation. Hence

$$\arc \sin x = \int_0^x \frac{dt}{\sqrt{1 - t^2}} \qquad -1 < x < 1.$$

The choice of the interval $[-\pi/2, \pi/2]$ on which $\sin x$ is $1:1$ is, of course, completely arbitrary. If we examine the graph of $\sin x$, this function is $1:1$ on the intervals $I_1 = [\pi/2, 3\pi/2]$, $I_2 = [3\pi/2, 5\pi/2]$, etc., and $I_{-1} = [-3\pi/2, -\pi/2]$, $I_{-2} = [-5\pi/2, -3\pi/2] \ldots$ (Figure 28). We may compute inverse functions for $\sin x$ when $x$ belongs to one of the intervals $I_1, I_2, \ldots$, and $I_{-1}, I_{-2}, \ldots$ very easily. Reflect Figure 28 about the line $y = x$ obtaining the curve shown in Figure 29.

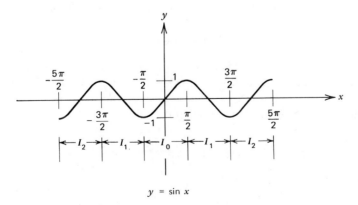

$y = \sin x$

**Fig. 28**

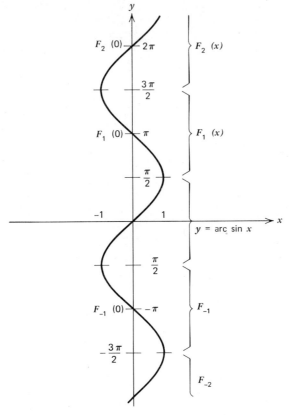

**Fig. 29** $F_k(x)$ is the inverse function for $x = \sin y$ if $y$ belongs to the interval $\left[\left(k-\frac{1}{2}\right)\pi, \left(k+\frac{1}{2}\right)\pi\right]$.

Let $F_1(x)$ be the inverse function for $x = \sin y$ when $y$ belongs to the interval $[\pi/2, 3\pi/2] = I_1$. Let $F_2(x)$ be the inverse function when $y$ belongs to $I_2 = [3\pi/2, 5\pi/2]$ etc. The domain of each of these inverse functions $F_k$ is the interval $-1 \leqslant x \leqslant 1$ and the range is the interval $I_k$. Moreover, if $-1 \leqslant x \leqslant 1$, then

(5) $$x = \sin F_k(x).$$

To determine formulas for the functions $F_k$ we differentiate (5). Applying the chain rule we get

$$1 = \cos F_k(x) F'_k(x).$$

Now for $k$ even, $\cos F_k(x) \geqslant 0$, hence $\cos F_k(x) = \sqrt{1 - \sin^2 F_k(x)} = \sqrt{1 - x^2}$. But for $k$ odd, $\cos F_k(x) \leqslant 0$. (Why?) Hence $F_k(x) = -\sqrt{1 - x^2}$. Therefore $F'_k(x) = 1/\sqrt{1 - x^2}$ if $k$ is even, and $F'_k(x) = -1/\sqrt{1 - x^2}$ if $k$ is odd. Now by the fundamental theorem of the calculus, it follows that

$$F_k(x) = \int_0^x \frac{dt}{\sqrt{1 - t^2}} + C \text{ if } k \text{ is even}$$

and

$$F_k(x) = -\int_0^x \frac{dt}{\sqrt{1 - t^2}} + C \text{ if } k \text{ is odd.}$$

Inspecting Figure 29 we see that $F_k(0) = k\pi$ for each value of $k$. Hence

$$F_k(x) = k\pi + \int_0^x dt/\sqrt{1-t^2} = k\pi + \text{arc sin } x \text{ if } k \text{ is even, and}$$

$$F_k(x) = k\pi - \int_0^x (1/\sqrt{1-t^2})\,dt = k\pi - \text{arc sin } x$$

if $k$ is odd. The function

$$F_0(x) = \text{arc sin } x = \int_0^x dt/\sqrt{1-t^2}$$

is sometimes called the principal inverse function for the sine, or in older books the "principal value" of the inverse function.

To determine an inverse function for the cosine we proceed similarly. First we choose an interval on which $\cos x$ is $1:1$. It is customary to take the interval $[0, \pi]$. Then the inverse function $F(y)$ for the cosine satisfies

$$y = \cos F(y) \qquad -1 \leqslant y \leqslant 1$$

and

$$x = F(\cos x) \qquad 0 \leqslant x \leqslant \pi.$$

We write $F(y) = \text{arc cos } y$. A sketch of the graph of $F(x) = \text{arc cos } x$ is easily obtained by reflecting the graph of $y = \cos x$ about the line $y = x$ and then interchanging the $x$ and $y$ axes (Figure 30). To compute $d(\text{arc cos } x)/dx$ we differentiate the equation

$$x = \cos (\text{arc cos } x)$$

obtaining

$$1 = -\sin (\text{arc cos } x) \frac{d}{dx} (\text{arc cos } x)$$

$$= -\sqrt{1 - \cos^2 (\text{arc cos } x)} \frac{d}{dx} (\text{arc cos } x)$$

$$= -\sqrt{1 - x^2} \frac{d}{dx} (\text{arc cos } x).$$

(Why is the positive sign taken for $\sqrt{1-x^2}$?) Therefore

(6) $$\frac{d}{dx} (\text{arc cos } x) = -\frac{1}{\sqrt{1-x^2}}.$$

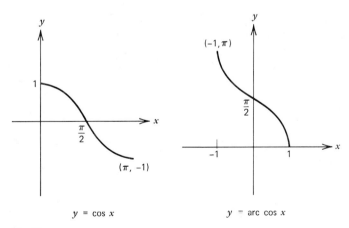

$$y = \cos x \qquad\qquad y = \text{arc cos } x$$

**Fig. 30**

Again by the fundamental theorem of the calculus

$$\text{arc cos } x = -\int_0^x \frac{dt}{\sqrt{1-t^2}} + C.$$

But since arc cos $0 = \pi/2$ (Why?), $C = \pi/2$ and we have

(7) $$\text{arc cos } x = \frac{\pi}{2} - \int_0^x \frac{dt}{\sqrt{1-t^2}}, -1 \leq x \leq 1.$$

If $s = $ arc cos $x$, then $s$ is the number between 0 and $\pi$ satisfying

$$\cos s = x.$$

Thus on the unit circle $s$ is the length of arc $PQ$ (Figure 31).

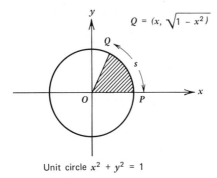

Unit circle $x^2 + y^2 = 1$

**Fig. 31**  If $Q$ is in the upper half plane,
$s = $ arc cos $x = $ length $\overset{\frown}{PQ}$.

If $A(x)$ is the area of the sector $POQ$, then we know from the discussion in Section 4.6 that

$$2A(x) = s = \text{arc cos } x.$$

Moreover, we have seen formula (7) before in the form

$$A(x) = \frac{\pi}{4} - \frac{1}{2} \int_0^x \frac{dt}{\sqrt{1-t^2}}.$$

(See Exercise 14, p. 167.)

If we restrict the cosine function to intervals $[\pi, 2\pi]$, $[2\pi, 3\pi]$, ... etc., in place of the interval $[0, \pi]$, then the cosine is $1:1$ and inverses may be determined by a procedure analagous to that used for sin $x$. We leave this to the exercises.

If we restrict the tangent function to the interval $-\pi/2 < x < \pi/2$, then tan $x$ is $1:1$ and an inverse $F(y) = $ arc tan $y$ can be defined. The sketch of the graph is Figure 32.
Using the same procedure as for the sine and cosine we may derive the formulas

(8) $$\frac{d}{dx} (\text{arc tan } x) = \frac{1}{1+x^2}$$

and

$$\text{arc tan } x = \int_0^x \frac{dt}{1+t^2}.$$

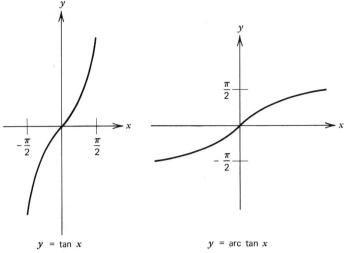

$y = \tan x$

$y = \text{arc } \tan x$

**Fig. 32**

The inverses for the remaining trigonometric functions are seldom encountered. Indeed, there is no general agreement as to which interval to take as the domain so that these functions will be $1:1$. However, if we select intervals on which the functions $\cot x$, $\sec x$, $\csc x$ are $1:1$, then we may construct inverses $F$, $G$, $H$ respectively for these three functions. Using the chain rule we may derive formulas for the derivatives of these three functions in exactly the same fashion as for $\sin x$ and $\cos x$. Indeed, if

$$x = \cot F(x), \qquad \text{then } F'(x) = -\frac{1}{1+x^2}.$$

If

$$x = \csc G(x), \qquad \text{then } G'(x) = \frac{\pm 1}{x\sqrt{x^2-1}}.$$

If

$$x = \sec H(x), \qquad \text{then } H'(x) = \frac{\pm 1}{x\sqrt{x^2-1}}.$$

The derivation of these results and the determination of the appropriate signs in the latter two cases are left to the exercises.

In terms of indefinite integrals, formulas (4), (6), and (8) become

$$(9) \qquad \int \frac{dx}{\sqrt{1-x^2}} = \text{arc } \sin x + C$$

$$(10) \qquad -\int \frac{dx}{\sqrt{1-x^2}} = \text{arc } \cos x + C$$

$$(11) \qquad \int \frac{dx}{1+x^2} = \text{arc } \tan x + C.$$

***Example 1.*** Compute $d[\text{arc } \cos (x+1)/(x-1)]/dx$. Applying the chain rule and formula (6) we have

$$\frac{d}{dx} \text{arc } \cos \left(\frac{x+1}{x-1}\right) = \frac{-1}{\sqrt{1-[(x+1)/(x-1)]^2}} \frac{d}{dx}\left(\frac{x+1}{x-1}\right)$$

$$= \frac{1-x}{2\sqrt{x}} \cdot \frac{(x-1)-(x+1)}{(x-1)^2} = \frac{1}{\sqrt{x}(x-1)}.$$

**Example 2.**   Compute $\int x/(1+x^4)dx$. If we compose arc tan $x$ with a function $f(x)$, then by the chain rule

$$\frac{d}{dx}(\text{arc tan} f(x)) = \frac{1}{1+f^2(x)} f'(x).$$

Hence

$$\int \frac{f'(x)dx}{1+f^2(x)} = \text{arc tan} f(x) + C.$$

In this example if we let $f(x) = x^2$, then $f'(x) = 2x$ and

$$\int \frac{2xdx}{1+x^4} = \text{arc tan} x^2 + C.$$

Hence

$$\int \frac{xdx}{1+x^4} = \frac{1}{2}\text{arc tan} x^2 + C.$$

### EXERCISES

1. Compute

   (a) $\text{arc cos}(0) - \text{arc sin}\left(\frac{\sqrt{2}}{2}\right)$

   (b) $\dfrac{\text{arc sin}(-1) + \text{arc cos}\left(\frac{1}{2}\right)}{\text{arc tan}(1)}$

   (c) $\text{arc sin}\left(-\frac{1}{2}\right) + \text{arc tan}\left(\frac{1}{\sqrt{3}}\right)$

2. Compute

   (a) $\lim\limits_{x\to 0} \dfrac{\sin x}{\text{arc sin}(\sin x/2)}$

   (b) $\lim\limits_{x\to 0} \dfrac{1-\cos x}{\text{arc cos}(\cos 4x)}$

3. Find $y'$ if

   (a) $y = \text{arc sin}\left(\dfrac{x-3}{2}\right)$

   (b) $y = \text{arc cos}(1+\sin 3x)$

   (c) $y = \text{arc tan}\left(\dfrac{x+1}{e^x}\right)$

   (d) $y = \cos(3+\text{arc tan} x)$

   (e) $y = \log(x-\text{arc sin} x)$

   (f) $y = \text{arc sin}(\sin x)$

   (g) $y = \text{arc cos}(1-e^{\sin x})$

4. Find $y'$ if $y = f(x)$ satisfies

   (a) $\text{arc tan}(x+y) = e^{x^2}$

   (b) $\text{arc sin}(x-y) = \tan(x+y)$

   (c) $\text{arc cos}(x+\sin y) = \log y$

   (d) $\text{arc tan}\dfrac{x}{y} + \log \sqrt{x^2+y^2} = 0$

5. If $y = \sin(a \text{ arc sin} x)$ show that

$$(1-x^2)y'' - xy' + a^2y = 0.$$

6. Show that $d(\text{arc tan} x)/dx = 1/(1+x^2)$. Show further that

$$\text{arc tan} x = \int_0^x 1/(1+t^2)dt.$$

   Hence, evaluate $\int_0^1 1/(1+t^2)dt$.

7. Restrict tan $x$ to the interval $\pi/2 < x < 3\pi/2$. Let $F$ be an inverse function for tan $x$ for $x$ in this interval. Sketch the graph of $F$ and determine a constant $C$ such that $F(x) = \int_0^x 1/(1+t^2)dt + C$.

8. Compute the following indefinite integrals.

(a) $\displaystyle\int \frac{dt}{\sqrt{1-4x^2}}$

(d) $\displaystyle\int \frac{e^x dx}{\sqrt{1-e^{2x}}}$

(b) $\displaystyle\int \frac{-xdx}{\sqrt{1-x^4}}$

(e) $\displaystyle\int \frac{dx}{x\sqrt{1-(\log x)^2}}$

(c) $\displaystyle\int \frac{dx}{a^2+x^2}, a \neq 0$

(f) $\displaystyle\int \frac{\cos x\, dx}{1+\sin^2 x}$

9. Let $\cos x$ be $1:1$ on some interval $I$. Let $F$ be an inverse function for $\cos x$ on this interval. Thus $F(\cos x) = x$ for $x \in I$ and $\cos (F(y)) = y$, $-1 \leqslant y \leqslant 1$. Show that $F$ is a primitive for either $1/\sqrt{1-x^2}$ or for $-1/\sqrt{1-x^2}$. Thus

$$F(x) = \pm \int 1/\sqrt{1-x^2}\ dx + C.$$

What are the possible values for $C$?

10. Determine an inverse function $F(y)$ for $y = \cot x$ if $0 < x < \pi$. Sketch the graph of this function, and write $F$ as an integral.

11. Let $F$ be the inverse function for $y = \csc x$ if $0 < x < \pi/2$. Sketch the graph of $F$ and determine $F'$.

12. Let $F$ be the inverse function for $y = \sec x$ if $0 \leqslant x < \pi/2$. Sketch the graph of $F$ and determine $F'$.

13. Let $F$ be the inverse function for $y = \sec x$ if $-\pi/2 < x \leqslant 0$. Sketch the graph of $F$ and determine $F'$.

## *4.9 Hyperbolic Functions

Certain combinations of exponential functions occur repeatedly in applications, especially in engineering. These are the so-called hyperbolic functions which have properties reminiscent of the trigonometric functions. They are defined as follows.

(1)
$$\sinh x = \frac{e^x - e^{-x}}{2}$$

(2)
$$\cosh x = \frac{e^x + e^{-x}}{2}$$

(3)
$$\tanh x = \frac{\sinh x}{\cosh x} = \frac{e^x - e^{-x}}{e^x + e^{-x}}$$

(4)
$$\coth x = \frac{\cosh x}{\sinh x}$$

(5)
$$\operatorname{csch} x = \frac{1}{\sinh x}$$

(6)
$$\operatorname{sech} x = \frac{1}{\cosh x}$$

The function $\sinh x$ is read "hyperbolic sine of $x$." The function $\cosh x$ is read "hyperbolic cosine of $x$," etc. The computation of the derivatives of the first two functions is immediate. Indeed,

$$\frac{d}{dx}(\sinh x) = \frac{d}{dx}\left(\frac{e^x - e^{-x}}{2}\right) = \frac{e^x + e^{-x}}{2} = \cosh x,$$

and

$$\frac{d}{dx}(\cosh x) = \frac{d}{dx}\left(\frac{e^x + e^{-x}}{2}\right) = \frac{e^x - e^{-x}}{2} = \sinh x.$$

The remaining formulas for the derivatives are left as exercises. Numerous identities are satisfied by these functions. For example,

$$[\cosh x]^2 - [\sinh x]^2 = 1$$

since

$$\frac{[e^x + e^{-x}]^2}{4} - \frac{[e^x - e^{-x}]^2}{4} = \frac{e^{2x} + 2 + e^{-2x} - [e^{2x} - 2 + e^{-2x}]}{4} = 1.$$

It immediately follows from the definitions that sinh $x$, and tanh $x$ are odd functions. That is,

$$\sinh (-x) = -\sinh x$$

$$\tanh (-x) = -\tanh x$$

and

$$\cosh x \text{ is an even function.}$$

That is,

$$\cosh (-x) = \cosh x.$$

Sketches of the graphs of these three functions are in Figure 33. The verification of further identities satisfied by these functions is left to the exercises. The connection between the hyperbolic functions and the hyperbola $x^2 - y^2 = 1$ is discussed in Exercise 7.

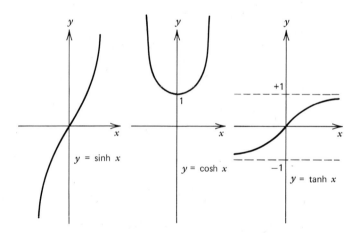

**Fig. 33**   The hyperbolic functions.

**EXERCISES**

1. Verify the following identities for the hyperbolic functions.

   (a) $1 - \tanh^2 x = \operatorname{sech}^2 x$

   (b) $\coth^2 x - 1 = \operatorname{csch}^2 x$

   (c) $\sinh^2 x/2 = \frac{1}{2}(\cosh x - 1)$

   (d) $\cosh^2 x/2 = \frac{1}{2}(\cosh x + 1)$

   (e) $\sinh (x + y) = \sinh x \cosh y$
   $+ \cosh x \sinh y$

   (f) $\cosh (x + y) = \cosh x \cosh y$
   $+ \sinh x \sinh y$

   (g) $\cosh x + \sinh x = e^x$

   (h) $(\cosh x + \sinh x)^n = \cosh nx + \sinh nx$

2. Show that

   (a) $d(\tanh x)/dx = \operatorname{sech}^2 x$

   (b) $d(\coth x)/dx = -\operatorname{csch}^2 x$

   (c) $d(\operatorname{csch} x)/dx = -\operatorname{csch} x \coth x$

   (d) $d(\operatorname{sech} x)/dx = -\operatorname{sech} x \tanh x$

3. The functions sinh $x$ and tanh $x$ are $1:1$ for all values of $x$. Hence, each of these

functions has an inverse, which we write $\sinh^{-1} x$ and $\tanh^{-1} x$. Sketch the graphs of these functions and show that

$$\frac{d}{dx}(\sinh^{-1} x) = \frac{1}{\sqrt{1+x^2}}$$

and

$$\frac{d}{dx}(\tanh^{-1} x) = \frac{1}{1-x^2}.$$

(Recall that if $F$ is an inverse for $f$ then $x = f(F(x))$. Differentiate this equation and apply the chain rule.)

4. Show that

   (a) $\sinh^{-1} x = \displaystyle\int_0^x \frac{dt}{\sqrt{1+t^2}}$

   and

   (b) $\tanh^{-1} x = \displaystyle\int_0^x \frac{dt}{1-t^2}.$

5. The function $f(x) = \cosh x$ is $1:1$ if $x \geqslant 0$ and is $1:1$ if $x \leqslant 0$. Let $F$ be an inverse function for $f$ if $x \geqslant 0$ and let $G$ be an inverse function for $f$ if $x \leqslant 0$. Sketch the graphs of $F$ and $G$ and compute $F'$ and $G'$. Write both $F$ and $G$ as definite integrals.

6. It can be shown that the function $y = a \cosh(x/a)$ describes the curve formed by a flexible cord or chain suspended from two points. This curve is called a catenary. Determine the area of the ordinate set for a catenary when $-a \leqslant x \leqslant a$.

*7. The relation between the hyperbolic sine, the hyperbolic cosine, and the hyperbola $x^2 - y^2 = 1$ is derived as follows. First sketch the portion of the graph of $x^2 - y^2 = 1$ that lies in the right half plane (Figure 34). If $\theta$ is $< POQ$, then we define

$$\sinh \theta = y$$
$$\cosh \theta = x.$$

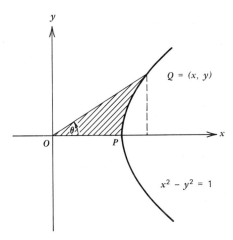

**Fig. 34**

These "geometric" hyperbolic functions have as their domains sets of angles, not sets of numbers. To show that these in reality are the same functions as those defined on p. 181, we must define a "hyperbolic measure" for the angle $\theta$ in the same way as radian measure of an angle was defined for the tirgonometric functions. We define this "hyperbolic measure" $m_\theta$ of the angle $\theta$ by the formula

$$m_\theta = 2 \cdot \text{area } POQ$$

where $POQ$ denotes the shaded region in Figure 34. We must then prove that

$$y = \sinh m_\theta$$

and

$$x = \cosh m_\theta.$$

We shall treat the case of the function $\cosh \theta$ first. Regard the area of $POQ$ as a function of $x$ where $Q = (x, y)$ and denote this function by $A(x)$. Develop a formula for $A(x)$ and show that $A'(x) = \frac{1}{2}(1/\sqrt{x^2 - 1})$.
From this, deduce that

$$2A(x) = \cosh^{-1} x,$$

and consequently

$$x = \cosh 2A(x) = \cosh m_\theta$$

if the point $Q = (x, y)$ is in the first quadrant. To derive the formula

$$y = \sinh m_\theta$$

follow the same procedure except that now the area $A$ of the sector $POQ$ should be thought of as a function of the $y$ coordinate of $Q$. Show then that

$$A'(y) = \frac{1}{2}\frac{1}{\sqrt{1 + y^2}}.$$

Hence

$$2A(y) = \sinh^{-1} y,$$

and

$$y = \sinh(2A(y)) = \sinh m_\theta.$$

# CHAPTER FIVE

# CONTINUOUS FUNCTIONS AND THEIR PROPERTIES

## 5.1 Introduction

In this chapter we shall derive two fundamental properties of continuous functions and discuss in detail the geometry of the graphs of such functions. In addition to deriving qualitative information that can be used to sketch the graphs of a continuous function, we shall obtain as a by-product a very valuable tool that we use to finally give a proof of the "vanishing derivative theorem." We shall be concerned with functions $f$ which are defined on a closed interval $[a, b]$ and which are continuous at each point $x_0 \in [a, b]$. Recall that a function is continuous at a point $x_0$, $a < x_0 < b$, if $\lim_{x \to x_0} f(x) = f(x_0)$. That is, for each $\epsilon > 0$ there is a corresponding $\delta > 0$ such that whenever $|x - x_0| < \delta$ it follows that $|f(x) - f(x_0)| < \epsilon$. We want the same condition to hold at the end points of the interval as well. For this we need the notion of a one-sided limit.

**Definition.** *Let $f$ be defined on the closed interval $[a, b]$; then we say that $f(x) \to L$ as $x$ approaches $b$ from the left if for each $\epsilon > 0$ there exists a corresponding $\delta > 0$ with the property that $|f(x) - L| < \epsilon$ whenever $b - \delta < x < b$. Similarly $f(x) \to M$ as $x$ approaches $a$ from the right if for each $\epsilon > 0$ there exists a corresponding $\delta > 0$ with the property that $|f(x) - M| < \epsilon$ whenever $a < x < a + \delta$. These two statements are abbreviated by*

$$\lim_{x \to b^-} f(x) = L \quad and \quad \lim_{x \to a^+} f(x) = M$$

*respectively.*

**Definition.** *If $f$ is defined on the closed interval $[a, b]$, then $f$ is continuous at $b$ if $\lim_{x \to b^-} f(x) = f(b)$. If $\lim_{x \to a^+} f(x) = f(a)$, then $f$ is said to be continuous at $a$.*

Note that the limit theorems, Theorem 1.10, hold for one-sided limits. The proofs are exactly the same. The fact that we are restricting the values of $x$ to lie to the left of $b$ on one hand and to the right of $a$ on the other hand causes no change in the arguments.

To begin our discussion of the properties of continuous functions we need a definition, valid for any function $f$ defined on a closed interval $[a, b]$, regardless of whether or not it is continuous.

**Definition.** *Let $f$ be defined on a closed interval $[a, b]$. Then we say $M$*

185

is a maximum *value for f on* [a, b] *if*

(1) $$M \geqslant f(x) \text{ for each } x \text{ in } [a, b]$$

*and*

(2) $$f(x_0) = M \text{ for some } x_0 \text{ in } [a, b]$$

*Similarly, m is a* minimum *value for f on* [a, b] *if*

(1') $$m \leqslant f(x) \text{ for each } x \text{ in } [a, b]$$

*and*

(2') $$f(x_0) = m \text{ for some } x_0 \text{ in } [a, b]$$

*If a function f has maximum and minimum values on* [a, b], *these numbers are called the* extreme values *of f on this interval.*

**Example 1.** For $x \in [-\sqrt{3}, \sqrt{3}]$, let $f(x) = x^3 - 3x$. The function $f$ has both a maximum and a minimum value on $[-\sqrt{3}, \sqrt{3}]$. Indeed, $f$ has a maximum value of 2 at $x = -1$ and a minimum value of $-2$ at $x = 1$ (Figure 1).

**Example 2.** For $x \in [-1, 1]$, let $f(x) = |x|$. This function, too, has both a maximum and a minimum value on its interval of definition. The number $1 = f(1) = f(-1)$ is the maximum value, and $0 = f(0)$ is the minimum (Figure 2).

**Example 3.** For $x \in [-2, 2]$, let

$$f(x) = x \qquad |x| < 1$$

$$f(x) = 0 \qquad 1 \leqslant |x| \leqslant 2.$$

This function has neither a maximum nor a minimum value on the interval $[-2, 2]$. Clearly, $-1$ should be the minimum value and $+1$ should be the maximum value, but these numbers are not values of the function $f$ (Figure 3).

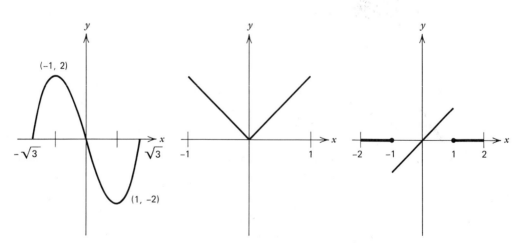

**Fig. 1**   Example 1: $f(x) = x^3 - 3x$.       **Fig. 2**   Example 2: $f(x) = |x|$.       **Fig. 3**   Example 3: $f(x) = 0, |x| \geqslant 1,$
$$= x, |x| < 1.$$

The two properties of functions that we wish to discuss are called the *extreme value property* and the *intermediate value property.*

**Definition.**   *A function f defined on the interval* [a, b] *has the* extreme value property *on* [*a*, *b*] *if f has both a maximum and a minimum value on* [*a*, *b*].

**Definition.**   *A function f defined on the interval* [*a*, *b*] *has the* intermediate value property *on* [*a*, *b*] *if, for each real number y between f*(*a*) *and f*(*b*), *there is at least one number x in the interval* [*a*, *b*] *for which f*(*x*) = *y*.

**THEOREM 5.1**   *If f is* continuous *on the closed interval* [*a*, *b*], *then f has the extreme value property on* [*a*, *b*].

**THEOREM 5.2**   *If f is* continuous *on the closed interval* [*a*, *b*], *then f has the intermediate value property on* [*a*, *b*].

Thus, if $f$ is continuous on the closed interval [*a*, *b*], $f$ takes on both a maximum and a minimum value somewhere on [*a*, *b*]. Furthermore, $f$ will take on each value $y$ between $f(a)$ and $f(b)$ at some point $x$ in [*a*, *b*].

To some extent, these properties seem self-evident. If, in addition to being continuous, $f$ is either decreasing or increasing, then $f$ obviously has the extreme value property and we have proved (Theorem 4.3) that $f$ has the intermediate value property. (See Figure 4.) If $f$ is piecewise monotone (Chapter 3), then both properties are still reasonably evident. However, there are continuous functions for which sketches of the graphs are quite inadequate, for example, functions with infinitely many oscillations close to a given point. For such functions, our geometric intuition is somewhat vague, and more than geometric arguments are necessary to establish Theorems 5.1 and 5.2.

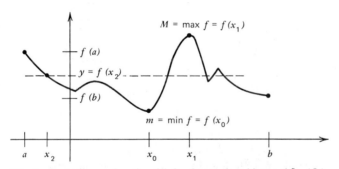

**Fig. 4**   A continuous function $f$ defined on a closed interval [*a*, *b*] has both the extreme value property and the intermediate value property.

Both Theorem 5.1 and Theorem 5.2 assert that there exists a number $x_0$ in the interval [*a*, *b*] with a certain property. In the first case, $f(x_0)$ is an extreme value of the function $f$. In the second, $f(x_0)$ is some number $y$ between $f(a)$ and $f(b)$. To prove these theorems, we must appeal to the completeness axiom for the real numbers to locate the desired point $x_0$. To show then that $f(x_0)$ is an extreme value or an intermediate value, we must use the continuity of $f$. Since the arguments are somewhat technical, they are contained in the appendix to this chapter. The material in the balance of this chapter does not depend on these arguments. The same construction also can be used to prove that every continuous function can be approximated by step functions. Hence, by Theorem 3.15, every continuous function is integrable.

**EXERCISES**

1. Determine which of the following functions defined on the interval $[-3, 3]$ satisfy the maximum value property, the minimum value property, and the intermediate value property. Sketch the graphs.

   (a) $f(x) = |x|$      $0 \leqslant |x| < 1$
   
            $= 1 - |x|$     $1 \leqslant |x| \leqslant 3$

   (b) $f(x) = 1 + x$    $-3 \leqslant x \leqslant 0$
   
            $= -4 - x$    $0 < x \leqslant 3$

   (c) $f(x) = \sin\dfrac{1}{x}$    $0 < |x| \leqslant 3$
   
            $= 0$        $x = 0$

   (d) $f(x) = \dfrac{1}{x+3}$    $-3 < x \leqslant 3$
   
            $= \dfrac{1}{6}$      $x = -3$

2. (a) If a function satisfies the extreme value property must it be continuous? Explain.

   (b) If a function satisfies the intermediate value property, must it be continuous? Explain.

3. (a) Does the extreme value property imply the intermediate value property? Explain.

   (b) Does the converse to (a) hold? Explain.

   For Exercises 4 to 7 you may assume the results of Theorems 5.1, 5.2 if they are applicable.

4. Define

   (a) $f(x) = \dfrac{\sin x}{x}$      $x \neq 0$
   
            $= 1$        $x = 0$

   (b) $f(x) = \dfrac{1 - \cos x}{x}$    $x \neq 0$
   
            $= 0$        $x = 0$

   Do these functions satisfy the extreme value or the intermediate value property on $[-\pi, \pi]$? Explain.

5. If $f(x) = \sin\left(\int_0^x \sqrt{1 + t^4}\, dt\right)$, does $f$ satisfy the extreme value or the intermediate value property on $[-\pi, \pi]$?

6. Suppose a continuous function $f$ is defined on the *open* interval $(a, b)$. Must $f$ satisfy the extreme value property? the intermediate value property? Explain.

7. Show that continuous 1:1 function defined on the interval $[a, b]$ is either increasing on $[a, b]$ or decreasing on $[a, b]$.

8. Show that $\lim_{x \to x_0} f(x)$ exists if the two limits $\lim_{x \to x_0^-} f(x)$, $\lim_{x \to x_0^+} f(x)$ exist and are equal.

## 5.2   Rolle's Theorem and the Mean Value Theorem

Next we investigate the role that the derivative plays in extreme value problems. If the function $f$ is continuous on $[a, b]$ and has a maximum value at an interior point $x_0$ of the interval, ($x_0$ is an interior point if $a < x_0 < b$), then it is geometrically evident that if $f$ has a tangent line at this point, that line is horizontal. A similar statement is true if $f$ has minimum value at an interior point. We incorporate these facts in the following theorem.

**THEOREM 5.3** *Let $f$ be defined on $[a, b]$ and suppose that $f$ has an extreme value at an interior point $x_0$ of $[a, b]$. Then if $f$ is differentiable at*

$x_0$, *it follows that*

$$f'(x_0) = 0.$$

**Proof.**    To prove this result we need one simple consequence of the definition of a limit. Namely, suppose $g$ is a function and suppose that $\lim_{x \to x_0} g(x) = L$ exists. Then if $L \neq 0$, it follows that there is an open interval $I$ containing $x_0$ such that if $x \in I$ and $x \neq x_0$ then $g(x)$ has the same sign as $L$. To see this, suppose $L > 0$, then by definition of a limit, for each $\epsilon > 0$ there is a corresponding $\delta$ so that whenever $0 < |x - x_0| < \delta$ it follows that $|g(x) - L| < \epsilon$. If we take $\epsilon = L/2$, say, and take the corresponding $\delta$, then if $x \in (x_0 - \delta, x_0 + \delta)$ and $x \neq x_0$ it follows that $|g(x) - L| < L/2$. In particular, $g(x) > L/2 > 0$. A similar proof works if $L < 0$.

To apply this fact to prove the theorem, let us suppose for definiteness that $f$ has its maximum value at the point $x_0$. By assumption $a < x_0 < b$. We shall show that it cannot be the case that $f'(x_0) > 0$ or that $f'(x_0) < 0$. Therefore, since $f$ is differentiable at $x = x_0$, it must follow that $f'(x_0) = 0$. To apply the above fact about limits, let $g(x)$ be the difference quotient for $f$. That is

$$g(x) = \frac{f(x) - f(x_0)}{x - x_0} = \frac{f(x_0 + h) - f(x_0)}{h} \qquad \text{when} \qquad x = x_0 + h.$$

Then, since

$$f'(x_0) = \lim_{x \to x_0} \frac{f(x) - f(x_0)}{x - x_0},$$

if $f'(x_0) > 0$, we must have a small interval about $x_0$ in which

(1)
$$\frac{f(x) - f(x_0)}{x - x_0} > 0.$$

But if $x > x_0$ in that interval, then (1) implies that $f(x) > f(x_0)$. (See Figure 5.) This contradicts the fact that $f(x_0)$ was the maximum value of $f$ on the interval $[a, b]$. Hence, it is impossible that $f'(x_0) > 0$. In a similar fashion, we show that $f'(x_0) < 0$ is also impossible. This theorem is illustrated in Figure 6.

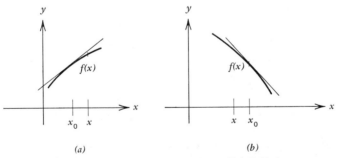

(a)                                      (b)

**Fig. 5**    (a) If $f'(x_0) > 0$, then for $x$ close to $x_0$, $\dfrac{f(x) - f(x_0)}{x - x_0} > 0$, hence $f(x_0)$ is not a maximum value for $f$.    (b) If $f'(x_0) < 0$, then for $x$ close to $x_0$, $\dfrac{f(x) - f(x_0)}{x - x_0} < 0$, hence $f(x_0)$ is not a maximum value.

Thus, for a continuous function $f$ defined on a closed interval $[a, b]$, the points at which $f$ takes on its extreme values belong to one of the following three categories:

1.  end points of $[a, b]$

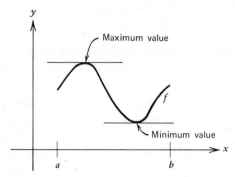

**Fig. 6**   The derivative of $f$ vanishes at interior extreme points.

2. points for which $f'(x) = 0$
3. points for which $f'(x)$ does not exist

Points satisfying (2) are called *stationary points* for $f$. Points satisfying (2) or (3) are called *critical points* for $f$. The points at which $f$ takes on an extreme value are called the *extreme points* for $f$.

**Example 1.**   Determine the extreme points for $f(x) = x^3 - x$ on the interval $[-1, 2]$. Since $f'(x) = 3x^2 - 1, f'(x) = 0$ if $x = \pm 1/\sqrt{3}$. Now

$$f\left(\frac{1}{\sqrt{3}}\right) = -\frac{2}{3\sqrt{3}} \quad \text{and} \quad f\left(-\frac{1}{\sqrt{3}}\right) = \frac{2}{3\sqrt{3}}.$$

Evaluating $f$ at the end points of the interval we have $f(-1) = 0$ and $f(2) = 6$. Therefore, $x = 2$, an end point, is the maximum point for $f$, and $x = 1/\sqrt{3}$ is the minimum point (Figure 7). The extreme values of $f$ on $[-1, 2]$ are 6 and $-2/3\sqrt{3}$.

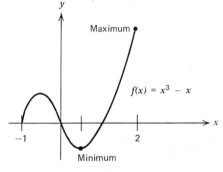

**Fig. 7**

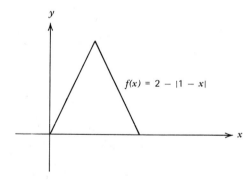

**Fig. 8**

**Example 2.**   Determine the extreme points of $f(x) = 2 - |1 - x|$ on the interval $[0, 2]$. Clearly, $f'(x) = \pm 1$ at each point $x$ where the derivative exists. Since $f'$ never vanishes, the extreme points must either be end points of the interval or the point $x = 1$ where $f'$ does not exist. If we compute the values of $f$ at these points we have $f(0) = 1, f(1) = 2$, and $f(2) = 1$. Therefore, $x = 1$ is the maximum point for $f$ and $x = 0, 2$ are the minimum points. The graph of $f$ is sketched in Figure 8.

Theorems 5.1 and 5.3 may be combined to yield the following result, first discovered by the French mathematician, Michel Rolle, in 1690.

**THEOREM 5.4 (Rolle's Theorem).**   *Let $f$ be continuous on the interval $[a, b]$ and assume that $f$ is differentiable at each interior point of this interval.*

*If, in addition, $f(a) = f(b)$, then at some point $x_0$, $a < x_0 < b$, it must follow that*

$$f'(x_0) = 0.$$

**Proof.** If $f$ is constant on $[a, b]$, then $f' = 0$, and the theorem certainly holds. So assume $f$ is not constant on the interval $[a, b]$. By Theorem 5.1, $f$ has both a maximum and a minimum value on $[a, b]$, since $f$ is continuous. However, since $f$ is not constant, one of these extremes must occur at an interior point $x_0$. Applying Theorem 5.3, we infer that $f'(x_0) = 0$ since $f$ is differentiable at each interior point.

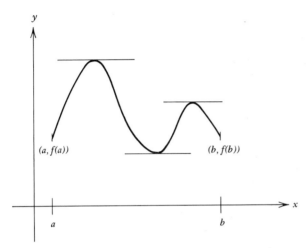

**Fig. 9**  If $f(a) = f(b)$, then $f'(x_0) = 0$ at some point $x_0$, $a < x_0 < b$.

Geometrically, Rolle's theorem asserts that if $f$ satisfies the hypotheses of Theorem 5.4, then at some point $x_0$, $a < x_0 < b$, the tangent line to the graph of $f$ is horizontal. That this fact should hold can be seen in Figure 9.

Suppose now that $f(a) \neq f(b)$. Then $f'$ need not vanish at any point, but if we draw a sketch analogous to Figure 9, it is apparent that at some interior point $x_0$, the tangent to the curve is parallel to the chord joining the two points $(a, f(a))$ and $(b, f(b))$. The slope of the tangent line is $f'(x_0)$. On the other hand, the slope of the chord is

$$\frac{f(b) - f(a)}{b - a}.$$

(See Figure 10.) Therefore, it appears that at some interior point $x_0$ of the interval $[a, b]$, the following formula holds:

$$f'(x_0) = \frac{f(b) - f(a)}{b - a}.$$

We shall now prove this result, which is called the mean value theorem of the differential calculus or just the mean value theorem.

**THEOREM 5.5 (Mean Value Theorem).** *Let $f$ be continuous on the interval $[a, b]$ and be differentiable at each interior point. Then there exists a point $x_0$, $a < x_0 < b$, such that*

$$f'(x_0) = \frac{f(b) - f(a)}{b - a}.$$

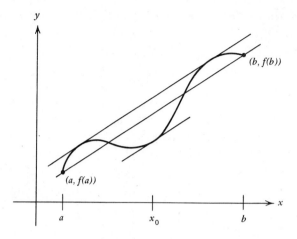

**Fig. 10**   Slope of tangent line $= f'(x_0) =$ slope of chord $= \dfrac{f(b) - f(a)}{b - a}$.

*Proof.*   To establish this formula, we must make an appeal to Rolle's theorem. We first must construct a function $g$ to which Rolle's theorem applies. To do this, let $l(x)$ be the straight line passing through $(a, f(a))$ and $(b, f(b))$ and let $g(x) = f(x) - l(x)$. (See Figure 11.) Certainly $g(a) = g(b) = 0$. In addition, $g$ is continuous on $[a, b]$ and differentiable at interior points since both $f$ and $l$ have that property. Applying Theorem 5.4 to $g$, we see that at some interior point $x_0$ of $[a, b]$

$$0 = g'(x_0) = f'(x_0) - l'(x_0).$$

But at all points $x$, $l'(x) = (f(b) - f(a)/b - a)$, since $l$ is the straight line passing through $(a, f(a))$ and $(b, f(b))$. Therefore, at the interior point $x_0$,

$$f'(x_0) = \frac{f(b) - f(a)}{b - a}$$

and we are finished.

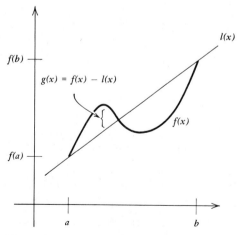

**Fig. 11**

The number $(f(b)-f(a))/(b-a)$ is called the *mean value of $f'$ on the interval* $[a, b]$. Thus, if

$$f'(x_0) = \frac{f(b)-f(a)}{b-a},$$

then $x_0$ is a point at which $f'$ assumes its mean value. If the function $f$ measures the displacement of a particle as a function of time and $f'$ measures the velocity, then the mean value theorem asserts that at some time $t_0$ the velocity of the particle equals the average velocity of the particle on the interval $[a, b]$.

**Example 3.**   If $f(x) = 3x^2 - x + 1$, find the point $x_0$ at which $f'$ assumes its mean value on the interval $[2, 4]$. Now

$$\frac{f(4)-f(2)}{4-2} = \frac{45-11}{2} = 17,$$

and $f'(x) = 6x - 1$. Therefore, setting $6x_0 - 1 = 17$, it follows that $x_0 = 3$, and this is the desired point.

The mean value theorem is an extremely important theoretical tool in the calculus. As a first illustration of this we use the result to prove the vanishing derivative theorem.

**THEOREM 5.6 (Vanishing Derivative Theorem).** *Suppose* $f'(x) = 0$ *for all points $x$ of the interval* $[a, b]$. *Then $f$ is constant on this interval.*

**Proof.**   Let $x < x < b$. We apply the mean value theorem to the function $f$ on the interval $[a, x]$. This implies that at some point $x_0$, $a < x_0 < x$,

$$f'(x_0) = \frac{f(x)-f(a)}{x-a}.$$

But since $f'(x_0) = 0$, it follows that $f(x) = f(a)$ at each point $x$. Hence, $f$ is constant.

The mean value theorem may be used to establish inequalities.

**Example 4.**   Show that $e^x \geq 1 + x$ for all real numbers $x$. For $x > 0$, we apply the mean value theorem to $f(x) = e^x$ on the interval $[0, x]$. Since $f'(x) = e^x$, we have

$$\frac{f(x)-f(0)}{x-0} = \frac{e^x-1}{x} = e^{x_0}$$

for some point $x_0$, $0 < x_0 < x$. But $x_0 > 0$ implies $e^{x_0} > 1$. Hence, $(e^x - 1)/x > 1$, or $e^x > x + 1$. Equality occurs if and only if $x = 0$. The argument for $x < 0$ proceeds similarly and is left as an exercise.

**Example 5.**   Show that

(2)
$$4 + \frac{2}{17} < \sqrt{17} < 4 + \frac{1}{8}.$$

To obtain this estimate, we apply the mean value theorem to $f(x) = \sqrt{x}$ on the interval $[16, 17]$. Then for some $x_0$, $16 < x_0 < 17$

(3)
$$\frac{1}{2}\frac{1}{\sqrt{x_0}} = \frac{\sqrt{17}-\sqrt{16}}{1} = \sqrt{17}-4.$$

We now must obtain two estimates on the size of $1/\sqrt{x_0}$. Since $16 < x_0 < 17$, $\frac{1}{16} > 1/x_0 > \frac{1}{17}$. Therefore $1/\sqrt{x_0} < 1/\sqrt{16} = \frac{1}{4}$. Combining this with (3), we infer that $\sqrt{17} = 4 + \frac{1}{2}\sqrt{x_0} < 4 + \frac{1}{8}$. To obtain the left-hand inequality in (2), observe that $x_0 > 16$ implies $\sqrt{x_0} > 4$. Hence, dividing this inequality by $x_0$, we have $1/\sqrt{x_0} > 4/x_0 > \frac{4}{17}$ since $1/x_0 > \frac{1}{17}$. Combining this with (3) we obtain

$$\sqrt{17} = 4 + \frac{1}{2}\frac{1}{\sqrt{x_0}} > 4 + \frac{1}{2} \cdot \frac{4}{17} = 4 + \frac{2}{17}$$

and we have both our estimates.

## EXERCISES

1. Determine the extreme values for the following functions on the specified intervals. Determine the points at which these extreme values are taken on.

   (a) $f(x) = 3x - 4$          $[1, 3]$
   (b) $f(x) = 2x^2 - 4x + 5$      $[0, 2]$
   (c) $f(x) = -x^2 + 6x + 1$     $[1, 4]$
   (d) $f(x) = x^2 - 8x + 4$      $[0, 3]$
   (e) $f(x) = 2 - |1 - x|$        $[0, 3]$
   (f) $f(x) = |2x + 1| - |x + 2|$    $[-3, 0]$
   (g) $f(x) = 2x^3 - 6x + 1$      $[-2, 2]$
   (h) $f(x) = 8x^3 - 9x^2 + 2$    $[-1, 1]$
   (i) $f(x) = x^4 - 2x^3 - 2x^2 + 1$   $[-2, 3]$
   (j) $f(x) = x^2 + 16/x$        $[1, 3]$

2. For the following functions and intervals determine the points at which the mean value of the derivative is taken on.

   (a) $f(x) = x^2 + 4x + 1$    $[1, 3]$
   (b) $f(x) = x^3 - x^2 + x$    $[0, 4]$
   (c) $f(x) = \sqrt{1 - x^2}$      $[0, 1]$
   (d) $f(x) = \log x$         $[1, t]$

3. Show that if $f(x)$ is a quadratic polynomial defined on the interval $[a, b]$, then $f'$ takes on its mean value at the midpoint of this interval.

4. If $f(x) = 1 - |x|$ on the interval $[-1, 1]$, then $f'(x)$ never vanishes on this interval. Why is this not a contradiction to Rolle's theorem?

5. Show that if $\lim\limits_{x \to x_0} f(x) = L$ and $L < 0$, then there is a small interval $I$ about $x_0$ such that for $x \in I, f(x) < L/2$. In particular, $f(x) < 0$ for $x \in I$.

6. Complete the proof of Theorem 5.3 by showing that $f'(x_0) < 0$ is also impossible.

7. Verify the following inequalities by appropriate applications of the mean value theorem.

   (a) $e^x \geq 1 + x$ all values of $x$

   (b) $\dfrac{x - 1}{x} < \log x < x - 1$ if $x > 1$

   (c) $|\sin x - \sin y| \leq |x - y|$ for all $x$ and $y$

   (d) $\dfrac{x}{1 + x^2} < \arctan x < x$ if $x > 0$

   (e) $|\arctan x - \arctan y| \leq |x - y|$ for all $x$ and $y$

8. Verify the following inequalities by the use of the mean value theorem.

   (a) $5 + \dfrac{5}{52} \leq \sqrt{26} \leq 5 + \dfrac{1}{10}$

   (b) $3 + \dfrac{1}{28} \leq \sqrt[3]{28} \leq 3 + \dfrac{1}{27}$

   (c) $2 + \dfrac{2}{165} \leq \sqrt[5]{33} \leq 2 + \dfrac{1}{80}$

9. Let $f$ be continuous on the interval $[a, b]$. By applying the mean value theorem to $F(x) = \int_0^x f(t)\, dt$ verify that

$$f(c)(b - a) = \int_a^b f(t)\, dt$$

for some point $c$ satisfying $a < c < b$. This formula is called the mean value formula of the integral calculus.

10. Using Exercise 9, show that

(a) $1 \leqslant \int_0^1 \sqrt{1 + x^2}\, dx \leqslant \sqrt{2}$

(b) $\dfrac{1}{2} \geqslant \int_1^2 \dfrac{dx}{1 + x^3} \geqslant \dfrac{1}{9}$

(c) $\dfrac{1}{\sqrt{7}} \leqslant \int_1^3 \dfrac{dx}{\sqrt{1 + x^3}} \leqslant \dfrac{2}{\sqrt{2}}$

11. Let $f(x)$ be a cubic polynomial. If $f$ has three distinct roots, how many roots must $f'$ have? Explain. If $f'$ has two distinct roots, how many roots must $f$ have? Explain.

12. Show that $x^3 - 3x + a$ can have at most one root in the interval $[-1, 1]$ regardless of the value of $a$.

13. Let $f$ be continuous on the interval $[a, b]$. Let $a = x_0 < x_1 < \cdots < x_n = b$ be a partition of $[a, b]$ and set $\Delta x_i = x_i - x_{i-1}$. Show that there exist points $x_i'$ in each interval $(x_{i-1}, x_i)$ such that

$$\int_a^b f(t)\, dt = f(x_1')\, \Delta x_1 + \cdots + f(x_n')\, \Delta x_n.$$

## 5.3 Relative Extremes and the First and Second Derivative Tests

To gain more qualitative information about the behavior of continuous functions it is useful to introduce the notion of relative extreme values.

*Definition. Let the continuous function $f$ be defined on an interval $I$ and let $x_0$ be an interior point. Then $f$ is said to have a relative maximum value at the point $x_0$ if there is an interval $[c, d] \subset I$ such that $c < x_0 < d$ and $f(x_0) \geqslant f(x)$ for all $x$ in $[c, d]$. The function $f$ has a relative minimum at $x_0$ if $f(x_0) \leqslant f(x)$ for all $x$ in some interval $[c, d]$ where $c < x_0 < d$ and $[c, d] \subset I$. Points at which $f$ has a relative maximum or minimum value are called relative extreme points, and the values of $f$ at these points are the relative extreme values. (See Figure 12.) The function $f$ will have a relative maximum at the left-hand end point $a$ of the interval $I$ if there is an interval $[a, c]$ such that*

$$f(a) \geqslant f(x) \text{ for all } x \text{ in } [a, c].$$

*Similarly, $f$ has a relative maximum at the right-hand end point $b$ if $f(b) \geqslant f(x)$ for all $x$ belonging to some interval $[c, b]$. Relative minimum values occurring at the end points $a$, $b$ are defined similarly. A relative maximum value that is the maximum value of $f$ over the entire interval $I$ is called the absolute maximum of $f$. An absolute minimum for $f$ is defined similarly.*

If $f$ has a relative extreme at an interior point $x_0$ and, in addition, $f$ is differentiable there, then Theorem 5.3 applies and we may conclude that $f'(x_0) = 0$. Thus the relative extremes of $f$ which are interior points occur either at points where $f' = 0$ or at points where $f'$ fails to exist, i.e., at the critical points of $f$. Next we wish to develop tests to determine which of these critical points are relative maximum points and which are relative minimum

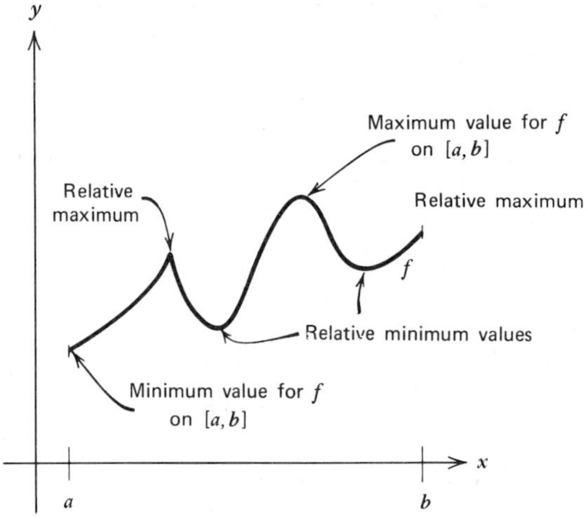

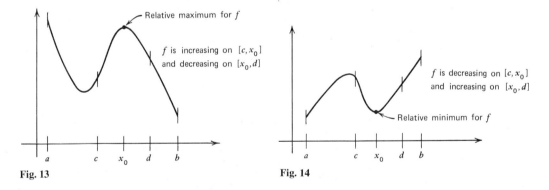

**Fig. 12**

points. These tests are simple consequences of the mean value theorem. They are applicable to end point extremes as well, but for the moment we shall confine our attention to interior points.

Suppose that on some interval $[c, d]$ in the domain of $f$, the function $f$ is increasing on the interval $[c, x_0]$ and decreasing on the interval $[x_0, d]$ where $c < x_0 < d$. (See Figure 13.) Then, clearly, $f$ has a relative maximum at $x_0$. Similarly, if $f$ is decreasing on $[c, x_0]$ and increasing on $[x_0, d]$, then $f$ has a relative minimum at $x_0$. This is illustrated by Figure 14.

**Fig. 13**   **Fig. 14**

Whether a function $f$ is increasing or decreasing on an interval can be determined by examining the sign of the first derivative as the next result shows.

**THEOREM 5.7**   *Let $f$ be continuous on the interval $[a, b]$ and assume that $f$ is differentiable at each interior point. Then $f$ is* increasing *on $[a, b]$ if $f'(x) > 0$ for each $x$, $a < x < b$; and $f$ is* decreasing *on $[a, b]$ if $f'(x) < 0$ for each $x$, $a < x < b$.*

**Proof.**   To show that $f$ is increasing, we must show that if $a \leqslant x_1 < x_2 \leqslant b$, then $f(x_1) < f(x_2)$. But applying the mean value theorem to $f$ on the interval $[x_1, x_2]$ we infer that

$$\frac{f(x_2) - f(x_1)}{x_2 - x_1} = f'(x_0)$$

for some point $x_0, x_1 < x_0 < x_2$. Since $f'(x) > 0$ for all values of $x$ in $(c, d)$ by assumption, we have

$$f(x_2) - f(x_1) = (x_2 - x_1) f'(x_0) > 0,$$

and $f$ is increasing. In a completely analogous fashion (Exercise 4) we establish that if $f'(x) < 0$ for all $x$, $a < x < b$, then $f$ is decreasing on $[a, b]$.

The application of Theorem 5.7 to the problem of determining relative extremes is often called the *first derivative test*. We summarize its use as follows.

(1) Determine the critical points for $f$ (stationary points and points at which $f'$ fails to exist).

(2) A critical point $x_0$ is a relative maximum point if, for some interval $c < x_0 < d$,

$$f'(x) > 0 \quad \text{if} \quad c < x < x_0$$

and

$$f'(x) < 0 \quad \text{if} \quad x_0 < x < d.$$

(3) A critical point $x_0$ is a relative minimum if, for some interval $c < x_0 < d$,

$$f'(x) < 0 \quad \text{if} \quad c < x < x_0$$

and

$$f'(x) < 0 \quad \text{if} \quad c < x < x_0$$

(4) A critical point will be neither a maximum nor a minimum if, for some interval $c < x_0 < d$,

$$f'(x) > 0 \quad \text{for} \quad c < x < d \quad x \neq x_0$$

or

$$f'(x) < 0 \quad \text{for} \quad c < x < d \quad x \neq x_0.$$

If the first condition in (4) holds, then $f$ is increasing on $[c, d]$; and if the second holds, $f$ is decreasing. We abbreviate these conditions by saying that $f$ has a relative maximum at $x_0$ if $f'$ changes sign from $+$ to $-$ at $x_0$ as $x$ moves from left to right, and $f$ has a relative minimum at $x_0$ if $f'$ changes sign from $-$ to $+$. If there is no change in sign of the derivative $f'$ as $x$ moves from the left of $x_0$ to the right of $x_0$, then $f$ does not have a relative extreme at $x_0$. These conditions are summarized in Figure 15.

| | $x < x_0$ | $x > x_0$ | |
|---|---|---|---|
| sign of | $+$ | $-$ | relative maximum |
| $f'$ for | $-$ | $+$ | relative minimum |
| $x$ close | $+$ | $+\rbrace$ | neither relative |
| to the | $-$ | $-$ | maximum nor |
| critical | | | relative minimum |
| point $x_0$ | | | |

**Fig. 15** First derivative test.

Theorem 5.7 may be used very effectively to sketch the graphs of functions. Indeed, often a reasonably accurate sketch of the graph of a function can be drawn if the relative extreme values are known, together with the intervals on which $f$ is either increasing or decreasing. We obtain the latter information by inspecting the sign of $f'$. We illustrate this with several examples.

***Example 1.***    Let $f(x) = 1/x + x$. Determine the relative extreme points for $f$ and the intervals on which $f$ is increasing or decreasing. Sketch the graph for $f$. Computing the derivative for $f$, we obtain

$$f'(x) = -\frac{1}{x^2} + 1 = \frac{x^2 - 1}{x^2} = \frac{(x-1)(x+1)}{x^2}.$$

Clearly, $f'(x)$ vanishes if $x = 1$ or $x = -1$. For $x < -1$ or $x > 1$, $f'(x) > 0$. If $-1 < x < 1$ and $x \neq 0$, then $f'(x) > 0$. Therefore, $f$ has a relative minimum at $x = -1$ and a relative minimum at $x = +1$. $f'$ does not exist at $x = 0$. The relative maximum value of $f$ is $f(-1) = -2$ and the relative minimum value is $f(1) = 2$. Note that $f$ has no absolute extremes. Applying Theorem 5.7, we see that $f$ is increasing if $x \leqslant 1$ or if $x \geqslant 1$. If $-1 \leqslant x < 0$ or $0 < x \leqslant 1$, $f$ is decreasing. Using this information we sketch the graph in Figure 16.

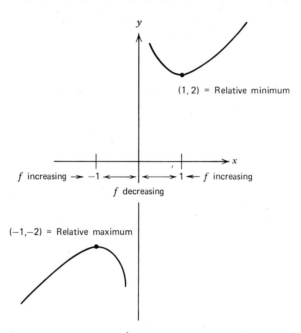

**Fig. 16**   Graph of $f(x) = \dfrac{1}{x} + x$.

***Example 2.***    Determine the relative extremes of $f(x) = \sqrt{|x|}$ and sketch the graph. The function $f$ is continuous everywhere. If $x > 0$ then $f'(x) = 1/2\sqrt{|x|}$; and if $x < 0$, then $f'(x) = -1/2\sqrt{|x|}$. The point $x = 0$ is a critical point since $f'(0)$ does not exist. Since $f'(x) < 0$ if $x < 0$ and $f'(x) > 0$ if $x > 0$, $f$ is decreasing if $x \leqslant 0$ and increasing if $x \geqslant 0$. Therefore, $f$ has a relative minimum at $x = 0$. Indeed, $x = 0$ is an absolute minimum. Does $f$ have an absolute maximum? The graph of $f$ is Figure 17.

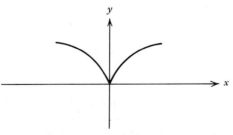

**Fig. 17**   Graph of $f(x) = \sqrt{|x|}$.

*Example 3.* Determine the relative extremes of $f(x) = 3x^4 - 4x^3$ and sketch the graph. Differentiating, we have $f'(x) = 12x^3 - 12x^2 = 12x^2(x-1)$. Therefore, $x = 0, 1$ are the stationary points of $f$. At $x = 1$, $f'$ changes sign from $-$ to $+$. Therefore, $f$ has a relative minimum at $x = 1$. However, at $x = 0$, $f'$ does not change sign. Indeed, for $x < 0$ and $0 < x < 1$, we have $f'(x) < 0$. Therefore, $f$ is decreasing for $x \leqslant 1$ and $x = 0$ is not an extreme point even though $f$ has a horizontal tangent at $x = 0$. For $x \geqslant 1$, $f$ is increasing. The graph of $f$ is Figure 18.

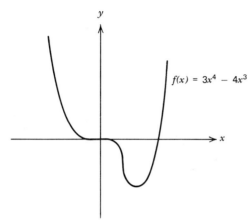

$f(x) = 3x^4 - 4x^3$

**Fig. 18**

The first derivative test applies to end point extremes as well. If on some interval $c < x < b$, $f'(x) > 0$, then $f$ has a relative maximum at the right end point $b$. If on some interval $a < x < c$, $f'(x) < 0$, then $f$ has a relative maximum at $a$. Analagous conditions may be given for end point minima. (See Exercise 7.)

If $x_0$ is a stationary point for $f$, then the sign of the second derivative at $x_0$ can be used to determine if $f$ has a relative maximum or a relative minimum at $x_0$.

**THEOREM 5.8** *Let $x_0$ be a stationary point for $f$ and assume that $f$ is differentiable for each $x$ in some interval $c < x < d$. Then if $f''(x_0) > 0$, it follows that $f$ has a relative minimum at $x_0$. If $f''(x_0) < 0$, then $f$ has a relative maximum at $x_0$. If $f''(x_0) = 0$, no information can be deduced.*

*Proof.* We use the same fact about limits that we used to prove Theorem 5.3. Namely, if $\lim_{x \to x_0} g(x) = L$ and $L \neq 0$, then there is a small interval $I$ about $x_0$ such that if $x \in I$ and $x \neq x_0$, then $g(x)$ has the same sign as $L$. Here we take

$$g(x) = \frac{f'(x) - f'(x_0)}{x - x_0} = \frac{f'(x)}{x - x_0}$$

since, by assumption, $f'(x_0) = 0$. Now, by definition, $\lim_{x \to x_0} g(x) = f''(x_0)$. Therefore, if $f''(x_0) > 0$, there is a small interval $I$ about $x_0$ such that for $x \in I \cap (c, d)$ and $x \neq x_0$, we have

$$\frac{f'(x)}{x - x_0} > 0.$$

But if $x > x_0$, this implies that $f'(x) > 0$, and if $x < x_0$, we have $f'(x) < 0$. Thus, by the first derivative test, we see that $x_0$ is a relative minimum point.

The fact that $f''(x_0) < 0$ implies that $x_0$ is a relative maximum point proceeds similarly and is left as an exercise. The following examples show that nothing can be inferred if $f''(x_0) = 0$.

**Example 4.**    Find the relative extremes of $f(x) = 3x^5 - 5x^3$. Now $f'(x) = 15x^4 - 15x^2 = 15x^2(x^2 - 1)$. Therefore, $x = 0, \pm 1$ are the stationary points for $f$. Computing the second derivative $f''$, we have

$$f''(x) = 60x^3 - 30x = 30x(2x^2 - 1).$$

$f''(1) = 30 > 0$. Therefore, $x = 1$ is a relative minimum point and $x = -1$ is a relative maximum since $f''(-1) = -30 < 0$. Now, $f''(0) = 0$, and to determine the behavior of $f$ at 0 we must use the first derivative test. Since $f'(x) < 0$ if $-1 < x < 1$ and $x \ne 0$, $f$ is decreasing on this interval. Therefore, $x = 0$ is not a relative extreme point. The graph of $f$ is Figure 19.

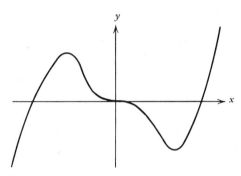

**Fig. 19**

**Example 5.**    If $f(x) = x^4$, $f'(x) = 4x^3$, $f''(x) = 12x^2$. Both $f'$ and $f''$ vanish at $x = 0$, yet $f$ has a minimum at $x = 0$. See Figure 20.

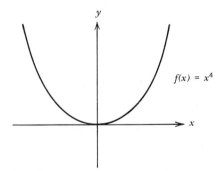

$f(x) = x^4$

**Fig. 20**

Examples 4 and 5 show that if $f''(x_0) = 0$ at a stationary point $x_0$, then $f$ may have a relative extreme value at $x_0$ but on the other hand it may not. For such functions one usually tries to ascertain the behavior of $f$ at $x_0$ by using the first derivative test.

It is somewhat cumbersome to always use the term "relative" extreme. Hence, in the future we shall drop this prefix and only use the terms, "extreme points," or "extreme values," etc. The word relative in this context is always to be understood. The maximum or minimum value of a function over its entire domain of definition will always be called the *absolute maximum or absolute minimum.*

**EXERCISES**

1. Locate the stationary points of the following functions and classify the extreme points using the first derivative test. Determine the intervals on which $f$ is increasing or decreasing. Use this information to sketch the graph of $f$.

(a) $f(x) = 4x^3 - 3x^2 + 2$

(b) $f(x) = 2x^3 - 15x^2 + 24x - 5$

(c) $f(x) = x^4 - 2x^2 + 1$

(d) $f(x) = 2(x-1)^3$

(e) $f(x) = 3x^4 - 16x^3 + 24x^2 + 3$

(f) $f(x) = (x+2)^4$

(g) $f(x) = x^4 - 2x^3 - 2x^2 - 1$

(h) $f(x) = 4(x+1)^7$

(i) $f(x) = x^5 + 5x^4 + 5x^3 + 1$

(j) $f(x) = 15x^7 - 63x^5 + 70x^3$

2. Locate the critical points of the following functions and classify the extreme points using either the first or second derivative test. Determine the intervals on which $f$ is increasing or decreasing and sketch the graph of $f$.

(a) $f(x) = x^2 + (4/x)$

(b) $f(x) = x^2 - (1/x^2)$

(c) $f(x) = x/(4-x^2)$

(d) $f(x) = 1 - |1-x|$

(e) $f(x) = (x-1)x^{1/3}$

(f) $f(x) = xe^{-x}$

3. Determine the relative and absolute extreme points of the following functions defined on the given intervals. Determine the values of the functions at these points as well.

(a) $f(x) = e^x$,   $[-1, 2]$

(b) $f(x) = xe^{-x}$,   $[-1, 2]$

(c) $f(x) = e^{-x^2}$,   $[-1, 1]$

(d) $f(x) = x^2 e^x$,   $[-3, 3]$

(e) $f(x) = \sin x - \cos x$,   $[0, 2\pi]$

(f) $f(x) = \sin^2 x$,   $[0, 2\pi]$

(g) $f(x) = e^x \sin x$,   $[0, 2\pi]$

(h) $f(x) = (\log^2 x/x)$,   $[1, 4]$

4. Complete the proof of Theorem 5.7 by showing that if $f'(x) < 0$ for all $x$ in the interval $x < x < b$, then $f$ is decreasing on this interval.

5. Show that if $x_0$ is a stationary point for $f$ and $f''(x_0) < 0$, then $x_0$ is a relative maximum point (Theorem 5.8).

6. (a) Let $f$ be defined on the interval $[a, b]$. Define the notion of a relative minimum for $f$ at the end points $a$ or $b$.
   (b) State an appropriate first derivative test for such end point minimum values.
   (c) Prove this test.

7. If the function $f$ is increasing on the interval $[a, b]$ and $f'$ exists for each $x$, $a < x < b$, must it follow that $f'(x) > 0$ for each such $x$? Explain.

8. Prove that if $f$ is increasing on $[a, b]$, and if $f$ is differentiable, for each $x$, $a < x < b$, then $f'(x) \geqslant 0$. (Suppose $f'(x) < 0$ and use the same fact about limits that was used to prove Theorem 5.3.)

9. Let $f$ be continuous on $[a, b]$. Suppose $f$ has a relative maximum at the points $x_0$, $x_1$ and assume that $x_0 < x_1$. Show that $f$ must have a relative minimum at some point $x_2$, $x_0 < x_2 < x_1$.

10. Let $f$ be continuous on the open interval $(a, b)$. Assume that $f$ has only one critical point $x_0$ in this interval and assume that this critical point is a relative maximum. Show that $f$ has an absolute maximum at the point $x_0$. (Since $f$ has only one critical point, $f$ is differentiable for all $x \neq x_0$. Now use Rolle's theorem and the intermediate value theorem.)

## 5.4   Applied Extremal Problems

The techniques of the previous section can be used for a variety of applied extremal problems. These problems usually take the following form. "Find the point $x$ at which a given function achieves its maximum or minimum." Although it is not explicitly stated, this usually means that we must find the absolute maximum or minimum for the function $f$. If $f$ is continuous and is defined on a closed interval, then we need only inspect the values of $f$ at the critical points and the end points of the interval to determine the extreme points.

However, most often the function $f$ is defined and continuous on an open interval, for example, the set of all real numbers or the set of all positive real numbers. For such functions the usual procedure is as follows. Determine the critical points for $f$. If there is only one critical point, then this point must be an absolute extreme for the function $f$ if it is a relative extreme (see Exercise 10, p. 000). Whether the point is a relative extreme can be determined by using the first or second derivative test. Often it is obvious on physical or other grounds that the point is a relative maximum or minimum.

If there is more than one critical point, more care must be exercised. First, $f$ may not have any absolute extremes. The relative extremes for $f$ may be determined among the critical points by the first or second derivative test. To determine which are the absolute extremes, one must usually determine the intervals on which $f$ is increasing or decreasing.

Often in physical problems one must exercise some care in constructing the right function to be maximized or minimized. A careful sketch illustrating the problem is usually the best technique for determining the appropriate function.

These matters are best illustrated by some examples.

**Example 1.** Find the positive number $x$ such that the sum of it and its reciprocal is a minimum. Here the function to be minimized is evident. It is

$$S(x) = x + \frac{1}{x}.$$

Differentiating we get $S'(x) = 1 - 1/x^2 = (x^2 - 1)/x^2$. Therefore, $S'(x) = 0$ if and only if $x = \pm 1$. Since we are only interested in positive values of $x$, $x = 1$ is the only stationary point of interest. Clearly, $S''(1) = 2 > 0$. Since $x = 1$ is the only critical point for $f$, the function must achieve its absolute minimum at $x = 1$. The graph for $S$ is sketched in Figure 21.

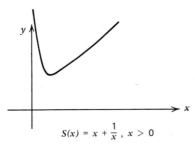

$$S(x) = x + \frac{1}{x}, \; x > 0$$

**Fig. 21**

**Example 2.** Find the dimensions of a cylindrical can of minimal surface area having a fixed volume. If $r$ denotes the radius and $h$ the height, then the surface area $S$ is given by

(1) $$S = 2\pi r^2 + 2\pi hr$$

while the volume

(2) $$V = \pi r^2 h.$$

Solving (2) for $h$ and substituting in (1) yields

$$S(r) = 2\pi r^2 + 2\pi r \cdot \frac{V}{\pi r^2}$$

$$= 2\pi r^2 + \frac{2V}{r}.$$

Differentiating, we get

$$S'(r) = 4\pi r - \frac{2V}{r^2}$$

(3)

$$= \frac{2}{r^2}(2\pi r^3 - V).$$

Therefore

(4)                    $S'(r) = 0$    if    $2\pi r^3 - V = 0.$

Hence, $r_0 = \sqrt[3]{V/2\pi}$ is the desired stationary point for $S$. The function $S$ has a minimum at $r_0 = \sqrt[3]{V/2\pi}$, since from (3) we see that $S'(r) < 0$ for $0 < r < r_0$ and $S'(r) > 0$ when $r > r_0$. The height corresponding to $r_0 = \sqrt[3]{V/2\pi}$ can now be determined from (2).

Actually in this problem the answer is neater if we determine the ratio $h/r$ yielding the minimum area. This is easy to do, for by (2) and (4)

$$V = \pi r^2 h = 2\pi r^3$$

if the surface area is a minimum. This means that

(5)                         $$\frac{h}{r} = 2$$

is the ratio of height to radius for minimum surface area.

Let us examine the mechanics of this problem. By (1) $S$ is expressed as a function of $h$ and $r$. A subsidiary equation (2) is then used to express $h$ as a function of $r$. After $S$ had been expressed as a function of $r$ alone, $S'$ was then computed and the stationary point found. It is important to note that the final ratio (5) can be determined without eliminating $h$ between equations (1) and (2) by exploiting the chain rule. Let us differentiate (1) and (2), treating $h$ as a function of $r$. Then, from (1), we have

$$S' = 4\pi r + 2\pi h + 2\pi r h'$$

and from (2), since $V$ is constant,

$$0 = 2\pi r h + \pi r^2 h' = \pi r(2h + rh').$$

Since $r \neq 0$, we have

(6)                         $2h + rh' = 0.$

Now $S' = 0$ if

(7)                     $2\pi(2r + h + rh') = 0.$

Eliminating the term $rh'$ between (6) and (7) we get

$$h = 2r.$$

This technique is very useful when the initial elimination of variables is difficult or impractical. (See Figure 22.)

**Example 3.**   Find the minimum distance from the point $(3, 0)$ to the parabola

(8)                          $y = x^2.$

If $D$ denotes the distance from $(3, 0)$ to a point $(x, y)$ on the parabola, then

(9)                      $D^2 = (x - 3)^2 + y^2.$

(See Figure 23.)

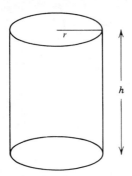

**Fig. 22**   A cylindrical can of fixed volume has minimum surface area if $h = 2r$.

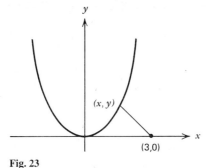

**Fig. 23**

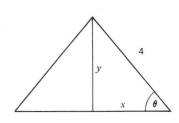

**Fig. 24**

We wish to minimize $D$, but it is more convenient to first differentiate (9) than to solve equation (9) for $D$. We have

$$2DD' = 2(x-3) + 2yy'.$$

Since $y = x^2$, $y' = 2x$. Consequently, $D' = (x-3+2x^3)/D$.

Now $D$ is always positive. Hence, $D' = 0$ if $x-3+2x^3 = 0$. By inspection we see that $x = 1$ is a root of this polynomial. Moreover, $2x^3+x-3 = (x-1)(2x^2+2x+3)$. Since for all values of $x$, $2x^2+2x+3 > 0$ (why?), it follows that, $x = 1$ is the only real root. Therefore, $D'(x) = 0$ if and only if $x = 1$. Moreover, since $D'(x) = ((x-1)(2x^2+2x+3))/D(x)$, $D'(x) < 0$ if $x < 1$ and $D'(x) > 0$ if $x > 1$. Therefore, $D(x)$ has an absolute minimum at $x = 1$. Since $D^2(1) = 4+1 = 5$, the minimum distance is $\sqrt{5}$.

*Example 4.*   What is the maximum area of an isoceles triangle if the two equal sides are 4 inches long. Calculations in this problem are simplified if the area $A$ is expressed as a function of the base angle $\theta$ (Figure 24). Indeed from the figure

$$A = \tfrac{1}{2}y \cdot 2x = 4 \sin \theta \, 4 \cos \theta$$
$$= 16 \sin \theta \cos \theta$$
$$= 8 \sin 2\theta.$$

Therefore, $A' = 16 \cos 2\theta$, and $A' = 0$ if $\cos 2\theta = 0$. Since $0 \leqslant \theta \leqslant \pi/2$, $\cos 2\theta = 0$ if $\theta = \pi/4$. This stationary point is clearly a maximum point for $A$ since $A(0) = A(\pi/2) = 0$. Thus the maximum area is 8 square inches.

*Example 5.*   Find the number $x$ such that its fourth power minus twice its second power plus one is a maximum. It appears that we must find the maximum value of the function

$$f(x) = x^4 - 2x^2 + 1.$$

Differentiating, we obtain $f'(x) = 4(x^3 - x) = 4x(x^2 - 1)$. Clearly, $x = 0, \pm 1$ are the stationary points. Moreover, by inspecting the sign of the first derivative on the intervals $x < -1, -1 < x < 0, 0 < x < 1$, and $x > 1$, it is clear that $x = \pm 1$ are relative minimum points and $x = 0$ is a relative maximum point. One is tempted to say that $f(0) = 1$ is the maximum value for $f$. This is obviously false, however, since $f(2) = 16 - 8 + 1 = 9 > 1$. The difficulty is that the function $f$ does not have a maximum over all values of $x$. Hence the problem as stated is incorrectly posed. This is clear when we graph $f$ (Figure 25).

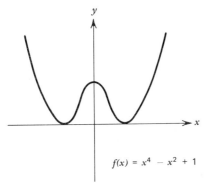

$$f(x) = x^4 - x^2 + 1$$

**Fig. 25**

## EXERCISES

1. Show that among all rectangles with given perimeter, a square has maximum area. Show that among all rectangles with given area, a square has minimum perimeter.

2. The sum of two positive numbers is 10. Find the numbers if (a) their product is a maximum, and (b) the sum of their squares is a minimum.

3. Find the minimum distance from the point $(1, 2)$ to the line $3x + 4y = 5$.

4. Determine the minimum distance from the point $(0, 1)$ to the hyperbola $x^2 - y^2 = 1$.

5. A right circular cylinder is inscribed in a sphere. Show that the lateral surface area will be a maximum if the height of the cylinder is twice the radius.

6. A rectangle is inscribed in a semicircle of radius $r$ with one side lying along a diameter. Determine the dimensions of the rectangle with maximum area.

7. What are the dimensions of the rectangle in Exercise 6 having maximum perimeter. What can you say about the minimum value of the perimeter?

8. A rectangle with base on the $x$ axis has its other two vertices on the parabola $y = 16 - x^2$. What are the dimensions of the rectangle with maximum area?

9. Determine the dimensions of the rectangle in Exercise 8 having maximum perimeter.

10. A farmer has $l$ feet of fencing. He wishes to fence three sides of a rectangular garden, the fourth side lying along a straight river. What should the dimensions be to maximize the area?

11. If the rectangular garden of problem 10 has fixed area $A$, what should the dimensions be to minimize the amount of fencing used to fence the three sides?

12. A man in a boat is 1 mile from a straight shore. He wishes to reach a point 10 miles down the coast. If he can row 1 mile an hour and walk 4 miles an hour, where should he land to reach his destination as quickly as possible?

13. The material used in making tops and bottoms of "pop-top" soft drink cans is twice as expensive as the material used to make the sides. Find the ratio of height to radius for minimum cost.

14. An isoceles triangle is circumscribed about a circle. What are the dimensions for minimum area?

15. An eastbound ship is sailing at a speed of 10 knots (a knot is 1 nautical mile

per hour). At noon the ship is 20 nautical miles north of a northbound ship sailing at 15 knots. Assuming the ships do not change course, when are they closest together, and what is the minimum distance?

16. Two points $P, Q$ on the same side of a straight line are at a distance of $a$ and $b$ units respectively from that line (Figure 26). Show that the sum of the distances from $P$ and $Q$ to the line $l$ will be a minimum if in Figure 26 the angles $\theta$ and $\varphi$ are equal. This is the Fermat principle in optics which states that the angle of incidence of a beam of light equals the angle of reflection. This may be derived by assuming that light reflected by the line $l$ will travel along a path that minimizes the time. Since the velocity of light is constant, this means that the light must travel over a path such that the sum of the distances from $P$ and $Q$ to the line $l$ is a minimum.

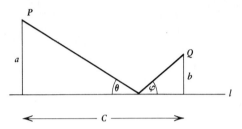

**Fig. 26**

17. Derive Snell's law of refraction which states that if $c_1, c_2$ are the velocities of light traveling through two media $M_1$ and $M_2$, then at the boundary $l$ between the media the light will be refracted according to the formula

$$\frac{\sin \alpha_1}{\sin \alpha_2} = \frac{c_1}{c_2}$$

where $\alpha_1, \alpha_2$ are the angles that the incident and refracted beams of light make with the normal to $l$. (See Figure 27.)
(Use the principle that light traveling from $P$ to $Q$ will follow a path that minimizes the time).

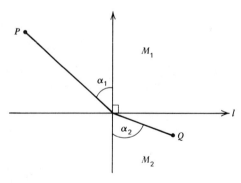

**Fig. 27**

18. Let $a_1, \ldots, a_n$ be $n$ real numbers. The number $(a_1 + \cdots + a_n)/n$ is called the arithmetic mean of the numbers $a_1, \ldots, a_n$. Show that the arithmetic mean of $n$ numbers $a_1, \ldots, a_n$ is the number $x$ that minimizes the sum of the squares of the distances from $x$ to the numbers $a_1, \ldots, a_n$.

19. A 27-foot flagpole is being carried down a corridor 8 feet wide. The corridor makes a right angle turn into a second corridor. How wide must the second

corridor be in order for the pole to go around the corner assuming that it remains horizontal?

20. Does there exist a number $x$ such that it minus its third power is a maximum or a minimum? Explain.

21. A right triangle is formed by the $x$ and $y$ axes and a line through the point $(27, 64)$. Find the minimum length of the hypotenuse of such a triangle.

22. A farmer has $l$ feet of fencing. He wishes to fence a rectangular garden using part or all of one side of a building $b$ feet long. How should he use the fencing to maximize the area of the garden? There are several possibilities depending on the relation between $b$ and $l$.

*23. How should a rectangular sheet of paper be folded so that one corner meets the opposite side and the length of the crease is a minimum?

*24. Let $a_1, \ldots, a_n$ be positive numbers. $M = (a_1 + \cdots + a_n)/n$ is called the arithmetic mean of these numbers and $m = \sqrt[n]{a_1 \cdots a_n}$ the geometric mean. An important inequality asserts that always

$$\sqrt[n]{a_1 \cdots a_n} \le \frac{a_1 + \cdots + a_n}{n}.$$

This inequality is proved by induction. It is obviously true for $n = 1$. To prove that the validity of the inequality for $n = k$ implies it for $n = k+1$, we investigate the behavior of

$$f(x) = \frac{(a_1 + \cdots + a_n + x)/(n+1)}{\sqrt[n+1]{a_1 \cdots a_n \cdot x}} \qquad \text{for } x > 0.$$

Show that $f$ has an absolute minimum at $x = (a_1 + \cdots + a_n)/n$ and the value of this minimum is

$$f\left(\frac{a_1 + \cdots + a_n}{n}\right) = \left(\frac{(a_1 + \cdots + a_n)/n}{\sqrt[n]{a_1 \cdots a_n}}\right)^{n/(n+1)}.$$

Use this fact to complete the proof of the inequality.
[*Hint.* Compute the derivative of $\log f(x)$].

25. Let $y = f(x)$ be a differentiable function and let $P = (x_0, y_0)$ be a point not lying on the graph of $f$. The distance between a point $Q = (x, y)$ on the graph of $f$ and the point $P$ is given by $D(x) = ((x - x_0)^2 + (f(x) - y_0)^2)^{1/2}$. Show that at each point $x$ at which $D(x)$ achieves a relative maximum or relative minimum, the line through $PQ$ is perpendicular to the tangent line to $f$ at $Q$ (Figure 28).

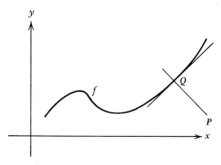

**Fig. 28**

## 5.5 Concavity and the Second Derivative

We can exploit the second derivative in order to gain more qualitative information about the graph of a continuous function. Suppose $f$ is differentiable on some interval $(a, b)$. Let $a < c < b$, and let $l(x)$ be the tangent line to $f$ at the point $(c, f(c))$.

**Definition.** *The graph of f is said to be* concave up *at the point c if there is an open interval I containing c such that for each x $\in$ I, x $\neq$ c,*

$$f(x) > l(x).$$

*If for each x $\in$ I, x $\neq$ c,*

$$f(x) < l(x),$$

*then the graph of f is said to be* concave down *at c.*

Thus the graph of $f$ is concave up at $c$ if the graph lies above the tangent line for $x$ close to $c$. It is concave down if it lies below the tangent line (Figure 29).

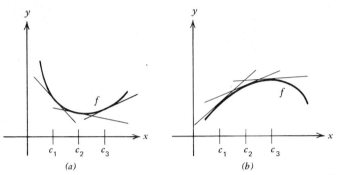

**Fig. 29**  *(a)* The graph of $f$ is concave up at $c_1, c_2, c_3$.  *(b)* The graph of $f$ is concave down at $c_1, c_2, c_3$.

**Definition.** *If the concavity of the graph of f changes at the point c, then c is called an* inflection point *for f.*

When we say that the concavity of the graph of $f$ changes at $c$, we mean one of two things. Either at all points in some interval $(d, c)$ the graph is concave up and at all points in some interval $(c, e)$ the graph is concave down, or just the reverse is true. (See Figure 30.)

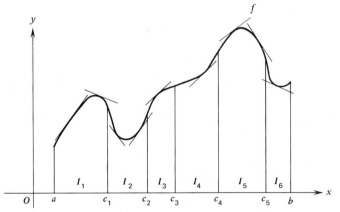

**Fig. 30**  $f$ is concave down on the intervals $I_1, I_3, I_5$ and concave up on $I_2, I_4, I_6$. The points $c_1, c_2, \ldots, c_6$ are all inflection points.

The sign of the second derivative of a function $f$ measures the sense of concavity for $f$. Indeed, if $f''(c) > 0$, then the graph of $f$ is concave up at $c$, and if $f''(c) < 0$ then the graph is concave down. If $c$ is an inflection point and $f''(c)$ exists, then $f''(c) = 0$. We establish these facts in the following theorem.

**THEOREM 5.9** *Let f be differentiable on $(a, b)$ and assume $f''(c)$ exists at the point $c, a < c < b$.*

(1) *If $f''(c) > 0$, the graph of f is concave up at $c$.*

(2) *If $f''(c) < 0$, the graph of f is concave down at $c$.*

(3) *If $c$ is an inflection point, then $f''(c) = 0$.*

**Proof.** All three of these results are consequences of the mean value theorem and simple facts about limits. If $l(x) = f'(c)(x - c) + f(c)$, then the graph of $l$ is the tangent line to $f$ at the point $c$. To show that the graph of $f$ is concave up at $c$, we must show that $f(x) - l(x) > 0$ for all $x \neq c$ in some interval $I$ containing $c$. (See Figure 31.) Now

$$(1) \qquad f(x) - l(x) = f(x) - f(c) - f'(c)(x - c).$$

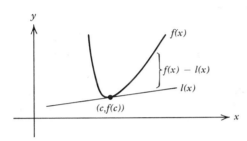

**Fig. 31**

We first apply the mean value theorem to $f$ either on the interval $[c, x]$ if $x > c$ or the interval $[x, c]$ if $x < c$. As a consequence, we claim that for some point $x_1$ between $c$ and $x$

$$f'(x_1) = \frac{f(x) - f(c)}{x - c}$$

or

$$f(x) - f(c) = f'(x_1)(x - c).$$

Substituting this in (1) we get

$$f(x) - l(x) = [f'(x_1) - f'(c)](x - c)$$
$$= \frac{f'(x_1) - f'(c)}{x_1 - c} \cdot (x_1 - c)(x - c).$$

Now if $f''(c) > 0$ (case 1), the difference quotient $(f'(x_1) - f'(c))/(x_1 - c)$ must be positive for all $x_1 \neq c$ in some interval $I$ containing $c$ as an interior point. (Why?) But if $x_1$ is between $c$ and $x$, then $(x - c)(x_1 - c) > 0$. Hence, for each $x$ in $I$ not equal to $c$

$$f(x) - l(x) > 0,$$

and thus the graph is concave up at $c$.

A similar argument shows that $f''(c) < 0$ implies the graph is concave down (Exercise 5). Part (3) of the theorem now follows as a consequence of parts (1) and (2) (Exercise 6).

Notice that a point $c$ at which $f''(c) = 0$ will be an inflection point if the sign of $f''$ changes at $c$. However, the vanishing of $f''(c)$ is not alone sufficient

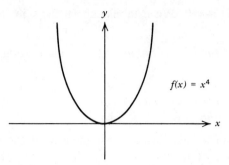

**Fig. 32**  $f$ is concave up at $x = 0$ even though $f''(0) = 0$.

for $c$ to be an inflection point. Indeed if $f(x) = x^4$, then $f''(0) = 0$. However, the graph of $f$ is concave up at $x = 0$. (See Figure 32.)

**Example 1.**   Let $f(x) = x^3 - 6x^2 + 9x$. Determine the inflection points for $f$ and the intervals of concavity of the graph. Differentiating, we get

$$f'(x) = 3x^2 - 12x + 9$$

and

$$f''(x) = 6x - 12 = 6(x - 2).$$

Clearly, $f''(x) > 0$ if $x > 2$, and $f''(x) < 0$ if $x < 2$. Therefore, the graph of $f$ is concave up for $x > 2$ and concave down for $x < 2$. Since the sign of $f''$ changes at $x = 0$, this point is an inflection point for $f$. To sketch the graph of $f$ we need only determine the intervals on which $f$ is increasing or decreasing. However, $f'(x) = 3(x^2 - 4x + 3) = 3(x - 1)(x - 3)$. Therefore, $f'(x) \geqslant 0$ if $x < 1$ or $x > 3$, and $f$ is increasing on these intervals. If $1 < x < 3$, $f'(x) < 0$ and $f$ is decreasing. The point $x = 1$ is a relative maximum for $f$, and $x = 3$ is a relative minimum. Figure 33 is a sketch of the graph.

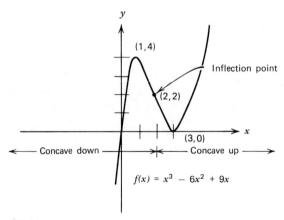

**Fig. 33**

**EXERCISES**

1. For the following functions determine the intervals of concavity and the points of inflection.
   (a) $f(x) = 2x^3 - 3x^2 + 3x + 1$
   (b) $f(x) = 3x^5 - 5x^3 + x - 3$
   (c) $f(x) = 40x^6 + 16x^5 - x^4$

(d) $f(x) = x/(1-x^2)$
(e) $f(x) = e^{-x^2}$
(f) $f(x) = \sqrt{1+x^2}$
(g) $f(x) = \log(1+x^2)$

2. The second derivative of the function $f(x) = x^{1/3}$ never vanishes, yet $f$ has an inflection point at $x = 0$. Why is this not a contradiction of Theorem 5.9?

3. Show that a polynomial of degree $n$ can have, at most, $n-2$ points of inflection.

4. If $f$ is a polynomial show that between any two consecutive stationary points there must exist an inflection point.

5. Prove part (2) of Theorem 5.9.

6. Prove part (3) of Theorem 5.9.

7. Let $f$ be differentiable on the open interval $(a, b)$ and assume the graph of $f$ is concave down on $(a, b)$. Show that $f$ can have, at most, one stationary point in $(a, b)$. Furthermore, show that if $f$ has a stationary point, this point must be an absolute maximum for $f$ on $[a, b]$. (Show that any stationary point for $f$ must be a relative maximum and then apply Exercise 10, p. 201).

### 5.6  Curve Sketching; Infinite Limits; Asymptotes

We have shown that the sign of the first derivative measures whether a function $f$ is increasing or decreasing, and the sign of the second derivative measures the sense of concavity of the graph of the function. We have also seen that both of these facts are extremely useful tools for sketching the graph of $f$. If we take advantage of any symmetry and also compute the intercepts of the graph, i.e., the points where the graph crosses the coordinate axes, then this, together with the information gained from the first and second derivative, is usually sufficient to draw a reasonable sketch of the graph of the function for small values of $x$. Furthermore, it is not necessary to plot large numbers of points. Indeed, just plotting points without knowing the general shape of the graph is usually worthless, since often many of the oscillations of the graph are completely overlooked.

If the function $f$ is defined for all real numbers or for all positive numbers then we have only to describe the behavior of $f$ for large values of $x$ in order to give a complete sketch of the graph. It is convenient to phrase this behavior in terms of limits.

*Definition.  Suppose $f$ is defined for all values of $x$. Then we say $f(x)$ tends to infinity as $x$ tends to infinity if, for each positive number $M$, there is a corresponding positive number $N$ such that whenever $x > N$ it follows that $f(x) > M$. We abbreviate this statement by writing $\lim_{x \to \infty} f(x) = \infty$ or by writing $f(x) \to \infty$ as $x \to \infty$.*

Note that both of these statements are abbreviations for the formal definition. We are not making an assertion that the infinity symbol $\infty$ is a real number.

We write $\lim_{x \to \infty} f(x) = -\infty$ if, for each $M > 0$, there is a corresponding $N > 0$ such that $f(x) < -M$ whenever $x > N$. For negative values of $x$ we have corresponding definitions of $\lim_{x \to -\infty} f(x) = \infty$ and $\lim_{x \to -\infty} f(x) = -\infty$. The function $f(x) = ax^n$ for $n$ a positive integer and $a \neq 0$ is an appropriate example to illustrate these ideas. We assert that $\lim_{x \to \infty} ax^n = \infty$ if $a > 0$ and $\lim_{x \to \infty} ax^n = -\infty$ if $a < 0$. (See Figure 34.) To verify the first of these statements let any positive number $M$ be given. Then we must construct a positive

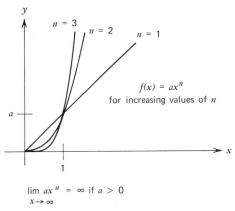

$$\lim_{x \to \infty} ax^n = \infty \text{ if } a > 0$$

**Fig. 34**

number $N$ such that whenever $x > N$ it follows that $ax^n > M$. The choice of $N$ is clear. Indeed $ax^n > M$ if $x > (M/a)^{1/n}$. Hence we let $N = (M/a)^{1/n}$. To verify the second statement is equally easy and we leave it as an exercise. For negative values of $x$, a similar argument shows that

$$\lim_{x \to -\infty} ax^n = \infty \qquad a > 0 \text{ and } n \text{ even}$$

$$\lim_{x \to -\infty} ax^n = -\infty \qquad a < 0 \text{ and } n \text{ even}$$

and

$$\lim_{x \to -\infty} ax^n = -\infty \qquad a > 0 \text{ and } n \text{ odd}$$

$$\lim_{x \to -\infty} ax^n = \infty \qquad a > 0 \text{ and } n \text{ odd}.$$

Now an arbitrary function $f$ defined for large values of $x$ does not necessarily get large as $x$ gets large. For example, consider $f(x) = 1/x$. We therefore need an appropriate definition of a finite limit for $f(x)$ as $x \to \infty$.

**Definition.** *We say* $\lim_{x \to \infty} f(x) = L$, *read limit of* $f(x)$ *equals* $L$ *as* $x$ *tends to* $\infty$, *if, for each positive number* $\epsilon$, *there exists a corresponding positive number* $N$ *such that whenever* $x > N$, *it follows that* $|f(x) - L| < \epsilon$. *We abbreviate this by saying* $f(x)$ *tends to* $L$ *as* $x$ *tends to* $\infty$, *or just* $f(x) \to L$ *as* $x \to \infty$. $\lim_{x \to -\infty} f(x) = L$ *is defined similarly.*

If $f$ and $g$ are two functions with finite limits as $x$ tends to $\infty$, then Theorem 1.10 applies and the limits of $f + g, f \cdot g$, and $f/g$ can be computed in the usual way. Indeed, if $\lim_{x \to \infty} f(x) = L$ and $\lim_{x \to \infty} g(x) = M$, then

$$\lim_{x \to \infty} (f(x) + g(x)) = L + M,$$
$$\lim_{x \to \infty} f((x) \cdot g(x)) = L \cdot M,$$

and

$$\lim_{\to \infty} \frac{f(x)}{g(x)} = \frac{L}{M} \qquad \text{provided that} \qquad M \neq 0.$$

However, if one or both of the functions have infinite limits, then Theorem 1.10 does not apply. We can make certain statements that follow immediately from the definition. Suppose $\lim_{x \to \infty} F(x) = \infty$ and $\lim_{x \to \infty} g(x) = M$. Then

(1)
$$\lim_{x \to \infty} (f(x) + g(x)) = \infty$$

(2)
$$\lim_{x\to\infty} f(x)\cdot g(x) = \infty \qquad \text{if} \qquad M > 0$$
$$= -\infty \qquad \text{if} \qquad M < 0$$

(3)
$$\lim_{x\to\infty}\frac{1}{f(x)} = 0.$$

The proofs of these statements are easy and we leave them as exercises.

Let us use these remarks to describe the behavior of a polynomial for large values of $x$. If $p(x) = a_n x^n + \cdots + a_0, a_n \neq 0$, then we assert that

(4)
$$\lim_{x\to\infty} p(x) = \infty \qquad \text{if} \qquad a_n > 0$$

and

(5)
$$\lim_{x\to\infty} p(x) = -\infty \qquad \text{if} \qquad a_n < 0.$$

To see this, write

$$p(x) = x^n\left(a_n + \frac{a_{n-1}}{x} + \cdots + \frac{a_0}{x^n}\right).$$

By (3) $a_k/x^k \to 0$ as $x \to \infty$. Therefore

$$\lim_{x\to\infty}\left(a_n + \frac{a_{n-1}}{x} + \cdots + \frac{a_0}{x^n}\right) = a_n \neq 0.$$

The desired result now follows from (2). We leave it as an exercise to show that (4) and (5) hold as $x \to -\infty$ provided that $n$ is even. If $n$ is odd, the signs of the limits are reversed.

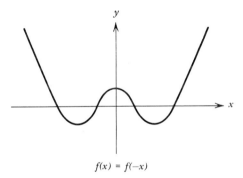

$f(x) = f(-x)$

**Fig. 35**   Graph of $f$ is symmetric about $y$ axis.

Symmetry properties of functions often simplify the problem of drawing an adequate sketch of the graph. If $f$ is even, i.e., $f(x) = f(-x)$, then the graph of $f$ is symmetric about the $y$ axis. (See Figure 35.) Thus, if the graph of $f$ is known for positive values of $x$, the graph for negative values can be obtained by reflecting the positive portion about the $y$ axis. The graph of $f$ is symmetric about the vertical line $x = a$ if

$$f(a+x) = f(a-x)$$

for all values of $x$. Again we may obtain the graph for $x < a$ from that for $x > a$ by reflecting about the line $x = a$.

If $f$ is odd, i.e., $f(x) = -f(x)$, then the graph for $x < 0$ can be obtained by first reflecting the graph for $x > 0$ about the $y$ axis and then reflecting about the $x$ axis. In this case we say that the graph of $f$ is *antisymmetric* about the

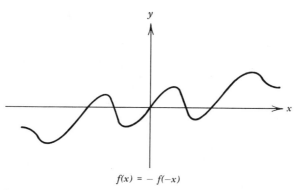

$$f(x) = -f(-x)$$

**Fig. 36**   Graph of $f$ is antisymmetric about $y$ axis.

$y$ axis. (See Figure 36.) A similar procedure can be used if the graph of $f$ is antisymmetric about the line $x = a$, i.e., $f(a+x) = -f(a-x)$ for all values of $x$.

Let us use these ideas to sketch the graph of a polynomial.

***Example 1.***   Sketch the graph of the polynomial $f(x) = 6x^5 - 5x^3 = x^3(6x^2 - 5)$. Clearly, $f(x) = -f(-x)$, so the graph is antisymmetric about the $y$ axis. The intercepts of $f$ are $x = 0, \pm\sqrt{\frac{5}{6}}$. If we compute $f'$ and $f''$, we obtain $f'(x) = 30x^4 - 15x^2 = 15x^2(2x^2 - 1)$ and $f''(x) = 120x^3 - 30x = 30x(4x^2 - 1)$. The derivatives $f'$ vanishes at $x = 0, \pm\sqrt{\frac{1}{2}}$. There is no extreme at $x = 0$ since $f'$ does not change sign there. The zeros of $f''$ are $x = 0, \pm\frac{1}{2}$. Each is an inflection point since $f''$ changes sign at each of the points. The function $f$ is concave up on the interval $x \geqslant \frac{1}{2}$ and down on the interval $0 \leqslant x \leqslant \frac{1}{2}$. By antisymmetry $f$ is concave up on $-\frac{1}{2} \leqslant x \leqslant 0$ and down if $x \leqslant \frac{1}{2}$. The point $(\frac{1}{2}, -1/\sqrt{2})$ is a relative minimum of the graph. Hence, by antisymmetry, $(-\frac{1}{2}, 1/\sqrt{2})$ is a relative maximum. Plotting the intercepts and the relative extremes and using the fact that $\lim_{x \to \infty} f(x) = \infty$ we have the following sketch of the graph (Figure 37).

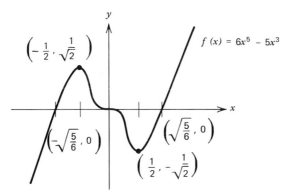

**Fig. 37**

Next we consider the graphs of rational functions, that is, the quotients $p/q$ of two polynomials. It is instructive to consider first the rational functions $f(x) = 1/(x-a)$ and $f(x) = 1/(x-a)^2$. Neither function has a limit as $x \to a$. However, we may describe the behavior of $f$ and $g$ near $x = a$ in terms of infinite limits.

***Definition.***   *Let $f$ be defined on an interval $(a, b)$ except possibly at the point $x_0$. Then $\lim_{x \to x_0} f(x) = \infty$ if, for each positive number $M$, there is a corresponding $\delta > 0$ such that $f(x) > M$ whenever $0 < |x - x_0| < \delta$.*

*If for each M there is a corresponding $\delta > 0$ such that $x_0 < x < x_0 + \delta$ implies $f(x) > M$, then we write $\lim_{x \to x_0^+} f(x) = \infty$. When this is the case we say that $f(x)$ tends to infinity as x approaches $x_0$ from the right. We say $f(x)$ tends to infinity as x approaches $x_0$ from the left, written $\lim_{x \to x_0^-} f(x) = \infty$ if for each M there is a corresponding $\delta > 0$ such that $f(x) > M$ whenever $x_0 - \delta < x < x_0$. The limits $\lim_{x \to x_0} f(x) = -\infty$, $\lim_{x \to x_0^+} f(x) = -\infty$, and $\lim_{x \to x_0^-} f(x) = -\infty$ are defined analogously.*

The rational functions $f(x) = 1/(x-a)$, and $f(x) = 1/(x-a)^2$ illustrate those ideas very well. Clearly, $\lim_{x \to a^+} 1/(x-a) = \infty$ and $\lim_{x \to a^-} 1/(x-a) = -\infty$. However, $\lim_{x \to a} 1/(x-a)^2 = \infty$. If $n$ is an odd integer $1/(x-a)^n$ behaves like $1/(x-a)$ for x near a. On the other hand if $n$ is an even integer $1/(x-a)^n$ behaves like $1/(x-a)^2$. The graphs of the functions $1/(x-a)$ and $1/(x-a)^2$ are sketched in Figure 38.

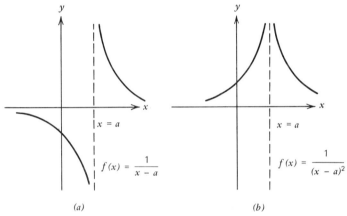

(a)                              (b)

**Fig. 38**   (a) Graph of $f(x) = \dfrac{1}{x-a}$.   (b) Graph of $f(x) = \dfrac{1}{(x-a)^2}$.

We note also the following analogues of (1), (2), and (3). If $\lim_{x \to a} f(x) = \infty$ and $\lim_{x \to a} g(x) = M$ then

(1')
$$\lim_{x \to a} (f(x) + g(x)) = \infty.$$

(2')
$$\lim_{x \to a} f(x) \cdot g(x) = \infty \qquad \text{if} \qquad M > 0$$
$$= -\infty \qquad \text{if} \qquad M < 0.$$

(3')
$$\lim_{x \to a} \frac{1}{f(x)} = 0.$$

These statements hold for one-sided limits as well. The verifications are immediate and are left as exercises.

To sketch the graph of an arbitrary rational function we need only determine the behavior of

$$f(x) = \frac{a_n x^n + \cdots + a_0}{b_m x^m + \cdots + b_0}$$

for large values of x. This is entirely determined by the ratio $n/m$ where $n$ is the degree of the polynomial in the numerator, and $m$ is the degree of the polynomial in the denominator. We assume $a_n \neq 0$ and $b_n \neq 0$. Then

$$f(x) = x^{n-m} \frac{(a_n + a_{n-1}/x + \cdots + a_0/x^n)}{(b_m + a_{m-1}/x + \cdots + b_0/x^m)}.$$

Hence, if $n = m$, then $f(x) \to a_n/b_n$ as $x \to \infty$. If $n < m$, then $f(x) \to 0$. If $n > m$, then $f(x) \to +\infty$ if $a_n/b_n > 0$, and $f(x) \to -\infty$ if $a_n/b_n < 0$. Similar remarks hold if $x \to -\infty$. The precise statements we leave as exercises.

**Example 2.** Sketch the graph of $f(x) = x^2/(x^2 - 1)$. Note first that $f$ is even. Hence, it suffices to sketch the graph for $x > 0$ and then reflect about the $y$ axis. Computing the first derivative we have

$$f'(x) = \frac{2x(x^2 - 1) - 2x^3}{(x^2 - 1)^2} = -\frac{2x}{(x^2 - 1)^2}.$$

By the first derivative test $x = 0$ is a relative maximum, and $f(0) = 0$. To determine the regions of concavity, we compute the second derivative, obtaining

$$f''(x) = \frac{-2(x^2 - 1)^2 + 4x(x^2 - 1)2x}{(x^2 - 1)^4} = \frac{6x^2 + 2}{(x^2 - 1)^3}.$$

Clearly, for $x > 0$, $f''(x) > 0$ if $x > 1$, and $f''(x) < 0$ if $0 < x < 1$. Therefore, the graph of $f$ is concave up if $x > 1$ and concave down if $0 < x < 1$. Now the degree of the numerator equals the degree of the denominator. Hence, by the above discussion, $\lim_{x \to \infty} f(x) = 1$. It only remains to determine the behavior of $f$ near $x = 1$. Now

$$\frac{x^2}{x^2 - 1} = \frac{x^2}{x+1}\frac{1}{x-1} \quad \text{and} \quad \lim_{x \to 1}\frac{x^2}{x+1} = \frac{1}{2}.$$

Therefore, by (2'), $\lim_{x \to 1^+} f(x) = \infty$ and $\lim_{x \to 1^-} f(x) = -\infty$. Combining these statements we have the sketch in Figure 39. Note we have plotted just one point, the relative maximum $(0, 0)$.

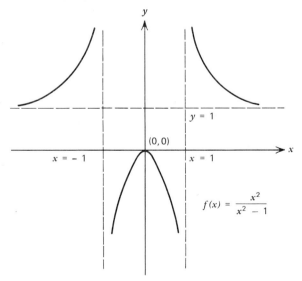

**Fig. 39**   Graph of $f(x) = \dfrac{x^2}{x^2 - 1}$.

If for a rational function

$$f(x) = \frac{a_n x^n + \cdots + a_0}{b_m x^m + \cdots + b_0}$$

we have $n \leqslant m+1$, then $f(x)$ behaves like a linear function for large values of $x$. To understand this, we must introduce the notion of an asymtote.

**Definition.**    *The linear function $l(x) = ax+b$ is called an* asymptote *for the function $f$ as $x \rightarrow \infty$ if $\lim\limits_{x \to \infty} f(x) - l(x) = 0$. The linear function $l$ will be an asymptote as $x \rightarrow -\infty$ if $\lim\limits_{x \to -\infty} f(x) - l(x) = 0$.*

In general two functions $f$ and $g$ are called asymptotic as $x$ tends to $\infty$ if $\lim\limits_{x \to \infty} f(x) - g(x) = 0$. A vertical line $x = a$ will be an asymptote for a function $f$ if either $\lim\limits_{x \to a^+} f(x)$ or $\lim\limits_{x \to a^-} f(x)$ is infinite.

A rational function

$$f(x) = \frac{p(x)}{q(x)} = \frac{a_n x^n + \cdots + a_0}{b_m x^m + \cdots + b_0},$$

where $p$ and $q$ are polynomials, will have the $x$ axis as an asymptote if $n = \text{degree } p < m = \text{degree } q$. If $n = m$, then $r$ will have the horizontal asymptote $l(x) = a_n/b_n$. If $n = m+1$, then we may divide $p$ by $q$ and write $r(x) = ax+b+s(x)$ where $\lim\limits_{x \to \infty} s(x) = \lim\limits_{x \to -\infty} s(x) = 0$. The line $l(x) = ax+b$ is a linear asymptote for $r$. If $n > m+1$, then $r(x)$ has no nonvertical linear asymptotes. The vertical asymptotes for a rational function $r$ correspond to the zeros of the denominator $q$.

**Example 3.**    Determine the asymptotes for the rational function $r(x) = (x^2-1)/(x+2)$ and sketch the graph. The vertical line $x = -2$ is a vertical asymptote and since $r(x) = x-2+3/(x+2)$, the line $l(x) = x-2$ is a linear asymptote. Computing the first derivative $r'$, we obtain that

$$r'(x) = 1 - \frac{3}{(x+2)^2} = \frac{x^2+4x+1}{(x+2)^2}.$$

Clearly $r' = 0$ if $x = -2 \pm 2\sqrt{3}$. By the first derivative test we conclude that $r$ has a maximum at $x = -2 - \sqrt{3}$ and a minimum at $x = -2 + \sqrt{3}$. Indeed $r(-2-\sqrt{3}) = -4 - 2\sqrt{3}$ and $r(-2+\sqrt{3}) = -4 + 2\sqrt{3}$. Figure 40 is a sketch of the graph.

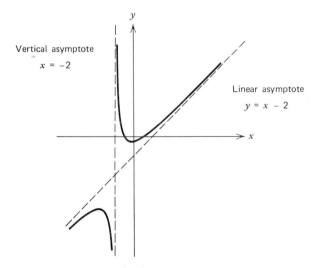

**Fig. 40**    Graph of $y = \dfrac{x^2-1}{x+2}$.

**EXERCISES**

1. Sketch the graphs of the following polynomials. Determine the intercepts, extremes, and inflection points where possible as well as the intervals on which the functions are increasing or decreasing and for which the sense of concavity is up or down. Discuss any symmetry present.

(a) $y = x^3$

(b) $y = 1 - 3x + 3x^2 - x^3$

(c) $y = 4x - x^3$

(d) $y = -x^4 + 2x^2 - 1$

(e) $y = -3x^4 + 16x^3 - 24x^2$

(f) $y = x^5 - x^3 - 2x$

(g) $y = 3x^5 - 30x^4 + 80x^3$

(h) $y = 15x^3 - x^5$

2. Sketch the graphs of the following rational functions. Determine the intercepts, extremes, etc., as in Exercise 1.

(a) $y = \dfrac{x}{x^2 - 1}$

(b) $y = \dfrac{x+2}{(x-1)^2}$

(c) $y = \dfrac{x}{x^2+1}$

(d) $y = \dfrac{x(x-1)}{x^2+4}$

(e) $y = \dfrac{x^2+1}{x^2-4}$

(f) $y = \dfrac{x^2+9}{x^2+1}$

(g) $y = \dfrac{x^2-4}{x+1}$

(h) $y = \dfrac{(x+1)^2}{(x+2)^2(x-1)}$

(i) $y = \dfrac{x}{x^3-1}$

(j) $y = \dfrac{x}{(x-1)^3}.$

3. Determine all the linear asymptotes for the following rational functions. Sketch the graph from a knowledge of the function and its asymptotes.

(a) $f(x) = \dfrac{x^2+1}{x-1}$

(b) $f(x) = \dfrac{x^2-2}{x+1},$

(c) $f(x) = \dfrac{2x^2}{x-1}$

(d) $f(x) = \dfrac{3x^2+x}{x-2}$

(e) $f(x) = \dfrac{x^2-4}{x^2-1}$

(f) $f(x) = \dfrac{x^2-2}{(x+4)^2}$

4. Define $f(x) \to \infty$ as $x \to -\infty$ and $f(x) \to -\infty$ as $x \to -\infty$.

5. Prove that if $a > 0$, then $ax^n \to \infty$ if $x \to \infty$. Prove that if $a < 0$, then $ax^n \to -\infty$ as $x \to \infty$.

6. Give criteria for determining when $ax^n \to \infty$ and when $ax^n \to -\infty$ as $x \to -\infty$. Prove the result.

7. For a real number $L$ define $\lim_{x \to -\infty} f(x) = L$.

8. Suppose $f(x) \to L$ and $g(x) \to \infty$ as $x \to \infty$. Prove that if $L > 0$, then $f(x)g(x) \to \infty$ and if $L < 0$, then $f(x)g(x) \to -\infty$ as $x \to \infty$. Show by an example that if $L = 0$ then $f(x) \cdot g(x)$ may or may not tend to $\infty$.

9. Analyze the behavior of a polynomial $p(x)$ as $x \to -\infty$ and prove your result.

10. For $n = 3, 4, \ldots$ analyze the behavior of $1/(x-a)^n$ as $x$ approaches $a$ from the left and from the right.

11. Analyze the behavior of a rational function $f(x) = p(x)/q(x)$ as $x \to -\infty$. Prove your results.

12. Show that as $x \to \infty$, $x^r \to \infty$ if $r$ is a positive rational number.

13. If $f(x) \to \infty$ as $x \to \infty$, show that $1/f(x) \to 0$. Is the converse of this statement valid? Namely, if $g(x) \to 0$ as $x \to \infty$, does $1/g(x) \to \infty$?

14. Prove statements $(1')$, $(2')$, and $(3')$.

## 5.7  More on Curve Sketching: Critical Points

We may use the analysis of the first and second derivative discussed in Sections 5.4 to 5.6 to sketch the graphs of transcendental functions.

*Example 1.*  Sketch the graph of $f(x) = 2 \sin x + \sin 2x$ locating all extremes and inflection points.

The function $f$ is periodic with period $2\pi$, so it suffices to consider $f$ defined on the interval $[0, 2\pi]$. Differentiating and applying the double angle formulas, we have

$$f'(x) = 2 \cos x + 2 \cos 2x$$
$$= 2[\cos x + 2 \cos^2 x - 1]$$
$$= 2(2 \cos x - 1)(\cos x + 1)$$

and

$$f''(x) = -2[\sin x + 2 \sin 2x]$$
$$= -2[\sin x + 4 \sin x \cos x]$$
$$= -2 \sin x(1 + 4 \cos x).$$

$f'(x) = 0$ if $2 \cos x - 1 = 0$ or $\cos x + 1 = 0$. For the former $x = \pi/3, 5\pi/3$, and for the latter $x = \pi$. Applying the first derivative test, it is easily seen that $f$ has a maximum at $\pi/3$ and a minimum at $5\pi/3$. Indeed, $f(\pi/3) = 3\sqrt{3}/2$ and $f(5\pi/3) = -3\sqrt{3}/2$. There is no extreme at $x = \pi$ since $f'$ does not change sign at this point. Now $f''(x) = 0$ if $\sin x = 0$, or $1 + 4 \cos x = 0$. For the former, $x = 0, \pi, 2\pi$, and for the latter $x = \text{arc cos}\frac{1}{4}$ and $x = 2\pi - \text{arc cos}\frac{1}{4}$. Since the second derivative changes sign at each of these points, they are all inflection points. Figure 41 is the sketch of the graph.

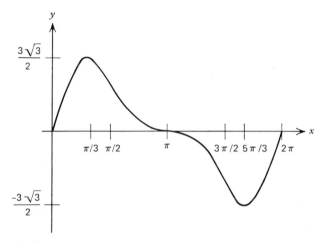

**Fig. 41**

**Example 2.**  Sketch the graph of $y = e^{-x^2}$, locating all extremes and points of inflection.

Note first that the graph is symmetric to the $y$ axis. If $y = e^{-x^2}$, then $y' = -2xe^{-x^2}$ and $y'' = e^{-x^2}[4x^2 - 2]$. The function has a stationary point at $x = 0$, which is easily seen to be a maximum since $y''(0) = -2$. The points $x = \pm\sqrt{\frac{1}{2}}$ are points of inflection, and the graph is concave up for $x \geq \sqrt{\frac{1}{2}}$ and $x \leq -\sqrt{\frac{1}{2}}$. It is concave down if $-\sqrt{\frac{1}{2}} \leq x \leq \sqrt{\frac{1}{2}}$. Since $e^x \geq 1 + x$, $e^x$ and $e^{x^2}$ tend to $\infty$ as $x \to \infty$. Hence, $\lim\limits_{x \to \infty} e^{-x^2} = 0$. A sketch of the graph is Figure 42.

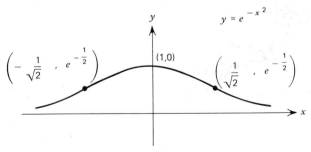

**Fig. 42**   Graph of $y = e^{-x^2}$.

***Example 3.***    Sketch the graph of $f(x) = e^{-x^2} \sin x$.

Note in this example that the function $f$ is the product of two functions for which the graphs are known. Hence, an adequate sketch can be made without further recourse to the calculus. Indeed, if we try to compute the stationary points for $f$, we note that they are the roots of the transcendental equation

$$2x = \cot x.$$

Without considerable computation we can only make crude estimates of these roots, and we do not gain any more information than we know already. Since $|\sin x| \leqslant 1$, it follows that

$$-e^{-x^2} \leqslant e^{-x^2} \sin x \leqslant e^{-x^2}.$$

Hence, an adequate sketch of the graph of $f$ is obtained by just squashing the graph of $\sin x$ by the factor $e^{-x^2}$. Figure 43 is the resulting sketch.

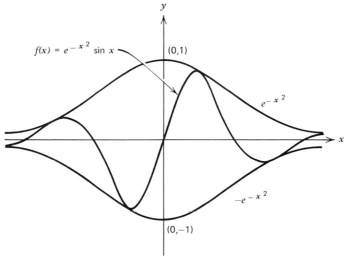

**Fig. 43**

***Example 4.***    Sketch the graph of $f(x) = xe^{-x^2}$.

Computing the first and second derivatives, we obtain

$$f'(x) = e^{-x^2}[1 - 2x^2]$$

and

$$f''(x) = e^{-x^2}[4x^3 - 6x].$$

Therefore, $f$ has a maximum and a minimum at the points $x = 1/\sqrt{2}$ and $x = -1/\sqrt{2}$ respectively. The inflection points are $x = 0, \pm\sqrt{\frac{3}{2}}$. To complete the discussion, we need to know the behavior of $xe^{-x^2}$ as $x \to \infty$. Since $e^{-x^2} \to 0$, we may infer nothing from the limit theorem. It is a fact, which we shall prove presently, that

(1) $$\lim_{x \to \infty} xe^{-x^2} = 0$$

Granting this for the moment, we sketch the graph in Figure 44.

We prove (1) by first establishing a fundamental fact about the behavior of the logarithm.

**THEOREM  5.10**    $\lim_{x \to \infty} (\log x)/x = 0$.

***Proof.***    We base the proof of this result on the known fact that

$$\lim_{x \to \infty} \frac{1}{\sqrt{x}} = 0$$

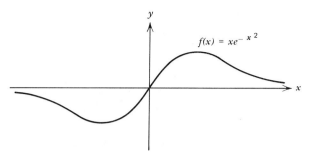

**Fig. 44**

(Exercise 12 of Section 5.6.) Now if $1 < t$, then $\sqrt{t} < t$; and hence $1/t < 1/\sqrt{t}$. Therefore, if $x > 1$

$$\log x = \int_1^x \frac{dt}{t} \leq \int_1^x \frac{dt}{\sqrt{t}} = 2\sqrt{t}\Big|_1^x$$

$$= 2(\sqrt{x} - 1) \leq 2\sqrt{x}.$$

Therefore, dividing by $x$ we have $0 \leq (\log x)/x \leq 2/\sqrt{x}$ and since $\lim_{x \to \infty} 2/\sqrt{x} = 0$, $\lim_{x \to \infty} (\log x)/x = 0$ as well.

To establish (1) we first prove that for any fixed positive number $r$,

(2) $$\frac{e^x}{x^r} \geq 1$$

if $x$ is large enough. From this we conclude that

(3) $$\lim_{x \to \infty} \frac{e^x}{x^r} = \infty$$

for each positive number $r$ since

$$\frac{e^x}{x^r} = \frac{e^x}{x^{r+1}} \cdot x \geq x$$

for $x$ large enough. Therefore, as $x \to \infty$, $e^x/x^r \to \infty$ as well. To establish (2), take the logarithm of both sides of inequality (2). Then (2) holds if and only if

(4) $$x - r \log x = \log\left(\frac{e^x}{x^r}\right) \geq \log 1 = 0.$$

For $x > 1$, (4) holds if and only if

$$\frac{x}{\log x} \geq r.$$

But by Theorem 5.10, $\lim_{x \to \infty} x/\log x = \infty$. Hence, for a given $r$ there is a constant $M$ such that whenever $x \geq M$, $x/\log x \geq r$. Hence, $x - r \log x \geq 0$ and consequently we may infer that for $x \geq M$, $e^x/x^r \geq 1$. Therefore, we have established (2), which in turn implies (3). Equation (3) expresses the very important fact that as $x \to \infty$, $e^x$ tends to $\infty$ faster than any fixed power of $x$.

As a corollary of Theorem 5.10 we may easily deduce the following limit relations. We leave the proofs as exercises.

(5) $$\lim_{x \to \infty} x^r \log x = \infty \text{ if } r \geq 0.$$

(6) $$\lim_{x \to \infty} x^r \log x = 0 \text{ if } r < 0.$$

(7) $$\lim_{x \to \infty} \frac{x^r}{e^x} = 0 \text{ for each real number } r.$$

(8) $$\lim_{x \to \infty} a^x x^r = \infty \text{ for each real number } r \text{ if } a \geq 1.$$

(9) $$\lim_{x \to \infty} a^x x^r = 0 \text{ for each real number } r \text{ if } 0 \leq a < 1.$$

We conclude our discussion of curve sketching with some remarks on critical points of functions. Recall that a point $a$ is a critical point for $f$ if $f'(a) = 0$ or if $f'$ does not exist at the point $a$. In the first instance the graph of $f$ has a horizontal tangent at $x = a$. If $f'$ fails to exist at the point $a$, then several things may happen. For instance, we may have

(10) $$\lim_{h \to 0^+} \frac{f(a+h) - f(a)}{h} = \pm\infty$$

and

(11) $$\lim_{h \to 0^-} \frac{f(a+h) - f(a)}{h} = \pm\infty.$$

If (10) and (11) hold, then we say that the graph has a vertical tangent at $x = a$. There are several possibilities depending on the sign of the difference quotient. If

$$\lim_{h \to 0^+} \frac{f(a+h) - f(a)}{h} = \infty = \lim_{h \to 0^-} \frac{f(a+h) - f(a)}{h},$$

then the difference quotient is positive if $h$ is close to zero and the graph behaves as in Figure 45.

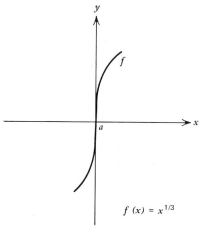

$$f(x) = x^{1/3}$$

**Fig. 45**

An example is $f(x) = x^{1/3}$. Note that

$$\frac{f(0+h) - f(0)}{h} = \frac{h^{1/3}}{h} = \frac{1}{h^{2/3}} > 0$$

if $h \neq 0$. As $h \to 0$, $1/h^{2/3} \to \infty$. However, if $\lim_{h \to 0^+} (f(a+h) - f(a))/h = \infty$ and $\lim_{h \to 0^-} (f(a+h) - f(a))/h = -\infty$, then for $h < 0$ the difference quotient is negative, and we have what is known as a cusp. The function $f(x) = \sqrt{|x|}$ is a typical example. (See Figure 46.)

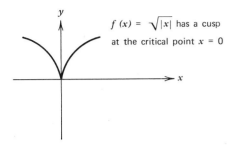

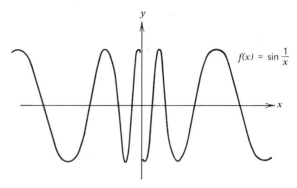

Fig. 46

Fig. 47

There are two further possibilities depending on the sign of the difference quotient. Examples of these are left to the exercises.

At a critical point that is not a stationary point the difference quotient need not become infinite. Indeed, we may have

$$\lim_{h \to 0^+} \frac{f(a+h) - f(a)}{h} = m$$

and

$$\lim_{h \to 0^-} \frac{f(a+h) - f(a)}{h} = M$$

with $M \neq m$. The function $f(x) = |x|$ is a familiar example that we have met many times. Here $m = 1$ and $M = -1$.

A third possibility is that the difference quotient may remain bounded but still may not approach a limit either from the right or from the left. A classic example of this phenomenon is the function

$$f(x) = x \sin \frac{1}{x} \quad \text{if} \quad x \neq 0$$
$$= 0 \quad \text{if} \quad x = 0.$$

This function is continuous for all values of $x$. If $x \neq 0$, this is obvious and if $x = 0$, then

$$\lim_{x \to 0} x \sin \frac{1}{x} = 0.$$

However

$$\frac{f(0+h) - f(0)}{h} = \frac{h \sin 1/h}{h} = \sin \frac{1}{h}.$$

Now in any interval about 0 the function $\sin 1/h$ will take on all values between $-1$ and 1. Hence, $\lim_{h \to 0} \sin 1/h$ does not exist, either from the right or from the left. Yet $|\sin 1/h| \leq 1$ for all values of $h$. (See Figures 47 and 48.)

### EXERCISES

1. Sketch the graphs of the following functions, locating extremes and inflection points.
   (a) $f(x) = \sin x + \cos x, 0 \leq x \leq 2\pi$
   (b) $f(x) = 2 \cos x + \cos 2x, 0 \leq x \leq 2\pi$
   (c) $f(x) = 5 \cos^3 x - 3 \cos x, 0 \leq x \leq 2\pi$
   (d) $f(x) = \sin x \cos^2 x, 0 \leq x \leq 2\pi$

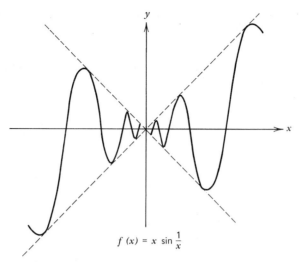

$$f(x) = x \sin \frac{1}{x}$$

**Fig. 48**

(e) $f(x) = xe^x$
(f) $f(x) = (1+e^x)^{-1}$
(g) $f(x) = x^2 e^{-x^2}$
(h) $f(x) = e^{-2x} - 2e^{-x}$
(i) $f(x) = \dfrac{\sin x}{x} \qquad x \neq 0$
$\qquad\quad = 0 \qquad\qquad x = 0$
(j) $f(x) = x \log x \qquad x > 0$
(k) $f(x) = e^x \cos x$
(l) $f(x) = e^{-x^2} \cdot \cos 2x$

2. Sketch the graphs of the following functions. Discuss the behavior at the critical points.

(a) $y = x^{1/3}(x+1)$
(b) $y = |x(x^2-1)|$
(c) $y = \sqrt{(1-x^2)^3}$
(d) $y = 1 + |\sin x|$
(e) $y = \sqrt{\dfrac{x+1}{x}}$

3. One or more functions are defined implicitly by the following equations. Sketch the graph of the equation.
(a) $y^2 = x(x-1)$
(b) $y^2 = x(x-1)^3$
(c) $y^2 = x^2(x^2-4)$
(d) $xy^2 = x^2 + 3x + 2$
(e) $x^{2/3} + y^{2/3} = 1$

4. Sketch the graph of
$$f(x) = x^2 \sin \frac{1}{x} \qquad x \neq 0$$
$$\qquad\quad = 0 \qquad\qquad x = 0.$$

Is $f$ differentiable at $x = 0$? Justify your answers. Does $f$ have a relative extreme at $x = 0$? Is $f'$ continuous at $x = 0$?

5. Prove that $\lim_{x \to \infty} xe^{-x^2} = 0$ as a consequence of (3).

6. Prove that $\lim_{x \to \infty} x^r \log x = \infty$ if $r \geq 0$ and that $\lim_{x \to \infty} x^r \log x = 0$ if $r < 0$.

7. Prove that $\lim_{x \to \infty} x^r/e^x = 0$ for each real number $r$.

8. Prove that $\lim_{x \to \infty} a^x x^r = 0$ for each real number $r$ if $0 \leq a < 1$.

# APPENDIX 1.  CONIC SECTIONS

**\*5.8   The Parabola**

We shall use the techniques for curve sketching developed in Chapter 5 to discuss the graphs of the conic sections. Geometrically these are the curves obtained by intersecting a right circular cone with a plane. If the plane is parallel to an edge of the cone, the resulting curve is called a *parabola*. If the plane is oblique to an edge of the cone and intersects only one nappe of the cone, then the curve is an *ellipse*. The ellipse will be a circle if the intersecting plane is perpendicular to the axis of the cone. If the intersecting plane cuts both nappes of the cone, the resulting curve is a *hyperbola*. See Figure 49.

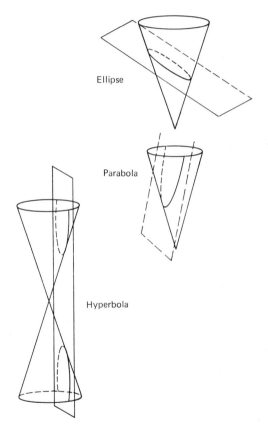

Ellipse

Parabola

Hyperbola

**Fig. 49**   The conic sections.

Unfortunately, it is awkward to base our discussion of these curves on their definition as plane sections of a cone. It is easier to define them as sets of points in the $(x, y)$ plane satisfying a given geometric condition. At the end of our discussion we shall show that the sections of a cone do indeed satisfy this condition. A set of points satisfying a geometric condition such as the one below is commonly called a *locus*.

**Definition.**   *A conic section is the set, or locus, of all points P in the plane such that the distance from P to a fixed point F is proportional to the distance from P to a fixed line l.*

The constant of proportionality $e$ is called the *eccentricity* of the conic. The point $F$ is the *focus* and the line $l$ is the *directrix*. The locus is an ellipse, parabola, or hyperbola, according as $0 < e < 1$, $e = 1$, $e > 1$. To derive an equation for the locus, let us assume that the directrix $l$ is parallel to the $y$ axis and that the focus $F$ has coordinates $(c_1, c_2)$. Let $P = (x, y)$ be a point on the locus. Let $D_1$ be the distance from $P$ to $F$, and $D_2$ be the distance from $P$ to the line $l$. (See Figure 50.)

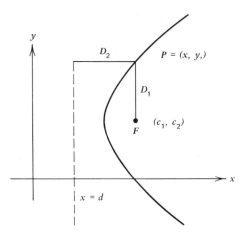

**Fig. 50**

Then $D_1 = \overline{PF} = \sqrt{(x-c_1)^2 + (y-c_2)^2}$. Now the distance from $P$ to the directrix $l$ is measured perpendicular to $l$. Hence, if the equation of the line $l$ is $x = d$, then $D_2 = |(x-d)|$. Thus, our geometric condition is just

$$(1) \qquad \frac{\sqrt{(x-c_1)^2 + (y-c_2)^2}}{|(x-d)|} = e.$$

Squaring both sides of (1), we obtain the equivalent form

$$(2) \qquad \frac{(x-c_1)^2 + (y-c_2)^2}{(x-d)^2} = e^2.$$

Let us consider first the case of the parabola, or the case when $e = 1$. To simplify equation (2), we assume that the focus $(c_1, c_2)$ is a point $(c, 0)$ on the $x$ axis and we assume further that the directrix has equation $x + c = 0$. This amounts to assuming $d = -c$. Then (2) becomes

$$\frac{(x-c)^2 + y^2}{(x+c)^2} = 1.$$

This reduces immediately to

$$(x-c)^2 + y^2 = (x+c)^2$$

or

$$(3) \qquad\qquad y^2 = 4cx.$$

A parabola with focus $(c, 0)$ and directrix the line $x + c = 0$ is said to be in *standard position*. It follows from the above that a point $(x, y)$ lies on a parabola in standard position if and only if $(x, y)$ satisfies equation (3). Furthermore a parabola is completely determined if its focus and directrix are specified.

To sketch the graph of (3) let us assume that $c > 0$. Then $y^2 = 4cx$ if and only if $y = \pm 2\sqrt{cx}$. The graph of $y = 2\sqrt{cx}$ is that portion of the parabola lying above the $x$ axis. If $y = -2\sqrt{cx}$, then we have the portion lying below the $x$ axis. If $y = 2\sqrt{cx}$, then $y' = 2\sqrt{c/x}$, and $y'' = -\sqrt{c}/2x^{3/2}$. Since $y' > 0$ and $y'' < 0$ for all $x > 0$, the graph is increasing and is concave down. Moreover, as $x$ tends to zero from the right the derivative $y'$ tends to $\infty$. Since $(x, y)$ satisfies (3) if and only if $(x, -y)$ satisfies (3), the graph of (3) is symmetric with respect to the $x$ axis. Hence the graph of (3) for $c > 0$ is given by Figure 51. For $c < 0$, we reflect about the $y$ axis, obtaining Figure 52.

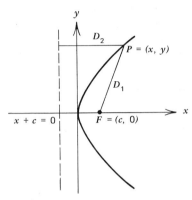

**Fig. 51**  Parabola for $c > 0$.     **Fig. 52**  Parabola for $c < 0$.

Next let us note that an arbitrary parabola with vertical directrix may be put in standard position if we change our coordinate axes appropriately. What we need to do is to translate the origin of our coordinate system to a new point in the plane. This device is called a *translation of coordinate axes* and is accomplished as follows.

Let $(h, k)$ be a point in the plane, and let $l_1$, $l_2$ be lines through $(h, k)$ parallel to the $x$ and $y$ axis respectively. Choose positive directions along $l_1$ and $l_2$ in the same direction as the positive directions along the $x$ and $y$ axes. If these lines are considered as new coordinate axes for the plane, then the coordinates of a point $P$ with respect to these new axes will be the lengths of the segments $O'P_1$ and $O'P_2$. Let us denote these distances by $x'$, $y'$ respectively. (See Figure 53.) If the coordinates of $P$ in the original

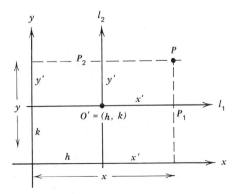

**Fig. 53**

coordinate system were $(x, y)$, then it is immediate that

(5)
$$x = h + x'$$

and

$$y = k + y'$$

or

(6)
$$x' = x - h$$

and

$$y' = y - k.$$

In short, if we translate our coordinate axes so that the origin in the new system is the point $O'$ with $(x, y)$ coordinates $(h, k)$, then the coordinates $(x', y')$ of any point $P$ with respect to the new axes are given by equation (6). The coordinates of $P$ with respect to the original coordinate system are, of course, $x$ and $y$.

To put a parabola with vertical directrix in standard position we need only translate the coordinate axes so that the origin is the point midway between the focus and the directrix. This point is the *vertex* of the parabola. If the focus has coordinates $(c_1, c_2)$ and the directrix is the line $x = d$, then the coordinates of the vertex are

$$\left( \frac{c_1 + d}{2}, c_2 \right).$$

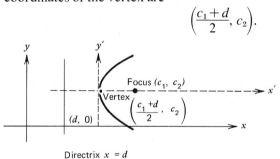

Directrix $x = d$

**Fig. 54**

(Figure 54.)

The constant $c$, which is the distance from the vertex to the focus, is then $(c_1 - d)/2$. If we translate the origin of our coordinate system to the point $((c_1 + d)/2, c_2)$ and set

$$x' = x - \frac{c_1 + d}{2}$$

$$y' = y - c_2,$$

then we have

$$x - c_1 = x' + \frac{c_1 + d}{2} - c_1 = x' - \frac{c_1 - d}{2} = x' - c,$$

$$y - c_2 = y',$$

and

$$x - d = x' + \frac{c_1 + d}{2} - d = x' + \frac{c_1 - d}{2} = x' + c.$$

By definition, a point $(x, y)$ *lies on the parabola if and only if*

$$\frac{(x - c_1)^2 + (y - c_2)^2}{(x - d)^2} = 1.$$

Using the three equations above, this is true if and only if

$$\frac{(x'-c)^2+y'^2}{(x'+c)^2}=1.$$

The latter reduces immediately to

$$y'^2=4cx'$$

which is the equation of the parabola in the $(x', y')$ coordinate system. In the original coordinate system, this equation is

$$(y-k)^2=4c(x-h)$$

where

$$(h, k)=\left(\frac{c_1+d}{2}, c_2\right)$$

are the coordinates of the vertex.

*Example 1.*  Let $(2, 1)$ be the focus of a parabola that has the $y$ axis as directrix. Determine an equation for the parabola and determine a translation of coordinate axes so that the parabola will be in standard position. What is the equation of the resulting parabola in the $(x', y')$ coordinate system?

The equation of the parabola in the $(x, y)$ system is

$$\frac{(x-2)^2+(y-1)^2}{x^2}=1$$

or

$$(x-2)^2+(y-1)^2=x^2.$$

This reduces to

$$-4x+4+y^2-2y+1=0$$

or

(7)     $$4x=y^2-2y+5.$$

To put the parabola in standard position we must translate our coordinate axes so that the origin is at the vertex of the parabola. This is the point midway between the focus and the directrix, which in this case is the point $(1, 1)$. Setting

$$x'=x-1$$

and

$$y'=y-1$$

we see that the $(x', y')$ coordinates of the focus are $(1, 0)$. Hence $c=1$ and the $x', y'$ equation of the parabola is

$$y'^2=4x'.$$

In the $(x, y)$ system, this is

(8)     $$(y-1)^2=4(x-1).$$

See Figure 55.

We may obtain (8) directly from (7), without any reference to the geometry of the parabola, by the simple device of completing the square on the right-hand side of (7). Indeed, if

$$4x=y^2-2y+5,$$

then

$$4x=y^2-2y+1+4=(y-1)^2+4.$$

Hence $(y-1)^2=4x-4=4(x-1)$ which is equation (8). We shall use this technique often in the following discussion.

If a parabola has horizontal directrix, then we say it is in standard position if $(0, c)$ is the focus and $y+c=0$ is the equation of the directrix. The equa-

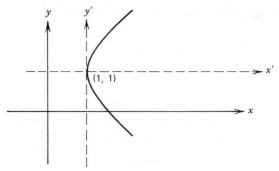

**Fig. 55**  Parabola with equation $y'2 = 4x'$  or  $(y-1)^2 = 4(x-1)$.

tion of the parabola is then

$$x^2 = 4cy.$$

If the directrix is horizontal but the parabola is not in standard position then, just as in the case of a vertical directrix, an appropriate translation of the coordinate axes will put the parabola in standard position. In both cases the origin is translated to the vertex of the parabola.

**EXERCISES**

1.  Find the equation of the parabola in standard position that satisfies the following conditions.
    (a)  focus is the point $(-3, 0)$
    (b)  directrix is the line $x+2=0$
    (c)  focus is the point $(0, 1)$
    (d)  directrix is the line $y=3$
    (e)  $(2, 3)$ lies on the parabola and the directrix is vertical
    (f)  $(1, 4)$ lies on the parabola and the directrix is horizontal
2.  Find the equations of the following parabolas. Determine the vertex and the equation for a translation of axes that will put the parabola in standard position.
    (a)  focus $(2, 1)$ directrix $x=3$
    (b)  vertex $(3, 2)$ focus $(5, 2)$
    (c)  focus $(2, 1)$ directrix $y=4$
    (d)  focus $(3, 2)$, directrix is vertical and $(5, 4)$ lies on the parabola
    (e)  vertex $(1, 3)$, directrix is horizontal and $(2, -4)$ lies on the parabola
    (f)  directrix is vertical and the points $(2, -2)$, $(5, -1)$, and $(14, 0)$ lie on the parabola
3.  For what value of $k$ is the line $16x-y+k=0$ tangent to the parabola $y=8x^2$?
4.  For what value of $k$ is the line $8y-x+k=0$ tangent to the parabola $(x-3)=4(y+1)^2$?
5.  If a line is drawn through the focus of a parabola parallel to the directrix, then that portion of this line cut off by the parabola is called the latus rectum of the parabola. Determine the area of the region bounded by the parabola $y^2=4cx$ and its latus rectum.
6.  The region in Exercise 5 is rotated about the $x$ axis. Sketch the region and determine the volume of the resulting solid.
7.  The region in Exercise 5 is revolved about the $y$ axis. Determine the volume of the resulting solid.
8.  Let $l$ be the line from a point $(x_1, y_1)$ on the parabola $y^2=4cx$ to the focus $(c, 0)$. Let $\alpha$ be the angle this line makes with the tangent to the parabola at the point $(x_1, y_1)$. (Figure 56.)

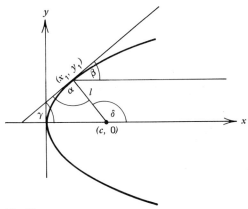

**Fig. 56**

Show that $\tan \alpha = 2c/y_1$. (Show that the slope of the tangent line is $2c/y_1$. There-fore, in Figure 56 $\tan \gamma = 2c/y_1$. Determine $\tan \delta$ and use the fact that $\alpha + \gamma = \delta$.)

9. In Exercise 8 let $\beta$ be the angle between the tangent to the parabola at $(x_1, y_1)$ and the horizontal line through the point $(x_1, y_1)$. Show that $\alpha = \beta$. This reflective property of parabolas has important applications. If a parabolic mirror is con-structed by rotating a parabola about its axis of symmetry and a light source is placed at the focus, then the light rays will be reflected by the mirror in rays all parallel to the axis of the parabola.

10. On page 226 the following statement is made. A point $(x, y)$ satisfies

(1)
$$\frac{\sqrt{(x-c_1)^2+(y-c_2)^2}}{|y-d|} = e$$

if and only if $(x, y)$ satisfies

(2)
$$\frac{(x-c_1)^2+(y-c_2)^2}{(y-d)^2} = e^2.$$

Clearly, if $(x, y)$ satisfies (1) it must satisfy (2). Justify the converse statement, i.e., show that if $(x, y)$ satisfies (2) it must satisfy (1).

## *5.9  The Ellipse

If $0 < e < 1$, then the locus of points $(x, y)$ satisfying

(1)
$$\frac{\sqrt{(x-c_1)^2+(y-c_2)^2}}{|x-d|} = e,$$

is an ellipse. We wish to define a standard position for an ellipse so that equation (1) takes a simple form. We assume first that the focus is at the point $(c, 0)$. Then (1) becomes

(2)
$$\frac{(x-c)^2+y^2}{(x-d)^2} = e^2$$

or

(3)
$$(x-c)^2+y^2 = (ex-de)^2.$$

Next we choose $d$ so that there is no term involving the first power of $x$ in (3). Since the coefficient of $x$ on the left-hand side of (3) is $-2c$ and on the right it is $-2de^2$, this means we must take $d = c/e^2$. Thus we say that the ellipse is in standard position if the focus is the point $(c, 0)$ and the directrix is the line $x = c/e^2$. Then equation (3) becomes

$$(x-c)^2 + y^2 = \left(ex - \frac{c}{e}\right)^2.$$

Multiplying this out, we obtain

$$x^2 + c^2 + y^2 = e^2x^2 + \frac{c^2}{e^2}$$

or

(4) $$x^2(1 - e^2) + y^2 = \frac{c^2}{e^2} - c^2.$$

If we set $a = c/e$ and $b = \sqrt{a^2 - c^2}$, then

$$b^2 = a^2 - c^2 = \frac{c^2}{e^2} - c^2 = a^2(1 - e^2).$$

Substituting $b^2 = (c^2/e^2) - c^2$ in equation (4), we obtain

$$x^2(1 - e^2) + y^2 = b^2$$

or

$$x^2\frac{(1 - e^2)}{b^2} + \frac{y^2}{b^2} = 1.$$

However, $a^2 = b^2/(1 - e^2)$, and consequently (4) reduces to the following simple equation

(5) $$\frac{x^2}{a^2} + \frac{y^2}{b^2} = 1.$$

This is the equation of an ellipse in standard position. Our argument shows that a point $(x, y)$ lies on an ellipse in standard position if and only if $(x, y)$ satisfies (5). Moreover, the equation of an ellipse (2) or (3) is completely determined if the focus, directrix, and eccentricity are known.

To sketch the graph of (5) we note first that the graph is symmetric in $x$ and $y$ since for each point $(x, y)$ on the graph $(\pm x, \pm y)$ also lies on the graph. Therefore, we may assume $x, y \geq 0$. The intercepts in the first quadrant are clearly $(a, 0)$ and $(0, b)$. If $y \geq 0$, then

$$y = \left[b^2\left(1 - \frac{x^2}{a^2}\right)\right]^{1/2} = \frac{b}{a}\sqrt{a^2 - x^2}.$$

Clearly, the domain of the function $f(x) = b/a\sqrt{a^2 - x^2}$ is the interval $[-a, a]$. Computing the first derivative, we obtain

$$f'(x) = -\frac{(b/a)x}{\sqrt{a^2 - x^2}}.$$

Since $f' < 0$ for $x > 0$, the function $f$ is decreasing for $x > 0$. Moreover, $f'(0) = 0$ and $\lim_{x \to a^-} f'(x) = \infty$. If we compute $f''$, we obtain

$$f''(x) = \frac{-ab}{(a^2 - x^2)^{3/2}}.$$

For all $x$ in $(-a, a)$, $f''(x) < 0$, hence the graph of $f$ is concave down. Figure 57 is the sketch of the graph of $f$ for $0 \leq x \leq a$. By symmetry we have the graph of (5), which is Figure 58.

Notice that the point $(-c, 0)$ is also a focus for the ellipse satisfying equation (5). The corresponding directrix is the line with equation $x+$

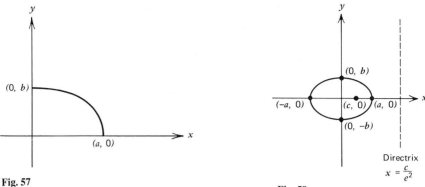

Fig. 57

Fig. 58

$(c/e^2) = 0$. This is clear since if we compute the equation satisfied by the ellipse with focus $(-c, 0)$ and directrix $x + (c/e^2) = 0$, we see that this is exactly equation (5). The point midway between the two foci of an ellipse is called the *center*. The line segment with end points $(a, 0)$, $(-a, 0)$ is called the *major axis* of the ellipse. The number $a$ is called the *length of the semimajor axis*. The line segment $(0, b)$, $(0, -b)$ is the *minor axis* of the ellipse and $b$ is the *length of the semiminor axis*. The points $(\pm a, 0)$ and $(0, \pm b)$ are the *vertices* of the ellipse. Notice that the foci always lie along the major axis and that $a > b$.

An ellipse with horizontal directrix will be in standard position if the foci are the points $(0, c)$ and $(0, -c)$ and the corresponding directrices are the lines $y - c/e^2 = 0$ and $y + c/e^2 = 0$. The equation of the ellipse is now $x^2/b^2 + y^2/a^2 = 1$ where as before $a = c/e$ and $b = \sqrt{a^2 - c^2}$. The major axis now lies along the $y$ axis. Since $a > b$, we note that $x^2/25 + y^2/16 = 1$ is the equation of an ellipse with major axis along the $x$ axis, whereas the major axis of the ellipse $x^2/9 + y^2/16 = 1$ lies along the $y$ axis. (See Figure 59.)

To put an arbitrary ellipse with horizontal or vertical directrix in standard position we need only locate the center $(h, k)$ of the ellipse. Then, if $x' = x - h$ and $y' = y - k$, the ellipse will have equation

(6)
$$\frac{(x-h)^2}{a^2} + \frac{(y-k)^2}{b^2} = 1$$

or

(7)
$$\frac{x'^2}{a^2} + \frac{y'^2}{b^2} = 1,$$

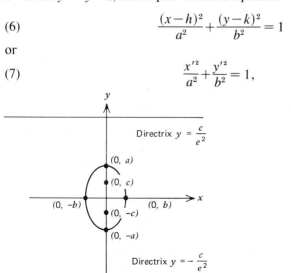

Fig. 59   Ellipse with vertical major axis.

if the directrix is vertical. The foci are the points $(h\pm c, k)$ and the directrices have equations $x - h \pm c/e^2 = 0$.

To determine the equation of an ellipse in the form (6) or (7), one may determine the center $(h, k)$ and the constants $a$, $b$ from given geometric information.

**Example 1.** Determine the equation of the ellipse with focus (5, 1), eccentricity $\frac{3}{5}$, and directrix $x = \frac{31}{3}$. We use the geometric information to determine linear equations involving the coordinates $(h, k)$ of the center and the constant $c$. First $(h+c, k) = (5, 1)$. Hence $k = 1$, and $h+c = 5$. Since the directrix has equation $x = \frac{31}{3}$, we obtain the equation

$$h + \frac{c}{e^2} = \frac{31}{3}$$

or

$$h + \frac{25}{9}c = \frac{31}{3} \qquad \text{since } e = \frac{3}{5}.$$

Thus we have two linear equations

$$9h + 25c = 93$$

and

$$h + c = 5.$$

The solution is $h = 2$, $c = 3$. Hence, $a = c/e = 5$ and $b = \sqrt{a^2 - c^2} = 4$. Therefore, the ellipse has equation

(8)
$$\frac{(x-2)^2}{25} + \frac{(y-1)^2}{16} = 1.$$

If we translate our coordinate system so that the origin is at the point (2, 1), then the ellipse is in standard position. Setting $x' = x - 2$ and $y' = y - 1$ equation (8) becomes

$$\frac{x'^2}{25} + \frac{y'^2}{16} = 1.$$

It is often simpler, however, to determine the equation of the ellipse directly from the geometric information and then complete the square to obtain (8). Since the ellipse in this example has focus (5, 1), eccentricity $\frac{3}{5}$ and directrix $x = \frac{31}{3}$, we see immediately from (2) that its equation is

(10)
$$(x-5)^2 + (y-1)^2 = \left(\frac{3}{5}\right)^2\left(x - \frac{31}{3}\right)^2.$$

Multiplying (10) out and simplifying, we obtain

$$16x^2 - 64x + 25y^2 - 50y = 311.$$

Completing the square in $x$ and $y$, we obtain

$$16(x^2 - 4x + 4) - 64 + 25(y^2 - 2y + 1) - 25 = 311.$$

Hence

$$16(x-2)^2 + 25(y-1)^2 = 311 + 64 + 25 = 400$$

or

$$\frac{(x-2)^2}{25} + \frac{(y-1)^2}{16} = 1$$

which is equation (8).

We use the fact that an ellipse in standard position satisfies the equation $x^2/a^2 + y^2/b^2 = 1$ to derive two important geometrical properties of the ellipse. The first is that the sum of the distances from a point $(x, y)$ on the ellipse to the foci $(\pm c, 0)$ is constant and equals $2a$. The second is that the

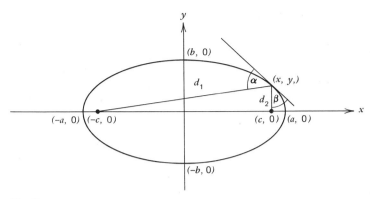

**Fig. 60**

lines joining the point $(x, y)$ on the ellipse to the foci make equal angles with the tangent to the ellipse at the point $(x, y)$. See Figure 60.

We may use the vanishing derivative theorem to verify the first of these properties. Let $f(x) = d_1 + d_2$ be the sum of the distances from $(x, y)$ to the foci. Then

$$f(x) = \sqrt{(x+c)^2+y^2} + \sqrt{(x-c)^2+y^2}.$$

Differentiate $f$ and show that $f'(x) = 0$. Hence $f(x)$ is constant. But $f(0) = 2\sqrt{b^2+c^2} = 2a$.

To verify the second property we must develop formulas for the tangents of the angles $\alpha$ and $\beta$ as functions of $x$. Since the angles $\alpha$ and $\beta$ are between $0$ and $\pi/2$, to prove $\alpha = \beta$ it suffices to show that $\tan \alpha = \tan \beta$. We leave the details of the verification of these two properties as exercises.

The first of these two properties can be used as an equivalent definition of an ellipse. (See Exercise 10.)

If in the equation

(11)
$$\frac{(x-h)^2}{a^2} + \frac{(y-k)^2}{b^2} = 1,$$

$a = b$, then (11) defines a circle with center $(h, k)$ and radius $a$. The equation of a circle with center at the origin, and radius $a$,

$$x^2 + y^2 = a^2,$$

is not a special case of equation (2), however, for if $a = b$ in (5), this implies that $e = 0$ since $b^2 = a^2(1 - e^2)$. Then $c = 0$, and (2) becomes $x^2 + y^2 = 0$.

**EXERCISES**
1. Determine the equation of the ellipse in standard position that satisfies the following conditions.
   (a) foci $(0, \pm 4)$, eccentricity $4/5$
   (b) foci $(\pm 3, 0)$, vertices $(\pm 5, 0)$
   (c) foci $(\pm 4, 0)$, vertices $(0, \pm 5)$
   (d) focus $(3, 0)$, directrix $x = 25/3$
   (e) vertex $(5, 0)$, directrix $x = 25/4$
   (f) vertex $(0, 3)$, focus on the $y$ axis, eccentricity $4/5$
2. Find the equations of the following ellipses. Determine the center of the ellipse and the equations for a translation of axes that will put the ellipse in standard position.
   (a) focus $(4, -2)$, directrix $x = 28/3$, eccentricity $3/5$

(b) focus $(5, 1)$, directrix $x = 29$, eccentricity $1/3$

(c) focus $(5, 7)$, directrix $y = 14$, eccentricity $3/4$

(d) vertices $(1, 8), (1-2), (5, 3), (-3, 3)$.

3. Find the equation of the tangent line to the ellipse $12x^2 + 4y^2 = 1$ at the point $(1/4, 1/4)$.

4. A point $(x_0, y_0)$ lies outside the ellipse $(x^2/a^2) + (y^2/b^2) = 1$ if $(x_0^2/a^2) + (y_0^2/b^2) > 1$. Show that $(4, 3)$ lies outside the ellipse $(x^2/25) + (y^2/9) = 1$. Determine the equations of the two lines through $(4, 3)$ that are tangent to this ellipse. (By implicit differentiation the slope of the tangent to the ellipse at the point $(x, y)$ satisfies $(x/25) + (yy'/9) = 0$. Use the condition that these tangent lines must pass through $(4, 3)$ to determine the points of tangency.)

5. Is it possible to determine lines passing through $(1, 1)$ that are tangent to the ellipse $(x^2/4) + (y^2/3) = 1$? Explain.

6. Determine the lines through $(2, 3)$ that are tangent to the ellipse $9x^2 + y^2 = 9$.

7. The ellipse $(x^2/a^2) + (y^2/b^2) = 1$ is revolved about the $x$ axis. Determine the volume of the resulting solid.

8. What is the volume of the solid obtained by rotating the ellipse of Exercise 7 about the $y$ axis?

9. Using the ideas discussed on p. 235, show that the sum of the distances from a point on an ellipse to the foci is constant and equals the length of the major axis.

10. Show that the locus of all points $P$ in the plane, such that the sum of the distances from $P$ to two fixed points $F_1, F_2$ is constant, defines an ellipse (Let $P = (x, y)$, $F_1 = (c, 0)$, $F_2 = (-c, 0)$, and denote the sum of the distances by $2a$. Show that the point $P$ satisfies

$$\sqrt{(x-c)^2 + y^2} + \sqrt{(x+c)^2 + y^2} = 2a$$

if and only if $x^2/a^2 + y^2/b^2 = 1$ where $b^2 = a^2 - c^2$.)

11. Let $l_1, l_2$ be two lines joining a point $(x, y)$ on the ellipse $x^2/a^2 + y^2/b^2 = 1$ with the foci. Let $\alpha, \beta$ be the angles that these lines make with the tangent to the ellipse at $(x, y)$ (Figure 61). Show that $\alpha = \beta$.

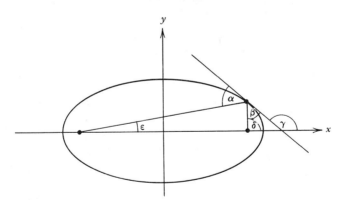

Fig. 61

(Introduce the angles $\epsilon$, $\delta$, $\gamma$ as in Figure 61. Show that $\alpha = \pi + \epsilon - \gamma$ and $\beta = \gamma - \delta$. Write the tangents of these angles as functions of $x$ and show that $\tan \alpha - \tan \beta = \tan (\epsilon - \delta) - \tan (\gamma - \delta) = 0$.)

This reflecting property of ellipses explains the "whispering gallery" effect of elliptical domes. If an ellipsoid is formed by rotating an ellipse about its major axis, and a sound source is placed at one focus, then all of the sound waves emanating from this point are reflected back to the other focus.

## *5.10    The Hyperbola

The locus of points $P$, such that the distance from $P$ to a fixed point $F$ divided by the distance from $P$ to a fixed line $l$ is a constant greater than one, is a *hyperbola*. If the focus $F = (c_1, c_2)$, and the directrix $l$ is vertical with equation $x = d$, then, as before, this geometric condition can be expressed by the following equation.

(1)
$$\frac{\sqrt{(x-c_1)^2 + (y-c_2)^2}}{|x-d|} = e > 1.$$

We say this hyperbola is in standard position if the focus $F = (c, 0)$ is on the $x$ axis and the directrix $l$ has equation $x = c/e^2$. Setting $(c_1, c_2) = (c, 0)$ and $d = c/e^2$, equation (1) has the following equivalent form:

$$(x-c)^2 + y^2 = e^2 \left(x - \frac{c}{e^2}\right)^2.$$

Simplifying, we obtain

$$x^2 + c^2 + y^2 = e^2 x^2 + \frac{c^2}{e^2}$$

or

(2)
$$(e^2 - 1)x^2 - y^2 = c^2 - \frac{c^2}{e^2}.$$

Just as with the ellipse we set $a = c/e$. Since $e > 1$, $c^2 - c^2/e^2 > 0$. Consequently, we set

$$b = \sqrt{c^2 - c^2/e^2} = \sqrt{c^2 - a^2}.$$

Substituting in (2), we obtain

$$(e^2 - 1)x^2 - y^2 = b^2$$

or

(3)
$$\frac{x^2}{a^2} - \frac{y^2}{b^2} = 1$$

since

$$b^2 = c^2 - \frac{c^2}{e^2} = \frac{c^2}{e^2}(1 - e^2) = a^2(1 - e^2).$$

Equation (3) is the equation of a hyperbola in standard position. Our argument shows that a point $(x, y)$ lies on a hyperbola in standard position if and only if $(x, y)$ satisfies (3).

A hyperbola with horizontal directrix and eccentricity equal to $e$ will be in standard position if the focus is at the point $(0, c)$ and the directrix is the line $y = c/e^2$. The equation of this hyperbola will then be

$$\frac{y^2}{a^2} - \frac{x^2}{b^2} = 1$$

where, as before, $a = c/e$ and $b = \sqrt{c^2 - a^2}$.

Let us now sketch the graph of (3). Since $(x, y)$ satisfies (3) if and only if $(\pm x, \pm y)$ does, the graph of (3) is symmetric about both the coordinate axes. Thus, if we obtain the graph of (3) for $x \geqslant 0$ and $y \geqslant 0$, we may obtain the entire graph of (3) by reflecting this portion about the $x$ and $y$ axes. Rearranging (3), we obtain

(4)
$$y^2 = \frac{b^2}{a^2}(x^2 - a^2).$$

Since $y^2 \geqslant 0$, a point $(x, y)$ in the first quadrant will satisfy (4) if and only if $x \geqslant a$. For such $x$

(5)
$$y = \frac{b}{a}\sqrt{x^2 - a^2}.$$

Computing the first and second derivative of thus function, we obtain

$$y' = \frac{b}{a}\frac{x}{\sqrt{a^2-x^2}}$$

and

$$y'' = \frac{-ab}{\sqrt{x^2-a^2}}.$$

For $x > a$, it follows that $y' > 0$ and $y'' < 0$. Hence the graph is increasing and is concave down. Moreover $y' \to \infty$ as $x \to a^+$.

To obtain a more accurate picture of the graph of (5) for large values of $x$, let us analyze the function $f(x) = b/a \sqrt{x^2-a^2}$ a little further. We may write

$$f(x) = \frac{b}{a}x\sqrt{1-a^2/x^2}.$$

Since $\sqrt{1-a^2/x^2} < 1$, $f(x)$ lies below the line $y = (b/a)x$. However, the vertical distance between the graph of $f$ and this line tends to zero as $x$ tends to $\infty$. To see this, let $d(x)$ be this vertical distance. Then

$$d(x) = \frac{b}{a}x - f(x)$$

$$= \frac{b}{a}(x - \sqrt{x^2-a^2})$$

$$= \frac{b}{a}(x - \sqrt{x^2-a^2})\left(\frac{x+\sqrt{x^2-a^2}}{x+\sqrt{x^2-a^2}}\right)$$

$$= \frac{ba}{x+\sqrt{x^2-a^2}}.$$

From this it is apparent that $\lim_{x\to\infty} d(x) = 0$. We now are in a position to sketch the graph of $f$. The complete graph of (3) is now obtained by reflecting Figure 62 about the coordinate axes. This is Figure 63.

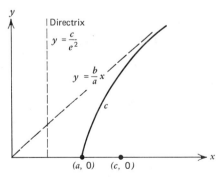

**Fig. 62**   Graph of $f(x) = \frac{b}{a}\sqrt{x^2-a^2}$.

The lines $y = (b/a)x$ and $y = -(b/a)x$ are called the *asymptotes* of the hyperbola. The graph of a hyperbola has two *branches*, i.e., two connected pieces. If the hyperbola is in standard position, one branch lies in the right half plane, the other in the left.

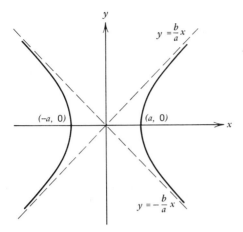

**Fig. 63** Graph of hyperbola $\dfrac{x^2}{a^2} - \dfrac{y^2}{b^2} = 1$.

***Example 1.*** A hyperbola has focus at the point $(5,0)$ and has accentricity $\frac{5}{4}$. If the hyperbola is in standard position, determine the equation and sketch the graph. Compute the asymptotes as well. Since the hyperbola is in standard position, the directrix $x = d = c/e^2 = \frac{16}{5}$. Furthermore, $a = c/e = 4$, and $b = (c^2 - a^2)^{1/2} = 3$. Therefore, the hyperbola has equation

$$\frac{x^2}{16} - \frac{y^2}{9} = 1.$$

The asymptotes are the lines $y = \pm\frac{3}{4}x$. The graph is Figure 64.

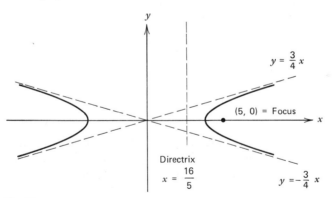

**Fig. 64**

The points $(\pm a, 0)$ on the hyperbola $x^2/a^2 - y^2/b^2 = 1$ are called the *vertices* of the hyperbola. Notice also that in our derivation of the equation of the hyperbola we could have replaced the focus $(c, 0)$ and directrix $x = c/e^2$ by the point $(-c, 0)$ and the line $x = -c/e^2$. Hence $(-c, 0)$ is also a focus for the hyperbola $x^2/a^2 - y^2/b^2 = 1$, and the line $x = -c/e^2$ is the corresponding directrix. The point midway between the vertices $(\pm a, 0)$ is called the *center* of the hyperbola.

An arbitrary hyperbola with vertical directrix is completely determined if the focus, directrix, and eccentricity are known. If the center $(h, k)$ is computed, then the hyperbola will have equation

(6) $$\frac{(x-h)^2}{a^2} - \frac{(y-k)^2}{b^2} = 1.$$

If we set $x' = x - h$ and $y' = y - k$, then this becomes $x'^2/a^2 - y'^2/b^2 = 1$, and the hyperbola is in standard position in the $(x', y')$ coordinate system.

**Example 2.**   A hyperbola with eccentricity $\sqrt{\tfrac{3}{2}}$ has the point $(2 + \sqrt{5}, 1)$ as focus and the line $x = 2 + \tfrac{4}{5}\sqrt{5}$ as directrix. Determine the center for the hyperbola and an equation for the hyperbola in the form

$$\frac{(x-h)^2}{a^2} - \frac{(y-h)^2}{b^2} = 1.$$

Determine the asymptotes and sketch the graph. Since the directrix is vertical, if $(h, k)$ is the center, then $(h + c, k)$ is the focus and $x = h + c/e^2$ is the directrix. Clearly, $k = 1$ and $h + c = 2 + \sqrt{5}$. Moreover, $h + c/e^2 = h + 4c/5 = 2 + \tfrac{4}{5}\sqrt{5}$. Eliminating $h$ between these equations we obtain $c = \sqrt{5}$ and $h = 2$. Since $a = c/e$, $a = \sqrt{5}/(\sqrt{\tfrac{5}{2}}) = 2$. Then since $b^2 = c^2 - a^2$ we have $b = 1$. The equation is then

(7)
$$\frac{(x-2)^2}{4} - \frac{(y-1)^2}{1} = 1.$$

The asymptotes have equation $y - k = \pm(b/a)(x - h)$ or $y - 1 = \pm\tfrac{1}{2}(x - 2)$. The graph is Figure 65.

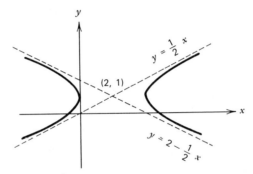

**Fig. 65**  Hyperbola  with  equation  $\dfrac{(x-2)^2}{4} -$ $\dfrac{(y-1)^2}{1} = 1.$

The technique of completing the square can be used to determine (7) directly from the definition of the hyperbola. From the knowledge of the focus, directrix, and eccentricity, we know from (1) that the equation of the hyperbola is

$$(x - (2 + \sqrt{5}))^2 + (y - 1)^2 = \frac{5}{4}\left(x - \left(2 + \frac{4}{5}\sqrt{5}\right)\right)^2.$$

This simplifies to

$$x^2 - 4x - 4y^2 + y = 4.$$

Completing the square, we obtain

$$(x^2 - 4x + 4) - 4 - 4(y^2 - 2y + 1) + 4 = 4$$

or

(8)
$$(x - 2)^2 - 4(y - 1)^2 = 4.$$

Dividing (8) by 4 yields (7).

Associated with the hyperbola $x^2/a^2 - y^2/b^2 = 1$ is another hyperbola having the lines $y = \pm(b/a)x$ as asymptotes. This is the hyperbola having equation

$$\frac{y^2}{b^2} - \frac{x^2}{a^2} = 1.$$

For $x \geqslant 0, y \geqslant 0$, this may be written

$$y = \frac{b}{a}\sqrt{x^2 + a^2}$$

or

$$y = \frac{b}{a}x\frac{\sqrt{1 + a^2}}{x^2}.$$

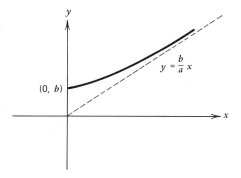

**Fig. 66**   Graph of $y = \frac{b}{a}\sqrt{x^2 - a^2}$.

Figure 66 is a sketch of the graph. If we sketch the graphs of $x^2/a^2 - y^2/b^2 = 1$ and $y^2/b^2 - x^2/a^2 = 1$ together, we obtain Figure 67.

The hyperbola $y^2/b^2 - x^2/a^2 = 1$ is called the *conjugate hyperbola* to the hyperbola $x^2/a^2 - y^2/b^2 = 1$. The focus of the former is the point $(0, c)$, where $c^2 = b^2 + a^2$. However, the eccentricity is $c/b$.

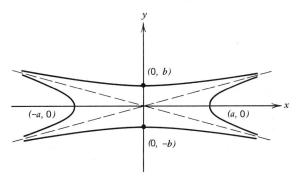

**Fig. 67**   Conjugate hyperbolas.

Just as with the ellipse, the hyperbola, may be characterized in terms of the foci alone. Indeed, a hyperbola is the locus of points such that absolute value of the difference of the distances from two fixed points is constant. If we denote the two fixed points by $(-c, 0)$ and $(c, 0)$ and the constant by $2a$, then this condition becomes

(9) $$\sqrt{(x + c)^2 + y^2} - \sqrt{(x - c)^2 + y^2} = \pm 2a.$$

We leave it as an exercise to verify that if $(x, y)$ lies on the hyperbola. $x^2/a^2 - y^2/b^2 = 1$ then $(x, y)$ satisfies (9). The right-hand side of (9) is positive if $(x, y)$ lies on the right-hand branch; it is negative if it lies on the left.

**EXERCISES**

1. For the following hyperbolas, determine the foci, vertices, directrices, asymptotes, and eccentricity and sketch the graphs.

   (a) $\dfrac{x^2}{25} - \dfrac{y^2}{16} = 1$

   (b) $\dfrac{x^2}{8} - \dfrac{y^2}{4} = 1$

   (c) $\dfrac{x^2}{9} - \dfrac{y^2}{16} = 1$

   (d) $4x^2 - 2y^2 = 1$

   (e) $x^2 - 2y^2 = -1$

   (f) $4x^2 - 3y^2 = -1$

2. Find the equation of the following hyperbolas in standard position if the following conditions are satisfied.
   (a) asymptotes $y = \pm 3x$, vertices $(\pm 4, 0)$
   (b) asymptotes $y = \pm 2x$, vertices $(0, \pm 5)$
   (c) focus $(+5, 0)$, directrix $x = 3$
   (d) focus $(0, -4)$, directrix $y + 2 = 0$
   (e) directrices $y = \pm \frac{32}{5}$, asymptotes $y = \pm \frac{4}{3}x$
   (f) asymptotes $y = \pm x$, passes through $(\sqrt{5}, 1)$.

3. Determine the equation for the following hyperbolas from the following geometric information. Write the equation in the form

$$\frac{(x - h)^2}{a^2} - \frac{(y - h)^2}{b^2} = \pm 1$$

   by completing the square. Determine the equations of the asymptotes.
   (a) focus $(2, 1)$, eccentricity 2, directrix $x = 0$
   (b) focus $(2, 1)$, eccentricity 2, directrix $y = -1$
   (c) focus $(3, 4)$, eccentricity 3, directrix $x + 1 = 0$
   (d) focus $(0, 0)$, eccentricity $\frac{3}{2}$, directrix $y = 5$

4. Determine the equation of the tangent and normal line to the hyperbola $4y^2 - x^2 = 1$ at the point $(1, 1/\sqrt{2})$.

5. Show that the tangent line to the hyperbola $x^2/a^2 - y^2/b^2 = 1$ at the point $(x_0, y_0)$ on the hyperbola satisfies the equation $x_0 x/a^2 - y_0 y/b^2 = 1$.

6. Find equations for the tangent lines to the hyperbola $x^2 - 4y^2 = 1$ passing through the point $(0, 2)$. Do there exist tangent lines to this hyperbola passing through the point $(2, 0)$? Explain.

7. Show that if $x_0^2 - y_0^2 < 1$, then there are two lines through $(x_0, y_0)$ that are tangent to the hyperbola $x^2 - y^2 = 1$. Sketch the hyperbola $x^2 - y^2 = 1$, and sketch the region $x_0^2 - y_0^2 < 1$.

8. Let $l$ be the line through the focus of a hyperbola parallel to the directrix. The chord on $l$ cut off by the hyperbola is called the *latus rectum* of the hyperbola. For the hyperbola $x^2/a^2 - y^2/b^2 = 1$, show that the length of the latus rectum is $2b^2/a$.

9. If $x^2/a^2 - y^2/b^2 = 1$, show that

$$\sqrt{(x + c)^2} - \sqrt{(x - c)^2 + y^2} = 2a \qquad \text{if} \qquad x > 0$$

   or

$$\sqrt{(x + c)^2 + y^2} - \sqrt{(x - c)^2 + y^2} = -2a \qquad \text{if} \qquad x < 0$$

   where $(\pm c, 0)$ are the foci of the hyperbola. (Use the ideas discussed on p. 235.)

10. Show that if $(x, y)$ satisfies either of the two equations in Exercise 9, then $x^2/a^2 - y^2/b^2 = 1$. (We may conclude from Exercise 9 and 10 that a hyperbola is the locus of points $P$ such that the differences of the distances from $P$ to two fixed points is constant.)

11. The region between a hyperbola and its latus rectum is revolved about the $x$ axis. What is the volume of the resulting solid if the equation of the hyperbola is $x^2/a^2 - y^2/b^2 = 1$? What is the volume of the solid formed by revolving the same region about the $y$ axis?

**\*5.11    The General Second Degree Equation**

In the previous three sections we showed that the equation for a parabola, ellipse, or hyperbola with vertical directrix could be written in the form

$$x - h = 4c(y - h)^2$$

$$\frac{(x-h)^2}{a^2} + \frac{(y-h)^2}{b^2} = 1$$

$$\frac{(x-h)^2}{a^2} - \frac{(h-k)^2}{b^2} = 1.$$

This could be accomplished by computing the equation of the conic from its focal directrix definition and then completing the square.

In this section we wish to apply this completion of the square technique to determine the graph of the general quadratic equation

(1)                    $$ax^2 + by^2 + cx + dy + e = 0.$$

We assume that $a$ and $b$ are not both zero. There are two cases to consider.

*Case I.*    One of the coefficients $a$ or $b$ is zero. Let us assume that $b = 0$ and $a \neq 0$. If we complete the square in $x$ in equation (1) we see immediately that (1) takes the form

(2)                    $$dy + e' = a(x - h)^2.$$

If $d \neq 0$, we may write (2) as

(3)                    $$d(y - k) = a(x - h)^2.$$

We know that the graph of (3) is a hyperbola with vertex at the point $(h, k)$. The parabola opens to the right if $a$ and $d$ have the same sign, and opens to the left if $a$ and $d$ have opposite signs.

If $d = 0$ in (2), then we may write (2) as

(4)                    $$(x - h)^2 = \frac{e'}{a}.$$

If $e'/a < 0$, the graph of (4) is empty. If $e'/a \geqslant 0$, the graph of (4) is the pair of vertical lines having equation $x = h \pm \sqrt{e'/a}$. A similar analysis holds if $a = 0$ and $b \neq 0$. Thus we see that if either $a$ or $b$ is zero but not both, and both $x$ and $y$ terms are present in (1), the graph is a parabola.

*Case II.*    Both of the coefficients $a$ and $b$ in (1) are different from zero. Then if we complete the square in (1) with respect to both $x$ and $y$, we see that (1) takes the form

(5)                    $$a(x - h)^2 + b(y - k)^2 = e'.$$

If $e' \neq 0$, we may divide by $e'$ obtaining

(6)                    $$a'(x - h)^2 + b'(y - k)^2 = 1.$$

If $a' = b'$ and both are positive, then the graph of (6) is a circle center at $(h, k)$ and radius $\sqrt{1/a'}$. If both $a'$, $b'$ are positive, then the graph of (6) is an ellipse. If $a'$, $b'$ have opposite signs, then the graph is a hyperbola. If $a'$, $b'$ are both negative, then clearly the graph of (6) is empty.

If, in (5), $e' = 0$, then if $a$ and $b$ have the same sign, the graph of (5) is just the point $(h, k)$. If $a$ and $b$ have opposite signs, then we may write (5) in the form

$$(7) \qquad (y-k)^2 = \left|\frac{b'}{a'}\right| (x-h)^2.$$

The graph of (7) is the pair of lines

$$y - k = \sqrt{\left|\frac{b'}{a'}\right|} (x-h)$$

$$y - k = -\sqrt{\left|\frac{b'}{a'}\right|} (x-h).$$

This analysis shows that the form that the graph of (1) takes is determined by the constants $a$ and $b$. The following are the three possibilities.

(i) Either $a = 0$ or $b = 0$ but not both. If the graph is not empty, it is either a parabola or a pair of lines (possibly coincident) parallel to one of the coordinate axes.

(ii) The constants $a$ and $b$ have the same sign and neither is zero. If the graph is not empty, it is an ellipse or a single point. If $a = b$, the ellipse is a circle.

(iii) The constants $a$ and $b$ have opposite signs and neither is zero. The graph is a hyperbola or a pair of intersecting lines.

Detailed information on the type of the graph can be obtained by completing the square in (1).

***Example 1.*** Determine the graphs of the following quadratic equations

$$(8) \qquad x + 1 - y^2 + 2y = 0$$

$$(9) \qquad x^2 + 2y^2 - 2x - 8y + 10 = 0$$

$$(10) \qquad 2x^2 - y^2 + 2y = 0.$$

In (1), one second-order coefficient is zero, but both $x$ and $y$ terms are present. Thus the graph of (8) is a parabola. If we complete the square, equation (8) becomes $x + 2 = (y-1)^2$. Hence the parabola has vertex at $(-2, 1)$ and opens to the right. In (9) the second-order terms have the same sign. Therefore, if the graph is not empty it is an ellipse or a single point. Completing the square (9) becomes $(x-1)^2 + 2(y-2)^2 = -1$. Thus the graph is empty. In (10) the coefficients of the quadratic terms have opposite sign, hence the graph is a hyperbola or a pair of crossed lines. Again completing the square, (10) becomes $(y-1)^2 - x^2 = 1$. Hence the graph of (10) is a hyperbola, center at $(0, 1)$ with foci along the $y$ axis. Figure 68 illustrates the graphs of (8) and (10).

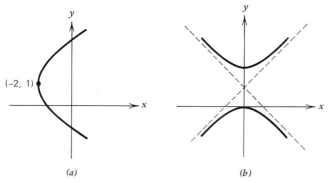

(a)                    (b)

**Fig. 68.** (a) Graph of $x + 1 - y^2 + 2y = 0$. (b) Graph of $2x^2 - y^2 + 2y = 0$.

Equation (1) is not the most general quadratic equation since there is no *xy* term present. To analyze the graph of the equation

(11) $$ax^2 + bxy + cy^2 + dx + cy + f = 0$$

we must introduce a rotation of coordinate axes to get rid of the *xy* term. This reduces (11) to an equation of the form (1). Hence the graph of (11) will be a conic section, a point, or a pair of lines. We omit a discussion of this problem here.

We close our discussion of conics with the determination of the focus and directrix of a conic defined as a section of a cone. The properties of conic sections have formed a subject of mathematical investigation for well over 2000 years. The treatise "The Conics" by Appolonius of Perga, written in the 3rd century B.C., was one of the high points of Greek geometry. Whether the focal directrix property of a conic was known to Appolonius is not clear. The first written account of this property comes in the commentaries of Pappus in the 3rd century A.D.

After the time of Pappus, there were no further important contributions until Renée Descartes' invention of analytic geometry in the 17th century. Then, however, many new results were discovered by Descartes, Pascal, and others. The following elementary and elegant construction of the focus and directrix was overlooked until around 1825 when it was discovered by two Belgian mathematicians, Adolphe Quetelet and Germinal Dandelin.

To construct the focus, directrix, and eccentricity of a section of a cone, we proceed as follows. Intersect a right circular cone $C$ with a plane $\pi_1$. Inscribe a sphere $S$ in the cone so that it is tangent to both the plane and the cone as in Figure 69. Draw a second plane $\pi_2$ through the circle of tangency of the sphere and the cone. We assert that the directrix of the conic is the line of intersection of the planes $\pi_1$ and $\pi_2$. The focus of the conic is the intersection of the sphere $S$ and the plane $\pi_1$. To establish the focal directrix property of the conic we must determine the eccentricity. Let $P$ be a point on the conic, and let $PO$ be the line from $P$ to the vertex $O$ of the cone.

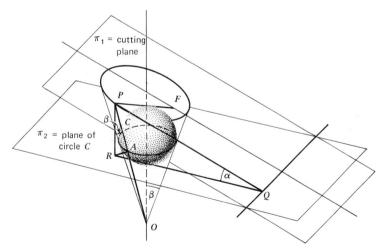

**Fig. 69**   Focal directrix property of conic sections. $\overline{FP} = \overline{PA}, \dfrac{\overline{PR}}{\overline{PA}} = \cos\beta, \dfrac{\overline{PR}}{\overline{PQ}} = \sin\alpha, \dfrac{\overline{FP}}{\overline{PQ}} = \dfrac{\sin\alpha}{\cos\beta} = e.$

This line intersects the circle of tangency of the sphere at a point $A$. Since $PF$ and $PA$ are tangents to a sphere drawn from a single point, we conclude that $\overline{PF} = \overline{PA}$. Next we drop a perpendicular from $P$ to the plane $\pi_2$. If this perpendicular intersects $\pi_2$ at $R$, then

$$(12) \qquad \qquad \frac{\overline{PR}}{\overline{PQ}} = \sin \alpha$$

where $\alpha$ is the angle between the planes $\pi_1$ and $\pi_2$.

If $\beta$ denotes one-half the central angle of the cone $C$, then for each choice of our point $P$, we have $< OPR = \beta$. Therefore

$$(12) \qquad \qquad \frac{\overline{PR}}{\overline{PA}} = \cos \beta.$$

Dividing (11) by (12) and using the fact that $PA = PF$ we conclude that

$$\frac{\overline{PF}}{\overline{PQ}} = \frac{\overline{PA}}{\overline{PQ}} = \frac{\overline{PA}}{\overline{PR}} \cdot \frac{\overline{PR}}{\overline{PQ}} = \frac{\sin \alpha}{\cos \beta}.$$

Therefore, $e = \sin \alpha / \cos \beta$, and we have established the focal directrix property of conic sections.

### EXERCISES
Identify and sketch the graphs of the following quadratic equations.

1. $x^2 - 2x + 4 - y = 0$.
2. $y^2 - 2x + 4y = 0$.
3. $x^2 + 2y^2 + x - y - 5 = 0$.
4. $3x^2 + y^2 - 12x + 2y + 20 = 0$.
5. $4x^2 + 4y^2 - 8x + 12y = 0$.
6. $4x^2 - 3y^2 + 8x - 5 = 0$.
7. $y^2 - 2x^2 + x - y + 2 = 0$.
8. $y^2 + y + 1 = 0$.
9. $x^2 - 2x - 3 = 0$.
10. $x^2 - 2y^2 - 2x - 16y - 28 = 0$.
11. $2x^2 + y^2 + 4x + 2y + 3 = 0$.
12. $x^2 - y^2 - x - y = 0$.

13. If in equation (1) the coefficients $a$ and $b$ are both positive or both negative, determine a condition on the coefficieats of (1) that will insure that the graph of (1) is nonempty.

14. If a section of a cone is an ellipse, then both foci may be determined by inscribing spheres in the cone above and below the cutting plane. Draw a sketch similar to Figure 69 and show that the sum of the distances from the foci to a point on the ellipse is constant. (Use the fact that tangents to a sphere from a fixed point must have the same length.)


## APPENDIX 2


*5.12   **Proofs of Extreme Value and Intermediate Value Theorems**

We wish to prove the following two theorems in this section.

**THEOREM 5.1**  *If $f$ is continuous on the closed interval $[a, b]$, then there exists points $x_1, x_2 \in [a, b]$ such that*

$$f(x_1) \geq f(x) \qquad \text{each} \qquad x \in [a, b]$$

and

$$f(x_2) \leq f(x) \qquad \text{each} \qquad x \in [a, b].$$

**THEOREM 5.2**  *Let $f$ be continuous on the interval $[a, b]$ and let $y_0$ be a number between $f(a)$ and $f(b)$. Then there exists a number $x_0$ in the* *interval $[a, b]$ satisfying*

$$f(x_0) = y_0.$$

Both these theorems assert that there exists a number $x$ with a certain property. Most assertions of this kind usually involve an appeal to the completeness axiom for the real numbers in some way or other. The two statements

1. there is a number $x$ such that $x^2 = 2$, and
2. there is a number $x$ such that $\log x = 1$

are examples of this. Indeed, they are both special cases of Theorem 5.2. The completeness axiom is the fundamental tool in the arguments we shall present and the student is advised to reread the discussion of this axiom in Chapter 1. We shall not try to give the shortest possible proofs but instead shall explore the problem somewhat.

The particular construction, involving the completeness axiom, which we shall use is interesting in its own right, so we shall discuss that first. Let $I_1$ be a closed interval $[a_1, b_1]$, and let $c_1$ be the midpoint. Let $J_1$ be the left-hand interval $[a_1, c_1]$ and $K_1$ the right-hand interval $[c_1, b_1]$ (Figure 70). Choose between $J_1$ and $K_1$ at random, say by flipping a coin. Heads,

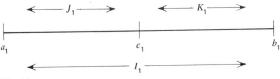

**Fig. 70**

we choose $J_1$; tails, we choose $K_1$. Let us call the chosen interval $I_2$. We now write $I_2 = [a_2, b_2]$ with $c_2$ the midpoint. Let $J_2$ and $K_2$ be the left-hand and right-hand subintervals, respectively, of $I_2$ If we get heads and choose $I_2 = J_1$, we would have the diagram shown in Figure 71.

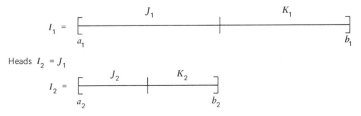

**Fig. 71**

Flip the coin again, choosing the left-hand interval $J_2$ if heads and the right-hand interval $K_2$ if tails. If we get tails, then we set $I_3 = K_2$ and repeat the process. If we continue this selection indefinitely, we have a decreasing sequence of closed intervals $I_1 \supset I_2 \supset \cdots \supset I_n \supset \cdots$. Figure 72 illustrates our decision process.

We assert that as a consequence of the completeness axiom there must be a number $m$ that belongs to *all* of the intervals $I_n$. There clearly can be no more than one such number $m$ because the length of the $n$th interval equals $(b_1 - a_1)/2^{n-1}$, which tends to zero as $n$ increases. To deduce that there is a number $m$ common to all the intervals $I_n$ we need only to con-

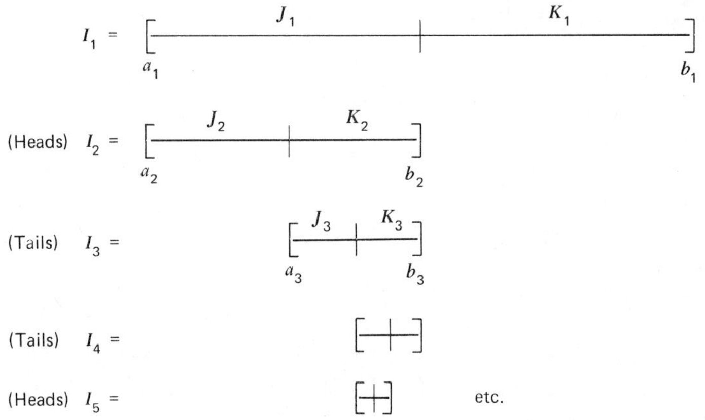

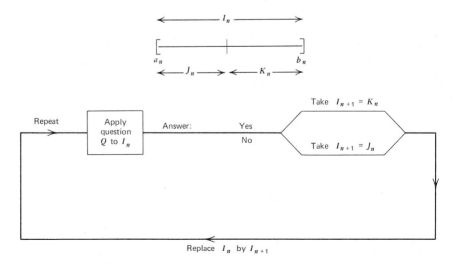

Fig. 72

struct two appropriate separated sets of real numbers $S$ and $T$. We do that as follows. Define $S = \{a_1, a_2, \ldots\}$ and $T = \{b_1, b_2, \ldots\}$. In other words, $S$ is the set of *left-hand* end points of the intervals $I_n$ and $T$ is the set of *right-hand* end points. It is evident that if $s \in S$ and $t \in T$, then $s \leqslant t$. Hence, $S$ and $T$ are separated. If $m$ is the number separating $S$ and $T$, then $m \geqslant s$ for each $s \in S$ and $m \leqslant t$ for each $t \in T$. In other words, $m \geqslant$ each left-hand end point $a_n$ and $m \leqslant$ each right-hand end point $b_n$. But this is just the statement

$$m \in I_n \qquad \text{for each } n.$$

Thus our assertion is established.

In applying this procedure, the intervals $I_n$ will not be chosen at random but rather as a result of some decision process. That is, a certain question $Q$ is asked at each stage of the construction. If the answer to the question is yes, choose $I_{n+1} = K_n$. If the answer is no, choose $I_{n+1} = J_n$. A repetitive process of this kind is called an *algorithm*. Indeed, we call this particular construction the *bisection algorithm*. A flow diagram can be drawn that illustrates this bisection algorithm very succintly. (Figure 73).

Fig. 73   Bisection algorithm.

The trick in applying the bisectiog algorithm to construct a number $m$ satisfying some desired property is to choose the correct question $Q$. For Theorem 5.2, this choice is not too hard. Let $f$ be a function defined and continuous on the closed interval $[a, b]$. For definiteness we assume $f(a) < f(b)$. Let $y_0$ be a number satisfying $f(a) < y_0 < f(b)$. To establish Theorem 5.2 we must prove that there is a number $x_0$, satisfying $a < x_0 < b$ and satisfying

$$f(x_0) = y_0.$$

We apply the bisection algorithm to construct $x_0$ in the following way. Let $I_1 = [a, b]$, and let $c$ be the midpoint. We set $J_1 = [a, c]$ and $K_1 = [c, b]$, as before and pose the following question.

$$\text{Is } f(c) \leqslant y_0? \quad Answer \quad \begin{array}{ll} \text{Yes:} & \text{Take } I_2 = K_1 \\ \text{No:} & \text{Take } I_2 = J_1 \end{array}$$

Replace $I_1$ by $I_2$ and repeat.

Thus at the $n$th stage if $c_n$ is the midpoint of $I_n$, we ask

$$\text{Is } f(c_n) \leqslant y_0? \quad Answer \quad \begin{array}{ll} \text{Yes:} & \text{Take } I_{n+1} = K_n \\ \text{No:} & \text{Take } I_{n+1} = J_n \end{array}$$

Replace $I_n$ by $I_{n+1}$ and repeat.

The net effect of this is to assure that if $I_n = [a_n, b_n]$ is the $n$th interval, then

(1) $$f(a_n) \leqslant y_0 \leqslant f(b_n).$$

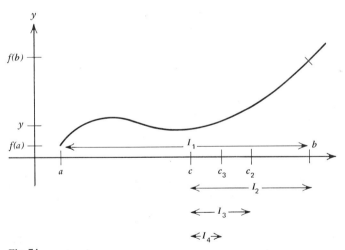

**Fig. 74**

Figure 74 illustrates the situation. We know by our discussion of the bisection algorithm that there is precisely one number $x_0$ common to all the intervals $I_n$. We must now prove that $f(x_0) = y_0$. To do that we must exploit the continuity of $f$. Suppose $f(x_0) < y_0$. Then set $\epsilon = y_0 - f(x_0)$. Choose $\delta$ so that

$$0 < |x - x_0| < \delta \text{ implies } |f(x) - f(x_0)| < \epsilon.$$

Now choose $n$ large enough so that $I_n = [a_n, b_n] \subset (x_0 - \delta, x_0 + \delta)$. Then $|f(a_n) - f(x_0)| < \epsilon$ and $|f(b_n) - f(x_0)| < \epsilon$. But this implies $f(a_n) < y_0$

and $f(b_n) < y_0$, which is impossible since it follows from our construction that for each integer $n$

$$f(a_n) \leq y_0 \leq f(b_n).$$

In a similar way we prove that $f(x_0) > y_0$ is also impossible. Thus, it must follow that $f(x_0) = y_0$ and we are finished.

The argument for Theorem 5.1 is similar but a bit more delicate. For definiteness we shall prove that if $f$ is continuous on the closed interval $[a, b]$, then there is a point $x_0 \in [a, b]$ such that $f(x_0) \geq f(x)$ for each $x \in [a, b]$. The problem now is this: What is the right question to ask so that the number $x_0$ produced by the bisection algorithm has the property that $f(x_0)$ is a maximum value for $f$ on $[a, b]$? To do this we must be able to show that if $I_n$ is the $n$th interval, then no point $x'$ outside $I_n$ has the property that

$$f(x') > f(x) \qquad \text{for each } x \text{ in } I_n.$$

Therefore, as before, set $I_1 = [a, b]$, and let $c$ be the midpoint of $I_1$. Let $J_1 = [a, c]$, $K_1 = [c, b]$, respectively. Then we ask: Is there a number $x' \in K_1$ such that $f(x') \geq f(x)$ for each $x \in J_1$?

$$\textit{Answer} \qquad \begin{array}{ll} \text{Yes:} & \text{Take } I_2 = K_1 \\ \text{No:} & \text{Take } I_2 = J_1 \end{array}$$

Replace $I_1$ by $I_2$ and repeat.

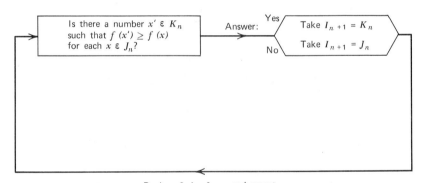

**Fig. 75**

Thus for the $n$th stage, if $J_n$, $K_n$ are the left- and right-hand halves of the interval $I_n$, we diagram the algorithm as in Figure 75. It is clear now that from this construction we may infer the following.

$$\bigstar \begin{cases} \text{For no point } x' \text{ outside } I_n \text{ is it possible that} \\ \qquad f(x') > f(x) \qquad \text{for each } x \in I_n. \end{cases}$$

Let $x_0$ be the point common to all the intervals $I_n$:

To prove $f(x_0)$ is a maximum value for $f$ on $[a, b]$, i.e., that $f(x_0) \geq f(x)$ for each $x \in [a, b]$, we use an indirect argument. Suppose this is not the case. That is, suppose there is a point $x_1$ such that

$$f(x_1) > f(x_0).$$

Then let $\epsilon = f(x_1) - f(x_0)$. Since $f$ is continuous, there is a corresponding number $\delta$ so that if $|x - x_0| < \delta$ then

$$|f(x) - f(x_0)| < \epsilon.$$

But from our construction there will be an integer $n$ so that all numbers $x$ in $I_n$ satisfy $x_0 - \delta < x < x_0 + \delta$. However, for this value of $n$, $f(x_1) > f(x)$ for all $x$ in $I_n$. But this violates condition ($\bigstar$) above. Therefore, our hypothesis that there is an $x_1$ for which $f(x_1) > f(x_0)$ must be false, and consequently the theorem is proved. Figure 76 illustrates the situation.

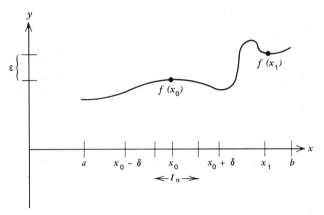

**Fig. 76** For all $x$ in $I_n |f(x) - f(x_0)| < \epsilon$. In particular, $f(x_0) < f(x_1)$.

**EXERCISES**

1. Starting with the interval $[1, 2]$, determine five successive intervals $I_2, \ldots, I_6$ in the determination of $\sqrt{2}$ by the bisection algorithm (Apply the argument used in the proof of Theorem 5.2 to the problem of finding an $x$ satisfying $x^2 = 2$.)
2. Answer Exercise 1 for $\sqrt{3}$.
3. Suppose the bisection algorithm is applied to the interval $[0, 1]$ and each time the left-hand interval is chosen we record 0 and each time the right-hand interval is chosen we record 1. If $x$ is the number common to all the intervals $I_n$, then associated with $x$ we have an infinite sequence of zeros and ones. For example

$$\frac{1}{3} = 0101 \ldots \text{etc.}$$

This sequence is called the binary expansion of $x$. Determine the first five terms in the binary expansion of
   (a) $\frac{1}{5}$
   (b) $\frac{2}{3}$
   (c) $\frac{1}{7}$
4. Construct an algorithm that would yield the decimal expansions of a number $x, 0 \leqslant x \leqslant 1$.
5. Show that in the proof of Theorem 5.2 that it is also impossible that $f(x_0) > y_0$.
6. Show that a continuous function on a closed interval has the minimum value property.
7. Let $I_1 \supset I_2 \supset \cdots \supset I_n \supset \cdots$ be a decreasing sequence of closed intervals. Using the completeness axiom, prove that there must always exist a point common to all the intervals $I_n$. If $I_n = [a_n, b_n]$ and $\lim_{n \to \infty} (b_n - a_n) = 0$, show that this point is unique. This result is often called the nested interval theorem.

**\*5.13   The Step Function Approximation Theorem; Uniform Continuity**

We wish to use the bisection algorithm technique to prove the following "step function approximation theorem." It asserts that a continuous function defined on a closed interval $[a, b]$ may be approximated to any degree of accuracy on this interval by a step function $h$.

**THEOREM 5.11**   *Let f be continuous on the closed interval* $[a, b]$. *Then for each* $\epsilon > 0$ *there exists a step function h satisfying*

$$|h(x) - f(x)| < \epsilon \qquad for\ all \qquad x \in [a, b].$$

It follows as a consequence of Theorem 5.11 that each continuous function is integrable on $[a, b]$. (See Chapter 3, p. 129.)

Before proving Theorem 5.11 we must establish a preliminary result. We have already noted (Exercise 7, p. 251) that if $I_n = [d_n, e_n]$ are closed intervals, which are nested, i.e., $I_1 \supset I_2, \ldots \supset I_n), \ldots$, and if $\lim_{n \to \infty} e_n - d_n = 0$, then there is precisely one point common to all the intervals $I_n$. What happens if we drop the assumption that the intervals are nested? It is fairly clear that we cannot assert that there is a point common to all the intervals. We leave it as an exercise to construct an example where this is the case.

However, if we assume that all the intervals $I_n$ are contained in a fixed interval $I$, then we can assert that there is a point $x_0 \in I$ such that infinitely many of the intervals $I_n$ are as close to $x_0$ as we wish. Let us state this result precisely. To avoid confusion with the intervals constructed in the bisection algorithm, we label these intervals $L_n$.

**LEMMA 5.12**   *For* $n = 1, 2, \ldots$, *let* $L_n = [d_n, e_n]$ *be closed intervals contained in the closed interval* $I = [a, b]$. *Assume that* $\lim_{n \to \infty} e_n - d_n = 0$. *Then there exists a point* $x_0 \in I$ *such that for each* $\epsilon > 0$, *infinitely many of the intervals* $L_n$ *are contained in the interval* $(x_0 - \epsilon, x_0 + \epsilon)$.

**Proof.**   We use the bisection algorithm to establish the existence of the point $x_0$. Let $c$ be the midpoint of the interval $[a, b]$. Let $J_1 = [0, c]$, $K_1 = [c, b]$. We ask the following question: Do infinitely many of the numbers $\{d_k\}$ belong to $J_1$? *Answer:* yes, take $I_2 = J_1$; no: take $I_2 = K_1$. Bisect $I_2$ and repeat. The flow diagram for this bisection algorithm is Figure 77.

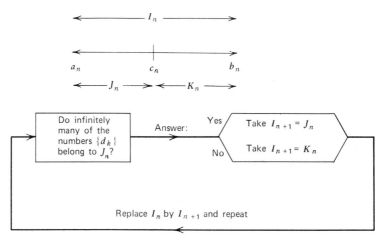

**Fig. 77**

At each stage of the construction infinitely many of the number $\{d_k\}$ must belong to either $J_n$ or $K_n$, so we may conclude that infinitely many of the numbers $\{d_k\}$ belong to each interval $I_n$. Let $x_0$ be the point common to the intervals $I_n$. Choose $\epsilon > 0$ and pick an interval $I_n$ such that $I_n \subset [x_0 - \epsilon/2, x_0 + \epsilon/2]$. By our construction, infinitely many of the numbers $\{d_k\}$ belong to $I_n$. But since $\lim_{k\to\infty} e_k - d_k = 0$, we may assert that infinitely many of the intervals $L_k = [d_k, e_k] \subset (x_0 - \epsilon, x_0 + \epsilon)$. Indeed these will be the intervals $[d_k, e_k]$ such that $d_k \in I_n$ and $|d_k - e_k| < \epsilon/2$. This completes the proof of the lemma.

To use this lemma to prove Theorem 5.11, we proceed as follows. Partition the interval $[a, b]$ into equal subintervals of width $(b - a)/2^n$. Then we may write $a = a_0 < a_1 < \cdots < a_n = b$ and $a_k - a_{k-1} = (b - a)/2^n$. Let $M_k$ and $m_k$ be the maximum and minimum values of $f$ on the interval $[a_{k-1}, a_k]$. Define the step function $h_n(x)$ by requiring that

$$h_n(x) = \frac{m_k + M_k}{2} \quad \text{if} \quad a_{k-1} \leq x < x_k.$$

If $\epsilon_n = \max_{k=1,\ldots,2^n} M_k - m_k$, then on each subinterval $[a_{k-1}, a_k]$

$$|f(x) - h_n(x)| \leq \frac{M_k - m_k}{2} < \epsilon_n.$$

Therefore, for all $x$ in $[a, b]$,

$$|f(x) - h(x)| < \epsilon_n.$$

Theorem 5.11 will be proved if we can show that $\lim_{n\to\infty} \epsilon_n = 0$. The construction of the approximating step function is illustrated in Figure 78.

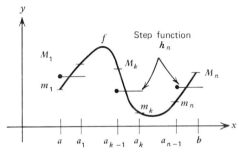

**Fig. 78**

To prove $\lim_{n\to\infty} \epsilon_n = 0$ we must use the lemma and exploit the continuity of $f$. For each value of $n$, $\epsilon_n = M_k - m_k$ for some value of $k = 1, 2, \ldots 2^n$. Let $L_n = [a_{k-1}, a_k]$ be the corresponding interval of width $(b - a)/2^n$. Notice that $\epsilon_{n+1} \leq \epsilon_n$ since each subinterval at the $n+1$st stage has half the length of a subinterval at the $n$th stage. By the lemma there is a point $x_0$ such that infinitely many of the intervals $L_n$ are as close to $x_0$ as we wish. However, $f$ is continuous at $x_0$. Hence, for each $\epsilon > 0$, there is a $\delta > 0$ such that if $|x - x_0| < \delta$ then $|f(x) - f(x_0)| < \epsilon$. But by the lemma there must be an interval $L_{n_0}$, in fact infinitely many, satisfying

$$L_{n_0} \subset (x_0 - \delta, x_0 + \delta).$$

Therefore, if $x, x' \in L_{n_0}$, it follows that

$$|f(x) - f(x')| < |f(x) - f(x_0)| + |f(x_0) - f(x')| < 2\epsilon.$$

Hence, $\epsilon_{n_0} = M_k - m_k = \max_{x \in L_{n_0}} f(x) - \max_{x \in L_{n_0}} f(x) < 2\epsilon$. Since $\epsilon_{n+1} \leq \epsilon_n$ we conclude that $\epsilon_n < 2\epsilon$ for $n \geq n_0$. This is just the assertion that $\lim_{n\to\infty} \epsilon_n = 0$, and the proof of Theorem 5.11 is complete.

In proving Theorem 5.11 we have in fact verified a very important property of continuous functions defined on *closed* intervals. We have shown that for each $\epsilon > 0$ there exists a $\delta > 0$ such that $|f(x) - f(x')| < \epsilon$ just so long as the points $x, x'$ satisfy $|x - x'| < \delta$. This is a much stronger condition than just the assertion that $f$ is continuous at each fixed point $x_0$. We give a formal definition of this property.

**Definition.** *A function $f$ is said to be* uniformly *continuous on a set $S$ of real numbers if, for each $\epsilon > 0$, there is a corresponding $\delta > 0$ such that $|f(x) - f(x')| < \epsilon$ whenever $x, x' \in S$ and $|x - x'| < \delta$.*

What we have demonstrated in the proof of Theorem 5.11 is the following.

**THEOREM 5.13** *If the function $f$ is continuous at each point of the closed interval $[a, b]$, then $f$ is uniformly continuous on $[a, b]$.*

Hints for the proof of this result based on what we have done in the proof of Theorem 5.11 are contained in the exercises. We also note that if a function $f$ is continuous at each point of an *open* interval $(a, b)$, then $f$ is not necessarily uniformly continuous on $(a, b)$. Examples illustrating this are discussed in the exercises.

**EXERCISES**

1. If open intervals $I_n = (a_n, b_n)$ are nested and $\lim_{n \to \infty} b_n - a_n = 0$, must there exist a point common to the intervals $I_n$? Explain.

2. Suppose the closed intervals $I_n = [a_n, b_n]$ are contained in a fixed interval $[a, b]$ and suppose $\lim_{n \to \infty} b_n - a_n = 0$. Provide an example that shows that there does not necessarily exist a point common to all of the intervals $I_n$.

3. Prove Theorem 5.13. (Suppose the theorem is false. Then for some $\epsilon > 0$, there is no $\delta > 0$ with the property that $|x - x'| < \delta$ implies $|f(x) - f(x')| < \epsilon$. Thus, if $\delta_n > 0$ we may find points $d_n, e_n$ satisfying $|d_n - e_n| < \delta_n$ and $|f(d_n) - f(e_n)| \geq \epsilon$. Assume $d_n < e_n$ and set $L_n = [d_n, e_n]$. By Lemma 5.12 there is a point $x_0$ in $[a, b]$ such that infinitely many of the intervals $L_n$ are arbitrarily close to $x_0$. Using the argument of Theorem 5.11, show that $f$ is not continuous at $x_0$. Since this is a contradiction, Theorem 5.12 is proved.)

4. The function $f(x) = 1/x$ is continuous on the open interval $(0, 1)$. Show that this function is not uniformly continuous on this interval (Show that it is impossible to find a $\delta$ so small that $|x - x'| < \delta$ implies $|f(x) - f(x')| < 1$.)

5. Show the function $f(x) = \sin \pi/x$ is not uniformly continuous on the interval $(0, 1)$. (Use the same idea as in Exercise 4.)

**CHAPTER SIX**

# FURTHER TECHNIQUES AND APPLICATIONS OF INTEGRATION

**6.1 Elementary Formulas**

In this chapter we shall consider in some detail the problem of determining a primitive or antiderivative $F$ for a given continuous function $f$.

If $F' = f$, then we write

$$F(x) = \int f(x)\,dx.$$

The symbol on the right is called the indefinite integral of $f$. Since any two primitives for $f$ differ by an additive constant, the symbol $\int f(x)\,dx$ is not uniquely determined. For example, if $f(x) = x^2$, then

$$\int x^2 dx = \frac{x^3}{3},$$

but also $\int x^2 dx = x^3/3 + 10$.

Both formulas are correct since

$$\frac{d}{dx}\left(\frac{x^3}{3}\right) = \frac{d}{dx}\left(\frac{x^3}{3} + 10\right) = x^2$$

Thus each primitive for a continuous function $f$ may be written

$\int f(x)\,dx + C$ for some constant $C$.

In the symbol $\int f(x)\,dx$, the function $f$ is referred to as the *integrand*.

To evaluate a *definite* integral $\int_a^b f$ we first must compute a primitive $F(x) = \int f(x)\,dx$. Then

$$\int_a^b f = F(x)\Big|_{x=a}^{x=b} = F(b) - F(a).$$

We may translate the differentiation formulas that we have encountered into formulas for primitives. The following is a list of the more important formulas.

| Function $f(x)$ | Primitive $F(x) = \int f(x)\,dx$ |
|---|---|
| (1) $x^r \; r \neq -1$ | $\dfrac{x^{r+1}}{r+1}$ |
| (2) $\dfrac{1}{x}$ | $\log|x|$ |
| (3) $e^x$ | $e^x$ |

255

| Function | Primitive |
|---|---|
| (4) $\sin x$ | $-\cos x$ |
| (5) $\cos x$ | $\sin x$ |
| (6) $\sec^2 x$ | $\tan x$ |
| (7) $\dfrac{1}{1+x^2}$ | arc tan $x$ |
| (8) $\dfrac{1}{\sqrt{1-x^2}}$ | arc sin $x$ |

Other formulas have been encountered to be sure but for the moment we shall concentrate on extending the above. To obtain new integration formulas or new formulas for primitives, we shall exploit the linearity of the indefinite integral

$$(9) \qquad \int kf(x)\,dx = k \int f(x)\,dx \qquad k \text{ a constant}$$

$$(10) \qquad \int (f(x)+g(x))\,dx = \int f(x)\,dx + \int g(x)\,dx$$

and the chain rule. This states that

$$\text{if } h(x) = f(g(x)), \qquad \text{then } h'(x) = f'(g(x))g'(x).$$

The analogous indefinite integral formula is

$$(11) \qquad f(g(x)) = \int f'(g(x))g'(x)\,dx.$$

We have met several instances of formula (11) in Chapter 4.

If $n \neq -1$ and $f'(x) = x^n$, then $f'(g(x))g'(x) = [g(x)]^n g'(x)$, and

$$(12) \qquad \frac{[g(x)]^{n+1}}{n+1} = \int [g(x)]^n g'(x)\,dx$$

If $f'(x) = 1/x$, then $f'(g(x))g'(x) = g'(x)/g(x)$, and

$$(13) \qquad \log |g(x)| = \int \frac{g'(x)}{g(x)}\,dx$$

(see p. 145).

Perhaps the central problem in finding formulas for indefinite integrals $\int f(x)\,dx$ is being able to recognize when a given function $h(x)$ is of the form

$$h(x) = f'(g(x))g'(x).$$

For then by (11), $\int h(x)\,dx = f(g(x))$. Sometimes this recognition is easy; sometimes it is not. In this and the following section we shall develop techniques for simplifying this recognition problem. However, there is no certain road to success. First, the student should train himself to spot the derivatives of the functions $F$ in the foregoing list. The best way to accomplish this is to work a large variety of examples.

**Example 1.**   Compute $\int 2x(x^2+1)^5\,dx$. Letting $g(x) = x^2+1$, we have $g'(x) = 2x$ if $h'(x) = x^5$, then

$$2x(x^2+1)^5 = g'(x)h'(g(x)).$$

Consequently

$$\int 2x(x^2+1)^5\,dx = \frac{(x^2+1)^6}{6}.$$

***Example 2.***   Compute

$$\int e^x \sin e^x \, dx.$$

Again, if $g(x) = e^x$, $g'(x) = e^x$; and if $h'(x) = \sin x$, then

$$e^x \sin e^x = g'(x)h'(g(x)).$$

Hence

$$\int e^x \sin e^x \, dx = -\cos e^x$$

Sometimes the fact that the function $h$ is of the form $h(x) = f'(g(x))g'(x)$ where $f$ and $g$ are known is pretty well disguised.

***Example 3.***   Compute

$$\int \frac{dx}{x^2 + 2x + 2}.$$

In this example we must complete the square in the denominator before it is in a form we can handle. But $x^2 + 2x + 2 = (x+1)^2 + 1$. Hence

$$\int \frac{dx}{x^2 + 2x + 2} = \int \frac{dx}{(x+1)^2 + 1}.$$

Setting $f'(x) = 1/(1+x^2)$ and $g(x) = x+1$, we have

$$\frac{1}{(x+1)^2 + 1} = f'(g(x))g'(x).$$

Hence from (7)

$$\int \frac{dx}{x^2 + 2x + 2} = \int \frac{dx}{(x+1)^2 + 1} = \text{arc tan } (x+1).$$

***Example 4.***   Compute $\int xe^{x^2} \, dx$. If we let $f(x) = e^x$ and $g(x) = x^2$, then

$$f'(g(x))g'(x) = 2xe^{x^2}$$

Therefore, $\int 2xe^{x^2} \, dx = e^{x^2}$. Hence by (9)

$$\int xe^{x^2} \, dx = \frac{2}{2}\int xe^{x^2} \, dx = \frac{1}{2}\int 2xe^{x^2} \, dx = \frac{1}{2}e^{x^2}.$$

The trick of writing an indefinite integral $\int h(x)\,dx$ as $1/k \int kh(x)\,dx$ for some appropriate nonzero constant $k$ is very useful. If $kh(x) = f'(g(x))g'(x)$, this is all that is necessary to compute $\int h(x)\,dx$. For then, by (11),

$$\int h(x)\,dx = \frac{1}{k}\int kh(x)\,dx = \int f'(g(x))g'(x)\,dx = \frac{1}{k}f(g(x)).$$

When integrals involve radicals such as $\sqrt{ax^2 + bx + c}$, it is usually necessary to complete the square before proceeding further.

***Example 5.***   Compute

$$\int \frac{dx}{\sqrt{15 - 4x - 4x^2}}.$$

Completing the square in the denominator we have

$$\int \frac{dx}{\sqrt{15 - 4x - 4x^2}} = \int \frac{dx}{\sqrt{16 - (1 + 4x + 4x^2)}}$$

$$= \int \frac{dx}{\sqrt{16 - (1 + 2x)^2}} = \frac{1}{4}\int \frac{dx}{\sqrt{1 - ((1 + 2x)/4)^2}}.$$

Now, if $g(x) = (1+2x)/4$ and $f'(x) = 1/(1-x^2)$, then

$$f'(g(x))g'(x) = \frac{1}{2} \frac{1}{\sqrt{1-((1+2x)/4)^2}}.$$

Hence

$$f(g(x)) = \text{arc sin}\left(\frac{1+2x}{4}\right)$$

$$= \int \frac{1}{2} \frac{dx}{\sqrt{1-((1+2x)/4)^2}}.$$

Therefore

$$\int \frac{dx}{\sqrt{15-4x-4x^2}} = \frac{1}{4} \int \frac{dx}{\sqrt{1-((1+2x)/4)^2}}$$

$$= \frac{1}{2} \int \frac{1}{2} \frac{dx}{\sqrt{1-((1+2x)/4)^2}} = \frac{1}{2} \text{arc sin}\left(\frac{1+2x}{4}\right).$$

Formula (13), $\log |f(x)| = \int f'(x)/f(x)\, dx$, has already been given in Chapter 4.

*Example 6.*   Compute

$$\int \frac{x}{1-x^2} dx.$$

If $f(x) = 1-x^2$, then $f'(x) = -2x$. Hence

$$\int \frac{x}{1-x^2} = -\frac{1}{2} \int \frac{-2x}{1-x^2}\, dx = -\frac{1}{2} \log |1-x^2|.$$

The fact that a function $h$ is of the form $f'/f$ may be disguised.

*Example 7.*   Compute $\int \tan x\, dx$.   Now   $\tan x = \sin x/\cos x = -f'(x)/f(x)$   if $f(x) = \cos x$. Therefore

$$\int \tan x\, dx = -\int \frac{-\sin x\, dx}{\cos x}$$

$$= -\log |\cos x|.$$

**EXERCISES**

1. Compute the following indefinite integrals. The integrands are all of the form $f'(g(x))g'(x)$ for known functions $f$ and $g$.

(a) $\displaystyle\int 2(2x-1)^4\, dx$

(b) $\displaystyle\int \frac{3x^2+1}{(x^3+x)^2} dx$

(c) $\displaystyle\int \frac{2x+1}{x^2+x} dx$

(d) $\displaystyle\int (x^4+x^3)^5(4x^3+3x^2)\, dx$

(e) $\displaystyle\int -2xe^{-x^2}\, dx$

(f) $\displaystyle\int e^{\sin x} \cos x\, dx$

(g) $\displaystyle\int \sin^3 x \cos x\, dx$

(h) $\displaystyle\int \tan x \sec^2 x\, dx$

(i) $\displaystyle\int \frac{e^x}{1+e^{2x}} dx$

(j) $\displaystyle\int \frac{(\text{arc tan } x)}{1+x^2} dx$

(k) $\displaystyle\int \frac{\log x}{x} dx$

(l) $\displaystyle\int e^x \sec^2 e^x\, dx$

(m) $\displaystyle\int \frac{dx}{\sqrt{2x-x^2}}$

(n) $\displaystyle\int \frac{2dx}{2+4x+4x^2}.$

2. Compute the following indefinite integrals. When multiplied by a suitable constant the integrands are of the form $f'(g(x))g'(x)$ for known functions $f$ and $g$.

(a) $\displaystyle \int \frac{dx}{4x+1}$

(h) $\displaystyle \int 2x\sqrt{1-3x^2}\,dx$

(b) $\displaystyle \int \frac{x}{(x^2+1)^3}\,dx$

(i) $\displaystyle \int \sqrt{x^2-x^4}\,dx$

(c) $\displaystyle \int \frac{(\log x^2)^3}{x}\,dx$

(k) $\displaystyle \int \cot 4x\,dx$

(d) $\displaystyle \int e^{\sin x^2} x \cos x^2\,dx$

(l) $\displaystyle \int \cos 2x \sin x \cos x\,dx$

(e) $\displaystyle \int \cos^3 4x \sin 4x\,dx$

(m) $\displaystyle \int \frac{x^2}{\sqrt{1-x^3}}\,dx$

(f) $\displaystyle \int \frac{x\,dx}{1+x^4}$

(n) $\displaystyle \int \frac{x}{4+3x^4}\,dx.$

(g) $\displaystyle \int x \tan x^2\,dx$

For integrals involving trigonometric functions the various trigonometric identities must often be exploited before the integral is in a form that can be evaluated.

**Example 8.**  Compute

$$\int \sin^2 x\,dx.$$

The integrand cannot be handled directly. However, $2\sin^2 x = 1 - \cos 2x$. Therefore

$$\int \sin^2 x\,dx = \int \left(\frac{1}{2} - \frac{1}{2}\cos 2x\right) dx$$

$$= \frac{x}{2} - \frac{1}{4}\sin 2x.$$

**Example 9.**  Evaluate $\int \sin x \cos 2x\,dx$.

From the addition formulas we know that

$$\sin (a+b) = \sin a \cos b + \cos a \sin b$$

and

$$\sin (a-b) = \sin a \cos b - \cos a \sin b.$$

Therefore, adding these two identities, we have

(14) $$\sin a \cos b = \frac{1}{2}\left[\sin (a+b) + \sin (a-b)\right].$$

Therefore

$$\sin x \cos 2x = \frac{1}{2}\left[\sin 3x + \sin (-x)\right]$$

$$= \frac{1}{2}\left[\sin 3x - \sin x\right],$$

and hence

$$\int \sin x \cos 2x\,dx = \frac{1}{2}\int \left[\sin 3x - \sin x\right] dx$$

$$= \frac{-1}{6}\cos 3x + \frac{1}{2}\cos x.$$

Identity (14) together with

(15) $$\sin a \sin b = \frac{1}{2}[\cos (a-b) - \cos (a+b)]$$

(16) $$\cos a \cos b = \frac{1}{2}[\cos (a-b) + \cos (a+b)]$$

enable us to handle all integrals of the form

$$\int \sin mx \cos nx \, dx, \int \sin mx \sin nx \, dx$$

$\int \cos mx \cos nx \, dx$ where $m$ and $n$ are constant and $m \neq n$.

Next we consider trigonometric integrals of the form $\int \sin^n x \cos^m x \, dx$, where $n$ and $m$ are positive integers. If one of these integers, for example, $m$, is odd, then the integral may be easily evaluated.

***Example 10.*** Evaluate $\int \sin^2 x \cos^3 x \, dx$.

Now

$$\sin^2 x \cos^3 x = \sin^2 x \cdot \cos^2 x \cdot \cos x$$
$$= \sin^2 x (1 - \sin^2 x) \cos x$$
$$= (\sin^2 x - \sin^4 x) \cos x.$$

Therefore

$$\int \sin^2 x \cos^3 x \, dx = \int \sin^2 x \cos x \, dx - \int \sin^4 x \cos x \, dx$$
$$= \frac{\sin^3 x}{3} - \frac{\sin^5 x}{5}.$$

Thus, to evaluate $\int \sin^n x \cos^{2k+1} x \, dx$, we write $\sin^n x \cos^{2k+1} x$ as $\sin^n x \cos^{2k} x \cos x = \sin^n x (1 - \sin^2 x)^k \cos x$. Since $\sin^n x (1 - \sin^2 x)^k$ is a polynomial in $\sin x$, we may multiply it out. The integral $\int \sin^n x \cos^{2k+1} x \, dx$ then is a sum of integrals of the form $a_l \int \sin^l x \cos x \, dx$. This is easily integrated to $a_l/(l+1) \sin^{l+1} x$. A similar technique handles integrals of the form $\int \cos^n x \sin^{2k+1} x \, dx$.

**EXERCISES**

3. Evaluate the following trigonometric integrals.

(a) $\int \cos^2 x \, dx$

(b) $\int \sin^2 2x \, dx$

(c) $\int \cos^3 x \, dx$

(d) $\int \cos^2 x \sin^3 x \, dx$

(e) $\int \tan x \sec^4 x \, dx$

(f) $\int \cot 3x \, dx$

(g) $\int \cos^3 x \sin^4 x \, dx$

(h) $\int \sin^5 x \cos^2 x \, dx$

(i) $\int \cos^2 x \sin 2x \, dx$

(j) $\int \sin 4x \cos 3x \, dx$

(k) $\int \sin x \sin 4x \, dx$

(l) $\int \cos 3x \cos x \, dx$

The foregoing method will not handle trigonometric integrals of the form $\int \sin^{2n} x \cos^{2m} x \, dx$. To evaluate such an integral, we first convert the integrand into a polynomial in even powers of $\cos x$ by substituting $(1 - \cos^2 x) =$

$\sin^2 x$. Thus $\sin^{2n} x \cos^{2m} x = (1 - \cos^2 x)^n \cos^{2m} x$. We have reduced the problem to that of handling integrals of the form

$$\int \cos^{2m} x \, dx.$$

But $\cos^{2m} x$ can be written

$$\cos^{2m} x = a_0 + a_1 \cos 2x + a_2 \cos 4x + \cdots + a_m \cos 2mx$$

by successive applications of the double angle formula. We illustrate this in the following example.

**Example 11.**   Compute $\int \cos^4 x \, dx$. Now $\cos^2 x = (\cos 2x + 1)/2$. Therefore, using this identity twice, we get

$$\cos^4 x = \frac{1}{4}[\cos 2x + 1]^2 = \frac{1}{4}[\cos^2 2x + 2 \cos 2x + 1]$$

$$= \frac{1}{4}\left[\left(\frac{\cos 4x + 1}{2}\right) + 2 \cos 2x + 1\right]$$

$$= \frac{\cos 4x}{8} + \frac{1}{2}\cos 2x + \frac{3}{8}.$$

Hence $\int \cos^4 x \, dx = \frac{1}{32}\sin 4x + \frac{1}{4}\sin 2x + (\frac{3}{8}x)$. For high powers of $\cos x$ this method is quite laborious. A simpler method is to use the following reduction formula, which will be derived in Section 6.3.

$$(17) \qquad \int \cos^n x \, dx = \frac{\cos^{n-1} x \sin x}{n} + \frac{n-1}{n}\int \cos^{n-2} x \, dx.$$

A similar formula holds for powers of the sine function,

$$(18) \qquad \int \sin^n x \, dx = -\frac{\sin^{n-1} x \cos x}{n} + \frac{n-1}{n}\int \sin^{n-2} x \, dx.$$

**EXERCISES**

4. Evaluate the following trigonometric integrals using the method of Example 11 or one of the reduction formulas.

(a) $\int \sin^4 x \, dx$

(b) $\int \sin^2 x \cos^2 x \, dx$

(c) $\int \sin^3 x \, dx$

(d) $\int \cos^2 x \sin^4 x \, dx$

(e) $\int \cos^6 x \, dx$

(f) $\int \sin^2 \frac{x}{2} \cos^4 \frac{x}{2} \, dx$

## 6.2   Integration by Substitution

The computation of definite integrals $\int_a^b f(x)\,dx$ and the finding of primitives for functions $f$ is often greatly simplified by substituting a new function $x = g(u)$ for $x$. The basic result that underlies this technique is the following substitution theorem.

**THEOREM 6.1**   *Let $f$ be continuous on $[a, b]$ and assume that the function $g$ has a continuous derivative on the interval $[c, d]$. If $g(c) = a$, $g(d) = b$,* and $a \leq g(u) \leq b$, whenever $c \leq u \leq d$, then,

(1) $$\int_a^b f(x)dx = \int_c^d f(g(u))g'(u)du.$$

***Proof.*** The proof of this result is very simple and depends only on the chain rule and the fundamental theorem of the calculus. Let $F$ be a primitive function for $f$. Then $F(b) - F(a) = \int_a^b f(x)dx$. Set $H(u) = F(g(u))$. Then $H'(u) = F'(g(u))g'(u) = f(g(u))g'(u)$.

Hence, by the fundamental theorem of the calculus,

$$H(d) - H(c) = \int_c^d f(g(u))g'(u)du.$$

But since $g(c) = a$ and $g(d) = b$,

$$H(d) - H(c) = F(g(d)) - F(g(c)) = F(b) - F(a) = \int_a^b f(x)\,dx,$$

and we are finished.

The trick in applying Theorem 6.1 is determining an appropriate function $x = g(u)$ so that the integral $\int_c^d f(g(u))g'(u)du$ can be easily evaluated.

***Example 1.*** Compute $\int_0^1 x\sqrt{1+2x}\,dx$. We try to choose a function $x = g(u)$ that simplifies the radical. If we set $1 + 2x = u$, then $x = (u-1)/2$. Since then $\sqrt{1+2x} = \sqrt{u}$, the substitution $x = g(u) = (u-1/2)$ looks promising. Now if $f(x) = x\sqrt{1+2x}$, $f(g(u)) = \sqrt{u}(u-1)/2$ and $g'(u) = 1/2$. Moreover $g(1) = 0$ and $g(3) = 1$. Therefore, $f(g(u))g'(u) = \sqrt{u}(u-1)/4$ and by (1) $\int_0^1 x\sqrt{1+2x}\,dx = \int_1^3 \sqrt{u}(u-1)/4\,du$. We can easily compute the latter integral. Indeed

$$\int_0^3 \left(\frac{u-1}{4}\right)\sqrt{u}\,du = \frac{1}{4}\left[\frac{2}{5}u^{5/2} - \frac{2}{3}u^{3/2}\right]_1^3 = \frac{2}{5}\sqrt{3} + \frac{1}{15}$$

To emphasize in (1) that $g(c) = a$ and $g(d) = b$, we write (1)

(1) $$\int_{g(c)}^{g(d)} f(x)dx = \int_c^d f(g(u))g'(u)du.$$

The actual mechanics of performing the substitution is virtually automatic if we use differentials that were introduced in Section 2.11. If $x = g(u)$, then $dx = g'(u)du$ and, as a result, $f(x)dx = f(g(u))g'(u)du$.

We may use this substitution technique to determine primitives for $f$ as well. Suppose $F(x) = \int f(x)dx$ is desired, and $H(u) = \int f(g(u))g'(u)du$ is known. Then if our substitution function $g$ is $1:1$ on some interval $[c, d]$ and $g'$ does not vanish on this interval, we may compute $F$ in the following way. Let $G$ be the inverse function for $g$. The function $G$ is defined on the interval $[g(c), g(d)]$, and if $x = g(u)$, then $u = G(x)$. Moreover

(2) $$g'(u)G'(x) = 1$$

(See Section 4.5, p. 160)

We assert that $P(x) = H(G(x))$ is a primitive for $f$ on the interval $[g(c), g(d)]$. To show this we must verify that $P'(x) = f(x)$. However, by applying the chain rule we have

$$P'(x) = H'(G(x))G'(x)$$
$$= H'(u)G'(x) \qquad \text{since } u = G(x),$$

$$= h(u)G'(x) \qquad \text{by definition of } H$$
$$= f(g(u))g'(u)G'(x) \qquad \text{by definition of } h$$
$$= f(g(u)) \cdot 1 \qquad \text{by equation (2)}$$
$$= f(x) \qquad \text{since } x = g(u).$$

Notice that in order to compute the primitive $F$ we had to impose the condition that $g$ is $1:1$. This was necessary in order to define the inverse $G$. However, to evaluate the definite integral (1) by our substitution technique we do not have to make this restriction.

In summary, to evaluate the definite integral $\int_a^b f(x)dx$ or to compute a primitive for $f$ by the substitution technique, first pick a function $x = g(u)$, so that if

$$f(x)dx = f(g(u))g'(u)du = h(u)du$$

then the indefinite integral $\int h(u)du$ can be evaluated. Then

$$\int_{a=g(c)}^{b=g(d)} f(x)dx = \int_c^d h(u)du$$

If $g$ is $1:1$ with inverse $G$, and $H(u) = \int h(u)\,du$, then $H(G(x)) = \int f(x)\,dx$.

The trick in applying this "integration by substitution" technique is, of course, picking the right function $x = g(u)$. We shall discuss many typical examples but there are no all-inclusive rules. Skill with this technique comes only with experience!

***Example 2.***   Evaluate $\int_0^1 x\sqrt{2-x}\,dx$, and compute $\int x\sqrt{2-x}\,dx$. We could use the substitution $u = 2-x$, but instead let us set $u = \sqrt{2-x}$. Then $x = 2-u^2$ and $dx = -2u\,du$. Consequently

$$x\sqrt{2-x}\,dx = (2-u^2)u\,(-2u)\,du$$

Moreover, if $u = \sqrt{2}$, then $x = 0$ and if $u = 1$, then $x = 1$. Hence, by (1),

$$\int_0^1 x\sqrt{2-x}\,dx = -2\int_{\sqrt{2}}^1 (2-u^2)u^2 du$$

$$= 2\int_1^{\sqrt{2}} (2-u^2)u^2 du = \frac{4}{3}u^3 - \frac{2u^5}{5}\Big|_{u=1}^{u=\sqrt{2}}$$

$$= \frac{16}{15}\sqrt{2} - \frac{14}{15}$$

To compute the indefinite integral $\int x\sqrt{2-x}\,dx$, we note that

$$\int x\sqrt{2-x}\,dx = -2\int (2-u^2)u^2\,du$$

$$= -2\int (2u^2 - u^4)\,du$$

$$= -\frac{4}{3}u^3 + \frac{2u^5}{5}$$

$$= -\frac{4}{3}(2-x)^{3/2} + \frac{2}{5}(2-x)^{5/2}.$$

This result may be checked immediately by differentiating the right-hand side. In general, if $P(x, \sqrt[n]{a+bx})$ is a polynomial in $x$ and $\sqrt[n]{a+bx}$, then the substitution $u = \sqrt[n]{a+bx}$ converts the integral $\int P(x, \sqrt[n]{a+bx})\,dx$ into an integral of the form $\int Q(u)\,du$ where $Q(u)$ is a polynomial in $u$. The integral $\int Q(u)\,du$ then can be evaluated immediately.

**Example 3.** Compute $\int \sqrt{1-x^2}\,dx$. In light of the previous example, we are tempted to try

$$u = \sqrt{1-x^2} \qquad \text{or} \qquad x = \sqrt{1-u^2}.$$

Then

$$dx = \frac{-u}{\sqrt{1-u^2}}\,du,$$

and

$$\sqrt{1-x^2}\,dx = \frac{-u^2}{\sqrt{1-u^2}}\,du.$$

Hence

$$\int \sqrt{1-x^2}\,dx = -\int \frac{u^2}{\sqrt{1-u^2}}\,du,$$

and the latter integral looks worse than the given one. The conclusion to be drawn is that we have probably made the wrong substitution. Let us try instead

$$x = \sin u.$$

Then

$$\sqrt{1-x^2} = \cos u,$$

and

$$dx = \cos u \, du.$$

Therefore

$$\sqrt{1-x^2}\,dx = \cos^2 u \, du.$$

Hence

$$\int \sqrt{1-x^2}\,dx = \int \cos^2 u \, du.$$

We have met this integral before. Using the double angle formulas and then substituting for $u$ in terms of $x$, we obtain

$$\int \cos^2 u \, du = \int \left(\frac{\cos 2u - 1}{2}\right) du = \frac{1}{4}\sin 2u - \frac{u}{2}$$

$$= \frac{1}{2}[\sin u \cos u - u]$$

$$= \frac{1}{2}[x\sqrt{1-x^2} - \text{arc} \sin x].$$

We can generalize this technique to handle any integral of the form

$$\int x^n \sqrt{(a^2 - x^2)^m}\,dx$$

where $m, n$ are positive integers by letting

$$x = a \sin u.$$

Then

$$\sqrt{a^2 - x^2} = a \cos u,$$

and

$$dx = a \cos u \, du.$$

Hence

$$x^n \sqrt{(a^2 - x^2)^m}\,dx = a^{n+m+1} \sin^n u \cos^{m+1} u \, du$$

and
$$\int x^n \sqrt{(a^2 - x^2)^m}\, dx = a^{n+m+1} \int \sin^n u \cos^{m+1} u\, du.$$

Methods for dealing with trigonometric integrals of the form $\int \sin^n u$ $\cos^m u\, du$ were discussed in the exercises of the previous section. The relation between the quantities $a, x$, and $u$ is illustrated in Figure 1.

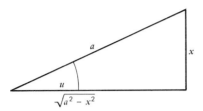

**Fig. 1**    $a \sin u = x. \; a \cos u = \sqrt{a^2 - x^2}.$

*Example 4.*    Compute

$$\int_0^2 x^2 \sqrt{4 - x^2}\, dx.$$

Here the appropriate substitution is $x = 2 \sin u$. Then $dx = 2 \cos u\, du$ and if $u = 0$, then $x = 0$. If $u = \pi/2$, then $x = 2$. Since, with this substitution,

$$x^2 \sqrt{4 - x^2}\, dx = 4 \sin^2 \cos^2 u\, du$$

$$= \sin^2 2u\, du = \frac{1}{2}[1 - \cos 4u]\, du,$$

we have

$$\int_0^2 x^2 \sqrt{4 - x^2}\, dx = \frac{1}{2} \int_0^{\pi/2} [1 - \cos 4u]\, du$$

$$= \frac{1}{2}\left[u - \frac{1}{4} \sin 4u\right]_0^{\pi/2} = \frac{\pi}{4}.$$

For integrals involving $\sqrt{x^2 + a^2}$ the substitution $x = a \tan u$ is suggested to exploit the fact that $1 + \tan^2 u = \sec^2 u$. For if $x = a \tan u$, then $\sqrt{x^2 + a^2} = a \sec u$. Similarly, if the integral involves $\sqrt{x^2 - a^2}$, take $x = a \sec u$. Then $\sqrt{x^2 - a^2} = a \tan u$. The relation between $a, x, u$ is given in Figure 2.

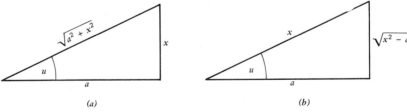

(a)                           (b)

**Fig. 2**    (a) $x = a \tan u; \sqrt{a^2 + x^2} = a \sec u.$    (b) $x = a \sec u; \sqrt{x^2 - a^2} = a \tan u.$

*Example 5.*    Compute $\int (1/\sqrt{x^2 - 1})\, dx$. Let $x = \sec u$, then $dx = \sec u \tan u\, du$, and $\sqrt{x^2 - 1} = \tan u$. Consequently, $(1/\sqrt{x^2 - 1})\, dx = \sec u\, du$ and $\int (1/\sqrt{x^2 - 1})\, dx$ $\int \sec du$. This latter integral we have not met before. To evaluate it we use the following "trick." Observe that

$$\sec u = \sec u \frac{\sec u + \tan u}{\sec u + \tan u}$$

$$= \frac{\sec^2 u + \sec u \tan u}{\tan u + \sec u} = \frac{f'(u)}{f(u)}$$

where $f(u) = \tan u + \sec u$. Since

$$\int \frac{f'(u)}{f(u)} \, du = \log |f(u)|,$$

we have

$$\int \sec u \, du = \log |\tan u + \sec u|.$$

Consequently

$$\int \frac{dx}{\sqrt{x^2 - 1}} = \log |\sqrt{x^2 - 1} + x|.$$

The integral $\int \csc u \, du$ may be evaluated in a similar fashion by multiplying numerator and denominator of the integrand by $\csc u - \operatorname{ctn} u$.

### EXERCISES

1. Evaluate the following definite integrals by making a substitution of the form $u = ax + b$ or $u = \sqrt[k]{ax + b}$.

(a) $\displaystyle \int_0^1 x\sqrt{1 + x} \, dx$

(d) $\displaystyle \int_{-1}^0 x\sqrt[3]{1 - 3x} \, dx$

(b) $\displaystyle \int_1^2 x^2 \sqrt{x - 1} \, dx$

(e) $\displaystyle \int_{-1}^2 x\sqrt{x + 2} \, dx$

(c) $\displaystyle \int_{-1}^1 (x + 2)\sqrt{4x + 5} \, dx$

(f) $\displaystyle \int_0^5 \frac{x \, dx}{\sqrt{3x + 1}}$

2. Evaluate the following definite integrals by making an appropriate trigonometric substitution.

(a) $\displaystyle \int_0^1 \sqrt{1 - x^2} \, dx$

(e) $\displaystyle \int_0^3 \frac{dx}{(9 + x^2)^2}$

(b) $\displaystyle \int_0^1 x\sqrt{1 - x^2} \, dx$

(f) $\displaystyle \int_{3/2}^3 \frac{\sqrt{9 - x^2}}{x} \, dx$

(c) $\displaystyle \int_0^2 \frac{x^2}{\sqrt{16 - x^2}} \, dx$

(g) $\displaystyle \int_{\sqrt{2}}^2 \frac{x^2}{\sqrt{x^2 - 1}} \, dx$

(d) $\displaystyle \int_0^1 \frac{dx}{\sqrt{(1 + x^2)^3}}$

3. Compute the following indefinite integrals.

(a) $\displaystyle \int x\sqrt{1 - x} \, dx$

(f) $\displaystyle \int (x^2 - 1)\sqrt{4x - 1} \, dx$

(b) $\displaystyle \int x^2 \sqrt{3x + 1} \, dx$

(g) $\displaystyle \int x^5 \sqrt{1 - x^3} \, dx$

(c) $\displaystyle \int x^2 \sqrt[3]{x + 4} \, dx$

(h) $\displaystyle \int \frac{(1 + x)^{3/4}}{x} \, dx$

(d) $\displaystyle \int x\sqrt[3]{x^2 - 2x + 1} \, dx$

(i) $\displaystyle \int \frac{dx}{x + x^{1/2}}$

(e) $\displaystyle \int \frac{x^2}{\sqrt{x + 1}} \, dx$

(j) $\displaystyle \int \frac{dx}{1 + x^{1/2}}$

4. Compute the following indefinite integrals by making an appropriate trigonometric substitution.

(a) $\displaystyle \int \sqrt{4 - 9x^2} \, dx$

(b) $\displaystyle \int x^2 \sqrt{1 - 4x^2} \, dx$

(c) $\displaystyle \int \frac{\sqrt{x^2-25}}{x^2}\,dx$

(g) $\displaystyle \int \frac{\sqrt{x^2-a^2}}{x^2}\,dx$

(d) $\displaystyle \int \sqrt{2x-x^2}\,dx$

(h) $\displaystyle \int \frac{dx}{\sqrt{x^2-4x}}$

(e) $\displaystyle \int \frac{dx}{(x^2+1)^2}$

(i) $\displaystyle \int \frac{dx}{x\sqrt{1+x^2}}$

(f) $\displaystyle \int \frac{dx}{(25+x^2)^{3/2}}$

5. Derive the formula

$$\int \csc u\,du = \log|\csc u - \operatorname{ctn} u|.$$

6. Find the area of the region common to the two circles $x^2+y^2=1$ and $x^2+y^2-2x=0$.

7. Find the area of the region to the right of the line $x=2$ and inside the ellipse $9x^2+16y^2=144$.

8. Find the area of the region in the right half plane lying to the right of the hyperbola $x^2-y^2=1$ and to the left of the line $x=4$.

9. Find the area of the region inside the ellipse $4x^2+9y^2=36$ and above the line $3y+4\sqrt{2}\,x=0$.

## 6.3. Integration by Parts

The product rule for differentiation,

(1)
$$(fg)' = fg' + gf'$$

can be immediately interpreted as an integration formula. Indeed, if we rewrite (1) using the indefinite integral notation, we have

$$f(x)g(x) = \int f(x)g'(x)\,dx + \int g(x)f'(x)\,dx$$

or

(2)
$$\int f(x)g'(x)\,dx = f(x)g(x) - \int g(x)f'(x)\,dx.$$

The technique of evaluating integrals by this formula is called *integration by parts*. To apply this method to the evaluation of an integral $\int h(x)\,dx$, we must choose appropriate functions $f$ and $g$ so that, first, $h(x) = f(x)g'(x)$ and, second, the integral

$$\int g(x)f'(x)\,dx$$

can be evaluated.

The integration by parts formula is often written using differentials. If $u = f(x)$ and $v = g(x)$, then $du = f'(x)\,dx$ and $dv = g'(x)\,dx$. Substituting in formula (2), we have the following equivalent formula

(3)
$$\int u\,dv = uv - \int v\,du.$$

***Example 1.*** Compute $\int x\cos x\,dx$. If $f(x) = x$, then $f'(x) = 1$. Hence, if we set $g'(x) = \cos x$ and $g(x) = \sin x$, we see that we can indeed compute the integral of $f'(x)g(x) = 1 \cdot \sin x$. Applying (2), we have

$$\int x\cos x\,dx = f(x)g(x) - \int g(x)f'(x)\,dx$$
$$= x\sin x - \int \sin x\,dx$$
$$= x\sin x + \cos x.$$

If the wrong choice is made for $f$ and $g'$, then no progress is made by using this method. In Example 1 if we had taken $f(x) = \cos x$ and $g'(x) = x$, then $f'(x) = -\sin x$ and $g(x) = x^2/2$ and

$$\int x \cos x \, dx = f(x)g(x) - \int g(x)f'(x) \, dx$$

$$= \frac{x^2}{2} \cos x + \int \frac{x^2}{2} \sin x \, dx.$$

The latter integral is more complicated than the one with which we started. Skill in choosing the appropriate functions $f$ and $g'$ comes only with practice.

**Example 2.**   Compute $\int xe^{-x} \, dx$. We use the differential notation. Let $u = x$, and $dv = e^{-x} \, dx$. Then $du = dx$, and $v = -e^{-x}$. Substituting in the formula $\int u \, dv = uv - \int v \, du$, we get

$$\int xe^{-x} \, dx = -xe^{-x} + \int e^{-x} \, dx$$

$$= -xe^{-x} - e^{-x} = -e^{-x}(x+1).$$

If $n$ is a positive integer, then integrals of the form $\int x^n \sin ax \, dx$, $\int x^n \cos ax \, dx$, and $\int x^n e^{ax} \, dx$ always can be evaluated by integration by parts. Repeated differentiation of $x^n$ reduces this function to a constant. Thus, after several applications of the integration by parts formula, we are left with integrals $\int \sin ax \, dx$, $\int \cos ax \, dx$, or $\int e^{ax} \, dx$, which can be immediately evaluated.

**Example 3.**   Compute $\int x^3 \cos 2x \, dx$. One application of the integration by parts formula yields

(4)   $$\int x^3 \cos 2x \, dx = \frac{x^2}{2} \sin 2x - \frac{3}{2} \int x^2 \sin 2x \, dx.$$

Repeating the process with the integral on the right, we obtain

$$\int x^2 \sin 2x \, dx = \frac{-x^2}{2} \cos 2x + \int x \cos 2x \, dx.$$

But

$$\int x \cos 2x \, dx = \frac{x}{2} \sin 2x - \frac{1}{2} \int \sin 2x \, dx$$

$$= \frac{x}{2} \sin 2x + \frac{1}{4} \cos 2x.$$

Substituting back in (4), we have

$$\int x^3 \cos 2x \, dx = \frac{x^3}{2} \sin 2x + \frac{3}{4} x^2 \cos 2x - \frac{3}{4} x \sin 2x + \frac{3}{8} \cos 2x.$$

If in (2) we set $g'(x) = 1$, then $g(x) = x$, and (2) becomes

(5)   $$\int f(x) \, dx = xf(x) - \int xf'(x) \, dx.$$

This device can be used to evaluate $\int \log x \, dx$.

**Example 4.**   Compute $\int \log x \, dx$. Setting $g'(x) = 1$ and $f(x) = \log x$, we obtain $g(x) = x$ and $f'(x) = 1/x$. Hence

$$\int \log x \, dx = x \log x - \int \frac{x}{x} \, dx$$

$$= x \log x - x.$$

Integrals of the form $\int e^{ax} \sin bx\, dx$, and $\int e^{ax} \cos bx\, dx$ can be evaluated by integrating by parts twice.

***Example 5.*** Evaluate $\int e^x \sin x\, dx$. Let $f(x) = e^x$ and $g'(x) = \sin x$. Then $f'(x) = e^x$ and $g(x) = -\cos x$, and we have

(6) $$\int e^x \sin x = -e^x \cos x + \int e^x \cos x\, dx.$$

At this point we are tempted to conclude that the wrong choice has been made for $f$ and $g'$. Let us, however, try again and integrate $\int e^x \cos x\, dx$ by parts with $f(x) = e^x$ and $g'(x) = \cos x$. Then $f'(x) = e^x$, and $g(x) = \sin x$. Hence

(7) $$\int e^x \cos x\, dx = e^x \sin x - \int e^x \sin x\, dx.$$

Substituting (7) in (6), we have

$$\int e^x \sin x\, dx = e^x(\sin x - \cos x) - \int e^x \sin x\, dx$$

or

(8) $$\int e^x \sin x\, dx = \frac{e^x(\sin x - \cos x)}{2}.$$

In the same way we may derive the formula

(9) $$\int e^x \cos x\, dx = e^x \left( \frac{\sin x + \cos x}{2} \right).$$

The reduction formulas (17) and (18) in Section 6.1 can be derived by integration by parts.

***Example 6.*** Show that for a positive integer $n$

$$\int \cos^n x\, dx = \frac{\cos^{n-1} x \sin x}{n} + \frac{n-1}{n} \int \cos^{n-2} x\, dx.$$

Set $f(x) = \cos^{n-1} x$ and $g'(x) = \cos x$. Then $f'(x) = -(n-1)\cos^{n-2} x \cdot \sin x$, and $g(x) = \sin x$. Hence integrating by parts we have

$$\int \cos^n x\, dx = \cos^{n-1} x \sin x + (n-1) \int \cos^{n-2} x \sin^2 x\, dx$$

$$= \cos^{n-1} x \sin x + (n-1) \int \cos^{n-2} x (1 - \cos^2 x)\, dx.$$

Therefore

$$(n-1+1) \int \cos^n x\, dx = \cos^{n-1} x \sin x + (n-1) \int \cos^{n-2} x\, dx,$$

or

$$\int \cos^n x\, dx = \frac{\cos^{n-1} x \sin x}{n} + \frac{n-1}{n} \int \cos^{n-2} x\, dx.$$

If in (2) the integrands are continuous on an interval $[a, b]$, which is true if $f'$ and $g'$ are continuous, we have an analagous definite integral formula.

$$\int_a^b f(x)g'(x)\, dx = f(b)g(b) - f(a)g(a) - \int_a^b g(x)f'(x)\, dx.$$

**EXERCISES**

1. Evaluate the following integrals using integration by parts one or more times.

(a) $\int x \cos 2x\, dx$　　　(c) $\int x^2 \sin 3x\, dx$

(b) $\int xe^{-2x}\, dx$　　　(d) $\int x^2 e^x\, dx$

(e) $\int x \log x \, dx$   (h) $\int e^{ax} \cos bx \, dx$

(f) $\int x(\log x)^2 \, dx$   (i) $\int \arcsin x \, dx$

(g) $\int x^{1/2} \log x \, dx$   (j) $\int \arctan x \, dx$

2. Compute the following integrals using integration by parts together with other techniques.

(a) $\int x \sin^2 x \, dx$   (d) $\int \tan^3 x \, dx$

(b) $\int x^2 \arctan x \, dx$   (e) $\int \tan^3 x \sec x \, dx$

(c) $\int \sec^3 x \, dx$

3. Derive the following reduction formulas.

(a) $\int \cos^n x \, dx = \dfrac{\cos^{n-1} x \sin x}{n} + \dfrac{n-1}{n} \int \cos^{n-2} x \, dx$

(b) $\int x^n e^{ax} \, dx = \dfrac{1}{a} x^n e^{ax} - \dfrac{n}{a} \int x^{n-1} e^{ax} \, dx$

(c) $\int x^m \log x)^n \, dx = \dfrac{x^{m+1}(\log x)^n}{m+1} - \dfrac{n}{m+1} \int x^m (\log x)^{n-1} \, dx \qquad m \neq -1$

4. Compute the following definite integrals.

(a) $\int_0^1 x e^x \, dx$   (e) $\int_0^{\pi/2} x^2 \cos x \, dx$

(b) $\int_1^2 \log x \, dx$   (f) $\int_0^{\pi/2} \sin^4 x \, dx$

(c) $\int_0^{\pi/2} e^x \sin x \, dx$   (g) $\int_0^{\pi/2} \cos^6 x \, dx$

(d) $\int_0^{\pi/2} x \sin 2x \, dx$   (h) $\int_0^1 \arcsin x \, dx$

## 6.4.   Integration by Partial Fractions

In this section we shall develop techniques for determining primitives for rational functions, that is, quotients of polynomials. First, if $f(x) = p(x)/q(x)$ where $p$ and $q$ are polynomials, we divide $p$ by $q$ obtaining

$$f(x) = s(x) + \frac{r(x)}{q(x)}$$

where both $s$ and $r$ are polynomials and the degree of $r$ is less than the degree of $q$. If the degree of $p$ is less than the degree of $q$ then, of course, $s = 0$, but in any case we can find the primitive for the polynomial $s$. Hence, we need only be concerned with rational functions for which the degree of the numerator is less than that of the denominator. So throughout this discussion we make that assumption. We shall also assume that the highest order coefficient of the demoninator $q$ is 1. The method we shall employ depends on the distribution of the roots of the denominator $q$. We distinguish three cases.

*Case I.*   The roots of $q$ are all real and distinct.
In this case if the degree of $q$ is $n$, then

$$q(x) = (x - a_1) \cdots (x - a_n).$$

It is a theorem of algebra that under this assumption on $q$, and assuming that the degree of $p$ is less than $n$, then $n$ real numbers $A_1, \ldots, A_n$ can be found so that

(1)
$$\frac{p(x)}{q(x)} = \frac{A_1}{(x-a_1)} + \cdots + \frac{A_n}{(x-a_n)}.$$

This is called the partial fraction decomposition of the rational function $p/q$. If we knew the numbers $A_1, \ldots, A_n$, the problem of finding a primitive for $p/q$ would be solved. For then

$$\int \frac{p(x)}{q(x)} dx = A_1 \int \frac{dx}{x-a_1} + \cdots + A_n \int \frac{dx}{x-a_1}$$

$$= A_1 \log |x-a_1| + \cdots + A_n \log |x-a_n|$$

$$= \log (|x-a_1|^{A_1} \cdots |x-a_n|^{A_n}).$$

The numbers $A_1, \ldots, A_n$ are determined in the following way. Observe that

$$\frac{A_1}{x-a_1} + \cdots + \frac{A_n}{(x-a_n)} = \frac{A_1(x-a_2) \cdots (x-a_n) + \cdots + A_n(x-a_1) \cdots (x-a_{n-1})}{(x-a_1) \cdots (x-a_n)}$$

$$= \frac{r(x)}{q(x)}$$

where $r$ is a polynomial of degree $n-1$. Since $p(x)/q(x) = r(x)/q(x)$ for all values of $x$, it follows that $p$ and $r$ are identical polynomials. Hence the coefficients of each power of $x$ in $p$ must be equal to the corresponding coefficient in $r$. However, the coefficients of $r$ are linear combinations of the unknowns $A_1, \ldots, A_n$. Hence, if we set these linear combinations of the unknowns $A_1, \ldots, A_n$ equal to the corresponding coefficient of $p$, we have $n-1$ linear equations which may be solved for the unknowns $A_1, \ldots, A_n$. This technique is called the method of undetermined coefficients. The following examples should clarify what is going on.

**Example 1.**  Determine the partial fraction decomposition of $(x+2)/(x^2-1)$ and compute $\int (x+2)/(x^2-1) \, dx$. We must determine numbers $A_1, A_2$ so that

$$\frac{x+2}{x^2-1} = \frac{A_1}{x-1} + \frac{A_2}{x+1}.$$

But

$$\frac{A_1}{x-1} + \frac{A_2}{x+1} = \frac{A_1(x+1) + A_2(x-1)}{x^2-1}$$

$$= \frac{(A_1+A_2)x + A_1 - A_2}{x^2-1}.$$

Therefore, if

$$\frac{x+2}{x^2-1} = \frac{(A_1+A_2)x + A_1 - A_2}{x^2-1},$$

the numerators are identical. Therefore, equating the coefficients of like powers of $x$, we have

$$A_1 + A_2 = 1 \quad \text{and} \quad A_1 - A_2 = 2.$$

From this we conclude $A_1 = \frac{3}{2}, A_2 = -\frac{1}{2}$. Therefore

$$\frac{x+2}{x^2-1} = \frac{3}{2}\frac{1}{x-1} - \frac{1}{2}\frac{1}{x+1}.$$

Hence

$$\int \frac{x+2}{x^2-1} dx = \frac{3}{2} \int \frac{dx}{x-1} - \frac{1}{2} \int \frac{dx}{x+1}$$

$$= \frac{3}{2} \log |x-1| - \frac{1}{2} \log |x+1|$$

$$= \log \left| \frac{(x-1)^3}{x+1} \right|^{1/2}.$$

*Example 2.*   Compute

$$\int \frac{x^3}{(x^2-4)(x+1)} dx.$$

Since the degree of the numerator of this fraction equals that of the denominator, we must first divide, obtaining

$$\frac{x^3}{(x^2-4)(x+1)} = 1 - \frac{x^2-4x-4}{(x^2-4)(x+1)}.$$

Next we obtain the partial fraction decomposition of

$$\frac{x^2-4x-4}{(x^2-4)(x+1)}.$$

Indeed

$$\frac{x^2-4x-4}{(x^2-4)(x+1)} = \frac{A_1}{x-2} + \frac{A_2}{x+2} + \frac{A_3}{x+1}$$

$$= \frac{A_1(x+2)(x+1) + A_2(x-2)(x+1) + A_3(x-2)(x+2)}{(x^2-4)(x+1)}$$

$$= \frac{(A_1+A_2+A_3)x^2 + (3A_1-A_2)x + 2A_1 - 2A_2 - 4A_3}{(x^2-4)(x+1)}$$

Therefore, equating coefficients for corresponding powers of $x$, we have

$$A_1 + A_2 + A_3 = 1$$
$$3A_1 - A_2 = -4$$
$$2A_1 - 2A_2 - 4A_3 = -4.$$

Solving for the unknowns, we infer that $A_1 = -\frac{2}{3}, A_2 = 2,$ and $A_3 = -\frac{1}{3}.$ Hence

$$\frac{x^3}{(x^2-4)(x+1)} = 1 + \frac{2}{3}\frac{1}{x-2} - \frac{2}{x+2} + \frac{1}{3}\frac{1}{x+1}.$$

Therefore

$$\int \frac{x^3\,dx}{(x^2-4)(x+1)} = \int dx + \frac{2}{3}\int \frac{dx}{x-2} - 2\int \frac{dx}{x+2} + \frac{1}{3}\int \frac{dx}{x+1}$$

$$= x + \frac{2}{3}\log|x-2| - 2\log|x+2| + \frac{1}{3}\log|x+1|$$

$$= x + \log\frac{|(x-2)^2(x+1)|^{1/3}}{|(x+2)^2|}.$$

**Case II.**   The roots of $q(x)$ are real but some of these are repeated.

A partial fraction decomposition for $p/q$ is still valid, assuming the degree of $p$ is less than the degree of $q$, but if the root $a$ has multiplicity $k$, that is, if $(x-a)^k$ is a factor of $q$, then in place of the term $A_1/(x-a)$ we have $k$ terms

$$\frac{A_1}{x-a} + \frac{A_2}{(x-a)^2} + \cdots + \frac{A_k}{(x-a)^k}.$$

For example

$$\frac{1}{(x-1)^2x} = \frac{A_1}{x-1} + \frac{A_2}{(x-1)^2} + \frac{A_3}{x}$$

and

$$\frac{1}{(x+1)^3(x-1)^2} = \frac{A_1}{x+1} + \frac{A_2}{(x+1)^2} + \frac{A_3}{(x+1)^3} + \frac{A_4}{x-1} + \frac{A_5}{(x-1)^2}.$$

The coefficients $A_k$ are determined in the same way as in Case I.

*Example 3.*   Compute $\int (1/(x-1)^2x)\,dx$. To determine the partial fraction decomposition of $1/(x-1)^2x$, we write

$$\frac{1}{(x-1)^2x} = \frac{A_1}{x-1} + \frac{A_2}{(x-1)^2} + \frac{A_3}{x}$$

$$= \frac{A_1(x-1)x + A_2x + A_3(x-1)^2}{(x-1)^2x}$$

$$= \frac{(A_1+A_3)x^2 + (-A_1+A_2-2A_3)x + A_3}{(x-1)^2x}$$

Therefore equating coefficients as before, we have

$$A_1 + A_3 = 0$$

$$-A_1 + A_2 - 2A_3 = 0$$

$$A_3 = 1$$

or $A_1 = -1, A_2 = 1$, and $A_3 = 1$. Hence

$$\int \frac{dx}{(x-1)^2x} = -\int \frac{dx}{(x-1)} + \int \frac{dx}{(x-1)^2} + \int \frac{dx}{x}$$

$$= -\log|x-1| - \frac{1}{x-1} + \log|x|$$

$$= \log\left|\frac{x}{x-1}\right| - \frac{1}{x-1}.$$

***Case III.***  The polynomial $q(x)$ has irreducible quadratic factors $x^2 + ax + b$.

A quadratic polynomial is called irreducible if it has no real roots. For the polynomial $x^2 + ax + b$, this will be the case if $a^2 - 4b < 0$. For each distinct irreducible quadratic factor of $q$ a term $(Ax+B)/(x^2+ax+b)$ occurs in the partial fraction decomposition of $p/q$. If the irreducible quadratic factor occurs with multiplicity $k$, then the following $k$ terms

$$\frac{A_1x+B_1}{x^2+ax+b} + \frac{A_2x+B_2}{(x^2+ax+b)^2} + \cdots + \frac{A_kx+B_k}{(x^2+ax+b)^k}$$

occur. For example

$$\frac{1}{(x^2+x+1)(x^2+1)} = \frac{Ax+B}{x^2+x+1} + \frac{Cx+D}{x^2+1},$$

and

$$\frac{1}{x(x^2+1)^3} = \frac{A}{x} + \frac{Bx+C}{x^2+1} + \frac{Dx+E}{(x^2+1)^2} + \frac{Fx+G}{(x^2+1)^3}.$$

The coefficients $A, B, C, \ldots$ are computed just as in the previous two cases.

We complete this discussion by considering integrals of the form

(2)  $$\int \frac{Ax+B}{(x^2+ax+b)^k} dx \qquad \text{where} \qquad k = 1, 2, \ldots.$$

If we complete the square in the denominator, we have $x^2 + ax + b = (x + (a/2))^2 + b - a^2/4$. Now $b - a^2/4 > 0$, so set $c^2 = b - a^2/4$, and let $u = x + a/2$. Then $du = dx$, and the integral (2) becomes

$$\int \frac{Cu+D}{(u^2+c^2)^k} du$$

where $C$ and $D$ are constants. To evaluate this integral note that

$$\int \frac{u\,du}{(u^2+c^2)^k} = \frac{1}{2}\frac{1}{1-k}(u^2+c^2)^{1-k} \qquad \text{if} \qquad k > 1,$$

$$= \frac{1}{2}\log|u^2+c^2| \qquad \text{if} \qquad k = 1.$$

and

$$\int \frac{du}{u^2+c^2} = \frac{1}{c} \text{ arc tan } \frac{u}{c}.$$

The integral $\int (1/(u^2+c^2)^k)\, du$ for $k > 1$ may be evaluated by the following reduction formula which may be derived by integrating by parts (Exercise 3).

(3)   $$\int \frac{du}{(u^2+c^2)^n} = \frac{u}{(2n-2)c^2(u^2+c^2)^{n-1}} + \frac{2n-3}{2n-2}\cdot\frac{1}{c^2}\cdot\int \frac{du}{(u^2+c^2)^{n-1}}.$$

We close this section with some observations on the problem of finding primitives. Polynomials, rational functions, trigonometric functions, exponential functions, and the inverses of such functions are called *elementary functions*. The indefinite integrals or primitives of the elementary functions that we have been discussing have all been elementary functions. This is by no means always the case. Many integrals that arise very naturally in applications cannot be expressed as finite combinations of elementary functions.

(4)   $$\int e^{-x^2}\, dx$$

(5)   $$\int \sqrt{1+x^3}\, dx$$

(6)   $$\int \sqrt{1-k^2 \sin^2 x}\, dx, \qquad k^2 \neq 1$$

are three such examples. Primitives for the functions $e^{-x^2}$, $\sqrt{1+x^3}$, $\sqrt{1-k^2 \sin^2 x}$, certainly exist, however. Since each function $f$ is continuous, the definite integral exists, and by the fundamental theorem of the calculus, $F(x) = \int_a^x f(t)\, dt$ is a primitive for $f$. If one is just interested in the numerical value of a definite integral $\int_a^b f(t)\, dt$ where $f$ is an elementary function, then it is really unimportant whether the primitive of $f$ is an elementary function. It is usually preferable to use a numerical approximation technique to estimate the integral $\int_a^b f$ directly rather than to attempt to find a formula for the primitive $F$. The numerical techniques for estimating definite integrals can be found in books on numerical analysis such as Introduction to Numerical Analysis, F. B. Hildebrand, McGraw-Hill Book Co., Chapter 3.

EXERCISES

1. Evaluate the following integrals using a partial fraction decomposition of the integrand.

(a) $\int \dfrac{dx}{x^2-1}$

(b) $\int \dfrac{x+1}{x^2-x+6}dx$

(c) $\int \dfrac{x^2}{x^2-x-2}dx$

(d) $\int \dfrac{dx}{2x^2+x-3}$

(e) $\int \dfrac{dx}{x^3-4x}$

(f) $\int \dfrac{x+1}{x^3}dx$

(g) $\int \dfrac{x^2-1}{x^3+2x^2-3x}dx$

(h) $\int \dfrac{x^3}{x^2-1}dx$

(i) $\int \dfrac{2x+4}{x^2(x-1)^2}dx$

(j) $\int \dfrac{dx}{x(x^2-1)^2}$

(k) $\int \frac{x^3-1}{(x+1)^3}dx$

(o) $\int \frac{x}{x^4-1}dx$

(l) $\int \frac{dx}{x^4-3x^2-4}$

(p) $\int \frac{x^3}{x^2+4}dx$

(m) $\int \frac{x^2+1}{x^3-1}dx$

(q) $\int \frac{x-1}{x^2(x^2+1)}dx$

(n) $\int \frac{x^2\,dx}{x^4+6x^2+5}$

(r) $\int \frac{x}{(x^2+1)^2}dx$

2. Evaluate the following integrals by any method.

(a) $\int \frac{x\,dx}{\sqrt{x-x^2}}$

(g) $\int \frac{x^2\,dx}{(x^2+1)^{3/2}}$

(b) $\int \frac{dx}{(x+1)\sqrt{x^2+2x-8}}$

(h) $\int x \arc \sin x\,dx$

(c) $\int \frac{\arc \tan x}{x^2}dx$

(i) $\int \frac{x\,dx}{(1-x^2)\sqrt{1+x^2}}$

(d) $\int \log\,(x+\sqrt{x^2+1})$

(j) $\int \frac{dx}{x\sqrt{1+x^2}}$

(e) $\int \frac{dx}{\sin^3 x}$

(k) $\int xe^{-\sqrt{x}}\,dx$

(f) $\int \tan^4 x\,dx$

(l) $\int \frac{\sqrt{x+1}+1}{\sqrt{x+1}-1}dx.$

3. Derive the reduction formula (3) of this section.

## 6.5    Further Applications of the Integral

We saw in Chapter 3 that the numerical solution $S$ to an applied problem associated with an interval $[a, b]$ could be written

$$S = \int_a^b f(x)\,dx$$

for an appropriate integrable function $f$ if two conditions were satisfied. The first condition is that the problem is additive on subintervals of $[a, b]$. This means that if $I(c, d)$ denotes the solution to the problem at hand on a subinterval $[c, d]$ of $[a, b]$ and $a < x_1 < \cdots < x_{n-1} < b$ is a partition of the interval $[a, b]$, then

$$S = I(a, b) = I(a, x_1) + \cdots + I(x_{n-1}, b)$$

Thus the solution to the problem on the whole interval is just the sum of the solutions on the subintervals of the partition. Area, work, and volumes of solids of revolution are examples of problems that are additive on an interval $[a, b]$.

The second condition concerns the determination of the integrable function $f$. We showed that $S = \int_a^b f$ if the following condition could be satisfied on each subinterval $[c, d]$. For each pair of constants $m, M$ satisfying $m \le f(x) \le M$, for all $x \in [c, d]$, it follows that

$$m(d-c) \le I(c, d) \le M(d-c)$$

If $f$ is assumed to be continuous on all of $[a, b]$, this is equivalent to the assertion that for each subinterval $[c, d]$ there exist points $x', x''$ in this interval such that

(1) $$f(x')(d-c) \leqslant I(c,d) \leqslant f(x'')(d-c).$$

Indeed we may take $f(x')$ to be the minimum value of $f$ on the interval $[c,d]$ and $f(x'')$ the maximum. If $I(a,b)$ represents the area of the ordinate set for a nonnegative continuous function $f$, then (1) just expresses the fact that the area of the strip through the interval $[c,d]$ is less than or equal to the area of a rectangle of height $f(x'')$ for some $x'' \in [c,d]$ and greater than or equal to the area of a rectangle of height $f(x')$. This is illustrated in Figure 3.

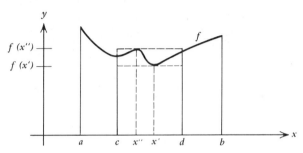

**Fig. 3** $f(x')(d-c) \leqslant I(c,d) \leqslant f(x'')(d-c)$ for some $x'$, $x'' \in [c,d]$.

In many applications the following situation is encountered. The problem under consideration is obviously additive on subintervals of $[a,b]$. Hence the solution $I(a,b)$ should be an integral. However, it may be impossible to verify (1). Often in place of one function $f$ there are two functions $f$ and $g$ defined on $[a,b]$. In place of condition (1), the best that one can verify is the following. On each subinterval $[c,d]$ of $[a,b]$

(2) $$f(x_1')g(x_2')(d-c) \leqslant I(c,d) \leqslant f(x_1'')g(x_2'')(d-c)$$

for some points $x_1', x_2', x_1'', x_2''$ in $[c,d]$. If $x_1' = x_2'$ and $x_1'' = x_2''$, then, of course, (2) is just inequality (1) and we may conclude immediately that

(3) $$I(a,b) = \int_a^b fg.$$

However, it may be impossible to achieve (2) if we take the points $x_1' = x_2'$ and $x_1'' = x_2''$. The question we must answer is: Can we assert that (3) holds just under the assumption that (2) holds for each subinterval $[c,d]$? The answer is yes, and it follows quite readily from our next result. Let us adopt the notation $\Delta x_k = x_k - x_{k-1}$ for consecutive points $x_{k-1}, x_k$ of a partition $a = x_0 < x_1 < \cdots < x_n = b$ of the interval $[a,b]$.

**THEOREM 6.2**  *Let $f$ and $g$ be continuous on the interval $[a,b]$. Then for each $\epsilon > 0$ there is a partition $a = x_0 < x_1 < \cdots < x_n = b$ such that for each pair of points $x_k', x_k''$ in the interval $[x_{k-1}, x_k]$, it follows that*

$$\left| \int_a^b f(x)g(x)\,dx - [f(x_1')g(x_1'')\Delta x_1 + \cdots + f(x_n')g(x_n'')\Delta x_n] \right| < \epsilon.$$

The proof of this result is sketched in the exercises at the end of this section. It depends on the fact that a continuous function on a closed interval $[a,b]$ may be approximated arbitrarily well on the whole interval $[a,b]$ by a step function $h$. In fact, the proof goes through verbatim for any pair of functions $f$ and $g$ with this property. The result also may be generalized from the product of two functions to any finite sum or product of continuous functions. However, we shall not make use of this fact.

It now follows quite readily from Theorem 6.2 that the answer $S$ to a problem that is additive on the subintervals of $[a, b]$ is given by $S = \int_a^b f(x)g(x)\,dx$ whenever (2) can be verified for each subinterval $[c, d]$. We leave the verification of that as an exercise also.

We illustrate the application of Theorem 6.2 by computing the volume of a solid of revolution by what is called the "shell method." Specifically, we wish to compute the volume of the solid formed by rotating about the $y$ axis the ordinate set of a positive continuous function defined on the interval $[a, b]$, $0 < a < b$. As always, let $a = x_0 < x_1 < \cdots < x_n = b$ be a partition of the interval $[a, b]$. If $V(a, b)$ denotes the volume of the solid and $V(x_{k-1}, x_k)$ denotes the volume of the shell formed by rotating the ordinate set of $f$ on the interval $[x_{k-1}, x_k]$, then certainly

$$V(a, b) = V(a, x_1) + \cdots + V(x_{n-1}, b)$$

Next we assert that $V(a, b) = \int_a^b f(x)g(x)\,dx$ where $f$ is the given function and $g$ is to be constructed. We verify that (2) holds for each subinterval $[c, d]$. To do this, we proceed as follows. Let $V(c, d)$ be the volume of the shell $S$ formed by rotating the ordinate set for $f$ on this interval. We construct two auxiliary shells $S_1, S_2$ such that $S_1$ is inside $S$ and $S_2$ contains $S$. Let $h$ be the minimum value of $f$ on the interval $[c, d]$, and $H$ the maximum value of $f$ on $[c, d]$. Then the inner shell $S_1$ has height $h$, inside radius $c$, and outside radius $d$. The volume $V_1$ of this shell is clearly given by

$$V_1 = \pi h(d^2 - c^2)$$

The outer shell has the same radii but height $H$. The volume $V_2$ is then

$$V_2 = \pi H(d^2 - c^2).$$

Since $S_1 < S < S_2$,

$$V_1 \leqslant V(c, d) \leqslant V_2.$$

(See Figure 4.)

Next write

$$V_1 = 2\pi h \left(\frac{d+c}{2}\right)(d-c)$$

and

$$V_2 = 2\pi H \left(\frac{d+c}{2}\right)(d-c)$$

Since $f$ is continuous, $h = f(x')$ and $H = f(x'')$ for some points $x', x''$ in $[c, d]$. Moreover, $(d+c)/2$ is a point $x'''$ in this interval. Indeed, it is the midpoint. Hence

$$V_1 = 2\pi x''' f(x')(d-c)$$

and

$$V_2 = 2\pi x''' f(x'')(d-c)$$

Thus, for each subinterval $[c, d]$, there are points $x', x'', x'''$ in $[c, d]$ such that $2\pi x''' F(x')(d-c) \leqslant V(c, d) \leqslant 2\pi x''' f(x'')(d-c)$. This is just condition

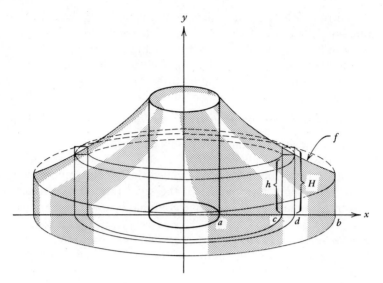

**Fig. 4**  Inside shell $S_1$ with height $h$ has volume $V_1 = \pi h(d^2 - c^2)$. Outside shell $S_2$ with height $H$ has volume $V_2 = \pi H(d^2 - c^2)$.

(2) where we are taking $g(x) = 2\pi x$. Therefore the desired volume is given by

(4)
$$V(a, b) = 2\pi \int_a^b xf(x)\, dx.$$

**Example 1.**  Compute the volume of the solid formed by rotating the ordinate set for $f(x) = x^2$ on the interval $[1, 3]$ about the $y$ axis. What is the volume of the solid formed by rotating this ordinate set about the line $x = -1$? (See Figure 5.)

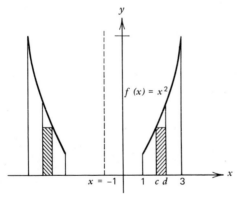

**Fig. 5**

For the first solid, formula (4) applies directly. Since $f(x) = x^2$,

$$V = V(1, 3) = 2\pi \int_1^3 x^3\, dx = 2\pi \frac{x^4}{4}\bigg|_1^3 = 40\pi.$$

To find the volume of the second solid, we compute the volume $V_1$ of the inside shell. This shell has inside radius $1 + c$ and outside radius $1 + d$. The height $h$ equals the minimum value of $f$ on the interval $[c, d]$. In this case $h = f(x')$ where $x'$ is the left-hand endpoint $c$. Thus

$$V_1 = 2\pi\left(1 + \frac{c + d}{2}\right) f(c)(d - c)$$

or

$$V_1 = 2\pi(1+x''')f(x')(d-c).$$

where $x'$, $x'''$ are two points in the interval $[c, d]$. If $V_2$ is the volume of the outer shell with height equal to the maximum value of $f$ on the interval $[c, d]$, then

$$V_2 = 2\pi(1+x''')f(x'')(d-c)$$

where $x''$, $x'''$ are two points in this interval. We conclude, therefore, that

$$V = 2\pi \int_1^3 (1+x)f(x)\,dx$$

$$= 2\pi \int_1^3 (1+x)x^2\,dx = 2\pi \left[\frac{x^3}{3}+\frac{x^4}{4}\right]_1^3$$

$$= \frac{172\pi}{3}.$$

As a final application of the definite integral we consider the problem of defining and computing the centroid of a region in the plane. If this region is thought of as a thin plate with equally distributed mass, then the centroid is the center of mass or balance point. One problem that confronts us is how to translate this physical definition into a mathematically useful fact.

If masses $m_1, \ldots, m_n$ are distributed at points $P_1 = (x_1, y_1), \ldots, P_n = (x_n, y_n)$, then the center of mass of this system is defined to be the point $P = (\bar{x}, \bar{y})$ such that

$$m_1(x_1 - \bar{x}) + \cdots + m_n(x_n - \bar{x}) = 0$$

and

$$m_1(y_1 - \bar{y}) + \cdots + m_n(y_n - \bar{y}) = 0.$$

Solving for $\bar{x}$ and $\bar{y}$, we obtain

(5)
$$\bar{x} = \frac{m_1 x_1 + \cdots + m_n x_n}{m_1 + \cdots + m_n}$$

and

(6)
$$\bar{y} = \frac{m_1 y_1 + \cdots + m_n y_n}{m_1 + \cdots + m_n}.$$

Notice that formula (5) implies that $\bar{x}$ is no larger than the largest of the numbers $x_1, \ldots, x_n$. This is clear because if $\hat{x}$ is the largest of these numbers, then

$$\bar{x} = \frac{m_1 x_1 + \cdots + m_n x_n}{m_1 + \cdots + m_n} \leqslant \frac{m_1 \hat{x} + \cdots + m_n \hat{x}}{m_1 + \cdots + m_n} = \hat{x}$$

Similarly, $\bar{x}$ is no smaller than the smallest of the numbers $x_1, \ldots, x_n$. A similar statement holds for $\bar{y}$. We may incorporate both of these statements in the following property $P$ of $(\bar{x}, \bar{y})$.

If all of the points $(x_i, y_i)$ lie on one side of a vertical or horizontal line $l$, then $(\bar{x}, \bar{y})$ lies on the same side.

To define the centroid $(\bar{x}, \bar{y})$ of the ordinate set for a nonnegative continuous function defined on an interval $[a, b]$, we take formulas (5) and (6) and property $P$ as our point of departure.

First we require that if a region in the plane lies entirely on one side of a horizontal or vertical line, then the centroid of this region lies on the same side of this line. Thus the $x$ coordinate of the centroid of the ordinate set for $f$ on an interval $[c, d]$ must lie within this interval. To see if the appropriate analogue of (5) holds, we ask the following question. Does there exist a point $\bar{x}$ in the interval $[a, b]$ such that for each partition $a = x_0 < x_1 < \cdots < x_n = b$ of the interval $[a, b]$

(7) $$A\bar{x} = A_1\bar{x}_1 + \cdots + A_n\bar{x}_n$$

where $A_k = \int_{x_{k-1}}^{x_k} f(x)\, dx$ is the area of the ordinate set on the interval $[x_{k-1}, x_k]$ and $\bar{x}_k$ is a point in this interval? Furthermore, the point $\bar{x}_k$ should depend only on the function $f$ and the interval $[x_{k-1}, x_k]$ (see Figure 6).

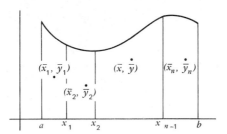

**Fig. 6** $A\bar{x} = A_1\bar{x}_1 + \cdots + A_n\bar{x}_n.$   $A\bar{y} = A_1\bar{y}_1 + \cdots + A_n\bar{y}_n.$
$A = \int_a^b f, A_k = \int_{x_{k-1}}^{x_k} f.$

By (7) the problem we are attempting to solve is clearly additive on subintervals of $[a, b]$. Therefore, the solution should be an integral. To see what form the integral should take, let us estimate one of the terms $A_k\bar{x}_k$.

Set $m_k = $ minimum value of $f$ on the interval $[x_{k-1}, x_k]$ and $M_k$ the maximum. Then $m_k = f(x_k')$ and $M_k = f(x_k'')$ for some points $x_k', x_k''$ in $[x_{k-1}, x_k]$. But then $\Delta x_k \leqslant A_k \leqslant M_k \Delta x_k$ (see Figure 7). Hence, on each subinterval $[x_{k-1}, x_k]$,

$$\bar{x}_k f(x_k') \Delta x_k \leqslant A_k \bar{x}_k \leqslant \bar{x}_k f(x_k'') \Delta x_k,$$

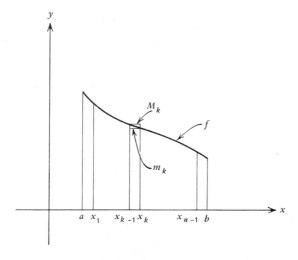

**Fig. 7**  $m_k(\Delta x_k) \leqslant A_k \leqslant M_k(\Delta x_k).$

if $\bar{x}_k \geqslant 0$. The inequalities are reversed if $\bar{x}_k < 0$. In either case for each interval $[x_{k-1}, x_k]$, there exist points $\bar{x}_k, x_k', x_k'' \epsilon [x_{k-1}, x_k]$ such that

$$\bar{x}_k f(x_k') \, \Delta x_k \leqslant A_k \bar{x}_k \leqslant \bar{x}_k f(x_k'') \, \Delta x_k$$

This is just condition (2) with $g(x) = x$. Hence we conclude that

$$A\bar{x} = \int_a^b xf(x) \, dx$$

or

(8)
$$\bar{x} = \frac{\int_a^b xf(x) \, dx}{\int_a^b f(x) \, dx}.$$

It now may be immediately verified that if we define $\bar{x}$ by formula (8), then (7) is satisfied where we define

$$\bar{x}_k = \frac{\int_{x_{k-1}}^{x_k} xf(x) \, dx}{\int_{x_{k-1}}^{x_k} f(x) \, dx} = \frac{\int_{x_{k-1}}^{x_k} xf(x) \, dx}{A_k}$$

To determine a formula for $\bar{y}$, we argue similarly. However, we must exercise a little care. Just as for the $x$ coordinate, if $a = x_0 < x_1 < \cdots < x_n = b$ is a partition of $[a, b]$, we require that

(9)
$$\bar{y}A = \bar{y}_1 A_1 + \cdots + \bar{y}_n A_n$$

where $\bar{y}_k$ is the $y$ coordinate of the centroid of the strip over the interval $[x_{k-1}, x_k]$. We must use property $P$ to make a preliminary estimate on the location of $\bar{y}_k$. If $m_k$ and $M_k$ denote the minimum and maximum values of $f$ on this interval we assert that

(10)
$$\frac{m_k}{2} \leqslant \bar{y}_k \leqslant \frac{M_k}{2}$$

See Figure 8.

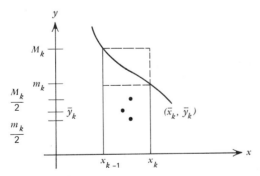

**Fig. 8** $\dfrac{m_k}{2} \leqslant y_k \leqslant \dfrac{M_k}{2}$.

To see this, note that the $y$ coordinate of the centroid of a rectangle of height $m_k$ is $m_k/2$. Denote the area of the rectangle by $B$. Let $C$ be the area of that portion of the ordinate set lying above the line $y = m_k$, and let $\hat{y}_k$ be the $y$ coordinate of the centroid of this region. By property $P$, $\hat{y}_k \geqslant m_k$. Since by (9)

$$\bar{y}_k = \frac{B\frac{m_k}{2} + C\hat{y}_k}{A_k} \geqslant \frac{B\frac{m_k}{2} + Cm_k}{A_k} \geqslant \frac{B\frac{m_k}{2} + C\frac{m_k}{2}}{A_k}$$

we conclude that $m_k/2 \leqslant \bar{y}_k$ since $B + C = A_k$. The proof that $\bar{y}_k \leqslant M_k/2$ proceeds similarly and is left as an exercise.

We only need now to estimate $A_k \bar{y}_k$. As in the case with $\bar{x}$

$$m_k(\Delta x_k) \leqslant A_k \leqslant M_k(\Delta x_k).$$

But $m_k = f(x_k')$ and $M_k = f(x_k'')$, where $x_k'$, $x_k''$ are points in the interval $[x_{k-1}, x_k]$. Using (10) we infer that

$$\frac{f^2(x_k')}{2} \Delta x_k \leqslant A_k \bar{y}_k \leqslant \frac{f^2(x_k'')}{2} \Delta x_k.$$

We conclude immediately from this that

$$A\bar{y} = \frac{1}{2} \int_a^b f^2(x)\, dx$$

or

(11)
$$\bar{y} = \frac{\dfrac{1}{2} \int_a^b f^2(x)\, dx}{\int_a^b f(x)\, dx}.$$

***Example 2.*** Determine the centroid of the ordinate set for $f(x) = x^2$ on the interval $[0, 1]$.

The area of this region $A = \int_0^1 x^2\, dx = \dfrac{1}{3}$, and

$$\int_0^1 xf(x)\, dx = \int_0^1 x^3\, dx = \frac{1}{4}.$$

Also

$$\frac{1}{2} \int_0^1 [f(x)]^2\, dx = \frac{1}{2} \int_0^1 x^4\, dx = \frac{1}{10}.$$

Therefore, $x = \left(\dfrac{1}{4}\right) \Big/ \left(\dfrac{1}{3}\right) = \dfrac{3}{4}$, and $\bar{y} = \left(\dfrac{1}{10}\right) \Big/ \left(\dfrac{1}{3}\right) = \dfrac{3}{10}$.

***Example 3.*** Determine the centroid of the region below the line $y = x$ and above the parabola $y = x^2$. See Figure 9. One must exercise a little care with this example. Analyzing the situation on a subinterval of a partition, we note that

$$\bar{x}_k[f(x_{k-1}) - g(x_k)]\Delta x_k \leqslant A_k \bar{x}_k \leqslant \bar{x}_k[f(x_k) - f(x_{k-1})]\Delta x_k$$

where $A_k$ is the area of the strip over the interval $[x_{k-1}, x_k]$.

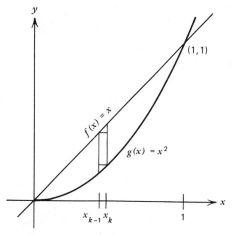

Fig. 9

Therefore

$$A\bar{x} = \int_0^1 x[f(x) - g(x)]\, dx$$

$$= \int_0^1 x[x - x^2]\, dx = \frac{x^3}{3} - \frac{x^4}{4}\Big|_0^1 = \frac{1}{12}$$

However

$$\bar{y}_k \leqslant \frac{f(x_k) + g(x_k)}{2}$$

and

$$\bar{y}_k \geqslant \frac{f(x_{k-1}) + g(x_{k-1})}{2}$$

Hence

$$A_k\bar{y}_k \leqslant \frac{f(x_k) + g(x_k)}{2}[f(x_k) - g(x_{k-1})]\Delta x_k$$

and

$$A_k\bar{y}_k \geqslant \frac{f(x_{k-1}) + g(x_{k-1})}{2}[f(x_{k-1}) - g(x_{k-1})]\Delta x_k.$$

Hence

$$A\bar{y} = \int_0^1 \left(\frac{f(x) + g(x)}{2}\right)(f(x) - g(x))\, dx$$

$$= \frac{1}{2}\int_0^1 [f^2(x) - g^2(x)]\, dx$$

$$= \frac{1}{2}\int_0^1 [x^2 - x^4]\, dx = \frac{1}{2}\left[\frac{x^3}{3} - \frac{x^5}{5}\right]_0^1 = \frac{1}{15}.$$

Since

$$A = \int_0^1 [x - x^2]\, dx = \frac{1}{6},$$

the coordinates of the centroid $(\bar{x}, \bar{y})$ are $\left(\frac{1}{2}, \frac{2}{5}\right)$.

## EXERCISES

1. Derive the formula for the volume of a sphere of radius $r$ by the shell method.
2. Derive the formula for the volume of a right circular cone of radius $r$ and height $h$ by the shell method.
3. The ellipse $x^2/4 + y^2/9 = 1$ is revolved about the $y$ axis. What is the volume of the resulting solid?
4. The ordinate set for $f(x) = \sin x$ on the interval $[0, \pi]$ is revolved about the $y$ axis. What is the volume of the resulting solid?
5. The region in the first quadrant above the hyperbola $xy = 1$ and below the line $2x + 2y = 5$ is revolved about the line $x + 2 = 0$. What is the volume of the resulting solid?
6. The ordinate set for $f(x) = e^x$ on the interval $[0, 2]$ is rotated about the line $x + 1 = 0$. Find the volume of the resulting solid.
7. The region inside the parabola $y^2 = x + 1$ and above the line $y = x - 1$ is revolved about the line $x = 3$. What is the volume of the resulting solid?
8. Determine the volume of a torus obtained by revolving a circle of radius $r$ about a line $R$ units away from the center of the circle. Assume $r < R$.
9. Show that the centroid of a rectangle is the intersection of the diagonals.
10. Locate the centroid of an isosceles triangle of height $h$.
11. Determine the centroid of the region $\{(x, y) : x^2 + y^2 \leqslant 1, y \geqslant 0\}$.
12. Determine the centroid of the region below $y = x^2$ and above $y = x^4$. Does the centroid lie in the region?
13. Determine the centroid of the ordinate set of $f(x) = \sin x$ on the interval $[0, \pi]$.

14. Determine the centroid of the ordinate set of $f(x) = e^{2x}$ on the interval $[0, 1]$.

15. Show that the centroid of a parallelogram is the intersection of the diagonals.

16. Locate the centroid of the upper half of the ellipse $x^2/a^2 + y^2/b^2 = 1$.

17. Determine the centroid of the region in Exercise 5.

18. Determine the centroid of the region in the first quadrant below the curve $x^{2/3} + y^{2/3} = 1$. (Note that by symmetry we may infer $\bar{x} = \bar{y}$.)

19. Prove that if the centroid $(\bar{x}, \bar{y})$ for the ordinate set of a nonnegative function is defined by equations (8) and (11), then equations (7) and (9) are satisfied.

20. In Figure 8, show that $\bar{y}_k \leqslant M_k/2$.

21. If $S$ is the solution to an additive problem of the interval $[a, b]$ and (2) can be verified for each subinterval $[c, d]$, show using Theorem 6.2 that $S = \int_a^b f(x)g(x)\, dx$.

*22. Prove Theorem 6.2 (We know that if $f$ is continuous on $[a, b]$ then $f$ can be approximated arbitrarily closely by a step function $p$. In particular, for each $\epsilon > 0$ there is a partition $a \leqslant x_0 < x_1 < \cdots < x_n = b$ such that if $x'_k$ are points in the intervals $[x_{k-1}, x_k]$ and $p(x) = f(x'_k)$, $x_{k-1} \leqslant x < x_k$, then $|f(x) - p(x)| < \epsilon$ for each $x \in [a, b]$ Show that if $f$ and $g$ are continuous and $\epsilon > 0$, then there is a partition of $[a, b]$ such that if $x'_k, x''_k \in [x_{k-1}, x_k]$ and we define $p(x) = f(x'_k)g(x''_k)$, $x_{k-1} \leqslant x < x_k$, then $|f(x)\, g(x) - p(x)| < \epsilon$ for each $x$ in $[a, b]$. From this conclude that

$$\left| \int_a^b fg - \int_a^b p \right| < \frac{\epsilon}{b-a}.$$

However

$$\int_a^b p = f(x'_1)g(x''_1)\Delta x_1 + \cdots + f(x'_n)g(x''_n)\Delta x_n.)$$

# CHAPTER SEVEN

# ELEMENTARY DIFFERENTIAL EQUATIONS

## 7.1 Introduction

In the preceding chapters we have encountered many equations in which the unknowns are functions rather than numbers. The functional equations

$$f(xy) = f(x) + f(y)$$

and

$$f(x+y) = f(x) \cdot f(y)$$

for the logarithm and exponential functions discussed in Chapter 4 are examples of this phenomenon. Another example is the problem of determining a function from an implicit equation. If we are given the equation

(1) $$g(x, y) = 0,$$

the function $y = f(x)$ is said to be defined implicitly by (1), or is a solution of (1), if for all values of $x$ in some interval $[a, b]$

$$g(x, f(x)) = 0.$$

A *differential equation* is an equation involving an unknown function $y = f(x)$ and some of its derivatives $y'$, $y''$, $y'''$, etc. For example,

(2) $$y''' + x^2 y' + \sqrt{1 - x^2 y} = 0$$

(3) $$y'' + \sin xy = 0$$

or

(4) $$y' + ky = 0.$$

The highest order derivative present in the equation is called the *order* of the differential equation. Thus the equations (2), (3), and (4) above have orders 3, 2, 1 respectively.

In this chapter we shall be concerned mainly with first order differential equations.

Rather than discuss the most general first order equation, we shall confine our attention to equations of the form

(5) $$y' = g(x, y)$$

where $g$ is a given function of two variables. A function $f$ is a solution to (5) on an interval $[a, b]$ if for all values of $x$ in this interval

$$f'(x) = g(x, f(x)).$$

285

We can obtain some pictorial information about the first order equation (5) in the following way. Through a point $(x_0, y_0)$ in the plane, draw a line segment having slope equal to $g(x_0, y_0)$. If there exists a solution $y = f(x)$ to equation (5) which passes through $(x_0, y_0)$, then what we have drawn is a section of the tangent line to $f$ at this point. The collection of numbers $\{g(x, y)\}$, each interpreted as the slope of a solution $f$ at the point $(x, y)$, is called the *direction field* for the equation (5).

If we attempt to draw the direction field for the equation $y' = y$, we have Figure 1. At each point $(x, y)$ we draw a line segment having slope equal to $y$. If enough of these lines are drawn, a rough picture of the graph of a solution $f$ can be obtained. We shall show presently that each solution to the equation $y' = y$ is given by $f(x) = ce^x$ where $c$ is a constant. The solutions $y = e^x$ and $y = -e^x$ passing through $(0, 1)$ and $(0, -1)$ are drawn in Figure 1.

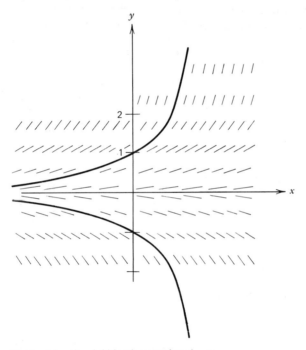

**Fig. 1**   Direction field for the equation $y' = y$.

Two fundamental problems are associated with the first order equation (5). First, we must specify conditions on the function $g$ that enable us to assert that solutions indeed exist on some interval $[a, b]$.

The second problem is to determine conditions that guarantee that there will be just one solution $f$ passing through a given point $(x_0, y_0)$ where $a \leq x_0 \leq b$. Theorems dealing with the first problem are called *existence theorems* for differential equations. Those dealing with the second are called *uniqueness theorems*. We shall not discuss the existence and uniqueness theorems for equation (5) but instead shall discuss special cases of (5) and associated techniques for the solution of the equation.

Suppose, for example, in (5) that the function $g$ does not depend on $y$ at all. Then (5) takes the form

(6) $$y' = g(x)$$

and a function $y = f(x)$ is a splution of (6) on the interval $[a, b]$ if and only if $f$ is a primitive for $g$. But if $g$ is continuous on this interval, the first fundamental theorem of the calculus states that

$$f(x) = \int_a^x g(t)\, dt$$

is a primitive for $g$. Moreover, by the second fundamental theorem, all primitives for $g$, hence all solutions to (6), are of the form

$$f(x) = \int_a^x g(t)\, dt + C$$

for some constant $C$. If the solution $f$ is to pass through the point $(x_0, y_0)$, then $y_0 = \int_a^{x_0} g(t)\, dt + C$, and $C$ is uniquely determined.

Thus, we may interpret the fundamental theorem of the calculus as an existence and uniqueness theorem for the differential equation

(6)
$$y' = g(x).$$

If $g$ is assumed to be continuous on the interval $[a, b]$, this theorem asserts that there is a unique solution $f$ to (6) on the interval $[a, b]$ passing through the point $(x_0, y_0)$ where $a \leqslant x_0 \leqslant b$.

The condition that $f(x_0) = y_0$, where $f$ is a solution to (5), is called an *initial condition* for the differential equation.

**EXERCISES**
For the following differential equations verify that the given functions are solutions. Specify an interval on which the given function is a solution.

1. $y' = 5y$; $y = e^{5x}$, $y = 2e^{5x}$.
2. $y'' = 9y$; $y = 4e^{3x}$, $y = e^{3x} + 2e^{-3x}$.
3. $xy' - xy^2 = y$; $y(x^2 - 2) = -2x$, $x^2 y + 2x = y$.
4. $y'' + 4y = 0$; $y = \sin 2x$, $y = 3 \cos 2x + 4 \sin 2x$.
5. $y''' - y'' + y' - y = 0$; $y = e^x \sin x$, $y = 4e^x \cos x - e^x \sin x$.
6. $xy' - y - 2x^3 = 0$; $y = x^3 + Cx$.

7. $xy' + 3y - e^x = 0$; $y = e^x \left( \dfrac{1}{x} - \dfrac{2}{x^2} + \dfrac{2}{x^3} \right) + C$.

Verify that if the function $y = f(x)$ is a solution of the following implicit equation, then $y$ is also a solution to the differential equation.

8. $y^2 + \log|y| = x^3 + 1$; $y' = \dfrac{3x^2 y}{2y^2 + 1}$.

9. $x^2 - y^2 = 10x$; $y' = \dfrac{x^2 + y^2}{2xy}$.

10. $\dfrac{1}{2} \log (x^2 + y^2) + \arctan \dfrac{y}{x} + 1 = 0$; $y' = \dfrac{y - x}{y + x}$.

11. $y^2 + 2\sqrt{1 - x^2} = 1$; $y' = \dfrac{x}{y\sqrt{1 - x^2}}$.

12. $\arctan y + \arcsin x = 1$; $y' = -\dfrac{1 + y^2}{\sqrt{1 - x^2}}$.

For arbitrary values of the constant $C$, $y = Cx^2$ defines a family of parabolas passing through the origin tangent to the $x$ axis. To find a differential equation satisfied by this family of functions, differentiate with respect to $x$ obtaining

$$y' = 2Cx.$$

Eliminating $C$ between the two equations yields $y' = 2y/x$, which is a differential equation satisfied by this family of parabolas.

Using this technique, determine differential equations satisfied by the following families of function. If there are two arbitrary constants, the equations must be differentiated twice to eliminate them.

13. $x^2 - y^2 = C$.
14. $y = Ae^{-x} + Be^{2x}$.
15. $y = x(A + e^x + Be^{2x})$.
16. $y = C + \sin(D - x)$.

Find a differential equation satisfied by the following families of curves.

17. All circles with centers on the $y$ axis.
18. All circles with centers on the line $y = x$.
19. All parabolas passing through the origin and having a vertical axis.
20. All parabolas with horizontal axis passing through $(0, 1)$.
21. Sketch the direction field for the differential equation $y' = x$. Superimpose the graph of the solution satisfying the initial condition $f(0) = 1$ on the sketch of the direction field.
22. Sketch the direction field for the differential equation $y' = -x^2$. Superimpose the graph of the solution satisfying $f(1) = 1$.
23. Sketch the direction field for the equation $y' = -xy$. Attempt to sketch the graph of the solution satisfying the initial condition $f(0) = 1$.
24. Determine all solutions to the following differential equations on the interval $-\infty < x < \infty$
   (a) $y' = x + e^{-x}$      (c) $y''' = e^{2x}$
   (b) $y'' = \sin x$
25. Show that $f(x) = \frac{1}{2} + Ce^{-2x}$ satisfies the differential equation $y' + 2y = 1$. Determine a solution satisfying the initial condition $f(1) = 1$. Determine a solution satisfying $f(x_0) = y_0$. If all solutions to the equation are known to be of the form $f(x) = \frac{1}{2} + Ce^{-2x}$, is the solution satisfying this initial condition unique?

## 7.2  An Equation for the Exponential Function; Separation of Variables

Consider the differential equation

(1) $$y' = ky$$

where $k$ is a constant. Let us determine all solutions to this equation. One trivial solution to (1) is the function identically equal to zero. However, any solution of (1) that never vanishes satisfies

$$\frac{y'}{y} = k$$

and vice versa. But we know that

$$\frac{d}{dx}(\log|y|) = \frac{y'}{y},$$

and by the fundamental theorem of the calculus if

$$\frac{d}{dx}(\log|y|) = k,$$

then

$$\log|y| = kx + C,$$

$C$ a constant. If we exponentiate both sides of this equation, we have

$$|y| = e^{kx+C} = e^C e^{kx}.$$

Thus, since $y$ must be continuous, either $y = e^C e^{kx}$ for all $x$, or $y = -e^C e^{kx}$ for all $x$. If we redefine our constant $C$, we can assert that if $y$ satisfies (1) and does not vanish, then

(3) $$y = Ce^{kx} \qquad \text{for some nonzero constant } C.$$

It is easily verified that all of the functions defined by (3) are solutions of (1). Conceivably, however, there are other solutions since, in order to solve (1), we had to assume that the solution $y$ never vanished. To show that all solutions of (1) are given by (3), we proceed as follows. Let $y$ be any solution of (1) and consider

$$g(x) = e^{-kx} y.$$

Differentiating, we get

$$g'(x) = -ke^{-kx}y + e^{-kx}y'$$
$$= e^{-kx}(-ky+y') = 0$$

since $y$ is assumed to be a solution of (1). Therefore, $g(x)$ must equal a constant $C$. Consequently, $C = e^{-kx}y$, or $y = Ce^{kx}$. Thus we have shown that $f$ is a solution to (1) on the interval $-\infty < x < \infty$ if and only if

$$f(x) = Ce^{-kx} \text{ for some constant } C.$$

Equation (1) states that the rate of change of the unknown function $y$ is proportional to $y$ itself. This phenomenon can be observed experimentally in many situations such as radioactive decay, population growth, and the like. We shall discuss these and other applications in the next section.

There are other special cases of the first order equation

(4) $$y' = g(x, y)$$

which can be handled by elementary means.

Suppose $g(x, y)$ can be written as the quotient $p/q$ where $p$ is a function only of $x$, and $q$ is a function only of $y$. The differential equation (4) then becomes

(5) $$y' = \frac{p(x)}{q(y)}$$

or

$$q(y)y' = p(x).$$

But if $P$ and $Q$ are primitives for $p$ and $q$, respectively, this implies

(6) $$Q(y) = P(x) + C.$$

To see this, differentiate (6) using the chain rule. Then

$$Q'(y)y' = P'(x),$$

or

$$q(y)y' = p(x)$$

since $Q'(y) = q(y)$ and $P'(x) = p(x)$. Thus, if $q(y) = q(f(x))$ does not vanish for $a \leqslant x \leqslant b$, then $y = f(x)$ is a solution of (5) if and only if $y$ is a solution of (6). Whenever a solution $y$ to a differential equation such as (5) has been determined up to an implicit equation such as (6), we say that $y$ is an *implicit* solution to (5). If it is possible to solve equation (6) for $y$ in terms of $x$, then the function $y = f(x)$ is said to be an *explicit* solution to the differential equation.

Whenever we have an equation that can be written

$$y' = \frac{p(x)}{q(y)},$$

we say that the "variables of the equation are separable." The process of collecting terms and writing the equation in the form

$$q(y)y' = p(x)$$

is called "separating the variables" of the equation. The Leibnitz notation $y' = dy/dx$ is convenient in this regard, since separating the variable just consists of writing all terms involving $y$ on one side of the equation and all terms involving $x$ on the other. Since the differential $dy = y'\,dx$, we may write

$$q(y)y' = p(x)$$

as

$$q(y)\,dy = p(x)\,dx.$$

Then the implicit equation for the solution $y$ is just

$$\int q(y)\,dy = \int p(x)\,dx + C.$$

*Example 1.*   Solve the differential equation

(7)
$$y' = \frac{x}{y^2\sqrt{1+x^2}}.$$

Separating the variables we get

$$y^2\,dy = \frac{x}{\sqrt{1+x^2}}\,dx.$$

Integrating we have

$$\int y^2\,dy = \int \frac{x\,dx}{\sqrt{1+x^2}} + C$$

or

(8)
$$\frac{y^3}{3} = \sqrt{1+x^2} + C.$$

Then functions defined implicitly by (8) are solutions of (7). We do not attempt to get explicit expressions for the solutions although that is possible in this example.

### EXERCISES
Solve the following differential equations. The solutions may be either explicitly or implicitly defined.

1. $y' + 10y = 0$.
2. $y' - x^2 y = 0$.
3. $e^x y' + xy = 0$.
4. $(\tan x)y' + y = 0$.
5. $y' + y + 1 = 0$.
6. $y' - a\frac{y}{x} = 0$   a constant.

7. $(x^2 + 1)yy' = 0$.
8. $y^2 y' - \cos^2 x = 0$.
9. $y' + 4y \sin 2x = 0$.
10. $(x^2 + 1)y' + y^2 + 1 = 0$.

The determination of a solution to a differential equation that satisfies certain initial conditions is called an *initial value problem*. Solve the following initial value problems.

11. $y' = \frac{y}{x}$, $y(1) = 2$.

12. $y'x \log x - y = 0$, $y(2) = \log 4$.

13. $y' - 2y \cot x = 0$, $y(\pi/2) = 2$.

14. $yy' e^{y^2} = x - 1$, $y(2) = 0$.

15. Suppose $p$ is continuous on $[a, b]$. Adapt the argument in the text for the equation $y' + ky = 0$ to show that all solutions $y = f(x)$ to the differential equation

(9) $$y' + p(x)y = 0$$

can be written

(10) $$y = f(x) = Ce^{-\int_a^x p(t)dt},$$

where $C$ is a constant. (Show first that if $y$ never vanishes and is a solution of (9), then $y$ satisfies (10). Show next that all solutions of (9) must be of this form. Note also that $C = f(a)$. Therefore, if $f(a) = 0$, the solution $y = f(x)$ is identically zero.)

## 7.3 Applications

Many physical phenomena can be described by differential equations for which the variables are separable. In particular, the equation

(1) $$\frac{y'}{y} = k$$

occurs very frequently. Suppose $y = y(t)$ describes a physical phenomenon as a function of time, and by experiment it can be verified that the rate of change of $y$ is proportional to $y$. Then $y$ satisfies equation (1), and, by the discussion in Section 7.2,

(2) $$y(t) = Ce^{kt}.$$

By substituting $t = 0$ in equation (2), we see that $C = y(0)$. To determine the constant $k$, additional information must be used. For example, if the value of $y$ is known at some time $t_1 \neq 0$, then this determines $k$. For if $y(t_1) = b$, then

$$b = Ce^{kt_1},$$

and hence

$$k = \frac{1}{t_1} \log\left(\frac{b}{C}\right).$$

If $y = y(t)$ denotes the amount of a radioactive substance present at time $t$, then physical experiments show that the rate of decay of the material does satisfy equation (1). Since the amount of the substance is decreasing, the constant $k$ in (1) is negative. It is then convenient to replace $k$ by $-k$, where $k$ is taken to be positive. We may then write

$$y(t) = y(0)e^{-kt}.$$

If $y(t_0)$ denotes the amount of material present at time $t_0$, and $y(t_0 + T)$ denotes the amount present after a time interval $T$, the ratio $y(t_0 + T)/y(t_0)$, or the percentage of the amount present at time $t_0$ that remains at time $t_0 + T$, does not depend on the starting time $t_0$ or the initial amount $y(0)$. This is clear since

(3) $$\frac{y(t_0 + T)}{y(t_0)} = \frac{y(0)e^{-k(t_0 + T)}}{y(0)e^{-kt_0}} = e^{-kT}.$$

Thus the "decay constant" $k$ may be determined if the time $T$ associated with a given value of the ratio (3) is known. If this ratio is $\frac{1}{2}$, then $T$ is called the *half-life*. Hence, if the half-life is known, this determines $k$. For instance, if the half-life is 10 years, then

$$\frac{1}{2} = e^{-10k}.$$

Taking the logarithms of both sides we have

$$k = \frac{1}{10} \log 2.$$

## EXERCISES

1. The half-life of radium is 1590 years. What percentage has disintegrated after 100 years?
2. Polonium has a half-life of 3.05 minutes. How long will it take for 90% of a given amount of polonium to disintegrate.
3. Scientists have observed that living matter contains, in addition to carbon, $C_{12}$, a fixed percentage of a radioactive isotope of carbon, carbon 14 or $C_{14}$. When the living material dies, the amount of $C_{12}$ present remains constant, but the amount of $C_{14}$ decreases with a half-life of 5550 years. To determine when the Lascaux caves in France were occupied, charcoal from these caves was examined and it was found that the amount of $C_{14}$ had decreased to 14.5% of the original amount. At what time were the Lascaux caves inhabited?
4. Between 1960 and 1967 the Norwegian archaeologist, Helge Ingstad, excavated several house sites on the northern coast of Newfoundland. Several old Norse artifacts were discovered at these sites. The charcoal from cooking pits was analyzed and the percentage of carbon 14 remaining in the charcoal was found to be 88.6%. What was the date of this Viking settlement in Newfoundland? (This date agrees with the dates in the Greenland saga for the voyages of Leif Eiriksson and others to "Vinland." For a fascinating account of the discovery of this settlement, see *Westward to Vinland* by Helge Ingstad, St. Martin's Press, New York, 1969.)
5. A crude mathematical assumption to explain population growth is to assume that the rate of change of the population is proportional to the population present. In 1880 the population of the United States was 50 million people; in 1900 it was 75 million. Assuming this rate of growth, how often is the population doubling? What would the population be in 1970?
6. If a body at temperature $T$ is immersed in a medium of temperature $T_0$, then assuming the temperature $T_0$ of the medium can be kept constant, it can be observed that the rate of change of $T$ is proportional to $(T - T_0)$. In other words, $T'(t) = k(T - T_0)$.

   A thermometer at 0°F is brought into a room and one minute later the thermometer reads 30°. At the end of another minute the thermometer reads 45°. What is the temperature of the room?
7. If a gas at low pressure is kept at constant temperature, then the volume and pressure are related by Boyle's law, which states that

$$V'(p) = -\frac{V}{p}.$$

   Solve this differential equation. If the pressure is doubled, what is the corresponding volume?
8. If a particle is falling freely near the surface of the earth and $y(t)$ denotes the height above the surface of the earth, then, assuming distance is measured in feet, and time in seconds, the motion is described by the differential equation $y''(t) = -32$. Suppose a cannon is fired at an angle $\alpha$ with an initial velocity $v_0$. Neglecting any frictional forces, show that the projectile follows a parabolic path. What should the angle be to maximize the distance that the projectile

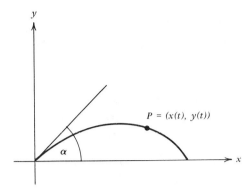

$P = (x(t), y(t))$

**Fig. 2**

travels? (Choose a coordinate system with the cannon at the origin and let $x(t)$, $y(t)$ denote the coordinates of the projectile at time $t$. (Figure 2.) Then $v_0 \sin \alpha$ is the initial vertical velocity and $v_0 \cos \alpha$ is the initial horizontal velocity. Set up and solve differential equations for the functions $x(t)$ and $y(t)$. To determine the angle for maximum distance, assume that the earth is flat.)

9. A dog is 100 yards away from a rabbit. The animals see each other and the rabbit begins to run at a constant speed in a straight line at right angles to the direction of the dog. At the same instant the dog begins to chase the rabbit. If the dog is always pointed at the rabbit and the distance between them does not change, determine a differential equation that describes the motion and solve this equation. (Choose an appropriate coordinate system.)

10. Often it is necessary to make a physical assumption in order to describe a physical phenomenon by a differential equation. This is the case in the following problem, credited to Professor Ralph Palmer Agnew. Sometime during a morning in February snow began falling at a heavy but constant rate. At noon the snow plows went out. By 1:00 *p.m.* one plow had cleared 8 miles of road but by 2:00 only 4 additional miles had been cleared by this plow. What time did it start snowing? (Show first that the rate the plow removes snow is proportional to the velocity of the plow multiplied by the depth of the snow. Then convince yourself that if the snow is not too deep, it is reasonable to assume that this rate of snow removal is constant. Let $y(t)$ denote the position of the plow at time $t$, and assuming $t = 0$ at noon, let $t_0$ be the time it started snowing. Set up an appropriate differential equation for $y$. Solve it and use the additional information to determine $t_0$.)

11. A certain mathematics professor is able to write calculus notes at a rate proportional to the number he has already written. The students in his class, however, can only read these notes at a constant rate. At the beginning of the year the professor hands the class 10 pages of notes and thereafter hands them out as he writes them. At the end of the fall quarter, one student finds himself 20 pages behind. At the end of the winter quarter, he is 70 pages behind. How far behind is the student at the end of the spring quarter?

## 7.4   First Order Linear Equations

An extremely important first order differential equation is the equation

(1) $$y' + p(x)y = q(x)$$

where $p$ and $q$ are assumed to be continuous. If $q = 0$, then the variables are separable, and the equation may be handled by the methods discussed in Section 7.2. Indeed, all solutions to

$$y' + p(x)y = 0$$

are given by

$$y = Ce^{-\int p(x)dx}$$

where $C$ is a constant (see Exercise 15 of Section 7.2). If $q$ is not identically zero, then we must proceed differently. The stratagem that we use is the observation that we may multiply (1) by a suitable positive function $P(x)$ so that the left-hand side is then the derivative of some recognizable function of $x$. Such a function $P(x)$ is called an *integrating factor* for the differential equation. Indeed, the appropriate function $P(x)$ here is just $e^{\int p(x)dx}$, since

(2)
$$e^{\int p(x)dx}[y' + p(x)y] = \frac{d}{dx}(ye^{\int p(x)dx}).$$

Formula (2) may be easily checked by differentiating the function $ye^{\int p(x)dx}$ and applying the chain rule. Thus to solve (1) we multiply

$$y' + p(x) \cdot y = q(x)$$

by $e^{\int p(x)dx}$ obtaining

$$e^{\int p(x)dx}[y' + p(x) \cdot y] = e^{\int p(x)dx} \cdot q(x).$$

But by (2) we infer that

$$\frac{d}{dx}(ye^{\int p(x)dx}) = e^{\int p(x)dx} q(x)$$

Hence by the fundamental theorem of the calculus

$$ye^{\int p(x)dx} = \int e^{\int p(x)dx} q(x)\, dx + C$$

or

(3)
$$y = e^{-\int p(x)dx}\left[\int e^{\int p(x)dx} q(x)\, dx + C\right].$$

Thus we have shown that if $y$ satisfies (1), then $y$ may be written in the form (3) for an appropriate choice of the constant $C$. Furthermore, it may be immediately checked by differentiation that each function defined by (3) is a solution of (1).

*Example 1.*    Solve

$$y' + y = x.$$

Here $p(x) = 1$ so $e^{\int p(x)dx} = e^x$. Multiplying the differential equation by this function we obtain

$$(y' + y)e^x = xe^x.$$

But

$$(y' + y)e^x = \frac{d}{dx}(ye^x).$$

Hence

$$\frac{d}{dx}(ye^x) = xe^x,$$

and therefore

$$ye^x = \int xe^x\, dx + C.$$

Integrating the right-hand side by parts, we get

$$ye^x = xe^x - e^x + C$$

or

(4)
$$y = (x - 1) + Ce^{-x}.$$

The constant $C$ may be determined by imposing an initial condition on the solution $y$. If $y(0) = 1$, for example, then substituting $x = 0$, $y = 1$ in (4) we obtain $1 = -1 + C$, or $C = 2$. Thus the solution to the initial value problem $y' + y = x$, $y(0) = 1$ is given by

$$y = (x - 1) + 2e^{-x}.$$

The equation

(1) $$y' + p(x)y = q(x)$$

is called a *linear equation* since all solutions of (1) are of the form

$$y = f(x) + Cg(x),$$

where $f(x)$ is one particular solution of (1) and $g(x)$ is a nonzero solution of $y' + p(x)y = 0$.

We have now completely solved the existence and uniqueness problem for first order linear differential equations by proving that if $p$ and $q$ are continuous for all values of $x$, then there exists a unique function $y = f(x)$ defined for all values of $x$ that satisfies (1) and passes through the point $(x_0, y_0)$. The proof of this result depends only on the fundamental theorem of the calculus.

**EXERCISES**

Solve the following first order linear equations by finding an integrating factor for the equation.

1. $y' - xy = 0$.
2. $y' + 2x^2 y = 0$.
3. $y' + (\sin x)y = 0$.
4. $y' - 2y = 1$.
5. $y' + 4xy = 2$.
6. $y' + xy = x$.
7. $y' - 4x^2 y = 2x^2$.
8. $y' - y = x^2 + 1$.

9. $y' = 4y + e^{2x}$.
10. $y' + (\cos x)y = \cos x$.
11. $xy' - y = x^3$.
12. $2xy' = \dfrac{4x^2 + 1}{x} - y$.
13. $2xy' - y = x^3 + x$.
14. $x(x + 1)y' + y = x(x + 1)^2 e^{-x^2}$.

15. In certain electrical circuits if $I(t)$ denotes the current at time $t$ and $E(t)$ denotes the electromotive force applied at time $t$, then the functions $E$ and $I$ are related by the linear differential equation

$$LI'(t) + RI(t) = E(t)$$

where $L$ and $R$ are positive constants. If $I(0) = 0$, and the electromotive force $E(t)$ is constant, say $E_0$, determine the current at time $t$. What is $\lim_{t \to \infty} I(t)$? Determine $I(t)$ if the initial current is zero, and the electromotive force $E(t) = E_0 \cos \omega t$.

16. In a certain chemical reaction, cane sugar in solution is converted into dextrose at a rate proportional to the amount of sugar present. If at time $t = 0$, 100 grams of sugar are in solution and after one half hour 10 grams have been converted, how much has been converted after 4 hours? Will all the sugar be converted at a certain time $t_0$? Explain.

17. In recent years there have been several reported instances of methyl-mercury poisoning caused by eating contaminated fish. It can be verified experimentally that the body eliminates methyl mercury at a rate proportional to the amount present with a half life of 70 days. If a fixed amount $m$ of methyl mercury is introduced into the body daily, determine a first order linear differential equation which describes the rate of change of the amount of methyl mercury present in the body. Solve the equation and determine the amount of methyl mercury present in the body after 100 days. What will be the maximum amount present

in the body? In 1968 the World Health Organization at a meeting in Stockholm set the maximum safe concentration of methyl mercury in the blood at 0.1 parts per million. Experiments show that the concentration of mercury in the blood ia about 40% of that in the total body. Certain types of fish caught off the California coast have been analyzed and have been found to contain a methyl mercury concentration of 1 part per million. How many pounds of such fish can a 150 lb. person eat per day and be sure of keeping his mercury concentration below 0.1 parts per million in the blood? (The data for this problem was supplied by Professor L. Sario.)

18. An equation such as

(5) $$f(x) = x + \int_1^x f(t)\, dt$$

is called an integral equation. The unknown here is, of course, the function $f$. Differentiate this equation and construct an appropriate differential equation. Solve this and then construct the unique function satisfying (5). Observe that if $f$ satisfies (5), then $f(1) = 1$.

Solve the following integral equations. There is exactly one solution to each equation. (See Exercise 18.)

19. $xf(x) = 1 + 2 \int_1^x f(t)\, dt \qquad x > 0.$

20. $f(x) = 2 + (1/x) \int_1^x f(t)\, dt \qquad x > 0.$

21. Show that each function defined by (3) satisfies the differential equation (1).

Certain first order differential equations for which the variables are not separable can be reduced to this case by means of a substitution or change of variable. We shall discuss one example of this phenomenon.

The first order differential equation $y' = f(x, y)$ is called *homogeneous* if for each real number $t \neq 0$,

$$f(tx, ty) = f(x, y).$$

For example,

$$y' = \sin\left(\frac{x+y}{x-y}\right) \qquad \text{and} \qquad y' = \frac{2xy}{y^2 - x^2}$$

are homogeneous differential equations. If $f(tx, ty) = f(x, y)$, then setting $t = 1/x$ we obtain $f(x, y) = f(1, y/x)$. Hence, $f$ is a function only of the ratio $y/x$. We may exploit this fact to obtain an implicit solution to the homogeneous equation (6) $y' = f(x, y)$. Define a new function $v = v(x)$ by $v = y/x$ where $y = y(x)$ is an unknown solution of (6). Then $y = xv$, and differentiating with respect to $x$ we have

$$y' = v + xv'.$$

Since the differential equation is homogeneous, $f(x, y) = f(1, y/x) = f(1, v)$. Set $g(v) = f(1, v)$. Then substituting in (6), we obtain

$$xv' + v = y' = f(x, y) = g(v).$$

Solving for $v'$, we get

(7) $$v' = \frac{g(v) - v}{x}.$$

The variables are separable in this equation, and thus any function $v = v(x)$ defined implicitly by

$$\int \frac{dv}{g(v) - v} = \int \frac{dx}{x} + C$$

is a solution of (7). Hence, $y = v \cdot x$ is a solution of (6).

In short, to obtain implicit solutions of a homogeneous equation $y' = f(x, y)$, make the substitution $y = vx$. The variables of the resulting equation are separable, and this equation may be solved directly.

***Example 2.*** Solve

(8)
$$y' = \frac{2xy}{y^2 - x^2}.$$

Equation (8) is homogeneous since for $t \neq 0$

$$f(tx, ty) = \frac{2txty}{t^2y^2 - t^2x^2} = \frac{2xy}{y^2 - x^2} = f(x, y).$$

The substitution $y = vx$ in (8) yields

$$xv' + v = y' = \frac{2xy}{y^2 - x^2} = \frac{2vx^2}{v^2x^2 - x^2} = \frac{2v}{v^2 - 1}.$$

Therefore

$$xv' = \frac{2v}{v^2 - 1} - v = \frac{3v - v^3}{v^2 - 1}$$

or

$$\int \frac{v^2 - 1}{3v - v^3} \, dv = \int \frac{dx}{x}.$$

Integrating, we get

$$-\frac{1}{3} \log |3v - v^3| = \log |x| + C.$$

Hence

$$\log |(3 - v^3)x^3| = C,$$

or

$$(3v - v^3)x^3 = C.$$

Substituting $v = y/x$ yields

$$\left(\frac{3y}{x} - \frac{y^3}{x^3}\right) x^3 = C$$

or

$$3x^2y - y^3 = C.$$

The desired solution $y = f(x)$ to the differential equation is defined implicitly by the above equation.

**EXERCISES**

Obtain implicit solutions to the following homogeneous differential equations.

22. $y' = \dfrac{y}{x} - 1.$

23. $y' = \dfrac{x^2 + 2y^2}{xy}.$

24. $y^2 + (x^2 - xy + y^2)y' = 0.$

25. $x^2y' + (xy + 2y^2) = 0.$

26. $xyy' = (x^2 - y^2).$

27. $x(x - y)y' + y^2 = 0.$

## 7.5 Simple Harmonic Motion

As our last example of a differential equation we consider the second order equation

(1)
$$y'' + k^2y = 0$$

where $k$ is a nonzero constant. If we interpret a solution $y = y(t)$ as measuring displacement of a particle moving along a line as a function of time $t$, then $y''(t)$ is the acceleration of this particle at time $t$. Equation (1) then states that this acceleration is proportional to the displacement and is in the opposite direction.

An example of a system, the motion of which is described by equation (1), is the oscillation of a weight suspended from a spring or elastic cord. We assume that the weight has been put into motion by an initial displacement from its equilibrium position or by an initial velocity. Also, we ignore all frictional forces. To see that the resulting motion is indeed described by equation (1), we proceed as follows. Choose a coordinate system so that the free end of the unstretched spring is at $y_0$. After the weight has been attached, we assume the weight is in the equilibrium position when $y = 0$, as in Figure 3.

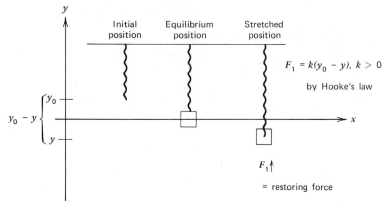

**Fig. 3**

Now we compute the forces acting on the weight. First there is the restoring force $F_1$ of the spring which, by Hooke's law, is proportional to the amount the spring has been stretched. Therefore, if the weight has been displaced $y$ units from the origin or equilibrium position, we have

$$F_1 = k(y_0 - y).$$

The constant $k$ is called the spring constant or stiffness constant. Indeed, $k > 0$ since the restoring force $F_1$ is directed upward. Next we have the gravitational force $F_2$ acting on the weight. $F_2 = -mg$, where $m$ is the mass of the weight and $g$ is the gravitational acceleration. If the mass is measured in pounds and the distance in feet, then $g = 32$. The gravitational force is directed downward, which explains the negative sign. Thus the net force acting on the weight is $F_1 + F_2$ which, by Newton's second law, must equal the product of the mass $m$ of the weight and its acceleration $y''$. Thus

(2)  $$my'' = F_1 + F_2 = k(y_0 - y) - mg.$$

At the equilibrium position $y = 0$ and $y'' = 0$. Hence, $ky_0 - mg = 0$. Combining this with (2), we have

$$my'' = -ky$$

or

$$my'' + ky = 0$$

which is the same as (1) since $k$ and $m$ are both positive.

Equation (1) is called a *linear* equation since, if $f$ and $g$ are solutions of (1), the linear combination $af + bg$, where $a$ and $b$ are arbitrary real numbers, is also a solution. To see this, set $h = af + bg$. Then

$$\begin{aligned} h'' + k^2 h &= af'' + bg'' + k^2(af + bg) \\ &= a(f'' + k^2 f) + b(g'' + k^2 g) \\ &= 0. \end{aligned}$$

Each of the terms in parentheses vanishes since both $f$ and $g$ are assumed to be solutions of (1).

To solve (1), we observe first that the functions $\sin kx$ and $\cos kx$ certainly are solutions of (1). Also, any linear combination $a \sin kx + b \cos kx$ is also a solution. It now follows from a general uniqueness theorem for linear differential equations that any solution $y$ to (1) must be of the form

$$(3) \qquad y(x) = a \sin x + b \cos x.$$

for appropriate constants $a$ and $b$. It is possible to give a direct proof of this fact without appealing to this general result by generalizing the uniqueness proof in Section 7.2 for the first order linear equation. However, since the argument is somewhat involved, it is given in the exercises.

To sketch the graph of the solution (3), it is convenient to rewrite (3) somewhat. Note first that since

$$\left(\frac{a}{\sqrt{a^2+b^2}}\right)^2 + \left(\frac{b}{\sqrt{a^2+b^2}}\right)^2 = 1,$$

there is a unique number $B$ satisfying $0 \leqslant B < 2\pi$ for which

$$\cos B = \frac{a}{\sqrt{a^2+b^2}}$$

and

$$\sin B = \frac{b}{\sqrt{a^2+b^2}}.$$

Hence from (3) we may write

$$y(x) = \sqrt{a^2+b^2}\left(\frac{a}{\sqrt{a^2+b^2}} \sin kx + \frac{b}{\sqrt{a^2+b^2}} \cos kx\right)$$

$$= \sqrt{a^2+b^2}\,(\cos B \sin kx + \sin B \cos kx)$$

$$= \sqrt{a^2+b^2}\,\sin (kx + B)$$

If we set $A = \sqrt{a^2+b^2}$, then the solution to (1) takes the form

$$(4) \qquad y(x) = A \sin (kx + B).$$

Knowing the graph of the sine function we may easily sketch the graph of (4). See Figure 4.

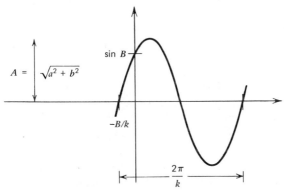

**Fig. 4** $y(x) = A \sin (kx + B)$.

Interpreting $y(x)$ as the motion of a particle satisfying equation (1), we see that since $\sin x$ is periodic of period $2\pi$, $\sin (kx + B)$ has a period of

$2\pi/k$. This we call the *period* of the motion. The reciprocal of the period, or $k/2\pi$, is called the *frequency* of the motion. Also, the fact that *for all* values of $x$, $|\sin x| \leq 1$ implies that the maximum value of $y$ is $A = \sqrt{a^2 + b^2}$. This quantity is called the *amplitude* of the motion. The independent variable $kx + B$ is often referred to as the *phase angle* of the motion. The number $B$, which is the value of this phase angle when $x = 0$, is called the *initial value* of the phase.

We have shown that any linear combination $y(x) = a \cos kx + b \sin kx$ can be also written $y(x) = A \sin (kx + B)$ for appropriate constants $A$ and $B$. If we let $B_1 = B - (\pi/2)$ then $\sin (kx + B) = \cos (kx + B_1)$. Hence the general solution to (1) may also be written

$$y(x) = A \cos (kx + B_1)$$

The functions $\sin (kx + B)$, $\cos (kx + B_1)$ are often called simple harmonics, hence the motion described by (1) is called *simple harmonic motion*.

The solution to (1) may be uniquely determined in two ways. One is to require that the solution satisfy the initial conditions $y(x_0) = y_0$ and $y'(x_0) = y_1$. Since there are two constants $a$ and $b$ to be determined, we now have two initial conditions. If we set

$$y_0 = y(x_0) = a \sin kx_0 + b \cos kx_0$$

$$y_1 = y'(x_0) = ka \cos kx_0 - kb \sin kx_0,$$

Then these equations may be solved for $a$ and $b$. We obtain

(5) $$a = ky_0 \sin kx_0 + y_1 \cos kx_0$$

(6) $$b = ky_0 \cos kx_0 - y_1 \sin kx_0.$$

Thus the initial value problem $y'' + k^2 y = 0$, $y(x_0) = y_0$, $y'(x_0) = y_1$ has the unique solution $y(x) = a \cos kx + b \sin kx$ where $a$ and $b$ are determined by (5) and (6).

The constants $a$ and $b$ in (3) may also be determined by requiring that the solution pass through two points $(x_0, y_0)$ and $(x_1, y_1)$. These are called boundary conditions on the solution $y$. The pair of equations that must be solved is then

$$y_0 = y(x_0) = a \cos kx_0 + b \sin kx_0$$
$$y_1 = y(x_1) = a \cos kx_1 + b \sin kx_1.$$

As long as $kx_0 - kx_1$ is not a multiple of $2\pi$ these equations may be uniquely solved for $a$ and $b$. The determination of the solution to the differential equation (1) satisfying the two boundary conditions is called a *boundary value problem*.

**Example 1.** Solve the initial value problem $y'' + 4y = 0$, $y(0) = 1$, and $y'(0) = -2$. What is the amplitude and period of the motion described by the solution to this problem, and what is the initial phase? If $y$ is a solution of $y'' + 4y = 0$, then

$$y = y(x) = a \sin 2x + b \cos 2x$$

and

$$y'(x) = 2a \cos 2x - 2b \sin 2x.$$

Setting $x = 0$ this implies

$$b = 1 \quad \text{and} \quad a = -1.$$

Hence, the amplitude $\sqrt{a^2 + b^2} = \sqrt{2}$. If we write $y(x)$ in the form $y(x) = \sqrt{2} \sin (2x + B)$, we may determine the initial phase $B$ from the initial conditions.

Since $y(0) = 1$ and $y'(0) = -2$, we have

$$\sin B = \frac{1}{\sqrt{2}}$$

and

$$\cos B = -\frac{1}{\sqrt{2}}.$$

Consequently, $B = 3\pi/4$. The period $2\pi/k = \pi$ since $k = 2$, and the appropriate solution is $y(x) = -\sin 2x + \cos 2x = \sqrt{2} \sin (2x + 3\pi/4)$. The solution is sketched in Figure 5.

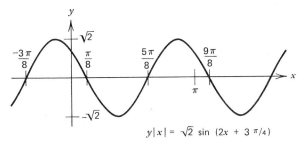

$$y|x| = \sqrt{2} \sin (2x + 3\pi/4)$$

**Fig. 5**   $y(x) = \sqrt{2} \sin (2x + 3\pi/4).$

*Example 2.*   Solve the boundary value problem $y'' + y = 0$, $y(0) = 1 = y(\pi/4)$. Determine the amplitude of the motion. The desired solution is $y(x) = a \sin x + b \cos x$ for appropriate constants $a$ and $b$. However, applying the boundary conditions we infer that

$$1 = y(0) = b$$

$$1 = y\frac{\pi}{4} = a\frac{\sqrt{2}}{2} + b\frac{\sqrt{2}}{2}$$

Therefore, $b = 1$ and $a = \sqrt{2} - 1$. The desired solution is $y(x) = (\sqrt{2} - 1) \sin x + \cos x$. The amplitude of the motion is $\sqrt{a^2 + b^2} = \sqrt{4 - 2\sqrt{2}}$.

## EXERCISES

1. Find the solution to the initial value problem $y'' + 4y = 0$, $y(\pi/4) = y'(\pi/4) = 1$. What is the amplitude and initial phase of the solution?
2. Find the solution to the equation $y'' + 5y = 0$ with initial phase $\pi/4$ and amplitude 10. What is the velocity of the motion at $x = \pi/3$?
3. Suppose a particle is moving in simple harmonic motion with an amplitude of 4 inches. When $y(t_0) = 2, y'(t_0) = 12\sqrt{3}$. What is the period of the motion?
4. Assume $y(t)$ is describing simple harmonic motion. If $y(t_0) = 3$ when $y'(t_0) = 4$, and $y(t_1) = 4$ when $y'(t_1) = 3$, determine the period of the motion.
5. Solve the boundary value problem $y'' + 2y = 0$, $y(\pi/4) = 2 = y(-\pi/4)$.
6. Solve the boundary value problem $y'' + y = 0$, $y(\pi/2) = 0$, $y(\pi) = 1$.
7. Solve the boundary value problem $y'' + 4y = 0$, $y(0) = 1$, $y(\pi/4) = -1$.
8. A weight attached to a spring causes it to stretch one foot. The spring is stretched one foot further and then released. What is the position and velocity of the weight one second later?
9. A weight attached to a spring causes it to stretch one foot. The weight is set in motion by an upward velocity of one foot per second. What is the amplitude and period of the motion?
10. Show that the differential equation

$$y'' + k^2 y = f(x)$$

has the solution

$$y(x) = \frac{1}{k} \int_a^x \sin k(x - t) f(t) \, dt.$$

*11. Show that all solutions to the differential equation (1) $y'' + k^2 y = 0$ are of the form $y(x) = a \sin kx + b \cos kx$ where $a$ and $b$ are constants. (We adapt the technique used in Section 7.2 for the equation $y' + ky = 0$. Suppose $y = y(x)$ is a solution of (1) on some interval $I$ on which $\cos kx$ does not vanish. Then define $u(x) = (y(x)/\cos kx)$. We assert that as a consequence $u(x) = a \tan kx + b$ for appropriate constants $a$ and $b$. This implies that

$$y(x) = u(x) \cdot \cos kx = a \sin kx + b \cos kx.$$

Differentiating $y(x) = u(x) \cos kx$ twice, we obtain

$$y''(x) = u''(x) \cos kx - 2ku'(x) \sin kx - k^2 u(x) \cos kx$$

and hence

$$y''(x) + k^2 y(x) = u''(x) \cos kx - 2ku'(x) \sin kx.$$

Since $y''(x) + k^2 y(x) = 0$, by assumption this implies

(7) $$u''(x) \cos kx - 2ku'(x) \sin kx = 0.$$

Multiplying (7) by $\cos kx$ yields

(8) $$u''(x) \cos^2 kx - 2ku'(x) \sin kx \cos kx = 0.$$

If we observe that the left-hand side of (8) is just

$$\frac{d}{dx}(u'(x) \cos^2 kx),$$

we have, by the fundamental theorem of the calculus,

$$u'(x) \cos^2 kx = c.$$

If $c = 0$, $u(x)$ must be constant and then $y(x) = b \cos kx$ for some constant $b$. If $c \neq 0$;

$$u'(x) = c \cdot \sec^2 kx.$$

Hence

$$u(x) = \frac{c}{k} \tan kx + d,$$

and in both cases $y(x) = a \sin kx + b \cos kx$ for appropriate constants $a$ and $b$. For an interval that includes a zero, say $x_0$, of $\cos kx$, we apply the above argument to the left of $x_0$ and to the right of $x_0$, obtaining

$$y(x) = a \sin kx + b \cos kx \qquad x < x_0$$
$$= a' \sin kx + b' \cos kx \qquad x > x_0$$

where conceivably $a \neq a'$, and $b \neq b'$. However, since $y(x)$ is continuous at $x_0$, we must have

$$y(x_0) = \lim_{x \to x_0^-} (a \sin kx + b \cos kx)$$
$$= a \sin kx_0$$

since $\cos kx_0 = 0$. Furthermore

$$y(x_0) = \lim_{x \to x_0^+} (a' \sin kx + b' \cos kx) = a' \sin kx_0.$$

Since $\sin kx_0 \neq 0$, we must have $a = a'$. On the other hand, since $y''(x_0)$ exists, $y'$ is continuous at $x_0$. Use this fact to show that $b = b'$. Thus, if $y(x)$ is a solution to (1) on any interval $I$, then

$$y(x) = a \sin kx + b \cos kx$$

for appropriate constants $a$ and $b$.)

*12. Verify that $e^{-kx}$ and $e^{kx}$ are solutions of the equation

(9) $$y'' - k^2 y = 0.$$

Verify that all linear combinations of these functions are solutions to (9). Show conversely that all solutions of (9) are linear combinations of $e^{-kx}$ and $e^{kx}$. (Adopt the argument in Exercise 11 for the equation $y'' + k^2 y = 0$ in the following way. Let $y$ be a solution of (9) and define $u = y \cdot e^{-kx}$ or $y = u e^{kx}$. Differentiate twice and obtain the second order equation

$$u'' e^{kx} + 2ku' e^{kx} = 0.$$

Multiply this equation by $e^{kx}$ and observe that

$$\frac{d}{dx}(u' e^{2kx}) = u'' e^{2kx} + 2ku' e^{2kx} = 0.$$

Use the fundamental theorem of the calculus to determine $u$ and hence $y$.)

In the remaining exercises you may assume that all the solutions to the equation $y'' - k^2 y = 0$ are given by $y = a e^{kx} + b e^{-kx}$.

13. A particle moves such that its acceleration is always twice its displacement. If at $t = 1$ the position and velocity equal 1, determine the displacement function $y(t)$ and sketch the graph.

14. A particle moves so that its displacement and acceleration are always equal. If $y(0) = 2$ and $\lim_{t \to \infty} y(t) = 0$, determine $y(t)$. What happens if in place of the assumption that $\lim_{t \to \infty} y(t) = 0$ we just assume that $y(t)$ remains bounded as $t \to \infty$?

15. Verify that $y = \cosh x$, and $y = \sinh x$ are solutions of $y'' - y = 0$. Are all solutions to this equation linear combinations of these two functions?

16. A particle moves so that its acceleration is always twice its displacement. If at $t = 0$, the displacement $y = 1$, determine the function $y(t)$ if for positive $t$ the velocity remains bounded.

17. Solve the initial value problem $y'' - 4y = 0$, $y(0) = y'(0) = 1$.

18. Solve the boundary value problem $y'' - y = 0$, $y(0) = e = y(1)$.

# CHAPTER EIGHT

# VECTORS AND VECTOR ALGEBRA

### 8.1 Definition of Vectors

One might say that real numbers were invented to obtain quantitative measurements of physical phenomena. We have met many instances of this: measurements of distance, speed, area, pressure, volume, to mention only a few. However, it is also apparent that in many situations one number alone is not going to be adequate to describe the phenomenon under consideration. Consider a space vehicle in flight. To specify the location of the rocket at a given time three numbers are required. Three more are necessary to specify the velocity. If we wish to specify the acceleration, we need three more. Thus to describe the position, velocity, and acceleration at time $t$, an array of nine numbers is needed.

To measure complicated physical phenomena in quantitative terms we are forced to use arrays of more than one number. Such an array is called a vector. To be precise, we define a *vector* to be an $n$-tuple of real numbers, that is, an ordered array of $n$ real numbers $(a_1, a_2, \ldots, a_n)$. The $i$th number entering in this array is called the $i$th *component* of the vector. When we say that a vector is an *ordered* array of real numbers, we mean that if $A = (a_1, \ldots, a_n)$ and $B = (b_1, \ldots, b_n)$ are two vectors, then $A = B$ if and only if $a_1 = b_1, a_2 = b_2 \ldots,$ and $a_n = b_n$. Thus the vector $A = (a_1, \ldots, a_n)$ is not to be confused with the set whose members are the numbers $a_1, \ldots, a_n$, which we write $\{a_1, \ldots, a_n\}$. The set of all such vectors will be written $V_n(R)$.

In our discussion, we shall always denote vectors by capital letters $A$, $B$, $C$, etc. and real numbers by small letters $a$, $b$, $c$, etc. (To further distinguish between vectors and real numbers the latter are often called *scalars*.) We shall formulate the definitions and theorems of vector algebra in terms of vectors from $V_n(R)$. However, for examples and calculations we shall assume for the most part that $n = 2$ or $n = 3$. We are, of course, limited to these cases when we give spatial interpretation to the notion of vectors.

Certain aspects of the algebra of real numbers extend readily to vectors. Without attempting to give any geometrical or physical motivation for the idea, it is reasonably apparent that addition between vectors in $V_n(R)$ can be readily defined. Indeed, if $A = (a_1, \ldots, a_n)$ and $B = (b_1, \ldots, b_n)$, we define the vector sum by the formula

$$A + B = (a_1 + b_1, a_2 + b_2, \ldots, a_n + b_n).$$

Thus the sum of two vectors is formed by adding the respective components. It is easily checked that vector addition has all the properties possessed by addition of real numbers. It is commutative and associative. There is a zero vector, $0 = (0, 0, \ldots, 0)$, and each vector has an inverse with respect to addition.

Indeed, if $A = (a_1, \ldots, a_n)$, then $-A = (-a_1, \ldots, -a_n)$. This implies that subtraction is always possible since

$$A - B = (a_1 - b_1, \ldots, a_n - b_n).$$

Since addition extends so readily to vectors, it is somewhat of a shock to learn that in general it is impossible to define a multiplication for vectors that possesses all the properties enjoyed by multiplication of real numbers. Indeed this is only possible when $n = 2$. (See Exercise 10, p. 306). However, it is possible to define a "multiplication" of a vector by a real number. This operation yields a vector as a result. The multiplication, called *scalar multiplication*, is defined by the following formula. If $a$ is a real number, or scalar, and $A = (a_1, \ldots, a_n) \in V_n(R)$, then we define

$$aA = (aa_1, aa_2, \ldots, aa_n).$$

We emphasize again that scalar multiplication is not a multiplication between vectors but is a multiplication of a vector by a real number or scalar that yields a vector as the product. If a vector $B$ can be written $B = aA$, then $B$ is called a *scalar multiple* of $A$. Thus $(2, -4, 6)$ is a scalar multiple of $(1, -2, 3)$ since

$$(2, -4, 6) = 2(1, -2, 3).$$

It may be easily checked that scalar multiplication satisfies the following distributive laws

(1) $$a(A + B) = aA + aB$$

(2) $$(a + b)A = aA + bA,$$

and the associative law

(3) $$(ab)A = a(bA).$$

If $k$ vectors $A_1, \ldots, A_k$ are given and there exist $k$ real numbers (scalars) $a_1, \ldots, a_k$ so that the vector $B$ can be written

$$B = a_1 A_1 + \cdots + a_k A_k,$$

then $B$ is called a *linear combination* of the vectors $A_1, \ldots, A_k$.

**Example 1.** Express $(-1, 2)$ as a linear combination of the vectors $(1, 2)$ and $(2, 3)$. We seek scalars $x_1, x_2$ so that

$$(-1, 2) = x_1(1, 2) + x_2(2, 3).$$

But

$$x_1(1, 2) + x_2(2, 3) = (x_1 + 2x_2, 2x_1 + 3x_2).$$

So

$$(-1, 2) = (x_1 + 2x_2, 2x_1 + 3x_2)$$

if and only if the components of the two vectors are equal. Consequently

$$-1 = x_1 + 2x_2$$

$$2 = 2x_1 + 3x_2.$$

This pair of linear equations can be immediately solved by elimination yielding $x_1 = 7$ and $x_2 = -4$. Thus

$$(-1, 2) = 7(1, 2) - 4(2, 3).$$

Finally, we define the *length* or *norm* $|A|$ of a vector $A$ by the formula

$$|A| = (a_1^2 + \cdots + a_n^2)^{1/2}$$

where
$$A = (a_1, \ldots, a_n).$$

It may be immediately checked that for a scalar $a$
$$|aA| = |a||A|$$

where $|a|$ is, of course, the absolute value of the scalar $a$.

**EXERCISES**

1. If $A = (-1, 4)$, $B = (2, 3)$, $C = (2, -1)$, compute $A + B$, $B - 2C$, $A - B + 2C$, $2(A + 3B)$, $|3A + 4B|$.

2. If $A = (3, 1, -2)$, $B = (2, 0, -1)$, $C = (1, 1, -2)$, compute $A - 2B$, $3(A + 3B)$, $2(B - A + C)$, $|2B - 4A + C|$.

3. Verify that vector addition possesses all the properties of addition of real numbers. That is, show that vector addition, is associative and commutative. Show that there is an identity for vector addition and that every vector has an additive inverse.

4. Show that scalar multiplication satisfies the distributive laws (1) and (2) as well as the associative law (3). Also verify the formula
$$|aA| = |a||A|.$$

5. By analogy with vector addition one is tempted to define the "product" of two vectors $A = (a_1, \ldots, a_n)$, $B = (b_1, \ldots, b_n)$, by the formula
$$A \ \# \ B = (a_1 b_1, a_2 b_2, \ldots, a_n b_n).$$

Verify that this "product" is commutative, associative, and distributive with respect to vector addition. Show, however, that not all the properties of multiplication of real numbers are possessed by this "product." (Show that the vector $I = (1, 1, \ldots, 1)$ is an identity with respect to the product $\#$, but not every nonzero vector $X$ has an inverse with respect to $\#$.)

6. Express $(2, -4)$ as a linear combination of $(1, 1)$ and $(2, 3)$.

7. Express $(1, -1, 2)$ as a linear combination of $(1, 0, -1)$, $(2, 2, 1)$, and $(2, -1, 3)$.

8. Can $(-1, 1)$ be expressed as a linear combination of $(1, 4)$ and $(-2, -8)$?

9. Can $(4, 1, 1)$ be expressed as a linear combination of $(1, -2, 1)$, $(2, 1, 1)$ and $(3, -1, 2)$?

10. A multiplication may be defined in $V_2(R)$ having all of the properties of multiplication of real numbers (that is, $V_2(R)$ provided with this multiplication is a field) in the following way. If $A = (a_1, a_2)$ and $B = (b_1, b_2)$, then we define (4) $A \cdot B = (a_1 b_1 - a_2 b_2, \ a_2 b_1 + a_1 b_2)$. Verify that the set $V_2(R)$ is a field when provided with this multiplication. This field is called the field of complex numbers. (The axioms for a field are given on p. 5. The verification of properties 1 to 4 is virtually automatic and may be omitted. To verify property 5, we must show that if $A \neq 0$, then $A$ has an inverse with respect to the multiplication defined by (4). To prove this, show first that $I = (1, 0)$ is an identity element with respect to the multiplication. Then to determine a formula for the inverse $A^{-1}$, solve the equation $A \cdot X = I$.)

### 8.2    Geometric Representation of Vectors

The geometric representation of vectors from $V_3(R)$ in space, and of vectors from $V_2(R)$ in the plane, is extremely useful both for giving spatial or geometric meaning to these algebraic ideas as well as for providing geometric models for applications. We construct the spatial representation for vectors in $V_3(R)$ as follows. First, construct a coordinate system in space consisting of three mutually perpendicular lines $l_1, l_2, l_3$ which intersect at a point that we take to be the origin. Next, choose positive directions along each of these lines. Which direction we take to be positive and in which order we label the lines is, of course, arbitrary. However, in this book we shall number the axes and choose directions as in Figure 1.

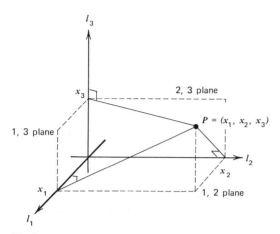

**Fig. 1**

The coordinates of $P$ will be the numbers corresponding to the points of intersection of the coordinate axes and lines through $P$ each perpendicular to the axes $l_1$, $l_2$, $l_3$. We shall label these axes the first, second, and third coordinate axes, rather than the $x$, $y$, $z$ axes. Hence the coordinates of a point $P$ will be written $(x_1, x_2, x_3)$ rather than $x$, $y$, $z$. The planes determined by pairs of the coordinate axes will be called the coordinate planes. We label the plane determined by the axes $l_1$, $l_2$ as the $1, 2$ plane or the $x_1, x_2$ plane. Similarly the $2, 3$ and $1, 3$ planes are determined by the axes $l_2$, $l_3$ and $l_1$, $l_3$, respectively.

Thus, to each point in space there corresponds a unique triple of real numbers $(x_1, x_2, x_3)$. Conversely, we assume that each triple of real numbers $(x_1, x_2, x_3)$ defines a unique point in space with these numbers $x_1, x_2, x_3$ as first, second, and third coordinates.

One might expect that the appropriate geometric representation of a vector $A = (a_1, a_2, a_3)$ would consist just of the point in space having coordinates $(a_1, a_2, a_3)$. This, unfortunately, is not general enough to give an adequate geometric interpretation to the operations we perform on vectors such as vector addition and scalar multiplication. Instead we represent the vector $A$ by a family of directed line segments. One line segment of this family consists of the segment drawn from the origin to the point with coordinates $(a_1, a_2, a_3)$. The remainder consist of all directed line segments having the same length and the same direction as the first one. This is illustrated in Figure 2. To indicate the direction of a directed line segment from a point $P$ to a point $Q$, we draw an arrow beginning at $P$ and ending at $Q$. If the directed line segment $PQ$ is a representation for the vector $A$, then the point $P$ is called the *initial point* of the representation and $Q$ is the *terminal point*. The points $Q$ and $P$ are also called the head and tail of the representation for $A$.

The length $|PQ|$ of the segment $PQ$ is given by

$$\sqrt{(q_1 - p_1)^2 + (q_2 - p_2)^2 + (q_3 - p_3)^2}$$

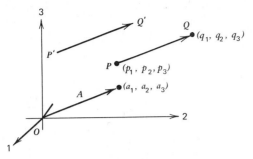

**Fig. 2**   Geometric representations for the vector $A$ consist of all directed line segments having the same length and the same direction as the line segment from the origin to the point with coordinates $(a_1, a_2, a_3)$.

and the length $|A|$ of the vector $A$ is given by $\sqrt{a_1^2 + a_2^2 + a_3^2}$. Hence, if $PQ$ is a representation for the vector $A$, we must have

(1) $$\sqrt{a_1^2 + a_2^2 + a_3^2} = \sqrt{(q_1 - p_1)^2 + (q_2 - p_2)^2 + (q_3 - p_3)^2}.$$

Next we must give mathematical meaning to the statement that the directed line segment $PQ$ has the same direction as the directed line segment from the origin to the point with coordinates $(a_1, a_2, a_3)$. We do this by introducing the notions of *direction angles* and *direction cosines* of a directed line segment.

Let us first consider a line segment from the origin to a point $P$ with coordinates $(p_1, p_2, p_3)$. The *direction angles* of the line segment $OP$ are the angles $\alpha_1, \alpha_2, \alpha_3$ between 0 and $\pi$ that this directed line segment makes with the positive first, second, and third coordinate axes. Since the length of this segment is $\sqrt{p_1^2 + p_2^2 + p_3^2}$, the cosines of each of these angles are given by the following formulas. (See Figure 3.)

$$\cos \alpha_1 = \frac{p_1}{\sqrt{p_1^2 + p_2^2 + p_3^2}}.$$

$$\cos \alpha_2 = \frac{p_2}{\sqrt{p_1^2 + p_2^2 + p_3^2}}.$$

$$\cos \alpha_3 = \frac{p_3}{\sqrt{p_1^2 + p_2^2 + p_3^2}}.$$

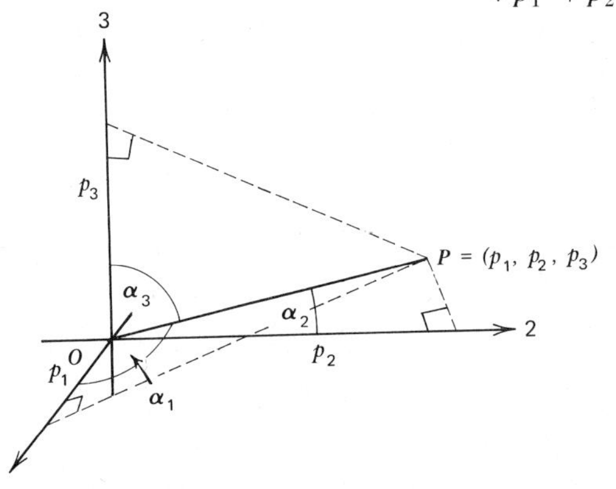

**Fig. 3**

The numbers $\cos \alpha_1$, $\cos \alpha_2$, and $\cos \alpha_3$ are called the *direction cosines* of the directed line segment $OP$. A line segment $OP'$ will have the same direction as $OP$ if and only if $OP'$ lies along $OP$. But this means that the direction angles of the two line segments are equal. Equivalently, their direction cosines are equal. (See Figure 4.)

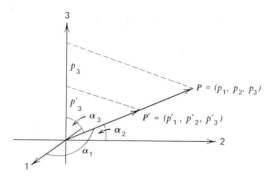

**Fig. 4**   Directed line segments emanating from the origin have the same direction if their direction cosines are equal.

To define the direction angles of an arbitrary directed line segment $PQ$, we translate our coordinate system so that the origin is at the point $P$. The new coordinate axes $l_1', l_2', l_3'$ emanating from $P$ have the same direction as the original first, second, and third coordinate axes. The direction angles of the segment $PQ$ are the angles between 0 and $\pi$ that this segment makes with the translated positive coordinate axes. Since the coordinates of the points $P$ and $Q$ are $(p_1, p_2, p_3)$ and $(q_1, q_2, q_3)$ respectively, the cosines of these angles are given by

$$\cos \alpha_1 = \frac{q_1 - p_1}{|PQ|} \qquad \cos \alpha_2 = \frac{q_2 - p_2}{|PQ|} \qquad \cos \alpha_3 = \frac{q_3 - p_3}{|PQ|}$$

where

$$|PQ| = \sqrt{(q_1 - p_1)^2 + (q_2 - p_2)^2 + (q_3 - p_3)^2}.$$

See Figure 5.

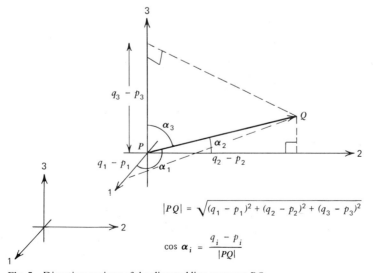

**Fig. 5**   Direction cosines of the directed line segment $PQ$.

Two directed line segments $PQ$ and $P_1Q_1$ have the same direction if, whenever one line segment, say $PQ$, is translated parallel to itself so that the point $P$ goes into the point $P_1$, the translate of $PQ$ will lie along $P_1Q_1$. (See Figure 6.) In terms of direction cosines, this just means that $PQ$ and $P_1Q_1$ will have the same direction if and only if the two segments have the same direction cosines.

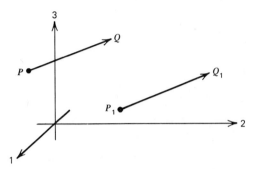

**Fig. 6**    Line segments having the same direction.

***Example 1.*** If $P = (2, 1, 3)$, $Q = (1, -1, 0)$, $P_1 = (1, 3, 4)$, and $Q_1 = (-1, -1, -2)$, show that the directed line segments $PQ$ and $P_1Q_1$ have the same direction. Let $\alpha_i$ be the direction angles for $PQ$ and $\beta_i$ be the direction angles for $P_1Q_1$. Clearly $|PQ| = \sqrt{14}$ and $|P_1Q_1| = \sqrt{56}$. Moreover,

$$\cos \alpha_1 = \frac{-1}{\sqrt{14}} \qquad \cos \alpha_2 = \frac{-2}{\sqrt{14}} \qquad \cos \alpha_3 = \frac{-3}{\sqrt{14}}$$

$$\cos \beta_1 = \frac{-2}{\sqrt{56}} = \frac{-1}{\sqrt{14}} \qquad \cos \beta_2 = \frac{-4}{\sqrt{56}} = \frac{-2}{\sqrt{14}} \qquad \cos \beta_3 = \frac{-6}{\sqrt{56}} = \frac{-3}{\sqrt{14}}.$$

Since the direction cosines are equal, the directed line segments have the same direction.

We define the *direction cosines* of a *vector A* to be just the direction cosines of the geometric representations for $A$.

For each value of $i$ these cosines may be written

$$(2) \qquad\qquad \cos \alpha_i = \frac{a_i}{|A|} = \frac{q_i - p_i}{|PQ|}$$

If $A = (a_1, a_2, a_3)$ and if the directed line segment $PQ$ is a representation for $A$. Notice that the direction cosines satisfy the following important identity:

$$\cos^2 \alpha_1 + \cos^2 \alpha_2 + \cos^2 \alpha_3 = \frac{a_1^2 + a_2^2 + a_3^2}{|A|^2} = 1.$$

Also, if we observe that $|A| = |PQ|$, then (2) implies that for each value of $i$, $a_i = q_i - p_i$, or equivalently $A = Q - P$. Thus, if the directed line segment $PQ$ is a geometric representation for the vector $A$, then the vectors $A, P, Q$ satisfy $A = Q - P$. See Figure 7. The converse of this statement is also true. Namely, if $A = Q - P$, then the segment $PQ$ is a representation for the vector $A$. We leave the verification of this as Exercise 8.

In practice, we only compute with the direction cosines of a vector $A$. Usually it is impractical and unnecessary to determine the angles $\alpha_i$. It is convenient to define the *direction* of a vector $A$ to be the vector

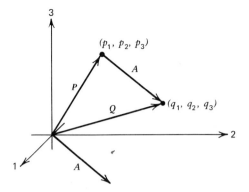

**Fig. 7** The directed line segment $PQ$ is a representation for the vector $A$ if and only if $A = Q - P$.

$(\cos \alpha_1, \cos \alpha_2, \cos \alpha_3)$ where $\cos \alpha_i$ are the direction cosines of $A$. Since

$$(\cos \alpha_1, \cos \alpha_2, \cos \alpha_3) = \left(\frac{a_1}{|A|}, \frac{a_2}{|A|}, \frac{a_3}{|A|}\right)$$

$$= \frac{1}{|A|}(a_1, a_2, a_3) = \frac{1}{|A|}A,$$

we see that the direction of a nonzero vector $A$ is just the vector $(1/|A|)A$, which we write $A/|A|$. Thus, two vectors $A$ and $B$ have the same direction if

$$\frac{A}{|A|} = \frac{B}{|B|}.$$

A direction, such as $\frac{A}{|A|}$, clearly has length equal to 1 since $\left|\frac{1}{|A|}A\right| = \frac{|A|}{|A|} =$

1. Such a vector is commonly called a *unit* vector to emphasize this fact. Thus "$B$ is a unit vector in the direction of $A$" and "$B$ is the direction of $A$" are synonymous statements. They both mean that

$$B = \frac{A}{|A|}.$$

The expression $\frac{A}{|A|}$ only makes sense for nonzero vectors; hence we adopt the position that the *zero vector has no direction.*

In space, the directions of the first, second, and third coordinate axes are given by the vectors $I_1 = (1, 0, 0)$, $I_2 = (0, 1, 0)$, and $I_3 = (0, 0, 1)$, respectively. The vectors $I_1, I_2, I_3$ are called the unit coordinate vectors, and we shall always reserve $I_k$ for this purpose. In $V_2(R)$ the unit coordinate vectors are $I_1 = (1, 0)$ and $I_2 = (0, 1)$. The direction of a vector $A = (a_1, a_2)$ is the vector $\frac{A}{|A|}$, and the two direction cosines of $A$ are $a_1/\sqrt{a_1^2 + a_2^2}$, and $a_2/\sqrt{a_1^2 + a_2^2}$. These are the cosines of the angles made by the vector $A$ with the first and second coordinate axes respectively. See Figure 8.

It is clear that the algebraic aspects of the above discussion are valid in general for vectors belonging to $V_n(R)$. If $A = (a_1, \ldots, a_n)$, then $\frac{A}{|A|}$ is the direction of $A$ and $\frac{a_k}{|A|}$ is the $k$th direction cosine for $A$. Since $-1 \leqslant \frac{a_k}{|A|} \leqslant 1$,

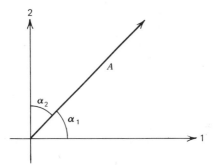

**Fig. 8** Direction angles of a vector in the plane.

there is a unique angle $\alpha_k$, satisfying $0 \leqslant \alpha_k \leqslant \pi$, and $\cos \alpha_k = \dfrac{a_k}{|A|}$. This angle $\alpha_k$ we define to be the $k$th direction angle of the vector $A$. The $k$th coordinate vector $I_k$ is the vector all of whose coordinates are zero except the $k$th which is 1. We note in passing that every vector $A$ in $V_n(R)$ can be written as a linear combination of the unit coordinate vectors $I_1, \ldots, I_n$. This is clear since $A = (a_1, \ldots, a_n) = a_1(1, 0, \ldots, 0) + \cdots + a_n(0, 0, \ldots, 0, 1) = a_1 I_1 + \cdots + a_n I_n$.

**EXERCISES**

1. Determine the direction cosines of the following directed line segments $PQ$ if the points $P$ and $Q$ have coordinates as follows.
    (a) $P = (2, 1, -1)$     $Q = (3, 2, 4)$
    (b) $P = (1, -1, 0)$     $Q = (4, 1, -1)$
    (c) $P = (4, 0, -1)$     $Q = (-1, 2, 1)$
2. Determine the direction of the vector $A$ if
    (a) $A = (3, 2, -1)$
    (b) $A = (4, -1, 2)$
    (c) $A = (0, -1, 2)$
3. Let $A = (2, -1, 0)$, $B = (1, 1, 5)$. Determine the direction of the vectors $A$, $B$, $A + 2B$, $A - B$.
4. Let $A = (2, -1)$, $B = (3, 4)$. Represent $A$ and $B$ geometrically as directed line segments emanating from the origin. Determine $A + B$, $A - B$, $3A + 2B$, and represent these vectors as directed line segments emanating from the origin. Determine the direction of these three vectors.
5. Let $A$ be the vector $(2, 3, 1)$. Determine the coordinates of a point $Q$ such that the directed line segment $PQ$ is a geometric representation for $A$ if $P$ has coordinates
    (a) $(2, -1, 0)$
    (b) $(3, 0, -2)$
    (c) $(4, 1, 3)$
6. Which of the directed line segments $PQ$ are geometric representations for the vector $A = (2, 3, -2)$ if the coordinates of $P$ and $Q$ are given by
    (a) $P = (2, 0, -1)$     $Q = (4, 1, 3)$
    (b) $P = (-2, 1, 0)$     $Q = (1, -1, -2)$
    (c) $P = (4, -2, 1)$     $Q = (6, 1, -1)$
7. Let $A$ be the vector $(4, 2, -1)$. Determine the coordinates of a point $P$ such that the directed line segment $PQ$ is a geometric representation for the vector $A$ if $Q$ has coordinates
    (a) $(2, -1, 3)$

(b) $(-1,-1,2)$

(c) $(-3,2,1)$

8. Let $A = (a_1, a_2, a_3)$, $P = (p_1, p_2, p_3)$, $Q = (q_1, q_2, q_3)$ be vectors from $V_3(R)$. If $A = Q - P$, show that the directed line segment $PQ$ is a geometric representation for the vector $A$.

9. If $A = (a_1, \ldots, a_n)$ is a nonzero vector of order $n$ with direction cosines $\cos \alpha_1, \ldots, \cos \alpha_n$, show that

$$\cos^2 \alpha_1 + \cdots + \cos^2 \alpha_n = 1.$$

10. Let $A$ and $B$ be nonzero vectors in $V_3(R)$. If $A = tB$ where $t$ is a positive scalar, show that $A$ and $B$ have the same direction.

11. If $A$ and $B$ are nonzero vectors in $V_3(R)$ having the same direction show that $A = tB$ for some positive scalar $t$.

12. Let $A$ and $B$ be nonzero vectors in $V_n(R)$. Show that they have the same direction if and only if one is a positive scalar multiple of the other.

*13. It was shown in 1877 by the German mathematician Frobenius that $V_n(R)$ can be made into a field only in the case $n = 2$. For $n = 4$, however, a multiplication can be defined in $V_4(R)$ having all the properties of multiplication in a field except that it is noncommutative. This multiplication, invented by the Irish mathematician W. R. Hamilton, is defined as follows. Let

$$X = (x_1, x_2, x_3, x_4) = x_1 I_1 + x_2 I_2 + x_3 I_3 + x_4 I_4$$

where

$$I_1 = (1,0,0,0), \ldots, I_4 = (0,0,0,1),$$

are the unit coordinate vectors in $V_4(R)$. We denote our multiplication by $X \cdot Y$ and postulate that for all vectors $X, Y, Z \in V_4(R)$ and real numbers $a$

(2) $$X \cdot (Y + Z) = X \cdot Y + X \cdot Z$$

(3) $$a(X \cdot Y) = (aX) \cdot Y = X \cdot aY.$$

Since each vector $X \in V_4(R)$ may be written $X = x_1 I_1 + \cdots + x_4 I_4$, we may infer from (2) and (3) that to define $X \cdot Y$ we need only define how the unit vectors $I_1, \ldots, I_4$ multiply together. Define

(4) $$I_1 \cdot I_k = I_k \cdot I_1 = I_k \quad \text{for} \quad k = 1, 2, 3, 4.$$

(5) $$I_2^2 = I_3^2 = I_4^2 = -I_1.$$

(6) $$\begin{cases} I_2 \cdot I_3 = -I_3 \cdot I_2 = I_4 \\ I_3 \cdot I_4 = -I_4 \cdot I_3 = I_2 \\ I_4 \cdot I_2 = -I_2 \cdot I_4 = I_3. \end{cases}$$

This multiplication is clearly noncommutative. Show that it is associative and show that if we define

$$X^{-1} = \frac{1}{|X|^2}(x_1, -x_2, -x_3, -x_4)$$

then

(7) $$X \cdot X^{-1} = X^{-1} \cdot X = I_1.$$

The set $V_4(R)$ provided with this multiplication is called the set of *quarternions*. For $n = 8$, a multiplication may be defined such that division is possible, i.e., so that (7) holds. However, in this case the multiplication is nonassociative as well as noncommutative. It was finally shown around 1960 by Raoul Bott, John W. Milnor, and others that only in these cases, $n = 2, 4, 8$, could a multiplication be defined in $V_n(R)$ for which division by nonzero vectors is always possible. This fact is regarded by many mathematicians as one of the truly outstanding accomplishments of 20th-century mathematics.

### 8.3   Geometric Vector Addition and Scalar Multiplication

We turn next to the geometric interpretation of scalar multiplication and vector addition for vectors in $V_3(R)$. If $B = tA$ for a scalar $t$, then $|B| = |tA| = |t||A|$, or the length of $B$ is the product of this length of $A$ and the absolute value of $t$. It can be immediately verified that if $t > 0$, then $B$ has the same direction as $A$, and if $t < 0$, then the direction of $B$ is the negative of that of $A$. On the other hand, the vectors $A$ and $B$ are parallel if $A$ and $B$ have the same direction or if $A$ and $B$ have opposite directions. Hence, we see that nonzero vectors $A$ and $B$ are parallel if and only if one is a nonzero scalar multiple of the other. See Figure 9.

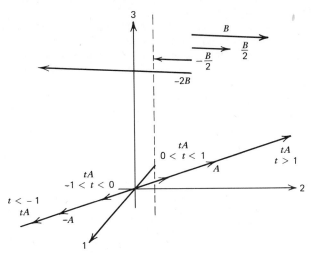

**Fig. 9**   Geometric representation of scalar multiplication.

The geometric interpretation of vector addition may be determined equally easily. Indeed, if $A$ and $B$ are represented as in Figure 10, then, $A + B$ is represented by the directed line segment from the origin to the fourth vertex of the parallelogram determined by the vectors $A$ and $B$. This follows because if $(c_1, c_2, c_3)$ is the fourth vertex of this parallelogram, then the line segment from $(a_1, a_2, a_3)$ to $(c_1, c_2, c_3)$ must be a representation for $B$ since the line segment has the same direction as $B$ as well as the same length. Therefore

$$B = C - A$$

or

$$C = A + B$$

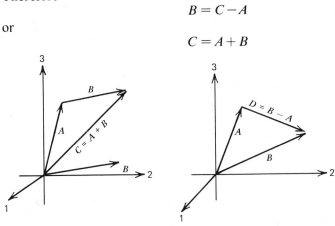

**Fig. 10**

When the vector $B$ is represented as the directed line segment from the point $(a_1, a_2, a_3)$ to the point $(c_1, c_2, c_3)$, we have noted that it is convenient to refer to the point $(a_1, a_2, a_3)$ as the initial point or tail of the vector $B$ and $(c_1, c_2, c_3)$ as the terminal point or head of the vector. These points are not uniquely defined, of course, but depend on the representation of $B$. Using this terminology, a very convenient recipe for performing vector addition geometrically can be given. Translate $B$ parallel to itself attaching the tail of $B$ to the head of $A$. The directed line segment from the tail of $A$ to the head of $B$ is then a representation for the vector $A + B$.

In the plane we have the following very useful characterization of non-parallel vectors.

**THEOREM 8.1**  *Let $A$ and $B$ be nonzero vectors in $V_2(R)$. Then each vector $X \in V_2(R)$ is a linear combination of the vectors $A$ and $B$ if and only if these vectors are not parallel.*

*Proof.*  We leave as an exercise the proof that if each vector $X$ can be written $X = sA + tB$, then $A$ and $B$ must be nonparallel. We shall give a geometric proof of the converse statement.

Let $A$ and $B$ be nonparallel vectors represented as directed line segments emanating from the origin. Then the collection of scalar multiples $\{sA\}$ and $\{tB\}$ determine lines through the origin, which we label $l_1$, $l_2$ (See Figure 11.)

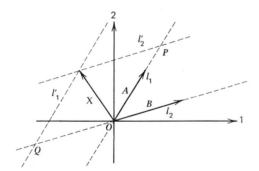

**Fig. 11**

Draw lines $l'_1$, $l'_2$ through the point with coordinates $(x_1, x_2)$ parallel to $l_1$ and $l_2$. The line $l'_2$ intersects $l_1$ at $P$ and $l'_1$ intersects $l_2$ at $Q$. But $OP$ is parallel to $A$. Hence, $OP = sA$ for some scalar $s$. Similarly, $OQ = tB$ for some scalar $t$. Since $X$ is the diagonal of the parallelogram determine by $OP$ and $OQ$, we infer that $X = sA + tB$.

To determine the scalars $s$ and $t$, we solve the equations

$$x_1 = sa_1 + tb_1$$

and

$$x_2 = sa_2 + tb_2$$

where $X = (x_1, x_2)$, $A = (a_1, a_2)$, and $B = (b_1, b_2)$. When each vector $X \in V_2(R)$ can be written as a linear combination of the vectors $A$ and $B$, we say that $A$ and $B$ *span* $V_2(R)$.

Much of the importance of vectors in applications stems from the fact that vector addition describes the way that quantities, which are represented by vectors, interact with one another. For example, the velocity of a particle moving in the plane is a vector in $V_2(R)$. It can be observed that if an airplane has a velocity $V$ and the wind has a velocity $V'$, then the resulting velocity of the airplane is the vector sum $V + V'$. Thus, if $V$ is 400 miles an hour northeast and $V'$ is 100 miles an hour southeast, then the resulting velocity $V + V'$ of the airplane can be constructed geometrically as in Figure 12.

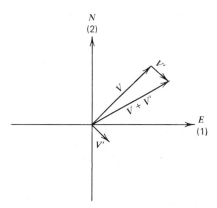

**Fig. 12**

Thus $V = 400(1/\sqrt{2}, 1/\sqrt{2})$ and $V' = 100(1/\sqrt{2}, -1/\sqrt{2})$. Hence

$$V + V' = \left(\frac{500}{\sqrt{2}}, \frac{300}{\sqrt{2}}\right) = 50\sqrt{2}(5, 3).$$

The direction of $V + V'$ is $1/\sqrt{34}\,(5, 3)$.

Another example is provided by forces acting on a particle in space. If the various forces acting on the particle are denoted by the vectors $F_1, \ldots, F_n$, then Newton's second law of motion states that the motion of the particle is described by the vector equation

$$mA = F_1 + \cdots + F_n$$

where $m$ is the mass of the particle, and $A$ is the acceleration. In this equation the mass $m$ is a scalar and the acceleration $A$ is a vector. Thus if the particle is at rest, the vector sum of the forces is zero. That is,

$$F_1 + \cdots + F_n = 0.$$

*Example 1.*  If $F_1 = (5, 6)$, and $F_2 = (2, -3)$ are vectors describing two forces acting on a particle $P$ in the plane, what is the third force $F_3$ acting on $P$ if the particle is at rest. By Newton's law

$$F_1 + F_2 + F_3 = 0.$$

Therefore

$$F_3 = -F_1 - F_2 = -(7, 3).$$

This is represented geometrically in Figure 13.

When we use vectors to describe physical phenomena it is convenient to make a choice between the many geometrical representations of the same vector. For example, suppose we are using vectors to describe a particle moving in the plane or in space. Suppose at time $t$ the coordinates of the particle are given by $(x_1, x_2, x_3)$. Then the directed line segment from the origin to the point with coordinates $(x_1, x_2, x_3)$ would be taken to represent the position vector $X = (x_1, x_2, x_3)$. Suppose for the same value of $t$ the

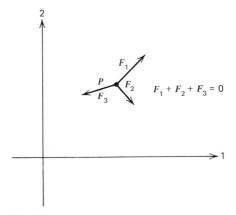

**Fig. 13**

velocity of the particle is given by $V = (v_1, v_2, v_3)$ and the acceleration by $A = (a_1, a_2, a_3)$. Then it is natural to take the geometric representation of these vectors to be directed line segments emanating from the point with coordinates $(x_1, x_2, x_3)$ since this is the position at which the velocity and acceleration are taking place. If for all values of $t$ the position of the particle is given by a curve, then the velocity and acceleration vectors would be represented by vectors attached or emanating from the appropriate points on the curve. Thus, if at time $t$, $X(t) = (x_1(t), x_2(t), x_3(t))$ is the position vector and $V(t) = (v_1(t), v_2(t), v_3(t))$ is the velocity and $A(t)$ the acceleration, these vectors would be drawn as in Figure 14.

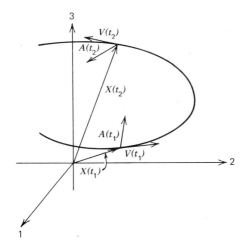

**Fig. 14** Representation of motion of a particle by vectors.

Similarly, if $F_1, F_2, F_3$ are forces acting on a particle whose position has coordinates $(x_1, x_2, x_3)$, the geometric representations of the vectors $F_1, F_2, F_3$ would be chosen as line segments emanating from this point.

**EXERCISES**

1. Let $A = (2, 1)$, $B = (3, -1)$. Represent $A$ and $B$ as directed line segments emanating from the origin. Represent geometrically the vectors $2A$, $-3A$, $A + B, A - B, A - 2B$.

2. Let $A = (3, 1)$, $B = (2, 1)$, $C = (0, 3)$. Represent geometrically the vectors $A - B + C$, $A + B - C$, and $2A - \frac{1}{2}B + 2C$.

3. Let $A = (1, 2, 3)$, $B = (1, -1, 1)$, $C = (2, 1, -3)$. Determine the vectors $A + B$, $2A - B$, $A + B - C$, and their respective directions. Represent these vectors as directed line segments.

4. Let $A = (2, 1)$, $B = (3, 4)$. For a real number $x$, let $C = A + xB$. Compute $C$ for three different values of $x$, and draw geometric representations as line segments with common initial point. What is the locus of the terminal points of all such vectors $C$?

5. Solve Exercise 4 if $A = (-1, 1)$, $B = (2, -3)$.

6. Let $A = (0, 1)$, $B = (-1, 2)$. What is the locus of all terminal points of the set of vectors $xA + (1 - x)B$ if $0 \leqslant x \leqslant 1$?

7. Let $A = (1, 2, 3)$, $B = (3, 1, 2)$. What is the locus of the set of terminal points of the vectors $xA + B$? What is the locus for the set of vectors $x(B - A) + A$ if $0 \leqslant x \leqslant 1$? Draw the loci involved.

8. Let $A = (1, 3)$, $B = (3, -1)$. Let $C = (c_1, c_2)$ be an arbitrary vector in the plane. Determine real numbers $x, y$ so that $C = xA + yB$. For $C = 0 = (0, 0)$, what are the values of $x$ and $y$? Represent this geometrically if $C = (2, 4)$.

9. Answer Exercise 8 if $A = (-1, 2)$, $B = (2, 3)$.

10. Let $A$ and $B$ be vectors in $V_2(R)$. If either $A$ or $B$ is zero, or if $A$ and $B$ are parallel, show that not every $X \in V_2(R)$ is a linear combination of $A$ and $B$.

11. Let $A = (1, 2, 3)$, $B = (1, 0, 1)$, $C = (-1, 1, 0)$. What is the locus of terminal points of vectors of the form $A + xB + yC$, $x, y$ real numbers? Can an arbitrary vector $D = (d_1, d_2, d_3)$ be written in this form? For what values of $d$ can the vector $D = (1, -1, d)$ be so represented?

12. Solve Exercise 11 if $A = (1, 0, -1)$, $B = (-1, 1, 1)$, $C = (0, 0, -1)$.

13. For $A, B, C$ in Exercise 11, can we find for an arbitrary vector $D = (d_1, d_2, d_3)$ real numbers $x, y, z$ so that
$$D = xA + yB + zC?$$

   If $D = (-1, 0, -1)$ what are the values of $x$, $y$, $z$? What are they for $D = (0, 0, 0)$?

14. Answer Exercise 13 for the vectors $A, B, C$ in Exercise 12.

15. An airplane is flying 500 miles per hour northwest. There is an easterly crosswind of 50 miles per hour. What is the resulting velocity of the airplane? What is the direction? (Choose a coordinate system so that the first and second coordinate axes correspond to the directions east and north respectively.)

16. Let $\pi$ be a regular polygon with center at the origin. Let $A_1, \ldots, A_n$ be the vectors from the origin to each of the vertices. Show (geometrically) that $A_1 + \cdots + A_n = 0$. (This is obvious if the number $n$ of vertices is even. If $n$ is odd, let $B = A_1 + \cdots + A_n$ and now rotate $B$ through a suitable angle.)

17. A 100-pound weight is suspended on a 5-foot flexible cord joining two pegs 2 feet apart. Determine the resulting forces at each peg if the coordinate system is chosen as in the following figure. Determine the forces if the right-hand peg is moved up 2 feet. Assume the weight is free to move along the cord. (See Figure 15.)

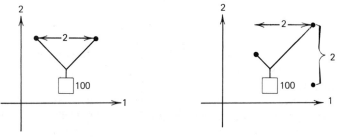

Fig. 15

## 8.4  The Inner Product of Two Vectors

One additional operation can be defined on vectors in $V_n(R)$ which has very useful geometric interpretations. This is the *inner product* of the vectors $A$ and $B$.

**Definition.**  *If* $A = (a_1, \ldots, a_n)$ *and* $B = (b_1, \ldots, b_n)$, *then the* inner product *of* $A$ *and* $B$, *written* $(A, B)$, *is defined by the formula*

$$(A, B) = a_1 b_1 + \cdots + a_n b_n.$$

For example if $A = (2, 1, 3)$ and $B = (-1, 1, -2)$, then $(A, B) = -2 + 1 - 6 = -7$. This inner product is an operation on vectors which *does not* yield a vector as a result but instead yields a scalar. It is often referred to as the *scalar product*, or *dot product*. If the latter terminology is used, then $(A, B)$ is written $A \cdot B$. We shall adhere to the parenthesis terminology, however.

The following algebraic properties of the inner product are readily established.

(1) $\qquad\qquad (A, B) = (B, A) \qquad\qquad$ (commutative law)

(2) $\qquad\quad (A, B + C) = (A, B) + (A, C) \qquad$ (distributive law)

If $c$ is a scalar, then

(3) $\qquad\qquad\qquad c(A, B) = (cA, B) = (A, cB).$

The verification of these three properties is immediate and is left as an exercise. The student should note that the associative law does not make sense for the inner product since $(A, B)$ is a scalar, not a vector. We note finally the following useful connection with the length of $A$:

(4) $\qquad\qquad (A, A) = a_1{}^2 + \cdots + a_n{}^2 = |A|^2.$

Therefore $(A, A) \geqslant 0$. Furthermore $(A, A) = 0$ if and only if $A = 0$.

For vectors in the plane or in space, the inner product of vectors $A$ and $B$ may be given an immediate geometric interpretation. Representing the vector $A - B$ as the directed line segment from the point with coordinates $(b_1, b_2, b_3)$ to the point with coordinates $(a_1, a_2, a_3)$ as in Figure 16, the law

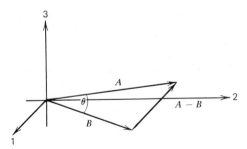

**Fig. 16**  $\cos \theta = \dfrac{(A, B)}{|A||B|}.$

of cosines gives the following formula for the angle $\theta$ between the vectors $A$ and $B$.

(5) $\qquad\qquad |A|^2 + |B|^2 - |A - B|^2 = 2|A||B| \cos \theta.$

But applying property (4) of the inner product we see that

(6)        $|A|^2+|B|^2-|A-B|^2 = (A,A)+(B,B)-(A-B,A-B).$

Using (1) to (3) we infer that

$$(A-B,A-B) = (A-B,A)-(A-B,B)    \text{(properties (2) and (3))}$$
$$= (A,A-B)-(B,A-B)    \text{(property (1))}$$
$$= (A,A)-(A,B)-(B,A)+(B,B)    \text{(by (2) and (3))}$$
$$= (A,A)-2(A,B)+(B,B)    \text{(by (1))}.$$

Therefore

$$(A,A)+(B,B)-(A-B,A-B) = 2(A,B).$$

Combining this with (5) and (6) we see that

$$(A,B) = |A||B| \cos \theta$$

or

(7)        $$\cos \theta = \frac{(A,B)}{|A||B|}.$$

For example, if $A = (1,2,3), B = (-1,-1,2)$, then the angle $\theta$ between them satisfies

$$\cos \theta = \frac{(A,B)}{|A||B|} = \frac{3}{\sqrt{14}\sqrt{6}} = \frac{3}{2\sqrt{21}}.$$

Consequently,

$$\theta = \text{arc} \cos \frac{3}{2\sqrt{21}}.$$

For vectors in $V_2(R)$ or $V_3(R)$, formula (7) shows that

$$-1 \leqslant (A,B)/|A||B| \leqslant 1.$$

A natural question to ask is does this formula hold in general for vectors in $V_n(R)$? It does, and the inequality is called the *Schwarz* inequality. It may be written in the following two equivalent forms:

$$|(A,B)| \leqslant |A||B|$$

or

$$(A,B)^2 \leqslant (A,A)(B,B).$$

Once this formula has been established for vectors in $V_n(R)$ it gives us the means for *defining* the angle between two such vectors $A$ and $B$. Indeed, we *define* $\theta$ to be that number $\theta$, satisfying $0 \leqslant \theta \leqslant \pi$, such that

(8)        $$\cos \theta = \frac{(A,B)}{|A||B|}.$$

This is legitimate once we have shown that

$$-1 \leqslant \frac{(A,B)}{|A||B|} \leqslant 1.$$

After this has been done, it can be immediately established that the law of cosines (5) holds for vectors in $V_n(R)$. The argument is exactly the same as the one that was used to prove (7) for vectors in space assuming the law of cosines.

This situation is very typical. A geometric fact that is proved for vectors in the plane or in space is taken to be the definition of a corresponding notion for vectors in $V_n(R)$. For example, if $A$ and $B$ are nonzero vectors in $V_3(R)$, then they are perpendicular if the angle $\theta$ between them is $\pi/2$. Applying (7) we have

$$0 = \cos \theta = (A, B)/|A||B|$$

or

$$(A, B) = 0.$$

We now use this fact as the definition of perpendicularity for vectors in $V_n(R)$.

*Definition.    If $A$ and $B$ are vectors in $V_n(R)$, then $A$ and $B$ are* orthogonal *if*

$$(A, B) = 0.$$

*If, in addition, $A$ and $B$ are different from zero, then $A$ and $B$ are called* perpendicular.

Another example of this is the following. We saw in Section 8.3 that nonzero vectors in space were parallel if and only if one was a nonzero scalar multiple of the other. For vectors in $V_n(R)$ we take this fact as the definition.

*Definition.    If $A$ and $B$ are nonzero vectors in $V_n(R)$, they are* parallel *if $A = tB$ for some nonzero scalar $t$.*

Note that under these definitions the notions of parallelism and perpendicularity do not apply to the zero vector. This is consistent with our position that the zero vector has no direction.

**EXERCISES**

1. Let  $A = (1, 2, 1)$,  $B = (2, -1, 0)$,  $C = (0, 1, -1)$.  Compute:  $(A, A)$,  $(A, B)$, $(C, A - B)$.
2. Using the vectors of Exercise 1, for which value of $t$ is $C$ perpendicular to $A + tB$? Represent the vectors involved as directed line segments.
3. If $A = (-1, 0, 2)$, $B = (1, -1, 1)$, $C = (-1, 2, 3)$, find a scalar $t$ so that $A$ is orthogonal to $B + tC$. Represent the vectors involved as directed line segments.
4. If $A = (4, 2, -4)$, $B = (2, -1, 0)$, $C = (0, 1, -1)$, find a scalar $t$ so that $A$ is parallel to $B + tC$. Draw the representation of the vectors involved. (*Hint.* Vectors $X$ and $Y$ are parallel if $X = sY$ for some nonzero scalar $s$.
5. Solve Exercise 4 if $A = (4, -1, 2), B = (2, 1, 1), C = (2, -1, 1)$.
6. If $A \neq 0$ and $(A, B) = (A, C)$ for vectors $B$ and $C$, does it follow that $B = C$? If not, give examples that satisfy the equation.
7. Let $A = (1, -1, 0)$, $B = (2, 0, -1)$. Find a nonzero vector $C$ orthogonal to both $A$ and $B$. (Write $C = (c_1, c_2, c_3)$ and apply the orthogonality relationship.) Are all such vectors $C$ parallel to each other?
8. Solve Exercise 7 if $A = (1, 1, 1), B = (2, 1, -1)$.
9. If $A = (-1, 1, 1)$, $B = (0, 0, 1)$, find all vectors $C$ of unit length perpendicular to both $A$ and $B$.
10. Solve Exercise 9 if $A = (1, 0, 1), B = (1, 1, 1)$.
11. If $A = (2, -1)$, $B = (1, 1)$, determine scalars $x$ and $y$ so that $xA + yB$ is perpendicular to $A$. Determine the scalars so that, in addition, $xA + yB$ has unit length. Represent the vectors involved as directed line segments.
12. Let $A = (1, -1, 0)$, $B = (-1, 0, 1)$. Solve Exercise 11 for these vectors.
13. Verify properties (1) to (3) for the inner product of vectors in $V_n(R)$.

14. Show that vectors $A$ and $B$ in $V_n(R)$ are orthogonal if and only if $|A+B|^2 = |A|^2 + |B|^2$ (theorem of Pythagoras).

15. Let $A$ and $B$ be vectors in $V_n(R)$. Using properties (1) to (4) of the inner product prove that

    (a) $|A+B|^2 - |A-B|^2 = 4(A,B)$.    (b) $|A+B|^2 + |A-B|^2 = 2\{|A|^2 + |B|^2\}$.

16. Let $A$ and $B$ be vectors in $V_3(R)$ and consider the parallelogram formed by $A$ and $B$. Interpret Exercise 15(b) as a geometric statement about this parallelogram.

17. Using Exercise 15(a), give a condition on the diagonals of a parallelogram that is necessary and sufficient for the parallelogram to be a rectangle.

18. A pirate captain, sailing along the coast of Florida, has recently plundered a Spanish galleon. He has a sizable amount of treasure that must be hidden to avoid detection. The captain chooses a stretch of coast with two large rocks about a quarter of a mile apart that will identify the spot. Inland between the rocks is a solitary oak tree. The captain gives the following directions for burying the treasure. Two men stand at the tree and march to the respective rocks pacing off the distances. They both then make right-angle turns inland and proceed to pace off a distance equal to what they have just walked. They then face each other and at a point half way between them the treasure is buried. Years later the pirate captain returns for his treasure. He finds the spot, guided by the two rocks, but if horrified to see that the oak has disappeared. Can he find his treasure without digging up the entire coastline? (It might help if the captain had studied vector algebra. Of course, it had not yet been invented, but you can ignore that difficulty.)

## 8.5   The Schwarz Inequality and the Triangle Inequality

**THEOREM 8.2 (Schwarz Inequality).** *Let $A$ and $B$ be vectors in $V_n(R)$. Then*

(1) $$(A,B)^2 \le (A,A)(B,B).$$

*Equality holds in the above expression if and only if one vector is a scalar multiple of the other.*

**Proof.**   If either $A$ or $B$ is the zero vector, then the theorem clearly holds. Suppose $B \ne 0$, and consider for a real number $x$ the function

$$p(x) = (A+xB, A+xB) = |A+xB|^2.$$

For all values of $x$, $p(x) \ge 0$, and if we expand the above inner product we get

(2) $$p(x) = (B,B)x^2 + 2(A,B)x + (A,A).$$

Therefore, $p(x)$ is a polynomial in $x$ of degree 2. If we complete the square on the right-hand side of (2), we obtain

$$p(x) = (B,B)\left[x^2 + \frac{2(A,B)x}{(B,B)} + \frac{(A,B)^2}{(B,B)^2}\right] - \frac{(A,B)^2}{(B,B)} + (A,A)$$

(3)

$$= (B,B)\left[x + \frac{(A,B)}{(B,B)}\right]^2 + \frac{(A,A)(B,B) - (A,B)^2}{(B,B)}$$

Set $x_0 = -(A,B)/(B,B)$. Then

(4) $$0 \le p(x_0) = 0 + \frac{(A,A)(B,B) - (A,B)^2}{(B,B)}.$$

Since $(B, B) > 0$, this implies

$$(A, A)(B, B) - (A, B)^2 \geq 0$$

or

$$(A, B)^2 \leq (A, A)(B, B).$$

Thus we have established (1). To finish the proof, suppose $(A, B)^2 = (A, A)(B, B)$. Then from (4)

$$p(x_0) = 0$$

where $x_0 = -(A, B)/(B, B)$. But $p(x_0) = |A + x_0 B|^2$. Hence $A + x_0 B = 0$ or $A$ is a scalar multiple of $B$. It remains only to show that if one vector is a scalar multiple of the other, then $(A, B)^2 = (A, A)(B, B)$. This can be immediately verified.

**COROLLARY.**   *Let $A$ and $B$ be nonzero vectors in $V_n(R)$. These vectors are parallel if and only if*

$$|(A, B)| = |A||B|$$

As an immediate consequence of the Schwarz inequality we have the following.

**THEOREM 8.3 (Triangle Inequality).** *Let $A$, $B$ be vectors in $V_n(R)$. Then*

$$|A + B| \leq |A| + |B|.$$

*Furthermore $|A + B| = |A| + |B|$ if and only if one vector is a nonnegative scalar multiple of the other.*

***Proof.***   Before beginning the proof let us observe that if we consider vectors in the plane or in space, then the triangle inequality just expresses the familiar fact that the sum of the lengths of two sides of a triangle exceeds the length of the third side. (Figure 17.)

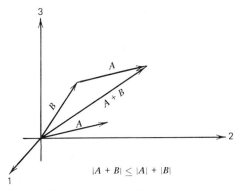

$$|A + B| \leq |A| + |B|$$

**Fig. 17**   The triangle inequality.

To establish the triangle inequality for vectors in $V_n(R)$ we perform the following computation.

$$\begin{aligned}
|A + B|^2 &= (A + B, A + B) \\
&= (A, A) + 2(A, B) + (B, B) \\
&\leq (A, A) + 2|(A, B)| + (B, B) \\
&\leq (A, A) + 2|A||B| + (B, B) \qquad \text{(Schwarz inequality)} \\
&= |A|^2 + 2|A||B| + |B|^2 = (|A| + |B|)^2.
\end{aligned}$$

Taking nonnegative square roots, we have

$$|A+B| \le |A|+|B|.$$

If we examine the above computation, we see that $|A+B| = |A|+|B|$ if and only if $(A,B) = |A||B|$. But by Theorem 8.2 this can hold if and only if one vector is a scalar multiple of the other. But if $B = tA$ and $(A,B) = |A||B|$, then $0 \le (A,B) = t(A,A)$. Therefore, if $A \ne 0$, it must follow that $t \ge 0$.

### EXERCISES

1. Let $A = (2,3,-1)$, $B = (-4,-6,2)$, $C = (1,1,-1)$, $D = (1,2,8)$. Which pairs of these vectors are orthogonal and which are parallel?

2. Determine the cosines of the angles of the triangle whose vertices are the points with coordinates $(2,1,3), (3,1,2), (2,-1,1)$.

3. For what values of $x$ is $(3,x,1)$ perpendicular to $(1,2,-1)$? Is it possible to choose $x$ so that these vectors are parallel?

4. Answer Exercise 3 if the vectors are $(x,2,-3)$ and $(-2,4,-6)$.

5. Deduce from the triangle inequality that if $A$ and $B$ are vectors in $V_n(R)$, then

$$||A|-|B|| \le |A \pm B| \le |A|+|B|.$$

   (Examine the proof of Theorem 1.4.)

6. Prove that if $A$ and $B$ are parallel vectors in $V_n(R)$, then

$$|(A,B)| = |A||B|.$$

7. Show by vector methods that a triangle inscribed in a semicircle is a right triangle. (See Figure 18.)

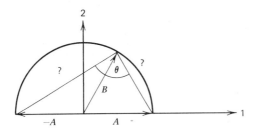

**Fig. 18**

8. Determine the cosines of the angles of the triangle in $V_4(R)$ whose vertices have coordinates $(1,2,-1,3),(2,1,4,0),(3,-1,2,2)$.

9. Let $A$, $B$, $C$ be vectors in $V_n(R)$. Assume $|A| = |B| = 1$ and $|C| = 4$. If $|A-B+C| = |A+2B+C|$ and the angle between $A$ and $B$ is $\pi/4$, determine the angle between $B$ and $C$.

10. Let $A$, $B$, $C$ be nonzero vectors and assume the angle between $A$ and $C$ equals that between $B$ and $C$. For what value of $t$ is $C$ perpendicular to

$$D = |B|A + tB?$$

11. Give an alternate proof of the Schwarz inequality by evaluating the polynomial $p(x) = |A+xB|^2$ at the point $x = -(A,A)/(A,B)$. (Of course, we must assume $(A,B) \ne 0$. What happens if $(A,B) = 0$?)

12. Let $A$ and $B$ be nonzero vectors in $V_n(R)$. For what value of $x$ does the quadratic polynomial $p(x) = |A+xB|^2$ take on its minimum value? What is the value of this minimum?

13. We have defined length or norm of a vector $A = (a_1, \ldots, a_n)$ by the following formula

$$|A| = (a_1^2 + \cdots + a_n^2)^{1/2}.$$

More properly, $|A|$ should be called the Euclidean length or norm of the vector $A$. The function $|A|$ is a real valued function defined on the vectors in $V_n(R)$, and we have shown that the following four properties hold for $|A|$

$$\text{(i) } |A+B| \leqslant |A|+|B|$$

$$\text{(ii) } |aA| = |a||A| \text{ for scalars } a$$

$$\text{(iii) } |A| > 0 \qquad \text{if} \qquad A \neq 0$$

$$\text{(iv) } |A| = 0 \qquad \text{if} \qquad A = 0.$$

Now any real valued function defined on the vectors in $V_n(R)$ that satisfies (i) to (iv) could be called a length or norm for the vector $A$. Such norms, if different from $|A|$, we shall call non-Euclidean. Verify that

$$||A|| = |a_1| + \cdots + |a_n|$$

and

$$|||A||| = \max(|a_1|, \ldots, |a_n|)$$

are two examples of non-Euclidean norms for the vectors of order $n$. (i.e., show that properties (i) and (iv) above hold for $||A||$ and $|||A|||$.) If we define $|A|_1 = |a_1 + \cdots + a_n|$, is this a non-Euclidean norm for $A$?

14. If $A$ is a vector in $V_2(R)$, then $\{A : |A| \leqslant 1\}$ is represented in the plane by the unit disc of radius 1. (Figure 19.)

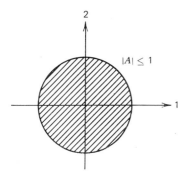

**Fig. 19**

Represent in the plane the sets

$$\{A \in V_2(R) : ||A|| \leqslant 1\} \qquad \text{and} \qquad \{A \in V_2(R) : |||A||| \leqslant 1\},$$

where $||A||$, $|||A|||$ are the non-Euclidean norms of Exercise 13.

## 8.6   Lines in $V_n(R)$

Let $A$ and $B$ be two vectors in $V_n(R)$ and assume $B \neq 0$. By the *line l through $A$ parallel to $B$* we shall mean the set of all vectors $X$ that can be written

$$(1) \qquad\qquad X = A + tB$$

for some scalar $t$. Equation (1) is called the vector equation of the line $l$.

To describe geometrically a line in $V_3(R)$, we represent all of the vectors $X$ satisfying (1) as directed line segments emanating from the origin. (See Figure 20.) Then a point $(x_1, x_2, x_3)$ in space lies on the line $l$ with equation (1) if and only if this point is a terminal point of a directed line segment representing one of the vectors $X$. With this choice of the representations for $X$ we see that a point with coordinates $(p_1, p_2, p_3)$ lies on the line $l$ if

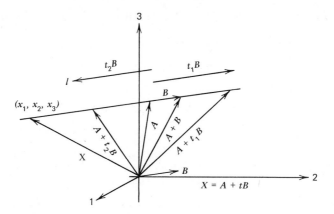

**Fig. 20**

and only if the vector $P = (p_1, p_2, p_3)$ satisfies the equation of the line. If we represent the vectors $A$ and $B$ in space, then the geometric representation of the line $l$ is illustrated in Figure 20.

If the vector $P$ satisfies the equation of the line, it is convenient to say that $P$ lies on the line. One should notice that this does *not* say that all points lying on the directed line segment from the origin to the point with coordinates $(p_1, p_2, p_3)$ lie on the line. It is clear that this happens if and only if the line passes through the origin, that is, if $A = 0$. In general only the terminal point of the directed line segment from the origin that represents $P$ lies on the line.

In addition to the vector equation (1), the line $l$ is also characterized by scalar equations obtained from (1) by equating the respective components. Thus in $V_3(R)$, equation (1) is equivalent to three scalar equations. For, if

$$X = (x_1, x_2, x_3) \qquad A = (a_1, a_2, a_3) \qquad B = (b_1, b_2, b_3),$$

Then $X = A + tB$ implies

(2)
$$\begin{cases} x_1 = a_1 + tb_1 \\ x_2 = a_2 + tb \\ x_3 = a_3 + tb_3 \end{cases}$$

These equations are often called the *parametric equations* of the line $l$.

Solving for the parameter $t$ we obtain the three equations

$$t = \frac{x_i - a_i}{b_i} \qquad i = 1, 2, 3,$$

Eliminating $t$ between these equations we have the following symmetric form of the equation for a line

$$\frac{x_1 - a_1}{b_1} = \frac{x_2 - a_2}{b_2} = \frac{x_3 - a_3}{b_3}.$$

If one of the components of $B$ is zero, say $b_1 = 0$, this pair of scalar equations becomes

$$x_1 = a_1; \frac{x_2 - a_2}{b_2} = \frac{x_3 - a_3}{b_3}.$$

The vector and scalar equations of the line $l$ may be determined in a variety of ways by specifying geometric conditions.

*Example 1.*  Determine the vector and scalar equation of the line passing through $A = (1, 2, -1)$ and $B = (2, 1, 1)$. Now the vector equation will be $X = A + tC$ for some vector $C$. But $B$ lies on the line, hence for some scalar $t_0$

$$B = A + t_0 C$$

or

$$C = \frac{B - A}{t_0}.$$

Thus we may take $C$ to be any vector parallel to $B - A = (1, -1, 2)$. In particular, if we take $C = B - A$, the vector equation becomes

$$X = (x_1, x_2, x_3) = A + t(B - A)$$
$$= (1, 2, -1) + t(1, -1, 2).$$

The resulting scalar equations are

$$x_1 = 1 + t$$
$$x_2 = 2 - t$$
$$x_3 = -1 + 2t.$$

Eliminating $t$ we have the symmetric equation of the line

$$x_1 - 1 = \frac{x_2 - 2}{-1} = \frac{x_3 + 1}{2}.$$

For vectors $A = (a_1, a_2)$ and $B = (b_1, b_2)$ in the plane, the scalar equation of the line $l$ through $A$ parallel to $B$ takes the form

$$\frac{x_2 - a_2}{b_2} = \frac{x_1 - a_1}{b_1}.$$

Writing this as

$$\frac{x_2 - a_2}{x_1 - a_1} = \frac{b_2}{b_1}$$

we see that the slope of the line is $b_2/b_1$.

*Example 2.*  Determine the vector equation of the line passing through $(1, 2)$ with slope 3. This line has scalar equation

$$\frac{x_2 - 2}{x_1 - 1} = 3$$

or

$$\frac{x_2 - 2}{3} = \frac{x_1 - 1}{1}.$$

Therefore the line passes through $(1, 2)$ and is parallel to $(1, 3)$. Hence the vector equation is

$$X = (1, 2) + t(1, 3).$$

If the lines $l_1$ and $l_2$ have vector equations $X_1 = A_1 + tB_1$ and $X_2 = A_2 + tB_2$, respectively, we say that these lines are perpendicular or parallel according to whether the vectors $B_1$ and $B_2$ are perpendicular or parallel.

Lines in the plane may be described in terms of vectors perpendicular to a fixed vector $N$. If $X = A + tB$ is the equation of a line $l$ in $V_2(R)$ and $B = (b_1, b_2)$, then let us set $N = (-b_2, b_1)$. It is clear that $B$ and $N$ are perpendicular since $(B, N) = -b_1 b_2 + b_1 b_2 = 0$. Moreover, if $X$ lies on the line $l$, then $X - A$ is perpendicular to $N$. This is clear since $(X - A, N) = (tB, N) = t(B, N) = 0$. Suppose conversely that $X$ is a vector satisfying $(X - A, N) = 0$. We assert that the vector $X$ lies on the line $l$. To see this, observe that since $B$ and $N$ are not parallel, we know by Theorem 8.1 that $X - A = tB + sN$ for some scalars $s$ and $t$. If we take the inner product of this equation with the vector $N$, we obtain

$$0 = (X - A, N) = (tB + sN, N)$$
$$= t(B, N) + s(N, N) = s(N, N)$$

since

$$(B, N) = 0.$$

However, $N \neq 0$. Hence $(N, N) \neq 0$, and therefore $s = 0$. Thus it follows that $X = A + tB$.

As a result we conclude that in $V_2(R)$ the line through $A$ parallel to $B$ may be characterized as the set of all vectors $X$, such that $(X - A, N) = 0$ where $N = (-b_2, b_1)$ and $B = (b_1, b_2)$. Hence we may speak of $l$ as the line through $A$ perpendicular to $N$. The vector equation of this line is then

$$(3) \qquad\qquad (X - A, N) = 0.$$

The vector $N$, or any nonzero scalar multiple of $N$, is called a *normal vector* to the line $l$. See Figure 21.

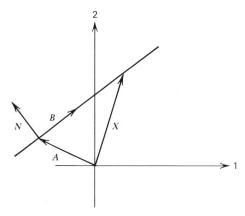

**Fig. 21**  The line through $A$ perpendicular to $N$ has the equation $(X - A, N) = 0$.

If $X = (x_1, x_2)$ and $N = (n_1, n_2)$, then we may write (3) as

$$(4) \qquad\qquad n_1 x_1 + n_2 x_2 = c \qquad \text{where } c = (N, A).$$

Thus if the equation of a line in the plane is written in the form (4), then $(n_1, n_2)$ is a normal vector to the line and the constant $c$ equals the inner product of $N$ with any vector lying on the line. For example, if the non-vertical straight line has equation $y = mx + b$, where $m$ is the slope and $b$ the $y$ intercept, then $(-m, 1)$ is a normal vector to this line. The constant $b$ is the inner product of the vector $(-m, 1)$ with any vector $X$ lying on the line.

*Example 3.*   A line $l$ in the plane passes through the points $(2, 1)$ and $(3, -2)$. Determine a normal vector to the line and the equation of the line in the form (4). The line clearly has equation

$$\frac{x_2 - 1}{x_1 - 2} = \frac{-3}{1}$$

or

$$3x_1 + x_2 = 7.$$

Therefore $(3, 1)$ is a normal vector to the line. The vector equation (1) to this line is

$$X = (2, 1) + t(-1, 3).$$

See Figure 22.

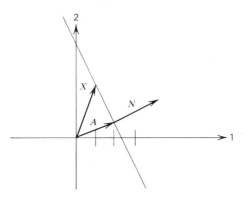

**Fig. 22**   $3x_1 + x_2 = 7$ is the equation of the line through $(2, 1)$ perpendicular to $(3, 1)$.

We turn next to the determination of the distance $d$ from the origin to the line $l$ through $A$ parallel to the vector $B$. This distance $d$ is defined to be the minimum length of all the vectors $X$ lying on the line $l$. This is, of course, just the minimum length of all vectors of the form $A + tB$. But by the Pythagorean theorem $|A + t_0 B|$ will be a minimum if $A + t_0 B$ is perpendicular to $B$. (See Figure 23.) To see this, note that if $(A + t_0 B, B) = 0$, then for any value of $t$

$$|A + tB|^2 = |A + t_0 B + (t - t_0)B|^2$$
$$= |A + t_0 B|^2 + |(t - t_0)B|^2$$
$$\geqslant |A + t_0 B|^2.$$

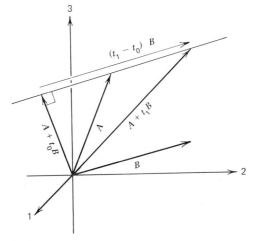

**Fig. 23**   $|A + t_0 B|$ will be a minimum if $A + t_0 B$ is perpendicular to $B$.

The second equality holds since $A + t_0 B$ and $(t - t_0)B$ are perpendicular. To compute the value of $d$, note that if $(A + t_0 B, B) = 0$, then

$$t_0 = -\frac{(A, B)}{(B, B)}.$$

Hence

$$d^2 = |A + t_0 B|^2 = (A, A) + 2t_0(A, B) + t_0^2(B, B)$$

$$= (A, A) - \frac{2(A, B)^2}{(B, B)} + \frac{(A, B)^2}{(B, B)}$$

$$= \frac{(A, A)(B, B) - (A, B)^2}{(B, B)}$$

and consequently

(5)
$$d = \frac{\sqrt{|A|^2|B|^2 - (A, B)^2}}{|B|}.$$

The numerator in (5) has a simple geometric description. If $\theta$ is the angle between the vectors $A$ and $B$, then we know from Section 8.4 that $(A, B) = |A||B| \cos \theta$. Therefore

$$\sqrt{|A|^2|B|^2 - (A, B)^2} = \sqrt{|A|^2|B|^2(1 - \cos^2 \theta)} = |A||B| \sin \theta.$$

Hence

(6)
$$d = |A| \sin \theta.$$

Thus the distance from the origin to the line through $A$ parallel to $B$ is just the product of the length of $A$ with the sine of the angle between $A$ and $B$. See Figure 24.

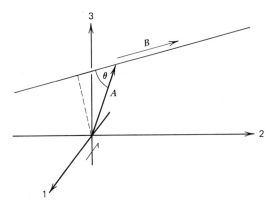

**Fig. 24**   The distance from origin to line $l$ is $|A| \sin \theta$.

This discussion is valid for lines in $V_n(R)$ as well, since we have used only the fundamental properties of the inner product to derive (5) and (6). For lines in the plane, the distance $d$ from the origin to a line $l$ may be expressed in terms of a normal vector $N$ to the line. Indeed, this distance $d$ will be the length of the vector $t_0 N$ where $t_0 N$ is chosen so that the terminal point of the line segment representing $t_0 N$ lies on the line $l$. (See Figure 25.) When this holds, $X = t_0 N$ must satisfy (3). Hence $(t_0 N - A, N) = 0$ where $A$ is some vector lying on the line $l$. Solving for $t_0$, we obtain

$$t_0 = \frac{(A, N)}{(N, N)}.$$

Therefore

$$d = |t_0 N| = \frac{|(A, N)|}{|N|}.$$

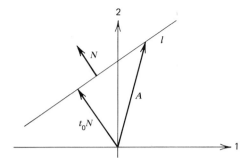

**Fig. 25** The distance from line $l$ to origin is $|t_0 N|$ where $t_0 N$ lies on $l$.

If the equation of $l$ has the form $n_1 x_1 + n_2 x_2 = c$, then $(A, N) = c$, and

$$d = \frac{|c|}{|N|} = \frac{|c|}{\sqrt{n_1{}^2 + n_2{}^2}}$$

### EXERCISES

1. Determine equations for the following lines in the plane in both the form $X = A + tB$ and $(X - A, N) = 0$.
   (a) line through $A = (1, -1)$ and parallel to $B = (2, 3)$
   (b) line through $A = (2, -3)$ and through $B = (-1, 2)$
   (c) line through $A = (4, 1)$ and perpendicular to $B = (1, -2)$
   (d) line through $A = (1, 3)$ and parallel to the line through $B = (-1, 4)$ and $C = (2, -1)$

2. Determine the vector equations of the medians of the triangle formed by $(1, 2)$, $(-1, 4)$, $(2, -1)$.

3. Let $A = (1, 3)$, $B = (2, 1)$, $C = (1, 4)$. Determine the vector equation of the line through $A$ bisecting the line segment between $B$ and $C$.

4. Determine normal vectors to the sides of the triangle with vertices $(3, 0)$, $(2, -1)$, and $(-3, 4)$.

5. Determine the distance from the origin to the line $y = 2x + 5$.

6. In space, determine vector and scalar equations of the lines defined by the following conditions.
   (a) through $A = (1, -1, 1)$ and parallel to $B = (2, 0, -3)$
   (b) through $A = (2, 1, -2)$ and through $B = (1, 2, 3)$
   (c) through $A = (1, -1, 1)$ and parallel to the line through $B = (1, 1, 1)$ and $C = (1, 4, 3)$

7. Show that the line through $A = (3, 5, -1)$ and $B = (4, 3, -1)$ is perpendicular to the vector $C = (2, 1, 1)$.

8. Show that the points $A = (0, 1, 3)$, $B = (-1, 0, 2)$, $C = (-1, 1, 4)$ are the vertices of a right triangle.

9. Show that the vectors $A = (5, 3, -2)$, $B = (4, 1, -1)$, $C = (2, -3, 1)$ are collinear.

10. Determine the vector equations of the lines defined by
    (a) $\dfrac{x - 5}{2} = \dfrac{y + 3}{-4} = \dfrac{z - 1}{6}$
    (b) $\begin{array}{l} 3x + 2y - 2z + 3 = 0 \\ 2x + y - 2z + 2 = 0 \end{array}$

11. What is the distance from the origin to the line through $(2, 1, -1)$ and $(1, -1, 3)$?

12. Show that the distance from the point with coordinates $(c_1, c_2, c_3)$ to the line through $A$ parallel to $B$ is given by

$$d = \frac{\sqrt{|A - C|^2 |B|^2 - (A - C, B)^2}}{|B|}$$

where $C$ is the vector $(c_1, c_2, c_3)$. Give a geometric interpretation of $d$ analogous to formula (6). (This distance will be the minimum length of the vectors $A + tB - C$. Using the same argument as in the text, show that this vector will have minimum length if $A + tB - C$ is perpendicular to $B$.)

13. If $l$ is a line in the plane with equation $c_1 x_1 + c_2 x_2 = d$, show that the distance from a point $(d_1, d_2)$ to $l$ is given by

$$\frac{|c_1 d_1 + c_2 d_2 - d|}{\sqrt{c_1^2 + c_2^2}}$$

(Make a sketch and convince yourself that this minimum distance will be the length of the vector $t_0 C$ if $t_0 C + D$ lies on the line $l$. $D$ is the vector $(d_1, d_2)$.)

14. Compute the distance from $(1, 2)$ to $2x + y + 4 = 0$.

15. What are the lengths of the altitudes of the triangle formed by $(1, -1, 1), (2, 3, 0)$, and $(1, 0, -1)$?

16. Determine the distance from $(1, -1, 2, 1)$ to the line through $(2, 0, 0, 1)$ and $(1, -1, 1, -1)$.

17. Let $l_1$ be the line through $(1, -1, 1)$ parallel to $(0, 1, 1)$ and $l_2$ be the line through $(2, 1, 1)$ parallel to $(1, -1, 0)$. What is the distance between these lines? (If $X_1$ lies on $l_1$ and $X_2$ on $l_2$, what is the geometric condition for $|X_1 - X_2|$ to be a minimum?)

## 8.7   Planes in $V_n(R)$

We showed in Theorem 8.1 that every vector $X$ in $V_2(R)$ is a linear combination of the vectors $A$ and $B$ if and only if $A$ and $B$ are *not* parallel. When every vector $X$ in $V_2(R)$ can be written

$$X = sA + tB$$

we say that $A$ and $B$ *span* $V_2(R)$.

If $A$ and $B$ are nonzero vectors in $V_n(R)$ that are not parallel, then in an analogous fashion we define the *plane spanned by $A$ and $B$* to be the set of vectors X which can be written

$$X = sA + tB.$$

where $s$ and $t$ are scalars. More generally, if $C$ is a third vector in $V_n(R)$, then the set of vectors

(1) $$X = C + sA + tB$$

will be called the plane *through $C$ spanned by $A$ and $B$*. Equation (1) is called the vector equation of this plane.

If the plane $\pi$ is in $V_3(R)$, then we represent the vectors $X$ satisfying (1) as line segments emanating from the origin. Then the representation of $\pi$ as a set of points will consist of all the points that are terminal points of the line segments representing the vectors $X$. Representing the vectors $X$ in this way we see that a point with coordinates $(p_1, p_2, p_3)$ lies on the plane if and only if the vector $P = (p_1, p_2, p_3)$ satisfies the equation of the plane.

Just as in the case of a line $l$ we say that a vector $X$ lies on the plane $\pi$ if $X$ satisfies the equation of the plane. Notice that this does *not* say that all points on the directed line segment representing $X$ lie on the plane. This will only be true if the plane $\pi$ passes through the origin. The geometric representation of the plane $\pi$ in space is given in Figure 26.

Planes in $V_3(R)$ may be characterized in much the same way as lines in $V_2(R)$. If $\pi$ is the plane through $C$ spanned by $A$ and $B$, we assert first that

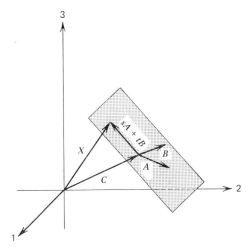

**Fig. 26** Plane $\pi$ with vector equation $X = C + sA + tB$.

there is a nonzero vector $N$ perpendicular to both $A$ and $B$. Second, this plane $\pi$ may be characterized as precisely those vectors $X$ such that $X - C$ is perpendicular to $N$. In terms of the inner product, this just means that

$$(2) \qquad\qquad (X - C, N) = 0.$$

As a result we may describe $\pi$ as the plane through $C$ perpendicular to $N$.

In order to construct this normal vector $N$ we must first derive two results on simultaneous linear equations. If $x_1, \ldots, x_n$ are unknowns and $a_1, \ldots, a_n, b$ are scalars, then

$$a_1 x_1 + \cdots + a_n x_n = b$$

is called a *linear equation* in the unknowns $x_1, \ldots, x_n$. If $b = 0$, then the equation is called a *homogeneous* linear equation. Now one or more homogeneous equations

$$a_1 x_1 + \cdots + a_n x_n = 0$$

always have one solution, namely, $x_1 = x_2 = \cdots = x_n = 0$. We call this the *trivial* solution to the equations. The problem is to determine conditions that guarantee that solutions exist to these homogeneous equations which are not identically equal to zero. Such a solution we call nontrivial. Rather than give the most general result at this time, we shall limit our consideration to what will be required in our discussion of planes.

First, one equation in two unknowns

$$(3) \qquad\qquad a_1 x_1 + a_2 x_2 = 0$$

certainly has a nontrivial solution. This is clear since if $a_1 = a_2 = 0$, then any pair of numbers $x_1, x_2$ is a solution. If one of the coefficients, say $a_1$, is different from zero, we may divide by it and obtain

$$(4) \qquad\qquad x_1 = -\frac{a_2}{a_1} x_2.$$

Thus, if $x_2$ is any nonzero real number and $x_1$ is then computed by (4), the resulting pair $(x_1, x_2)$ is a nontrivial solution to (3). Notice that this solution to (3) is not unique. Since in (4) we may take $x_2$ to be any nonzero real number, there must be infinitely many nontrivial solutions to (3).

We may now extend this result by the familiar process of elimination of unknowns.

**THEOREM 8.4**   *Two homogeneous linear equations in three unknowns*

(5)
$$a_1x_1 + a_2x_2 + a_3x_3 = 0$$

$$b_1x_1 + b_2x_2 + b_3x_3 = 0$$

*always have a nontrivial solution* $x_1, x_2, x_3$.

Before proving this result, let us consider an example. To find a solution to

(6)
$$2x_1 - x_2 + x_3 = 0$$

$$x_1 - 4x_2 + 3x_3 = 0$$

different from the solution $x_1 = x_2 = x_3 = 0$, solve the first equation for $x_3$ obtaining

(7)
$$x_3 = -2x_1 + x_2.$$

Substitute this in the second equation, obtaining

$$x_1 - 4x_2 + 3(-2x_1 + x_2) = 0$$

or

$$-5x_1 - x_2 = 0$$

then

(8)
$$x_2 = -5x_1.$$

Now for any nonzero value of $x_1$, $x_2$ is determined by (8) and $x_3$ by (7). For example, if $x_1 = 1$, then $x_2 = -5$ and $x_3 = -7$. There are, of course, infinitely many nontrivial solutions to (6). Each corresponds to a nonzero value of $x_1$ in (8).

To prove Theorem 8.4 observe first that if all of the coefficients $a_1, a_2, a_3$, $b_1, b_2, b_3$ are zero, then any triple $x_1, x_2, x_3$ is a solution of (5). Therefore, assume one of the coefficients, for example, $a_1$, is different from zero. Then, dividing by $a_1$, we have

(9)
$$x_1 = -\frac{a_2}{a_1} x_2 - \frac{a_3}{a_1} x_3.$$

If we now substitute this in the second equation of (5), the unknown $x_1$ is eliminated, and we have one equation

(10)
$$c_1x_2 + c_2x_3 = 0$$

in the two unknowns $x_2, x_3$.

We know by what we proved on page 333 that there is a nontrivial solution to (10). Hence, if we then determine $x_1$ by (9), we have a nontrivial solution to (5). In fact there must exist infinitely many nontrivial solutions to (5).

It is readily apparent that we may extend this result to three equations in four unknowns. Indeed, using induction, the result may be extended to $n$ equations in $n+1$ unknowns. We state this result as a theorem and leave the proof as an exercise.

**THEOREM 8.5**  *The n homogeneous linear equations in the $n+1$ unknowns $x_1, \ldots, x_{n+1}$ defined by*

$$a_{1,1}x_1 + \cdots + a_{1,n+1}x_{n+1} = 0$$
$$a_{2,1}x_1 + \cdots + a_{2,n+1}x_{n+1} = 0$$
$$\cdot \qquad \qquad \cdot$$
$$\cdot \qquad \qquad \cdot$$
$$\cdot \qquad \qquad \cdot$$
$$a_{n,1}x_1 + \cdots + a_{n,n+1}x_{n+1} = 0$$

*always have a nontrivial solution, that is, a solution different from the solution $x_1 = x_2 = \cdots = x_{n+1} = 0$.*

We now apply these results to describe the plane through $C$ spanned by $A$ and $B$. (See Figure 27.) First we assert that there is a nonzero vector $N = (n_1, n_2, n_3)$ perpendicular to both $A = (a_1, a_2, a_3)$ and $B = (b_1, b_2, b_3)$.

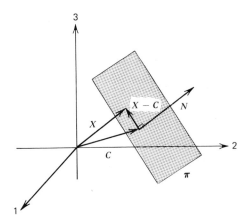

3

2

1

**Fig. 27**  Plane through $C$ perpendicular to $N$.

This just means that
$$(N, A) = n_1 a_1 + n_2 a_2 + n_3 a_3 = 0$$
and
$$(N, B) = n_1 b_1 + n_2 b_2 + n_3 b_3 = 0.$$

But by Theorem 8.4 this pair of homogeneous linear equations must have a nontrivial solution $(n_1, n_2, n_3) = N$.

Next observe that $(N, A) = 0 = (N, B)$ implies $(N, sA + tB) = 0$ for any scalar $s$ and $t$. Hence, if $X$ is a vector in the plane through $C$ spanned by $A$ and $B$, then
$$X - C = sA + tB$$
and consequently

(11)                              $(X - C, N) = 0.$

We now must show that every vector $X$ in $V_3(R)$ satisfying (11) lies in the plane through $C$ spanned by $A$ and $B$. This is tantamount to showing that every vector $Y$ perpendicular to $N$ must be a linear combination of $A$ and $B$. To establish this we must first show that $N, A, B$ together span all of $V_3(R)$. That is, we must show that any vector $Y$ in $V_3(R)$ may be written

(12)                              $Y = rN + sA + tB.$

Now by Theorem 8.5 there exist scalars $u, v, w, x$ not all zero such that

(13) $$uY + vN + wA + xB = 0$$

since this vector equation represents three homogeneous equations, one for each component, in the four unknowns $u, v, w, x$. By Theorem 8.5 these equations must have a nontrivial solution. If in equation (13) $u \neq 0$, we may divide by it, solve for $Y$, and obtain (12). Let us show that it cannot be the case that $u = 0$ in equation (13). We argue by contradiction. If $u = 0$ then from (13) we have

(14) $$vN + wA + xB = 0.$$

Taking the inner product of (14) with $N$ we obtain

$$0 = (0, N) = (vN + wA + xB, N)$$
$$= v(N, N) + w(A, N) + x(B, N) = v(N, N),$$

since $(A, N) = (B, N) = 0$. But since $N \neq 0$, this implies that $v = 0$, or that

$$wA + xB = 0.$$

This is clearly impossible since $A$ and $B$ are assumed to be nonparallel. We therefore have proved that $u \neq 0$ and have established (12).

To prove that $(Y, N) = 0$ implies $Y = sA + tB$, write

$$Y = rN + sA + tB.$$

If we take the inner product of this equation by $N$, we have

$$0 = (Y, N) = (rN + sA + tB, N)$$
$$= r(N, N) + s(A, N) + t(B, N) = r(N, N).$$

Since $N \neq 0$, this implies $r = 0$. Hence $Y = sA + tB$ and we are finished.

We have now shown that any vector $X - C$ satisfying $(X - C, N) = 0$ may be written

$$X - C = sA + tB$$

for appropriate scalars $s$ and $t$. Thus the plane through $C$ perpendicular to $N$ is precisely the plane through $C$ spanned by $A$ and $B$. We note also that equation (12) shows that all vectors $M$ perpendicular to the plane $\pi$ are parallel to $N$. This is, of course, geometrically clear, and an arithmetic proof may be easily given. We leave this verification for the exercises.

The vector $N$, or any nonzero scalar multiple of $N$, is called a *normal vector* for the plane $\pi$. Two planes $\pi_1$ and $\pi_2$ in $V_3(R)$ are said to be parallel or perpendicular according to whether their normal vectors $N_1$ and $N_2$ are parallel or perpendicular.

The vector equation (2) for the plane $\pi$ through $C$ perpendicular to $N$ may be translated into a scalar equation by expanding the inner product. Thus

$$0 = (X - C, N) = (x_1 - c_1)n_1 + (x_2 - c_2)n_2 + (x_3 - c_3)n_3$$

or

$$x_1n_1 + x_2n_2 + x_3n_3 = c_1n_1 + c_2n_2 + c_3n_3 = c.$$

Conversely, the scalar equation

(15) $$x_1n_1 + x_2n_2 + x_3n_3 = c$$

defines the plane through the points $(c/n_1, 0, 0)$, $(0, c/n_2, 0)$, $(0, 0, c/n_3)$, perpendicular to $(n_1, n_2, n_3)$. Indeed, any vector lies in the plane $\pi$ if and only if its components satisfy (15). For example,

$$3x_1 - 4x_2 + x_3 = 5$$

is a scalar equation of the plane perpendicular (normal) to the vector $(3, -4, 1)$ and passing through $(5/3, 0, 0)$.

The numbers $c/n_1$, $c/n_2$, $c/n_3$ are the intercepts of the plane, that is the coordinates of the points of intersection of the plane with the first, second, and third coordinate axis, respectively. See Figure 28.

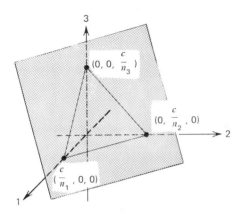

**Fig. 28** Intercepts of a plane.

**Example 1.** Determine a normal vector for the plane $\pi$ through $C = (2, 1, 1)$ spanned by $A = (1, -1, 1)$ and $B = (2, 0, 1)$. Determine a scalar equation for this plane. $N = (n_1, n_2, n_3)$ will be a normal vector for $\pi$ if $(N, A) = (N, B) = 0$. This implies

$$n_1 - n_2 + n_3 = 0$$

and

$$2n_1 + n_3 = 0.$$

Setting $n_1 = 1$, we have $n_3 = -2$ and $n_2 = -1$. Therefore, $N = (1, -1, -2)$ is one such normal vector. All others are, of course, parallel to $N$. Now $C = (2, 1, 1)$ is assumed to lie in the plane. So translating the vector equation $(X - C, N) = 0$ into scalar form, we have

$$(x_1 - 2)1 + (x_2 - 1)(-1) + (x_3 - 1)(-2) = 0$$

or

$$x_1 - x_2 - 2x_3 + 1 = 0.$$

This is a scalar equation for the plane $\pi$ as is any scalar multiple of this equation.

**Example 2.** Determine two vectors $A$ and $B$ that span the plane determined by the scalar equation

$$2x_1 - x_2 - x_3 = 5.$$

Now $(2, -1, -1)$ is a normal vector for this plane. Hence, any two nonparallel (necessarily nonzero) vectors $A$ and $B$ that are perpendicular to $(2, -1, 1)$ will span the plane $\pi$. Two such vectors may be easily computed by inspecting the equation

$$2x_1 - x_2 - x_3 = 0$$

which must be satisfied by both of these vectors. One choice is $A = (1, 2, 0)$, and $B = (1, 0, 2)$. They both are perpendicular to $N$ and are clearly not parallel.

*Example 3.*   Determine a scalar equation for the plane

$$\pi = \{(1, 3, 2) + s(2, 1, 1) + t(0, 1, -1)\}.$$

This plane has vector equation

$$X = (1, 3, 2) + s(2, 1, 1) + t(0, 1, -1).$$

Since $X = (x_1, x_2, x_3)$, this vector equation is equivalent to three scalar equations

$$x_1 = 1 + 2s$$

(16)   $$x_2 = 3 + s + t$$

$$x_3 = 2 + s - t.$$

These scalar equations are often called *parametric* equations of the plane. The scalar equation for the plane may be determined as in Example 1 by computing a vector normal to the vectors $(2, 1, 1)$ and $(0, 1, -1)$. This equation also may be determined by eliminating the parameters $s$ and $t$ in the equations (16). Indeed, solving the first equation for $s$, we obtain

$$s = \frac{x_1 - 1}{2}.$$

Solving for $t$ in the second equation, we obtain

$$t = x_2 - 3 - s = x_2 - 3 - \frac{x_1 - 1}{2}.$$

Substituting these values of $s$ and $t$ in the third equation, we obtain

$$x_3 = 2 + \frac{x_1 - 1}{2} - \left( x_2 - 3 - \frac{x_1 - 1}{2} \right),$$

or

$$x_1 - x_2 - x_3 + 4 = 0$$

which is the desired scalar equation of the plane.

For $n > 3$ the set of vectors $X - C$ in $V_n(R)$ satisfying

(10)   $$(X - C, N) = 0$$

cannot be spanned by two nonparallel vectors $A$ and $B$. Hence for $n > 3$, equation (10) does not determine a plane. It can be shown, however, that there are $n - 1$ vectors $A_1, \ldots, A_{n-1}$ which span this set of vectors. For $n > 3$, the set of vectors $X$ satisfying (10) is often called the *hyperplane* through $C$ perpendicular to $N$.

Next we determine a formula for the distance from a point to a plane in space. In $V_3(R)$, the distance from a point with coordinates $(a_1, a_2, a_3)$ to the plane $\pi$ described by (10) is clearly the minimum of the quantity

$$|X - A|$$

where $X$ ranges over all the vectors of the plane. This minimum will be obtained when $X - A$ is normal to the plane or, in other words, when $X - A = tN$ for some scalar $t$. But since $X$ lies in the plane, $(X - C, N) - 0$. Combining these two facts, we have

$$(A + tN - C, N) = (X - C, N) = 0$$

Solving for $t$ we obtain

$$t = \frac{(C - A, N)}{|N|^2}.$$

Thus the distance from $A$ to $\pi$ is

$$|tN| = |t\|N| = \frac{|(C-A, N)|}{|N|}.$$

See Figure 29.

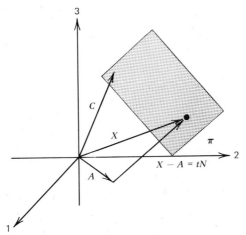

**Fig. 29**    If $X = A + tN$ lies in $\pi$, and $N$ is normal to the plane $\pi$, then $|tN|$ is the distance from the point with coordinates $(a_1, a_2, a_3)$ to the plane $\pi$.

*Example 4.*    Determine the distance from $(1,-1,2)$ to the plane with scalar equation $2x_1 + x_2 - x_3 = 5$. This plane has normal vector $N = (2, 1, -1)$ and $C = (0, 5, 0)$ lies in the plane. Therefore, letting $A = (1, -1, 2)$ we have $C - A = (-1, 6, -2)$. The required distance is

$$\frac{|(C-A, N)|}{|N|} = \frac{6}{\sqrt{6}} = \sqrt{6}.$$

For $n > 3$ the same procedure determines the distance from a point with coordinates $(a_1, \ldots, a_n)$ to the hyperplane in $V_n(R)$ through $C$ perpendicular to $N$.

**EXERCISES**

1. If the plane $\pi$ has scalar equation
$$2x_1 - x_2 + 3x_3 = 5,$$
   determine
   (a) the intercepts of the plane
   (b) a normal vector to the plane of unit length (length $= 1$)
   (c) the vector equation of the plane
   (d) two vectors that span the plane
2. Let $\pi$ be the plane through $(1, 1, 3)$ perpendicular to $(1, -1, 1)$. Determine
   (a) a scalar equation for this plane
   (b) two vectors that span the plane
   (c) the intercepts of the plane.
   (d) the distance from the origin to the plane.
3. Determine the cartesian equation of the plane passing through the terminal points of the vectors $A = (-1, 2, 4)$, $B = (2, 1, 0)$, $C = (3, -1, 2)$. Determine a unit normal vector to this plane. Determine two vectors spanning the plane as well as the distance from $(4, 1, 1)$ to the plane.
4. Let $\pi = \{C + sA + tB: \ s \text{ and } t \text{ real numbers}\}$, where $C = (-1, 1, 2)$, $A = (2, 1, -2)$, $B = (1, 1, 3)$. Which of the following vectors lie in the plane $\pi$?
   (a) $(-1, 2, 0)$    (b) $(3, 4, 6)$    (c) $(3, -2, 7)$    (d) $(0, 1, -3)$

5. Determine a scalar equation for the plane $\pi = \{C + sA + tB: s \text{ and } t \text{ real num-}$ bers$\}$ if $C = (2, 3, 7), A = (1, 0, -1), B = (1, -1, 2)$.

6. Determine scalar equations for the two planes parallel to the plane through $C = (1, -1, 2)$ spanned by $A = (1, 1, 5)$ and $B = (2, 0, 1)$ and at a distance 5 from this plane.

7. Let $\pi_1, \pi_2$ be two planes and let $A, B$ be two vectors belonging to the intersection of these planes. Prove that the line through $A$ and $B$ is contained in $\pi_1 \cap \pi_2$.

8. Determine the distance from the point $(-1, 2, 1)$ to the line of intersection of the planes defined by

$$2x_1 - x_2 + x_3 = 2$$

and

$$x_1 + x_2 - 3x_3 = 1.$$

9. Determine the equation of the plane parallel to $2x_1 + x_2 - x_3 = 5$ if the point with coordinates $(3, 1, 1)$ is equidistant from the two planes.

10. Find the distance between the parallel planes
   (a) $2x_1 - 3x_2 + x_3 = 5$ $\qquad 4x_1 - 6x_2 + 2x_3 = 1$
   (b) $x_1 - x_2 + 3x_3 = 2$ $\qquad -x_1 + x_2 - 3x_3 = 6$

11. Prove that if a plane $\pi$ in $V_3(R)$ has scalar equation $ax_1 + bx_2 + cx_3 = d$, then the distance between the origin and this plane is

$$\frac{|d|}{\sqrt{a^2 + b^2 + c^2}}$$

12. Prove that if a plane $\pi$ in $V_3(R)$ has scalar equation $ax_1 + ax_2 + cx_3 = d$, then the distance from the point $(x_1, x_2, x_3)$ to the plane $\pi$ is given by

$$\frac{|d - ax_1 - bx_2 - cx_3|}{\sqrt{a^2 + b^2 + c^2}}.$$

13. Determine a scalar equation for the plane containing the point $(1, -2, 1)$ as well as the line through $(1, 4, 2)$ parallel to the vector $(3, 1, 7)$.

14. In $V_4(R)$ if $\pi$ is the hyperplane through $C = (c_1, c_2, c_3, c_4)$, perpendicular to $N = (n_1, n_2, n_3, n_4)$, determine a scalar equation to this hyperplane. (Argue by analogy with the case $n = 3$.) If $C = (1, -1, 2, 0)$ and $N = (1, 0, -1, -1)$, what is the distance from the origin to this hyperplane?

15. Determine a nontrivial solution to the equations

$$x_1 - 2x_2 + x_3 = 0$$

$$2x_1 + x_2 - 3x_3 = 0.$$

16. Determine a nontrivial solution to the equations

$$2x_1 + x_2 - x_3 + 3x_4 = 0$$

$$x_1 - x_2 + x_3 - 2x_4 = 0$$

$$2x_1 - x_2 - x_3 + x_4 = 0.$$

17. Using Theorem 8.4., prove that three homogeneous linear equations in four unknowns always possess a nontrivial solution.

18. Using induction and Theorem 8.4, prove that $n$ homogeneous linear equations in $n + 1$ unknowns always have a nontrivial solution.

19. Deduce from Exercise 18 that $k$ homogeneous linear equations in $n$ unknowns always has a nontrivial solution provided that $n > k$.

20. Use the fact that two homogeneous equations in three unknowns always possess a nontrivial solution to prove that two nonparallel vectors $A$ and $B$ will span $V_2(R)$.

21. Using the fact that one homogeneous equation in two unknowns always has a nontrivial solution, show that if $A$ and $B$ span $V_2(R)$, then $A$ and $B$ are not parallel.

22. Let $N$ be a normal vector to the plane through $C$ spanned by $A$ and $B$. If $M$ is a vector perpendicular to $A$ and $B$, show that $M$ is parallel to $N$. (We must show $M = rN$ for some scalar $r$. By the discussion on pp. 335–6 the vectors $A$, $B$, $N$ span $V_3(R)$. Hence $M = rN + sA + tB$. By taking the inner product of this equation with the vectors $A$, $B$, and $N$, show that $s = t = 0$ and $r = (M, N)/|N|^2$.)

## 8.8 The Vector Product

Although it is impossible to define a multiplication between vectors in $V_3(R)$ for which division is always possible, a multiplication that has many useful and interesting properties can be defined. This multiplication is called the *vector product or cross product* and it is written $A \times B$. We first require that the cross product of parallel vectors always vanishes. Hence, for the unit coordinate vectors in particular, we define

(1) $$I_1 \times I_1 = I_2 \times I_2 = I_3 \times I_3 = 0.$$

Second, if $I$ and $J$ are orthogonal unit vectors, then $I \times J$ is to be a unit vector orthogonal to both $I$ and $J$. There are two choices for $I \times J$ satisfying this requirement. We take $I \times J$ to be the vector $K$ such that if the triple $I$, $J$, and $K$ are rotated so that $I$ lies along $I_1$, and $J$ lies along $I_2$, then $K = I \times J$ will lie along $I_3$. See Figure 30.

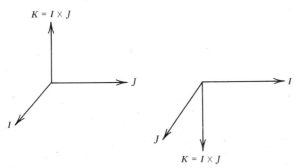

**Fig. 30**  Orientation of the vectors $I, J, I \times J$.

This geometric condition immediately gives us a multiplication table for the unit coordinate vectors. Accordingly, we define

$$I_1 \times I_2 = -I_2 \times I_1 = I_3$$
(2) $$I_2 \times I_3 = -I_3 \times I_2 = I_1$$
$$I_3 \times I_1 = -I_1 \times I_3 = I_2.$$

We extend the cross product to all vectors $A, B \in V_3(R)$ by requiring, in addition, that the product $\times$ satisfy the distributive law

(3) $$A \times (B + C) = A \times B + A \times C,$$

and that for each real scalar $a$,

(4) $$a(A \times B) = (aA) \times B = A \times (aB).$$

To obtain a formula for $A \times B$ we need only multiply out the expression

$$A \times B = (a_1 I_1 + a_2 I_2 + a_3 I_3) \times (b_1 I_1 + b_2 I_2 + b_3 I_3)$$

using (1) to (4). This yields

$$A \times B = (a_1 I_1 + a_2 I_2 + a_3 I_3) \times (b_1 I_1 + b_2 I_2 + b_3 I_3)$$

(5)

$$= (a_2 b_3 - a_3 b_2) I_1 + (a_3 b_1 - a_1 b_3) I_2 + (a_1 b_2 - a_2 b_1) I_3.$$

We note a few properties of this cross product. First, $A \times B$ always defines a vector in $V_3(R)$. Second, (2) shows that the cross product is noncommutative. It is in nonassociative as well since, by (1) and (2),

$$I_1 \times (I_1 \times I_2) = I_1 \times I_3 = -I_2.$$

On the other hand, $(I_1 \times I_1) \times I_2 = 0$. Furthermore there is no identity since it is easy to check (Exercise 6) that if $A \times A = A$, then $A = 0$.

Formula (5) is most easily remembered and used if determinant notation is introduced. Recall from high school algebra that a $2 \times 2$ determinant is defined by the formula

$$\begin{vmatrix} a_1 & a_2 \\ b_1 & b_2 \end{vmatrix} = a_1 b_2 - a_2 b_1.$$

A $3 \times 3$ determinant is defined by the formula

(6)
$$\begin{vmatrix} a_1 & a_2 & a_3 \\ b_1 & b_2 & b_3 \\ c_1 & c_2 & c_3 \end{vmatrix} = a_1 b_2 c_3 - a_1 b_3 c_2 + a_2 b_3 c_1 - a_2 b_1 c_3 + a_3 b_1 c_2 - a_3 b_2 c_1.$$

We may simplify the right-hand side of this expression by using $2 \times 2$ determinants as follows. By (6)

(7)
$$\begin{vmatrix} a_1 & a_2 & a_3 \\ b_1 & b_2 & b_3 \\ c_1 & c_2 & c_3 \end{vmatrix} = a_1(b_2 c_3 - b_3 c_2) - a_2(b_1 c_3 - b_3 c_1) + a_3(b_1 c_2 - b_2 c_1)$$

$$= a_1 \begin{vmatrix} b_2 & b_3 \\ c_2 & c_3 \end{vmatrix} - a_2 \begin{vmatrix} b_1 & b_3 \\ c_1 & c_3 \end{vmatrix} + a_3 \begin{vmatrix} b_1 & b_2 \\ c_1 & c_2 \end{vmatrix}.$$

Formula (7) is called the *Lagrange expansion* of a $3 \times 3$ determinant. It may perhaps best be remembered in the following way. Each term $a_1, a_2, a_3$ in the first row is multiplied by the $2 \times 2$ determinant obtained by striking out the first row and the column containing the appropriate coefficient $a_1, a_2, a_3$. The resulting terms

$$a_1 \begin{vmatrix} b_2 & b_3 \\ c_2 & c_3 \end{vmatrix}, \quad a_2 \begin{vmatrix} b_1 & b_3 \\ c_1 & c_3 \end{vmatrix}, \quad a_3 \begin{vmatrix} b_1 & b_2 \\ c_1 & c_2 \end{vmatrix},$$

are then multiplies by $1, -1, 1$, respectively, and added together.

Now observe that formula (5) just states that

$$A \times B = \begin{vmatrix} a_2 & a_3 \\ b_2 & b_3 \end{vmatrix} I_1 - \begin{vmatrix} a_1 & a_3 \\ b_1 & b_3 \end{vmatrix} I_2 + \begin{vmatrix} a_1 & a_2 \\ b_1 & b_2 \end{vmatrix} I_3.$$

Hence it is quite reasonable to adopt the terminology

(8)
$$A \times B = \begin{vmatrix} I_1 & I_2 & I_3 \\ a_1 & a_2 & a_3 \\ b_1 & b_2 & b_3 \end{vmatrix},$$

where, in accord with (7), we define

$$\begin{vmatrix} I_1 & I_2 & I_3 \\ a_1 & a_2 & a_3 \\ b_1 & b_2 & b_3 \end{vmatrix} = \begin{vmatrix} a_2 & a_3 \\ b_2 & b_3 \end{vmatrix} I_1 - \begin{vmatrix} a_1 & a_2 \\ b_1 & b_3 \end{vmatrix} I_2 + \begin{vmatrix} a_1 & a_2 \\ b_1 & b_2 \end{vmatrix} I_3.$$

Thus if $A = (2, -1, 1)$, and $B = (-1, 1, 2)$, then

$$A \times B = \begin{vmatrix} I_1 & I_2 & I_3 \\ 2 & -1 & 1 \\ -1 & 1 & 2 \end{vmatrix} = -3I_1 - 5I_2 + I_3 = (-3, -5, 1).$$

It follows immediately from the definition of a determinant that if two rows of a determinant are interchanged, then the sign is reversed. Using (8) we immediately conclude that the vector product is anticommutative. That is,

$$A \times B = -B \times A.$$

The determinant notation gives a very concise formula for the inner product of $A \times B$ with a third vector $C = (c_1, c_2, c_3)$. If

$$A \times B = \begin{vmatrix} a_2 & a_3 \\ b_2 & b_3 \end{vmatrix} I_1 - \begin{vmatrix} a_1 & a_3 \\ b_1 & b_3 \end{vmatrix} I_2 + \begin{vmatrix} a_1 & a_2 \\ b_1 & b_2 \end{vmatrix} I_3$$

and

$$C = c_1 I_1 + c_2 I_2 + c_3 I_3,$$

then

$$(C, A \times B) = \begin{vmatrix} a_2 & a_3 \\ b_2 & b_3 \end{vmatrix} c_1 - \begin{vmatrix} a_1 & a_3 \\ b_1 & b_3 \end{vmatrix} c_2 + \begin{vmatrix} a_1 & a_2 \\ b_1 & b_2 \end{vmatrix} c_3 = \begin{vmatrix} c_1 & c_2 & c_3 \\ a_1 & a_2 & a_3 \\ b_1 & b_2 & b_3 \end{vmatrix}.$$

Thus if $A = (0, 1, -1)$, $B = (1, 1, 1)$, and $C = (2, -1, -1)$,

$$(C, A \times B) = \begin{vmatrix} 2 & -1 & -1 \\ 0 & 1 & -1 \\ 1 & 1 & 1 \end{vmatrix} = 2 \begin{vmatrix} 1 & -1 \\ 1 & 1 \end{vmatrix} + \begin{vmatrix} 0 & -1 \\ 1 & 1 \end{vmatrix} - \begin{vmatrix} 0 & 1 \\ 1 & 1 \end{vmatrix} = 4 + 1 + 1 = 6.$$

The inner product $(C, A \times B) = (A \times B, C)$ is called the *scalar triple product* of the vectors $A$, $B$, and $C$. It follows immediately from the definition that a determinant has value zero if two rows are the same. Using this fact we deduce that

(9) $$(A, A \times B) = (B, A \times B) = 0.$$

Hence, if $A \times B \neq 0$, the vector $A \times B$ is perpendicular to both $A$ and $B$.

Next we investigate what it means for $A \times B$ to be equal to the zero vector. We note first that if $A$ and $B$ are parallel, then $A \times B = 0$. This is clear since, if $A$ and $B$ are parallel, then $B = aA$ for some scalar $a$ and

$$A \times B = A \times (aA) = a(A \times A) = a \begin{vmatrix} I_1 & I_2 & I_3 \\ a_1 & a_2 & a_3 \\ a_1 & a_2 & a_3 \end{vmatrix} = 0.$$

This discussion shows that if $A, B \in V_3(R)$ and one vector is a scalar multiple of the other, then $A \times B = 0$. The converse to this statement is also valid. The proof requires the following identity which is credited to Lagrange.

**THEOREM 8.6** *If $A, B \in V_3(R)$, then*

$$|A \times B|^2 = |A|^2 |B|^2 - (A, B)^2.$$

*Proof.* The identity results from the following calculation. Letting $A = (a_1, a_2, a_3)$, $B = (b_1, b_2, b_3)$, we have

$$|A|^2 |B|^2 - (A, B)^2 = (a_1{}^2 + a_2{}^2 + a_3{}^2)(b_1{}^2 + b_2{}^2 + b_3{}^2) - (a_1 b_1 + a_2 b_2 + a_3 b_3)^2.$$

Observe that each of the squares in $(a_1b_1 + a_2b_2 + a_3b_3)^2$ cancels with an appropriate term of $(a_1^2 + a_2^2 + a_3^2)(b_1^2 + b_2^2 + b_3^2)$. Hence we have

$$|A|^2|B|^2 - (A, B)^2 = a_1^2(b_2^2 + b_3^2) + a_2^2(b_1^2 + b_2^2) + a_3^2(b_1^2 + b_2^2)$$
$$- 2(a_1a_2b_1b_2 + a_1a_3b_1b_3 + a_2a_3b_2b_3)$$
$$= (a_1b_2 - a_2b_1)^2 + (a_1b_3 - a_3b_1)^2 + (a_2b_3 - a_3b_2)^2$$
$$= \begin{vmatrix} a_1 & a_2 \\ b_1 & b_2 \end{vmatrix}^2 + \begin{vmatrix} a_1 & a_3 \\ b_1 & b_3 \end{vmatrix}^2 + \begin{vmatrix} a_2 & a_3 \\ b_2 & b_3 \end{vmatrix}^2 = |A \times B|^2.$$

**COROLLARY.** $A \times B = 0$, *if and only if one vector is a scalar multiple of the other.*

**Proof.** By Theorem 8.6, $A \times B = 0$ if and only if $|(A, B)| = |A||B|$. But by the Schwarz inequality, Theorem 8.2, this is true if and only if one vector is a scalar multiple of the other.

If $A$ and $B$ are vectors and one is a scalar multiple of the other, then it is easy to see that this is equivalent to the assertion that there exist scalars $a$ and $b$ not both equal to zero such that $aA + bB = 0$. (The student should convince himself that these two assertions are equivalent.) When either of these assertions hold for vectors $A$ and $B$, these vectors are said to be *linearly dependent.* Thus the corollary to Theorem 8.6 says that $A \times B = 0$ if and only if $A$ and $B$ are linearly dependent. We shall discuss linear dependence further in the next section.

Next we turn our attention to some geometric interpretations of Lagrange's identity. First recall that if $B \neq 0$ and $l$ is the line through $A$ parallel to $B$, then the distance $d$ from the origin to $l$ is given by the formula

$$d = \frac{\sqrt{|A|^2|B|^2 - (A, B)^2}}{|B|}$$

(p. 330). In view of Theorem 8.6 this may be written

$$d = \frac{|A \times B|}{|B|}.$$

Second, if $A$ and $B$ are nonzero vectors and $\theta$ is the angle between them, then

$$|A \times B|^2 = |A|^2|B|^2 - (A, B)^2$$
$$= |A|^2|B|^2 - |A|^2|B|^2 \cos^2 \theta$$
$$= |A|^2|B|^2 \sin^2 \theta.$$

Therefore

$$|A \times B| = |A||B| \sin \theta,$$

and it is easily verified that this is the area of the parallelogram formed by $A$ and $B$. (See Figure 31.)

Thus if $A$ and $B$ are not parallel, $A \times B$ is perpendicular to the plane $\pi$ determined by $A$ and $B$ and has length $|A \times B| = |A||B| \sin \theta$. To give a complete geometric description of $A \times B$, it only remains to determine the direction of $A \times B$. This direction is, of course, reversed if we interchange the order of $A$ and $B$.

The direction of $A \times B$ is determined by the choice of the initial coordinate axes. Fixing one axis, for example $I_3$, to be vertical, the axes $I_1, I_2$ may be

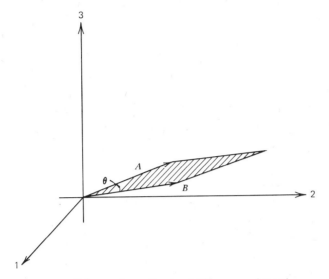

**Fig. 31** Parallelogram formed by $A$ and $B$ has area $= |A \times B|$.

specified in essentially two different ways. By different, we mean that one choice cannot be transformed into the other by a combination of rotations and translations. These two sets of axes are called right hand and left hand, respectively. (Figure 32.)

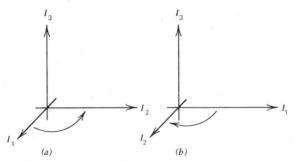

**Fig. 32** (a) Right hand coordinate system. (b) Left hand coordinate system.

If we have chosen a right-hand coordinate system, as has been the convention in this book, then the direction of $A \times B$ is given by the right-hand rule. That is, if the direction of rotation from $A$ to $B$ is given by the clasped fingers of the *right* hand, then the direction of $A \times B$ is in the direction of the thumb of the right hand. (Figure 33.) If we had chosen a left hand coordinate system, then the direction of $A \times B$ would be just the reverse. The student may object to this description of the orientation of a coordinate system in three space as highly nonmathematical. It is, but it is not worthwhile at this point to erect the necessary machinery to put these ideas on a firm mathematical foundation.

We summarize the properties of the vector product in the following theorem.

**THEOREM 8.7** *Let $A, B, C \in V_3(R)$ and $a \in R$. If $A \times B$ denotes the vector product of $A$ and $B$, then*

(i) $A \times (B+C) = A \times B + A \times C$

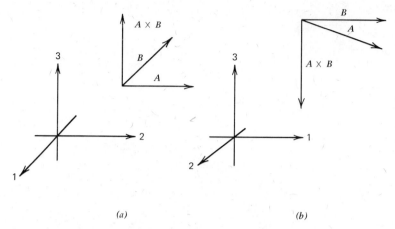

**Fig. 33** (*a*) Direction of $A \times B$ in right-hand coordinate system. (*b*) Direction of $A \times B$ in left-hand coordinate system.

(ii) $a(A \times B) = (aA) \times B = A \times (aB)$
(iii) $(A \times B) = -(B \times A)$
(iv) $A \times B = 0$, *if and only if A and B are linearly dependent.*

**EXERCISES**

1. If $A = (2, -1, 3), B = (1, -1, -1), C = (1, 2, -1)$, compute
   (a) $A \times (B \times C)$                     (f) $(A, B \times C)$
   (b) $(A \times B) \times C$                     (g) $(B, C \times A)$
   (c) $(A \times A) \times B$                     (h) $(A - B, A \times B)$
   (d) $A \times (A \times B)$                     (i) $(A \times B, A \times C)$
   (e) $(A - B) \times (B - C)$

2. Using the vector product determine a normal vector to the planes spanned by
   (a) $(2, -1, 3)$ and $(4, 6, -2)$
   (b) $(-1, -1, 2)$ and $(2, -1, 1)$
   (c) $(1, 0, 1)$ and $(1, 1, 1)$

3. Using the vector product, determine a normal vector for the plane passing through
   (a) $(1, -1, 1), \ (2, 2, -1), \ (1, 0, 1)$
   (b) $(2, -1, 3), \ (1, -1, -1), \ (1, 2, -1)$
   (c) $(2, -1, 0), \ (1, 3, 4), \ (0, -1, 5)$

4. Let $P, Q, R$ be three points in space. Determine a formula for the area of the triangle determined by these points that involves the vector product of appropriate vectors.

5. Using the vector product, find the area of the triangle determined by the following triples of points. (Use Exercise 4.)
   (a) $(1, 2, -3), \ (1, 1, 1), \ (2, 0, -1)$
   (b) $(2, 0, 1), \ (-1, 5, 2), \ (1, 1, -1)$
   (c) $(3, 2, 1), \ (-1, 1, 0), \ (0, -1, 1)$

6. Show that if $A \times A = A$, then $A = 0$. Also show that $0 \times A = 0$ for each $A \in V_3(R)$. Using these two facts, deduce that there is no identity element for the vector product.

7. Using Lagrange's identity, deduce a necessary and sufficient geometric condition on nonzero vectors $A$ and $B$ so that

$$|A \times B| = |A||B|.$$

Prove the condition.

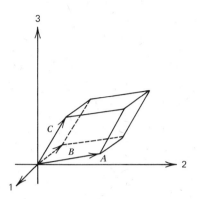

**Fig. 34**

8. Vectors $A$, $B$, $C$ in $V_3(R)$ determine a parallelepiped in space as in Figure 34. The volume of this parallelepiped is clearly the product of the area of the base and the altitude. Assuming that the base is determined by the vectors $A$ and $B$, show that the altitude equals

$$\frac{|(A \times B, C)|}{|A \times B|}.$$

Hence the volume of the parallelepiped is given by $|(A \times B, C)|$. (What is the cosine of the angle between $C$ and a normal to the plane determined by $A$ and $B$?)

9. Let $A = (1,3,4)$, $B = (2,-1,1)$, $C = (1,1,-2)$. For what value of $t$ is the volume of parallelepiped determined by $A$, $B$, $tC$ equal to 1?

10. Show directly from the definition of $A \times B$ that

$$A \times B = -B \times A.$$

11. Derive the following properties of the scalar triple product.
   (a) $(A \times B, C) = (A, B \times C)$
   (b) $(A, B \times C) = -(B, A \times C)$
   (c) $(A \times B, C) = -(C, B \times A)$
   (d) $(A \times B, C) = (B \times C, A)$.

12. If $A, B, C, D \in V_3(R)$, prove that
   (a) $A \times (B \times C) = (A, C)B - (A, B)C$
   (b) $A \times (B \times C) + B \times (C \times A) + C \times (A \times B) = 0$
   (c) $(A \times B, C \times D) = (A, C)(B, D) - (A, D)(B, C)$.
   (Derive (a) directly from the definition; (b) and (c) now follow from (a).)

13. Let $L$ be the line through the points with coordinates $(a_1, a_2, a_3)$ and $(b_1, b_2, b_3)$. Show that the distance from the point $(c_1, c_2, c_3)$ to $l$ is given by

$$\frac{|(C-A) \times (A-B)|}{|A-B|}$$

where $A$, $B$, and $C$ are the vectors $(a_1, a_2, a_3)$, $(b_1, b_2, b_3)$, and $(c_1, c_2, c_3)$.

14. Let $A$ and $B$ be nonparallel vectors in $V_3(R)$. Show that if real numbers $a, b, c$ satisfy

$$aA + bB + cA \times B = 0$$

then $a = b = c = 0$. (Compute the inner product of this equation with the vector $A \times B$.)

15. Let $A, B, C \in V_3(R)$ and suppose $(A, B \times C) \neq 0$. Show that if real numbers $a$, $b$, $c$ exist satisfying $aA + bB + cC = 0$, then $a = b = c = 0$. (Compute the vector product of the equation $aA + bB + cC = 0$ with one vector, then the inner product of the resulting equation with another vector.)

### *8.9   Linear Independence of Vectors

Theorem 8.1 states that two vectors $A$ and $B$ span $V_2(R)$ if and only if $A$ and $B$ are not parallel. Let us consider the analogous question in $V_3(R)$. The problem is this: What are necessary and sufficient conditions on three vectors $A$, $B$, $C$ in $V_3(R)$ so that each vector $D \in V_3(R)$ can be written

$$D = x_1 A + x_2 B + x_3 C$$

for appropriate scalars $x_1, x_2, x_3$? What we need is an appropriate generalization of the notion of parallelism.

If $A$ and $B \in V_n(R)$, then $A$ and $B$ are parallel if the vectors are nonzero and if one is a scalar multiple of the other. The latter condition means that there exist scalars $a$, $b$ not both zero so that

$$aA + bB = 0.$$

When this condition holds, we have said that $A$ and $B$ are *linearly dependent*. The only difference between parallelism and linear dependence for two vectors is that parallelism only applies to nonzero vectors, whereas linear dependence applies to the zero vector as well. It is this notation of linear dependence that we shall generalize.

**Definition.**   *Vectors $A_1, \ldots, A_k$ in $V_n(R)$ are said to be* linearly dependent *if there exist scalars $a_1, \ldots, a_k$ not all equal to zero such that*

$$a_1 A_1 + \cdots + a_k A_k = 0.$$

*If no such scalars different from zero exist, then the vectors $A_1, \ldots, A_k$ are called* linearly independent.

Thus, to verify that vectors $A_1, \ldots, A_k$ are linearly independent one must prove that if scalars $a_1, \ldots, a_k$ exist so that

$$a_1 A_1 + \cdots + a_k A_k = 0,$$

then it must follow that

$$a_1 = a_2 = \cdots = a_k = 0.$$

**Example 1.**   The unit coordinate vectors $I_1 = (1, 0, \ldots, 0)$, $I_2 = (0, 1, 0, \ldots, 0)$, $\ldots, I_n = (0, 0, \ldots, 0, 1)$ in $V_n(R)$ are linearly independent since if $a_1 I_1 + \cdots + a_n I_n = 0$, then $(a_1, \ldots, a_n)$ is the zero vector. Hence, $a_1 = a_2 = \cdots = a_n = 0$.

**Example 2.**   If one of the vectors $A_1, \ldots, A_k$ is the zero vector, then these vectors are linearly dependent. For if $A_j = 0$, then

$$0 \cdot A_1 + 0 \cdot A_2 + \cdots + 1 \cdot A_j + \cdots + 0 \cdot A_k = 0,$$

and not all the scalars in the above equation are equal to zero.

**Example 3.**   The vectors $A = (1, -1, 1)$, $B = (-1, 4, 2)$, $C = (1, 2, 3)$ are linearly dependent since

$$2A + B - C = 0.$$

**Example 4.**   Two vectors $A$ and $B$ in $V_n(R)$ are linearly independent if and only if they are not parallel. (Why?)

**Example 5.**   If $A_1, \ldots, A_k$ are nonzero vectors in $V_n(R)$ that are pairwise orthogonal, that is, $(A_i, A_j) = 0$ if $i \neq j$, then $A_1, \ldots, A_k$ are linearly independent. This follows easily since if

(1)  $$a_1 A_1 + \cdots + a_k A_k = 0,$$

then taking the inner product of the equation with $A_1$, we get

$$0 = a_1(A_1, A_1) + a_2(A_1, A_2) + \cdots + a_k(A_1, A_k) = a_1(A_1, A_1).$$

Since $A_1 \ne 0$, this implies $a_1 = 0$. Taking the inner product of (1) successively with the vectors $A_2, \ldots, A_k$, we deduce that

$$a_2(A_2, A_2) = 0 = a_3(A_3, A_3) = \cdots = a_k(A_k, A_k).$$

Since

$$(A_k, A_k) = |A_k|^2 \ne 0$$

we have

$$a_2 = \cdots = a_k = 0.$$

**Example 6.** In $V_3(R)$ if $A$ and $B$ are linearly independent, then $A$, $B$, $A \times B$ are linearly independent (Exercise 12, Section 8.8).

Whenever a vector $B$ can be written as a linear combination of vectors $A_1, \ldots, A_k$, we say that $B$ is in the *span* of the vectors $A_1, \ldots, A_k$. It follows immediately from the definition of linear independence that vectors $A_1, \ldots, A_k$ are linearly independent if no vector $A_i$ in this set is in the span of any of the other vectors of the set. We leave the verification of this fact as an exercise.

**THEOREM 8.8**   *Three vectors A, B, C will spaa $V_3(R)$ if and only if they are linearly independent.*

Furthermore, the scalar triple product is precisely the tool that determines if the vectors are linearly independent.

**THEOREM 8.9**   *Vectors A, B, C in $V_3(R)$ are linearly independent if and only if*

$$(A, B \times C) \ne 0.$$

To prove 8.8 we need to know that any set of four vectors in $V_3(R)$ is linearly dependent. This follows readily from Theorem 8.5 but since the proof of 8.5 was left as an exercise we state and give the proof of the result we need.

**LEMMA 8.10**   *Any set of four vectors A, B, C, D in $V_3(R)$ is linearly dependent.*

**Proof.**   We seek scalars $x_1, x_2, x_3, x_4$ not all zero satisfying

$$x_1 A + x_2 B + x_3 C + x_4 D = 0.$$

Writing $A = (a_1, a_2, a_3)$ etc., the above vector equation gives rise to three scalar equations

(2)
$$x_1 a_1 + x_2 b_1 + x_3 c_1 + x_4 d_1 = 0$$
$$x_1 a_2 + x_2 b_2 + x_3 c_2 + x_4 d_2 = 0$$
$$x_1 a_3 + x_2 b_3 + x_3 c_3 + x_4 d_3 = 0.$$

If all the vectors $A$, $B$, $C$, $D$ are zero, the set is certainly linearly dependent. So assume one vector, $A$ say, is nonzero. Therefore, one of the real numbers $a_1$, $a_2$, $a_3$ is different from zero. Assume $a_3 \ne 0$. Then

(3)
$$x_1 = -\frac{1}{a_3}(x_2 b_3 + x_3 c_3 + x_4 d_3).$$

Substitute this value of $x_1$ in the first two equations of (2). This yields two equations

$$x_2b_1' + x_3c_1' + x_4d_1' = 0$$

(4)

$$x_2b_2' + x_3c_2' + x_4d_2' = 0$$

with three unknowns $x_2, x_3, x_4$ and certain new constants $b_i', c_i', d_i', i = 1, 2$. However, the statement that two homogeneous linear equations in three unknowns always possess a nontrivial solution is just Theorem 8.4. Therefore (4) must have a nonzero solution $x_2, x_3, x_4$. Defining $x_1$ by (3) we have a nonzero solution to (2), and hence we have shown that the vectors $A, B, C, D$ are linearly dependent.

We are now ready to prove Theorem 8.8. First we show that if $A, B, C$ are linearly independent, then $A, B, C$ span $V_3(R)$. If $D \in V_3(R)$, we know that $A, B, C, D$ are linearly dependent. Hence there are scalars, $x_1, x_2, x_3, x_4$ not all zero such that

(5)

$$x_1A + x_2B + x_3C + x_4D = 0.$$

Now since $A, B, C$ are linearly independent, it follows that in equation (5) $x_4 \neq 0$. (Why?) Thus dividing by $x_4$ we have

$$-\frac{x_1}{x_4}A - \frac{x_2}{x_4}B - \frac{x_3}{x_4}C = D.$$

Consequently $A, B, C$ span $V_3(R)$.

To finish the proof of 8.8 we must show that if $A, B, C$ are linearly dependent, then these vectors do not span $V_3(R)$. So suppose

$$x_1A + x_2B + x_3C = 0$$

and at least one of the numbers $x_1, x_2, x_3$ is not zero. Assume $x_3 \neq 0$. Then

(6)

$$C = -\frac{x_1}{x_3}A - \frac{x_2}{x_3}B.$$

We saw in Section 8.7 that if $A$ and $B$ were any two vectors in $V_3(R)$ there exists a nonzero vector $N$ orthogonal to $A$ and $B$. Indeed, if $A$ and $B$ are linearly independent, we may take $N = A \times B$. But $(N, A) = (N, B) = 0$ implies by (6) that $(N, C) = 0$. This vector $N$ is not in the span of $A, B, C$ for if $N = xA + yB + zC$, then

$$(N, N) = x(A, N) + y(B, N) + z(C, N) = 0.$$

This is impossible since $N \neq 0$. Thus we have proved that $A, B, C$ do not span $V_3(R)$.

We now turn to the proof of Theorem 8.9. First let us prove that if $(A, B \times C) \neq 0$, then $A, B, C$ are linearly independent. Suppose therefore

(7)

$$aA + bB + cC = 0.$$

We must show $a = b = c = 0$. Taking the cross product of (7) with $C$ we have since $C \times C = 0$

(8)

$$aA \times C + bB \times C = 0.$$

Taking the inner product of (8) with $A$, we obtain

$$a(A, A \times C) + b(A, B \times C) = 0.$$

Since $(A, A \times C) = 0$ and $(A, B \times C) \neq 0$, this implies that $b = 0$. In a similar fashion we deduce that $a = c = 0$ (Exercise 10).

To complete the proof of 8.9 we must show that if $(A \times B, C) = 0$, then $A, B, C$ are linearly dependent. If $A \times B = 0$, then as proved in Section 8.8, $A$ and $B$ are linearly dependent. Hence $A, B, C$ certainly are linearly dependent. If $A \times B \neq 0$, then $A \times B$ is a normal vector to the plane through the origin spanned by $A$ and $B$. Moreover, we showed in Section 8.7 that any vector $C$ perpendicular to this normal vector must lie in that plane. This, however, is just the statement that $A, B, C$ are linearly dependent, and the proof of Theorem 8.9 is complete.

If $A, B, C$ are linearly independent in $V_3(R)$ and $D$ is a fourth vector, then using the scalar triple product we may determine scalars $x_1, x_2, x_3$ so that

(9)
$$x_1 A + x_2 B + x_3 C = D.$$

If we write this vector equation as three scalar equations,

(10)
$$a_1 x_1 + b_1 x_2 + c_1 x_3 = d_1$$
$$a_2 x_1 + b_2 x_2 + c_2 x_3 = d_2$$
$$a_3 x_1 + b_3 x_2 + c_3 x_3 = d_3$$

then this amounts to exhibiting the solution to this triple of linear equations. Indeed, taking the vector product of (9) by $B$ and then taking the inner product of the resulting equation by $C$, we obtain first

$$x_1 A \times B + x_3 C \times B = D \times B$$

since $B \times B = 0$. Hence

$$x_1 (A \times B, C) = (D \times B, C)$$

or

$$x_1 = \frac{(D \times B, C)}{(A \times B, C)}.$$

In a similar fashion we may obtain

$$x_2 = \frac{(A \times D, C)}{(A \times B, C)}$$

and

$$x_3 = \frac{(A \times B, D)}{(A \times B, C)}.$$

Using the determinant notation for the scalar triple product, we have

$$x_1 = \frac{\begin{vmatrix} d_1 & d_2 & d_3 \\ b_1 & b_2 & b_3 \\ c_1 & c_2 & c_3 \end{vmatrix}}{\begin{vmatrix} a_1 & a_2 & a_3 \\ b_1 & b_2 & b_3 \\ c_1 & c_2 & c_3 \end{vmatrix}} \qquad x_2 = \frac{\begin{vmatrix} a_1 & a_2 & a_3 \\ d_1 & d_2 & d_3 \\ c_1 & c_2 & c_3 \end{vmatrix}}{\begin{vmatrix} a_1 & a_2 & a_3 \\ b_1 & b_2 & b_3 \\ c_1 & c_2 & c_3 \end{vmatrix}} \qquad x_3 = \frac{\begin{vmatrix} a_1 & a_2 & a_3 \\ b_1 & b_2 & b_3 \\ d_1 & d_2 & d_3 \end{vmatrix}}{\begin{vmatrix} a_1 & a_2 & a_3 \\ b_1 & b_2 & b_3 \\ c_1 & c_2 & c_3 \end{vmatrix}}$$

The determination of the scalars $x_1, x_2, x_3$ by the above formulas is called *Cramer's rule* for solving the linear equations (9). Note that each of the numerators in the above fractions are obtained by replacing the first, second, and third row respectively in

$$(A \times B, C) = \begin{vmatrix} a_1 & a_2 & a_3 \\ b_1 & b_2 & b_3 \\ c_1 & c_2 & c_3 \end{vmatrix}$$

by the components of the vector $(d_1, d_2, d_3)$.

***Example 7.*** Show that the vectors $A = (1, -1, 1), B = (0, 4, 3)$, and $C = (-1, 1, 0)$ are linearly independent. If $D = (1, 1, 1)$ determine $x_1, x_2, x_3$ so that $x_1 A + x_2 B + x_3 C = D$. To show that the vectors are linearly independent compute

$$(A \times B, C) = \begin{vmatrix} 1 & -1 & 1 \\ 0 & 4 & 3 \\ -1 & 1 & 0 \end{vmatrix} = 4 \neq 0.$$

By Cramer's rule

$$x_1 = \frac{\begin{vmatrix} 1 & 1 & 1 \\ 0 & 4 & 3 \\ -1 & 1 & 0 \end{vmatrix}}{4} = -\frac{2}{4} = -\frac{1}{2}$$

$$x_2 = \frac{\begin{vmatrix} 1 & -1 & 1 \\ 1 & 1 & 1 \\ -1 & 1 & 0 \end{vmatrix}}{4} = \frac{2}{4} = \frac{1}{2}$$

$$x_3 = \frac{\begin{vmatrix} 1 & -1 & 1 \\ 0 & 4 & 3 \\ 1 & 1 & 1 \end{vmatrix}}{4} = -\frac{6}{4} = -\frac{3}{2}.$$

Therefore $D = -\dfrac{A}{2} + \dfrac{B}{2} - \dfrac{3C}{2}.$

**EXERCISES**

1. Determine which of the following sets of vectors are linearly independent.
   (a) $A = (1, 1, 0), B = (0, 1, 1), C = (1, 0, 1)$
   (b) $A = (1, 1, -1), B = (0, 2, 3), C = (2, 0, -5)$
   (c) $A = (1, -1, 3), B = (2, 2, 1), C = (1, -2, 0)$
   (d) $A = (1, 1, 0), B = (0, 2, 3), C = (1, -1, -3)$
2. For what values of $t$, if any, are the following sets of vectors linearly dependent?
   (a) $A = (-1, 1, t), B = (2, -1, 3), C = (1, 1, -1)$
   (b) $A = (1, 2t, t^2), B = (-1, 2, 3), C = (0, -1, 1)$
   (c) $A = (1, -t, t), B = (t, -1, 1), C = (1, 1, -1)$
3. For the following choices of $A, B, C, D$ determine scalars $x_1, x_2, x_3$ by Cramer's rule so that $x_1 A + x_2 B + x_3 C = D$ or explain why this rule is not applicable.
   (a) $A = (-1, 1, 4), B = (2, 0, -3), C = (1, 1, 1), D = (-1, 1, 0)$
   (b) $A = (1, 4, 3), B = (-1, 1, 2), C = (3, 7, 4), D = (1, 0, -1)$
   (c) $A = (-1, 0, 1), B = (2, 3, 5), C = (2, -1, 0), D = (1, 1, 1)$
   (d) $A = (1, 2, 3), B = (-1, 4, 2), C = (-1, 2, 1), D = (0, 0, 1).$
4. For the following choices of $A, B, C$, determine scalars $x_1, x_2, x_3$ by Cramer's rule so that

$$x_1 A + x_2 B + x_3 A \times B = C$$

   or explain why this rule is not applicable.
   (a) $A = (1, 4, 3), B = (-1, 1, 2), C = (1, 0, -1)$
   (b) $A = (-1, 1, 1), B = (2, -2, -2), C = (-1, 0, 1)$
   (c) $A = (4, 3, 2), B = (-1, 1, 1), C = (0, 0, 1)$
   (d) $A = (2, -1, 1), B = (6, -3, 3), C = (4, -2, 2)$
5. If $A, B, C$ are nonzero vectors that are orthogonal, prove that they span $V_3(R)$.
6. If the zero vector in $V_n(R)$ can be written in more than one way as a linear

combination of vectors $A_1, \ldots, A_k$ prove that the vectors $A_1, \ldots, A_k$ are linearly dependent.

7. If one vector $B$ in $V_n(R)$ can be written in more than one way as a linear combination of vectors $A_1, \ldots, A_k$, prove that $A_1, \ldots, A_k$ are linearly dependent.

8. Let $A_1, \ldots, A_k$ be vectors in $V_n(R)$. If no vector in this set is in the span of the remaining vectors, show that the vectors $A_1, \ldots, A_k$ are linearly independent.

9. Let $A, B, C \in V_3(R)$. If each vector $D$ in the span of $A, B, C$ can be written in precisely one way as a linear combination of $A, B, C$, prove that $A, B, C$ span $V_3(R)$.

10. Prove that any set of $k$ vectors in $V_3(R)$ is linearly dependent if $k > 3$.

11. Finish the proof of Theorem 8.8 by showing that if $(A, B \times C) \neq 0$, and $aA + bB + cC = 0$, then $a = b = c = 0$.

12. Derive the formulas for $x_2$ and $x_3$ in Cramer's rule if

$$x_1 A + x_2 B + x_3 C = D.$$

13. Show that

$$\begin{vmatrix} a_1 & a_2 & a_3 \\ b_1 & b_2 & b_3 \\ c_1 & c_2 & c_3 \end{vmatrix} = \begin{vmatrix} a_1 & b_1 & c_1 \\ a_2 & b_2 & c_2 \\ a_3 & b_3 & c_3 \end{vmatrix}.$$

Hence, Cramer's rule may be written

$$x_1 = \frac{(D \times B, C)}{(A \times B, C)} = \frac{\begin{vmatrix} d_1 & b_1 & c_1 \\ d_2 & b_2 & c_2 \\ d_3 & b_3 & c_3 \end{vmatrix}}{\begin{vmatrix} a_1 & b_1 & c_1 \\ a_2 & b_2 & c_2 \\ a_3 & b_3 & c_3 \end{vmatrix}}.$$

What are the analagous formulas for $x_2$, $x_3$?

The following exercises sketch the analogue of Theorem 8.8 for $V_n(R)$. Namely that $A_1, \ldots, A_n$ span $V_n(R)$ if and only if they are linearly independent. The proof follows the case $n = 3$.

14. Prove Theorem 8.5. That is, prove that for each positive integer $n$ the $n$ equations in $n + 1$ unknowns

$$a_{1,1}x_1 + \cdots + a_{1,n+1}x_{n+1} = 0$$
$$a_{2,1}x_1 + \cdots + a_{2,n+1}x_{n+1} = 0$$
$$\vdots \qquad \vdots \qquad \vdots$$
$$a_{n,1}x_1 + \cdots + a_{n,n+1}x_{n+1} = 0$$

always have a solution $x_1, \ldots, x_{n+1}$ not identically zero. (Use induction and Theorem 8.4.)

15. Prove that $n + 1$ vectors in $V_n(R)$ are linearly dependent.

16. Prove that if $A_1, \ldots, A_n$ are linearly independent vectors in $V_n(R)$, then these vectors span $V_n(R)$.

17. Prove that if $A_1, \ldots, A_n$ are linearly dependent, there is a nonzero vector $B \in V_n(R)$ orthogonal to $A_1, \ldots, A_n$.

18. Prove that if $B \neq 0$ and is orthogonal to $A_1, \ldots, A_n$, then $B$ is not in the span of $A_1, \ldots, A_n$.

19. Prove that if $A_1, \ldots, A_n$ are linearly dependent, they do not span $V_n(R)$.

## CHAPTER NINE

# CURVES IN THE PLANE AND SPACE

### 9.1 Vector-Valued Functions

In Chapter 2 we applied the ideas of the calculus to the study of motion along a line. If $f(t)$ denotes the position of a particle on the line at time $t$, then the scalar function $f'(t)$ represents the velocity and $f''(t)$ the acceleration. In this chapter we wish to generalize this problem and consider the question of describing the motion of a particle that is moving arbitrarily either in the plane or in space.

There are several problems that arise. First, we cannot describe even the position of this particle with a scalar valued function. If at each point $t$, the position of the particle has coordinates $x_1(t), x_2(t), x_3(t)$, then instead of a scalar valued function we need a position function whose values are vectors $(x_1(t), x_2(t), x_3(t))$. Next, the velocity and acceleration of the particle will also be vectors, and if we consider these quantities as functions of time we have two more "vector-valued functions" to be discussed. To describe these functions we shall exploit heavily the vector techniques developed in the previous chapter.

Thus, to describe the motion of a particle in the plane or in space we shall be concerned with functions mapping intervals on the real line into $V_2(R)$ or $V_3(R)$, and we call such a function a *vector-valued* function. We have met one example of such a function already, the line through a vector $A$ and parallel to a vector $B$. This function is defined by the formula

$$X(t) = A + tB.$$

If an arbitrary vector-valued function $X$ is written in component notation

$$X(t) = (x_1(t), x_2(t), x_3(t)),$$

then the three real-valued functions $x_1$, $x_2$, $x_3$ are called the *component functions* of the vector function $X$. Just as for real-valued functions the symbol $X(t)$ properly denotes the value of the vector function $X$ at the point $t$. However, we often abuse this terminology and speak of the vector-valued function $X(t)$.

We may define the concepts of continuity and differentiability for vector-valued functions in the same way as for scalar-valued functions.

*Definition.* $\lim_{t \to t_0} X(t) = L$ *if for each* $\epsilon > 0$ *there is a corresponding* $\delta > 0$ *such that whenever* $0 < |t - t_0| < \delta$ *it follows that* $|X(t) - L| < \epsilon$. *The quantity* $|X(t) - L|$ *is of course, the length of the vector* $X(t) - L$.

*Definition.* *The vector-valued function* $X$ *is* continuous *at the point* $t_0$ *if*

$$\lim_{t \to t_0} X(t) = X(t_0).$$

If $L = (l_1, l_2, l_3)$ and $X(t) = (x_1(t), x_2(t), x_3(t))$, then it is easy (see Exercise 10) to verify that $\lim_{t \to t_0} X(t) = L$ if and only if

$$\lim_{t \to t_0} x_i(t) = l_i \qquad i = 1, 2, 3.$$

**Definition.**  *If $X$ is a vector-valued function, then $X$ is differentiable at $t$ if*

$$\lim_{h \to 0} \frac{1}{h} [X(t+h) - X(t)] = X'(t)$$

*exists.*

Just as for real-valued functions the above limit is called the derivative of $X$ at $t$, and $X$ is differentiable at $t$ if and only if the component functions are differentiable at $t$. Indeed

$$X'(t) = (x_1'(t), x_2'(t), x_3'(t)).$$

When we studied curves in the plane as the graphs of real-valued functions, certain difficulties were encountered. One was that many curves were not the graphs of functions, for example, circles, ellipses, etc. To describe these curves we were forced to use more than one function to describe the graph. Even to discuss the unit circle we had to consider the two functions $f_1(x) = \sqrt{1 - x^2}$ and $f_2(x) = -\sqrt{1 - x^2}$. These two functions had the added disadvantage that they were not differentiable at $x = \pm 1$. Hence, we could not use them to discuss the vertical tangents to the circle at these points.

These and other difficulties may be avoided if we change our point of view with respect to curves. Instead of thinking of a curve as a set of points in the plane or in space, we shall define a curve as a vector-valued function. The precise definition we shall use is the following.

**Definition.**  *A curve in $V_2(R)$ or $V_3(R)$ is a continuous vector-valued function defined on an interval $I$ of real numbers and having values in $V_2(R)$ or $V_3(R)$. The range of values of the function $X$, that is, the set of those vectors*

$$\{X(t) : t \in I\},$$

*is called the* trace *or* locus *of the curve.*

The motion of a particle in the plane or in space is a very helpful model to have in mind when considering curves. For example, if a horse is running around a circular race track, then the curve that describes this motion is the position function $X(t) = (x_1(t), x_2(t))$ as a function of time. If the race-track is centered at the origin, then the trace of the curve satisfies the equation $x_1^2 + x_2^2 = r^2$. In many ways it is quite natural to take the position function of the particle to be the curve rather than the set of points that is its trace. For example, if the horse runs once around the track, twice around the track, or once around in an opposite direction, these motions are all different. Hence, we need different curves, that is, different vector-valued functions to describe each motion.

Curves as we have defined them are often called parametrized curves. The word parameter refers to the independent variable $t$ of the function $X$. The student should not confuse the trace of a curve with its graph. The latter is the set of pairs

$$\{(t, X(t)) : t \in I\}.$$

We shall seldom have occasion to consider the graph of a curve.

If $X$ is a curve, then for each $t \in I$, the vector $X(t)$ is called the radius or position vector to the curve at the point $t$. We always represent this vector as the directed line segment from the origin of our coordinate system to the point with coordinates $(x_1(t), x_2(t), x_3(t))$. This is illustrated in Figure 1.

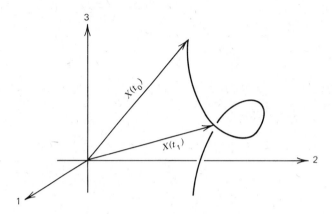

**Fig. 1** Vector-valued function = curve in space.

Just as for real-valued functions the derivative $X'(t_0)$ enables us to define the direction of the curve at the point $X(t_0)$. To show this, let $t_1 < t_0 < t_2$, and consider the vectors $X(t_2) - X(t_0)$ and $X(t_0) - X(t_1)$, which we assume are different from zero.

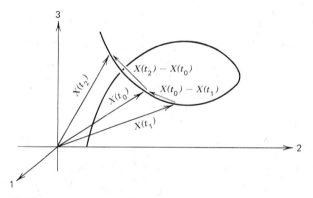

**Fig. 2**

Referring to Figure 2 the directions of these two vectors are given by the unit vectors

$$\frac{X(t_2) - X(t_0)}{|X(t_2) - X(t_0)|} \quad \text{and} \quad \frac{X(t_0) - X(t_1)}{|X(t_0) - X(t_1)|}$$

respectively. If

$$(1) \qquad \lim_{t_2 \to t_0^+} \frac{X(t_2) - X(t_0)}{|X(t_2) - X(t_0)|} \quad \text{and} \quad \lim_{t_1 \to t_0^-} \frac{X(t_0) - X(t_1)}{|X(t_0) - X(t_1)|}$$

both exist and are equal, then the two-sided limit defines a unit vector, called the *unit tangent* to the curve at $X(t_0)$, which we write $T(t_0)$. We define the direction of the curve at $X(t_0)$ to be this vector $T(t_0)$. Whenever the limit fails to exist we say the curve has no direction at $t_0$. We represent the unit tangent as a directed line segment emanating from the point with coordinates $(x_1(t_0), x_2(t_0), x_3(t_0))$. Thus to indicate the direction of the curve at the point $(x_1(t_0), x_2(t_0), x_3(t_0))$ we attach the unit tangent vector to the trace of the curve at this point. See Figure 3. The *tangent line* to the curve at the point $X(t_0)$ is

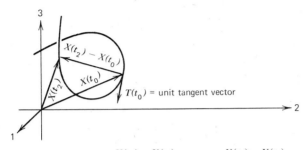

**Fig. 3**   $T(t_0) = \lim\limits_{t_2 \to t_0} + \dfrac{X(t_2) - X(t_0)}{|X(t_2) - X(t_0)|} = \lim\limits_{t_1 \to t_0} - \dfrac{X(t_0) - X(t_1)}{|X(t_0) - X(t_1)|}.$

then defined to be the line through the point with coordinates $(x_1(t_0), x_2(t_0), x_3(t_0))$, which is parallel to the vector $T(t_0)$. This line has the vector equation

$$Y(t) = X(t_0) + tT(t_0).$$

We may use the difference quotient $(X(t+h) - X(t))/h$ to express the limit relations (1) very concisely. Indeed if $t_2 > t_0$ and $t_2 = t_0 + h$, then

$$\frac{X(t_2) - X(t_0)}{|X(t_2) - X(t_0)|} = \frac{1/h}{1/|h|} \cdot \frac{[X(t_0 + h) - X(t_0)]}{|X(t_0 + h) - X(t_0)|} = \frac{1/h[X(t_0 + h) - X(t_0)]}{|1/h[X(t_0 + h) - X(t_0)]|}$$

If $t_1 < t_0$ and we set $t_1 = t_0 + h$, then $h < 0$. Consequently,

$$\frac{X(t_0) - X(t_1)}{|X(t_0) - X(t_1)|} = \frac{-1/h}{1/|h|} \frac{[X(t_0) - X(t_0 + h)]}{|X(t_0) - X(t_0 + h)|}$$

$$= \frac{1/h[X(t_0 + h) - X(t_0)]}{|1/h[X(t_0 + h) - X(t_0)]|}$$

Thus the unit tangent $T(t_0)$ to the curve $X$ at the point $X(t_0)$ is defined by the formula

$$T(t_0) = \lim_{h \to 0} \frac{(1/h)[X(t_0 + h) - X(t_0)]}{|(1/h)[X(t_0 + h) - X(t_0)]|}.$$

if this limit exists.

It is an easy exercise to prove that if $X$ is differentiable at the point $t_0$, that is, if

$$X'(t_0) = \lim_{h \to 0} \frac{X(t_0 + h) - X(t_0)}{h}$$

exists, and if in addition $X'(t_0) \neq 0$, then

$$T(t_0) = \frac{X'(t_0)}{|X'(t_0)|}.$$

Thus in this case the tangent line to the curve at the point $X(t_0)$ may be written

$$Y(t) = X(t_0) + tX'(t_0).$$

For emphasis we state these facts as a theorem.

**THEOREM 9.1**  *If the vector-valued function $X$ is differentiable at the point $t_0$, and if $X'(t_0) \neq 0$, then the unit tangent $T(t_0)$ to the curve at $X(t_0)$ exists. Indeed*

$$T(t_0) = \frac{X'(t_0)}{|X'(t_0)|}.$$

The tangent line to the curve $X$ at the point $X(t_0)$ has equation

$$Y(t) = X(t_0) + tX'(t_0).$$

The proof of Theorem 9.1 is easy and is left as Exercise 9.

If $X'(t_0) = 0$ or even if $X'(t_0)$ fails to exist, the curve may still have a direction $T(t_0)$ at the point $t_0$. Illustrations of what may happen in these cases are contained in Examples 6 and 7 and Exercise 15. However, we wish to avoid this pathology, so in the future we shall only consider those curves for which $X'(t)$ exists and is different from zero at each point $t$. Such curves we call *regular curves*.

If we interpret $X(t)$ as the position of a particle moving in space as a function of time, then the vector $X'(t)$ is called the *velocity vector*, and the second derivative $X''(t)$ is called the *acceleration vector*. This is reasonable since $X'(t)$ measures time rate of change of position, which is what we mean by velocity. Similarly $X''(t)$ measures time rate of change of velocity. The length of the velocity vector, $|X'(t)|$, is called the *speed* of the particle. We shall use these terms, velocity, speed, and acceleration to describe $X'(t)$, $|X'(t)|$, and $X''(t)$ for any curve or vector-valued function $X$.

We have noted earlier that when the radius vector $X(t)$ is represented by a directed line segment, this line segment is chosen with initial point at the origin. The terminal point then has coordinates $(x_1(t), x_2(t), x_3(t))$ and lies on the trace of the curve. However, we represent the vectors $X'(t)$, $X''(t)$ differently. The directed line segments representing these vectors are chosen to have the point with coordinates $(x_1(t), x_2(t), x_3(t))$ as their initial point. Thus we may say that $X'(t)$ and $X''(t)$ or, more correctly, their geometric representations, are attached to the curve $X$ at the point with coordinates $X(t)$. See Figure 4.

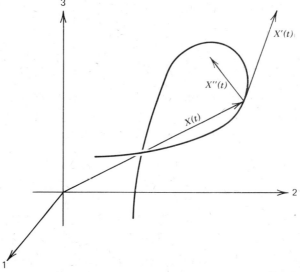

**Fig. 4**   $X'(t), X''(t)$ are attached to the curve at the terminal point of the vector $X(t)$.

***Example 1.*** Let $X(t) = (\cos t, \sin t)$. This vector-valued function describes a curve in the $(x_1, x_2)$ plane. The trace is, of course, a circle. The curve $X$ is differentiable for every value of $t$. Indeed

$$X'(t) = (-\sin t, \cos t).$$

Since for all values of $t$, $|X'(t)| = 1$, the unit tangent $T(t) = X'(t)$. The acceleration $X''(t) = (-\cos t, -\sin t) = -X(t)$. Thus $X''(t)$ is always directed toward the origin. The tangent line at $X(\pi/4)$ is given by

$$Y(t) = \frac{1}{2}(\sqrt{2}, \sqrt{2}) + t(-\sqrt{2}, \sqrt{2}).$$

Also the vertical tangent line to the curve at $X(0)$ has equation

$$Y(t) = (1, 0) + t(0, 1).$$

See Figure 5.

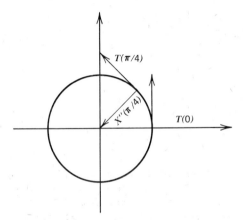

**Fig. 5**  $X(t) = (\cos t, \sin t)$.

Note that the vectors $X(t)$, $X'(t)$ are always perpendicular since

$$(X(t), X'(t)) = -\cos t \sin t + \cos t \sin t = 0.$$

***Example 2.*** The trace of the curve

$$X(t) = (a \cos t, b \sin t),$$

is an ellipse. To see this, set $X(t) = (x_1, x_2)$. Then

$$x_1 = a \cos t \quad \text{and} \quad x_2 = b \sin t,$$

and

$$\frac{x_1^2}{a^2} + \frac{y_1^2}{b^2} = \frac{a^2}{a^2} \cos^2 t + \frac{b^2}{b^2} \sin^2 t = 1.$$

To obtain an equation for the trace of a curve in the plane one must eliminate $t$ between the two equations

$$x_1 = x_1(t) \quad \text{and} \quad x_2 = x_2(t)$$

as was done in the above example. Usually this can be done in only the most favorable cases. For example, to determine the trace of the curve $X(t) = (t+1, t^2-2)$, set $x_1 = t+1$ and $x_2 = t^2-2$. From the first equation we have $t = x_1 - 1$.

Hence $x_2 = (x_1 - 1)^2 - 2 = x_1^2 - 2x_1 - 1$. If we use $(x, y)$ in place of $(x_1, x_2)$, the equation for the trace of the curve is

$$y = x^2 - 2x - 1.$$

This is, of course, a parabola.

***Example 3.***   If $y = f(t)$ is a real-valued continuous function defined on an interval, then the graph of the scalar function $f$ defines a curve in the plane. The appropriate vector-valued function is just

$$X(t) = (t, f(t)).$$

The velocity vector $X'(t) = (1, f'(t))$ and the acceleration $X''(t) = (0, f''(t))$. Thus, for such a curve the acceleration is always directed along the second coordinate axis. If $f''(t) < 0$, the trace is concave down. If $f''(t) > 0$, the trace is concave up. See Figure 6.

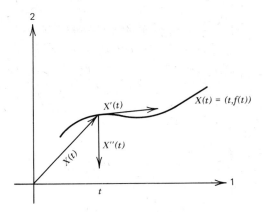

**Fig. 6**

***Example 4.***   Different curves may have the same trace. For example, if the interval $I = [0, 2\pi]$ and $X(t) = (\cos t, \sin t)$, then the trace of this curve is the unit circle. The trace of $Y(t) = (\cos 2t, \sin 2t)$ as well as $Z(t) = (\cos t, -\sin t)$ is also the unit circle, but these three curves are all different. The circle is traversed once by $X$, and twice by $Y$. The curve $Z$ traverses the circle once but in a direction opposite to that of $X$.

***Example 5.***   The space curve $X(t) = (\cos t, \sin t, t)$ is called a *circular helix*. The trace of $X$ for $t > 0$ lies above the unit circle in the 1, 2 plane, and is illustrated in Figure 7.

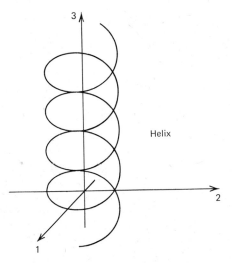

**Fig. 7**

The next example shows that a curve $X$ may have a direction at a point $t_0$ even though $X'(t_0) = 0$.

**Example 6.** If $X(t) = (t^3, t^6)$, then the trace of $X$ is the parabola $x_2 = x_1^2$. (See Figure 8.) $X'(t) = (3t^2, 6t^4)$ and $X'(0) = (0,0)$. For $t = 0$, we cannot determine the direction of the curve from the formula $T(0) = X'(0)/|X'(0)|$. Nevertheless, the curve has a direction at $t = 0$, since

$$T(0) = \lim_{h \to 0} \frac{(1/h)[X(h) - X(0)]}{|(1/h)[X(h) - X(0)]|} = \lim_{h \to 0} \frac{(1/h)(h^3, h^6)}{|(1/h)(h^3, h^6)|} = \lim_{h \to 0} \frac{h^2(1, h^3)}{|h^2(1, h^3)|}$$

$$= \lim_{h \to 0} \frac{(1, h^3)}{|(1, h^3)|} = \lim_{h \to 0} \frac{1}{(1 + h^6)^{1/2}} (1, h^3) = (1, 0).$$

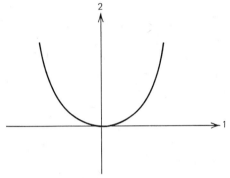

**Fig. 8** Trace of $X(t) = (t^3, t^6)$.

On the other hand, if $X'(t_0) = 0$, the curve may not have a direction at $t_0$. This is illustrated by the next example.

**Example 7.** If $X(t) = (t^2, t^4)$, then the trace of $X$ still satisfies $x_2 = x_1^2$ and $X'(0) = (0,0)$. However this curve has no direction at $t = 0$ since

$$\lim_{h \to 0} \frac{(1/h)[X(h) - X(0)]}{|(1/h)[X(h) - X(0)]|} = \lim_{h \to 0} \frac{(1/h)[(h^2, h^4)]}{|(1/h)(h^2, h^4)|} = \lim_{h \to 0} \frac{h(1, h^2)}{\sqrt{h^2 + h^6}} = \lim_{h \to 0} \frac{h}{|h|} \frac{(1, h^2)}{\sqrt{1 + h^4}}$$

does not exist. In this case the trace of $X$ is that portion of the parabola $x_2 = x_1^2$ lying in the first quadrant. See Figure 9.

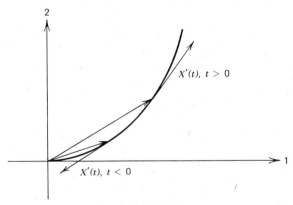

**Fig. 9** $X(t) = X(-t)$, $X'(t) = -X'(-t)$.

Note that for $t \neq 0$, $X(t) = X(-t)$ yet $X'(t) = (2t, 4t^3) = -X'(-t)$. Therefore, at $t = 0$, there appears to be an instantaneous reversal of direction of the curve $X$. The mathematical explanation of this is just that the curve has no direction at $t = 0$.

*Example 8.*   If a curve has a corner or a cusp at some point, then the curve has no direction at this point. For example, let

$$X(t) = (t, |t|), \quad -1 \leqslant t \leqslant 1.$$

It is an easy exercise to check that

$$\frac{(1/h)[X(h) - X(0)]}{1/|h||X(h) - X(0)|}$$

does not have a limit as $h \to 0$. The trace of this curve is Figure 10.

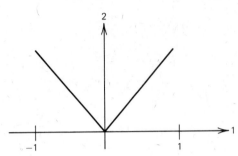

**Fig. 10**   $X(t) = (t, |t|)$ has no direction at $t = 0$.

### EXERCISES

1. Let $X(t) = (4 \cos t, 2 \sin t)$. Determine the vector equation of the tangent line to $X$ at $t = 0$, and $t = \pi/4$. What is the trace of the curve?
2. Let $X(t) = (t - 3, t^2 - 4)$. Determine the vector equation to the tangeat line when $t = 1$. For what values of $t$ are the radius vector and the acceleration orthogonal?
3. Let $X(t) = (t^2, \sin t, \cos t)$. Determine $X'(t)$ and the vector equation of the tangent line to $X$ at $t = \pi/4$.
4. Let $X(t) = (e^t, e^{-t}, t)$. Determine $X'(t)$ and the vector equation of the tangent line at $t = 0$.
5. If $X(t) = (\cos t, \sin t, t)$, determine the cosine of the angle between the radius vector and the tangent vector as a function of $t$.
6. If $X(t) = (2 \cosh t, \sinh t)$, what is the tangent line to the curve when $t = 1$? What is the trace of the curve?
7. Let $X(t) = (t^5, t)$ and $Y(t) = (t, t^{1/5})$. Show that these curves have the same trace. Which curve is differentiable when $t = 0$? Which curve has a direction when $t = 0$?
8. Let $X(t) = (x_1(t), x_2(t), x_3(t))$. Show that if $\lim_{t \to t_0} X(t) = L$, then $\lim_{t \to t_0} |X(t)| = |L|$.
9. Using Exercise 8 show that if $X$ is differentiable at $t_0$ and if $X'(t_0) \neq 0$, then the unit tangent $T(t_0)$ is given by

$$T(t_0) = \frac{X'(t_0)}{|X'(t_0)|}.$$

10. If $X(t) = (x_1(t), x_2(t), x_3(t))$, and $L = (l_1, l_2, l_3)$, prove that $\lim_{t \to t_0} X(t) = L$ if and only if $\lim_{t \to t_0} x_i(t) = l_i$ for $i = 1, 2, 3$.
11. Using Exercise 10 show that $X$ is continuous if and only if the component functions $x_1$, $x_2$, $x_3$ are continuous.
12. If $\lim_{t \to t_0} X(t) = L$ and $\lim_{t \to t_0} Y(t) = M$, show that
    (a) $\lim_{t \to t_0} aX(t) = aL$
    (b) $\lim_{t \to t_0} X(t) + Y(t) = L + M$
    (c) $\lim_{t \to t_0} (X(t), Y(t)) = (L, M)$

13. If a differentiable vector function $X(t)$ has constant length, show that the radius vector $X(t)$ and the velocity $X'(t)$ are always orthogonal.
14. If a particle moves in space with constant speed, what can you say about the direction of the acceleration?
15. Show that the curve $X(t) = (t, t^{1/3})$ is not differentiable at the origin, yet this curve has a direction at $t = 0$. Sketch the trace of $X$. (Since $f(t) = t^{1/3}$ is not differentiable at $t = 0$, the vector function $X$ is not either. However, show that for $h \neq 0$

$$\frac{X(h) - X(0)}{|X(h) - X(0)|} = \frac{1}{(h^2 + h^{2/3})^{1/2}} (h, h^{1/3}).$$

Conclude from this that $T(0) = (0, 1)$. Hence $X$ has a direction for $t = 0$.)

## 9.2   Differentiation Formulas

We saw in the last section that if $X(t) = (x_1(t), x_2(t), x_3(t))$, then the derivative $X'(t)$ of this vector-valued function was the vector-valued function defined by

(1) $$X'(t) = (x_1'(t), x_2'(t), x_3'(t)).$$

Just as for scalar functions, in addition to the notation $X'$ we use $DX$ and $dX/dt$ to denote derivatives. Thus

$$X'(t) = DX(t) = \left(\frac{dX}{dt}\right)(t) = \lim_{h \to 0} \frac{X(t+h) - X(t)}{h}.$$

If $Y(t) = (y_1(t), y_2(t), y_3(t))$, it follows immediately from (1) that $D(X + Y)(t) = DX(t) + DY(t)$ and $D(aX)(t) = aDX(t)$, since these properties hold for the derivatives of scalar-valued functions.

It is important that we have analogues of the product formula for derivatives as well. We have three products to consider. Let $X(t)$, $Y(t)$ be vector functions and $f(t)$ be a scalar function. Then we may form

(2) $$g(t) = (X(t), Y(t)) = (X, Y)(t) \qquad \text{(inner product)}$$

(3) $$Z(t) = f(t)X(t) = (fX)(t) \qquad \text{(scalar multiplication)}$$

(4) $$W(t) = X(t) \times Y(t) = (X \times Y)(t) \qquad \text{(vector product)}.$$

If the functions $X$, $Y$, and $f$ are differentiable, then it may be immediately verified that

(5) $$g'(t) = (X, Y)'(t) = (X(t), Y'(t)) + (X'(t), Y(t))$$

(6) $$Z'(t) = (fX)'(t) = f(t)X'(t) + f'(t)X(t)$$

(7) $$W'(t) = (X \times Y)'(t) = X(t) \times Y'(t) + X'(t) \times Y(t).$$

Using the $d/dt$ notation for derivatives, these formulas may be written

$$\frac{d}{dt}(X, Y) = \left(X, \frac{dY}{dt}\right) + \left(\frac{dX}{dt}, Y\right)$$

$$\frac{d}{dt}(fX) = f\frac{dX}{dt} + \frac{df}{dt}X$$

$$\frac{d}{dt} X \times Y = X \times \frac{dY}{dt} + \frac{dX}{dt} \times Y.$$

Each of these formulas may be verified in a straightforward way from the definition of the inner product, scalar multiplication, and vector product, respectively. These verifications are left to the exercises.

We note one application of (6). For each value of $t$, formula (6) implies that $(fX)'(t)$ is a linear combination of the vectors $X(t)$, $X'(t)$. Thus if the vectors $X(t)$, $X'(t)$ are linearly independent, $(fX)'(t)$ lies in the plane spanned by $X(t)$ and $X'(t)$.

**EXERCISES**

1. Compute $D(X, Y)(t)$, $D(fX)(t)$, $D(fY)(t)$, and $D(X \times Y)(t)$ if
   (a) $X(t) = (\cos t, \sin t, t)$
   $Y(t) = (t, t^2, t^3)$
   $f(t) = e^t$

   (b) $X(t) = \left(\dfrac{1-t}{1+t}, t, \dfrac{1+t}{1-t}\right)$
   $Y(t) = (e^t, e^{-t}, 1)$
   $f(t) = (1+t)^2$

   (c) $X(t) = (\log t, 1, t)$
   $Y(t) = \left(\dfrac{1}{t}, e^{t^2}, 1\right)$

   $f(t) = \sin t$

2. Verify that for differentiable vector functions $X$ and $Y$,
$$D(X + Y)(t) = (DX)(t) + (DY)(t) \qquad \text{and} \qquad D(aX)(t) = a(DX)(t).$$

3. Verify that for differentiable vector functions $X$ and $Y$ and a differentiable scalar function $f$ it follows that
$$(X, Y)'(t) = (X(t), Y'(t)) + (X'(t), Y(t)).$$
$$(fX)'(t) = f(t)X'(t) + f'(t)X(t).$$
$$(X \times Y)'(t) = X(t) \times Y'(t) + X'(t) \times Y(t).$$

4. Show that if for all values of $t$ in the interval $I$, the vectors $X(t)$ and $X'(t)$ are orthogonal, then the vector $X$ has constant length. If the curve lies in $V_3(R)$, determine an equation in $x_1, x_2, x_3$ satisfied by the trace of the curve.

5. The unit tangent vector $T(t)$ to the curve $X(t)$ satisfies
$$T(t) = \frac{1}{|X'(t)|} X'(t) \qquad \text{if} \qquad X'(t) \neq 0.$$

   Show that for each $t$, $T'(t)$ is a linear combination of $X'(t)$ and $X''(t)$.

6. Show that if the unit tangent $T$ to the curve $X$, and its derivative $T'$ do not vanish at the point $t$, then the acceleration $X''(t)$ lies in the plane spanned by $T(t)$ and $T'(t)$.

7. Show that if $X'(t) \neq 0$, then the unit tangent $T(t)$ to the curve at the point $t$ and its derivative $T'(t)$ are orthogonal.

**9.3 The Normal and Osculating Planes for a Curve**

From now on we confine our attention to regular curves $X$. These are differentiable curves for which $X'$ never vanishes. Hence, as we saw in Section 9.1, the unit tangent $T(t)$ is always given by

$$T(t) = \frac{X'(t)}{|X'(t)|}$$

There are two planes passing through a point $X(t)$ on a curve which are useful in studying the geometry of the curve. These are the normal and osculating planes to the curve. The first of these, the normal plane, is the plane through the point with coordinates $(x_1(t), x_2(t), x_3(t))$ that is perpendicular to the tangent vector $T(t)$. We call this plane the *normal plane to the curve at the point t.* (See Figure 11.)

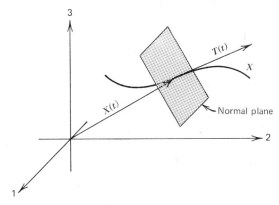

**Fig. 11**   Normal plane to curve is plane through the point with coordinates $X(t)$, which is perpendicular to the vector $T(t)$.

Since the curve is regular, both $T(t)$ and $X'(t)$ are normal vectors to this plane at the point $t$. Scalar equations for the normal plane may readily be determined.

**Example 1.**   Determine the equation of the normal plane to $X(t) = (t^2, t^3, t^4)$ when $t = 1$. Clearly $X'(t) = (2t, 3t^2, 4t^3)$ and $X'(1) = (2, 3, 4)$. Therefore the desired plane is the plane through $(1, 1, 1)$ perpendicular to $(2, 3, 4)$. The scalar equation for this plane is

$$2(x_1 - 1) + 3(x_2 - 1) + 4(x_3 - 1) = 0$$

or

$$2x_1 + 3x_2 + 4x_3 = 9.$$

The *osculating plane* to the curve $X$ at the point $t$ is the plane through the point with coordinates $(x_1(t), x_2(t), x_3(t))$, which is spanned by the unit tangent $T(t)$ and its derivative $T'(t)$ provided that $T'(t) \neq 0$. One remark is in order here. For this definition to make sense we must know that the vectors $T(t)$ and $T'(t)$ are not parallel. (If they are parallel they do not span a plane.) However, the tangent vector $T(t)$ has constant length. Indeed,

$$1 = |T(t)|^2 = (T(t), T(t)).$$

Differentiating this expression with respect to $t$ and applying (5) of Section 9.2, we obtain

$$0 = (T(t), T'(t)) + (T'(t), T(t)) = 2(T(t), T'(t)).$$

Therefore, $(T(t), T'(t)) = 0$ and the vectors $T(t)$ and $T'(t)$ are orthogonal. Hence, if $T'(t) \neq 0$, the vectors $T(t)$ and $T'(t)$ are linearly independent, which is the same thing as saying that they are not parallel.

The result of the above argument is quite useful. Hence we express it as a theorem.

**THEOREM 9.2**  *If the differentiable vector function $X(t)$ has constant length, then $X(t)$ and $X'(t)$ are orthogonal.*

Whenever the derivative $T'$ of the unit tangent $T$ is different from zero, we define

$$N(t) = \frac{T'(t)}{|T'(t)|}.$$

This vector, which is, of course, perpendicular to the unit tangent vector, is called the *principal normal* to the curve. See Figure 12.

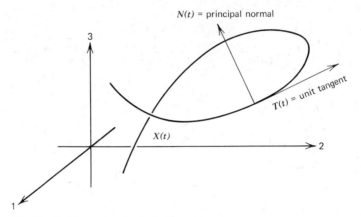

**Fig. 12**   Unit tangent and principal normal.

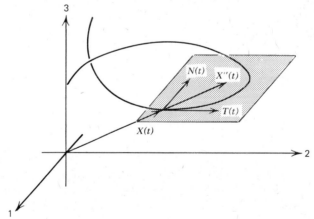

**Fig. 13**   Osculating plane at $t$ is plane through $X(t)$ spanned by the vectors $T(t)$ and $N(t)$.

Thus the osculating plane to the curve at the point $t$ is the plane through $X(t)$, which is spanned by the unit tangent $T(t)$ and the principal normal $N(t)$. See Figure 13.

Let us show next that the acceleration $X''(t)$ is always a linear combination of the unit tangent and principal normal. Hence the vector $X(t) + X''(t)$ always lies in the osculating plane. This result is an immediate consequence of our differentiation formulas. To see this, let $v(t)$ denote the *speed* of the curve. Hence $v(t) = |X'(t)|$, and

$$T(t) = \frac{X'(t)}{v(t)}$$

Therefore

$$X'(t) = v(t)T(t).$$

Differentiating and applying (6) of Section 9.2, we get

$$X''(t) = v'(t)T(t) + v(t)T'(t).$$

But

$$N(t) = \frac{T'(t)}{|T'(t)|} \quad \text{or} \quad T'(t) = |T'(t)|N(t)$$

Hence

(1) $$X''(t) = v'(t)T(t) + v(t)|T'(t)|N(t)$$

Since $v'(t)$ and $v(t)|T'(t)|$ are scalars, this shows that the acceleration $X''(t)$ is a linear combination of the vectors $T(t)$ and $N(t)$ as required.

The scalar functions $v'(t)$ and $v(t)|T'(t)|$ are called the *tangential* and *normal* components of the acceleration, respectively. Since $T(t)$ and $N(t)$ are orthogonal unit vectors, we may derive formulas for the tangential and normal components of $X''(t)$ by computing the inner product of $X''(t)$ with $T(t)$ and $N(t)$, respectively. Indeed

$$(X''(t), T(t)) = v'(t)$$

and

$$(X''(t), N(t)) = v(t)|T'(t)|.$$

**Example 2.**   Determine the tangential and normal components of the acceleration of the curve $X(t) = (t, t^2)$ as functions of time for $t > 0$. Clearly

$$X'(t) = (1, 2t) \quad \text{and} \quad X''(t) = (0, 2),$$

and $T(t) = [1/(1+4t^2)^{1/2}](1, 2t)$. Therefore the tangential component of $X''(t)$ is given by

$$v'(t) = (X''(t), T(t)) = \frac{4t}{(1+4t^2)^{1/2}}.$$

For the normal component it is often easier to compute $v(t)|T'(t)|$ directly rather than to first compute $N(t)$ and then evaluate the inner product $(X''(t), N(t))$. To compute $v(t)|T'(t)|$, differentiate

$$T(t) = \left( \frac{1}{(1+4t^2)^{1/2}}, \frac{2t}{(1+4t^2)^{1/2}} \right)$$

obtaining

$$T'(t) = \left( \frac{-4t}{(1+4t^2)^{3/2}}, \frac{2}{(1+4t^2)^{3/2}} \right).$$

Therefore

$$v(t)T'(t) = \left( \frac{-4t}{(1+4t^2)}, \frac{2}{(1+4t^2)} \right) = \frac{2}{1+4t^2}(-2t, 1)$$

and

$$v(t)|T'(t)| = \frac{2}{(1+4t^2)}(1+4t^2)^{1/2} = \frac{2}{(1+4t^2)^{1/2}}.$$

Another way to compute $v(t)|T'(t)|$ is to first write $T'(t)$ as a linear combination of $X'(t)$ and $X''(t)$. This is accomplished by differentiating $T(t)$. Indeed, since $T(t) = (1/v(t))X'(t)$

$$T'(t) = -\frac{v'(t)}{v^2(t)} X'(t) + \frac{1}{v(t)} X''(t).$$

Hence

$$v(t)T'(t) = -\frac{v'(t)}{v(t)} X'(t) + X''(t).$$

In the example all the terms on the right-hand side are known. Substituting we obtain

$$v(t)T'(t) = -\frac{4t}{(1+4t^2)}(1, 2t) + (0, 2)$$

$$= \left(\frac{-4t}{(1+4t^2)}, \frac{-8t^2}{(1+4t^2)}\right) + (0, 2) = \frac{2}{1+4t^2}(-2t, 1).$$

Hence the normal component

$$v(t)|T'(t)| = \frac{2}{(1+4t^2)^{1/2}}$$

as before.

We know from our discussion of planes in Section 8.7 that the osculating plane at $t$ may also be described as the plane through the point $X(t)$ perpendicular to an appropriate normal vector. Since the plane is spanned by the vectors $T(t)$ and $N(t)$, this normal vector is just $T(t) \times N(t)$. Hence the osculating plane is the plane through $X(t)$ normal to $T(t) \times N(t)$.

Another normal vector to the osculating plane is given by $X'(t) \times X''(t)$. This may be readily seen by computing the cross product of (1) with $X'(t)$. Indeed

$$(2) \qquad X'(t) \times X''(t) = v(t)T(t) \times [v'(t)T(t) + v(t)|T'(t)|N(t)]$$

$$= v^2(t)|T'(t)|T(t) \times N(t),$$

since $T(t) \times T(t) = 0$.

***Example 3.*** Determine a scalar equation for the osculating plane to the helix $X(t) = (\cos t, \sin t, t)$ at $t = \pi/4$. Determine the vectors $T(t)$, $N(t)$, $T(t) \times N(t)$ as functions of time. Now $X'(t) = (-\sin t, \cos t, 1)$, and $X''(t) = (-\cos t, -\sin t, 0)$. Therefore

$$X'(t) \times X''(t) \begin{vmatrix} I_1 & I_2 & I_3 \\ -\sin t & \cos t & 1 \\ -\cos t & -\sin t & 0 \end{vmatrix} = (\sin t, -\cos t, 1).$$

Therefore at $t = \pi/4$ the osculating plane has a normal vector $(\sqrt{2}/2, -\sqrt{2}/2, 1)$. A scalar equation to the osculating plane at $t = \pi/4$ is thus given by

$$\left(x_1 - \frac{1}{\sqrt{2}}\right)\frac{\sqrt{2}}{2} + \left(x_2 - \frac{1}{\sqrt{2}}\right)\left(-\frac{\sqrt{2}}{2}\right) + \left(x_3 - \frac{\pi}{4}\right) = 0.$$

Simplfying we have

$$\frac{x_1}{2} - \frac{x_2}{2} + \frac{x_3}{\sqrt{2}} = \frac{\pi}{4\sqrt{2}}$$

or

$$x_1 - x_2 + \sqrt{2}x_3 = \frac{\pi\sqrt{2}}{4}.$$

To compute $T(t)$, $N(t)$, $T(t) \times N(t)$, observe that

$$T(t) = \frac{X'(t)}{|X'(t)|} = \frac{1}{\sqrt{2}}(-\sin t, \cos t, 1),$$

and

$$T'(t) = \frac{1}{\sqrt{2}}(-\cos t, -\sin t, 0).$$

Hence

$$N(t) = \frac{T'(t)}{|T'(t)|} = (-\cos t, -\sin t, 0).$$

Note that in this example the speed $v(t)$ is constant. Hence the acceleration $X''(t)$ is a scalar multiple of the principal normal. Indeed $X''(t) = N(t)$. Thus

$$T(t) \times N(t) = \frac{1}{\sqrt{2}} X'(t) \times X''(t) = \frac{1}{\sqrt{2}} (\sin t, -\cos t, 1).$$

Since the osculating plane at $t = \pi/4$ is the plane through $(1/\sqrt{2}, 1/\sqrt{2}, \pi/4)$ spanned by $T(\pi/4) = (-\frac{1}{2}, \frac{1}{2}, 1/\sqrt{2})$ and $N(\pi/4) = (-1/\sqrt{2}, -1/\sqrt{2}, 0)$, this plane has vector equation

$$X(t, s) = \left(\frac{1}{\sqrt{2}}, \frac{1}{\sqrt{2}}, \frac{\pi}{4}\right) + t\left(-\frac{1}{2}, \frac{1}{2}, \frac{1}{\sqrt{2}}\right) + s\left(-\frac{1}{\sqrt{2}}, -\frac{1}{\sqrt{2}}, 0\right).$$

The vector $T(t) \times N(t)$ is called the *unit binormal* to the curve at the point $X(t)$ and is written $B(t)$. The three vectors $T(t)$, $N(t)$, and $B(t)$ are mutually perpendicular unit vectors and several vector functions that describe the motion of the curve can be conveniently expressed as linear combinations of these three vectors. For example, it can be shown (see Exercise 16) that the trace of a curve $X(t)$ will lie in a fixed plane in space if and only if the binormal $B(t)$ is a constant vector.

**EXERCISES**

1. Let $X(t) = (t, 2t^2, 3t)$.
   (a) Determine a scalar equation for the normal plane to this curve at $t = 1$.
   (b) Determine two vectors spanning the normal plane when $t = 1$.
   (c) Determine a scalar equation for the normal plane $\pi$ to this curve at the point at which the tangent vector is orthogonal to $(1, 1, 1)$

2. Let $X(t) = (\cos t\pi, \sin t\pi, t^2 - 1)$.
   (a) Determine a scalar equation for the normal plane to this curve when $t = 2$.
   (b) For what values of $t$ does the normal plane pass through the origin?
   (c) Determine the scalar equations for the normal planes for the values of $t$ in part (b)

3. Show that the normal plane to a curve is spanned by the principal normal and binormal to the curve for each value of $t$.

4. Show that if the normal planes to a curve are always parallel, the curve is a straight line.

5. Determine the equations of the normal planes to the "twisted cubic" $X(t) = (t, t^2, t^3)$ at the points where the curve pierces the plane $2x_1 + x_2 - x_3 = 0$.

6. Determine the acceleration vector $X''(t)$ and its normal and tangential components for the following vector functions at the specified values of $t$.
   (a) $X(t) = (t^2, 2t)$, $t = 2$
   (b) $X(t) = (2 \sin t, 2 \cos t)$, $t = 3\pi/4$
   (c) $X(t) = (e^t, e^{-t})$, $t = 0$
   (d) $X(t) = (3 \sin t, 3(1 - \cos t))$ when $t = \pi/2$

7. A particle in the first quadrant moves along the parabola $y^2 = 8x$ such that for all values of $t$ $x'(t) = 1$. If $X(0) = (2, 4)$ determine the vector function $X(t) = (x(t), y(t))$ describing this motion. Determine the tangential component of the acceleration as a function of time.

8. The trace of a curve $X$ in the plane satisfies the equation $4y = x^3$. If $X(0) = (0, 0)$ and for all values of $t$, $y'(t) = 2$, determine $X'(1)$, $X''(1)$, and the normal and tangential components of the acceleration when $t = 1$.

9. The foot of a ladder 30 feet long slides away from a wall at a rate of 6 feet per second. If the top of the ladder rests against the wall, determine a vector function describing the motion of the midpoint of the ladder. What are the normal and tangential components of the acceleration when the foot of the ladder is 20 feet from the wall?

10. Determine the equations of the osculating planes to the following curves at the indicated points.
    (a) $X(t) = (t, t^2, t^3)$, $t = 1$
    (b) $X(t) = (t, 2t, t^2)$, $t = 2$

(c) $X(t) = (t - \sin t, 1 - \cos t, 4 \sin t/2)$, $t = \pi$

(d) $X(t) = (t \cos t, t \sin t, t)$, $t = \pi/2$

11. Determine the unit tangent, principal normal, and binormal to the following curves as functions of time.

(a) $X(t) = (\cos t, \sin t, t)$

(b) $X(t) = (e^t, e^{-t}, t)$

12. Show that if $X(t) \times X'(t)$ is independent of time, then the curve $X$ lies in a plane through the origin. If the curve $X$ lies in a plane through the origin, does it follow that $X(t) \times X'(t)$ is constant?

13. Show that the following curves lie in a plane through the origin and determine the equation of that plane (see Exercise 12).

(a) $X(t) = (\cos t, \sin t, \cos t)$.

(b) $Y(t) = (e^t, e^{-t}, \cosh t)$.

(c) $Z(t) = (e^t, e^{-t}, \sinh t)$.

14. Show that for a regular curve $X$, the velocity and acceleration are linearly independent if and only if $N(t) \neq 0$. (Use equation (2).)

15. Suppose a particle moves in space in such a way that its acceleration is always a scalar multiple of its position. In other words, for each value of $t$, $X''(t) = f(t)X(t)$ where $f(t)$ is a scalar-valued function of $t$. Show that the particle moves in a plane through the origin. Find the equation of this plane if $X(0) = (1, 1, 1)$ and $X'(0) = (1, -1, 1)$. (Differentiate $X(t) \times X'(t)$.)

16. Suppose a regular curve $X$ has the property that $N(t)$ is continuous and never vanishes. Prove that the trace of the curve lies in a plane if and only if the binormal $B(t)$ is a constant vector. (If the trace of the curve $X$ lies in a plane, then $(X(t) - X(t_0), A) = 0$ for some nonzero vector $A$. Show then that $B(t) = \pm A/|A|$. Hence $(B(t), A/|A|) = \pm 1$. Since this inner product must be continuous it follows that for all $t$, $B(t) = A/|A|$ or $B(t) = -A/|A|$. Conversely if $B(t) = B$ is constant, show that for all $t$, $(X(t) - X(t_0), B) = 0$.)

17. If the trace of a curve lies in a fixed plane in space, then the curve is called a *plane curve*. Using Exercise 16, show that the following curves are all plane curves, and determine the equation of the plane containing the curve.

(a) $X(t) = (\sin t + 1, \sin 3t, 2 \sin^3 t)$

(b) $Y(t) = (e^t + 2, 1 - e^{-t}, \cosh t)$

## 9.4  Plane Curves in Polar Coordinates

The description of certain curves in the plane is often simplified if in place of rectangular coordinates $(x_1, x_2)$ for points in the plane we use polar coordinates. Recall that the polar coordinates $(r, \theta)$ of a point with rectangular coordinates $(x_1, x_2)$ are defined by the following formulas.

$$x_1 = r \cos \theta$$

(1)

$$x_2 = r \sin \theta$$

In order for these formulas to define $r$ and $\theta$ uniquely in terms of $x_1$ and $x_2$, we must restrict the values of $r$ and $\theta$. It is customary to take $r \geq 0$ and $0 \leq \theta < 2\pi$. Then $r = \sqrt{x_1^2 + x_2^2}$ is the length of the radius vector from the origin to the point with coordinates $(x_1, x_2)$. The angle $\theta$, called the *polar angle* of the radius vector, is the angle that the radius vector makes with the positive $x_1$ axis. This angle is measured in a counterclockwise direction. See Figure 14.

Now, however, if we consider $x_1$ and $x_2$ as being defined in terms of $r$ and $\theta$ by equation (1), then we may allow both $r$ and $\theta$ to take on all real values. Thus, if $f$ is a function and we set $r = f(\theta)$, then the pair of numbers $(r, \theta)$

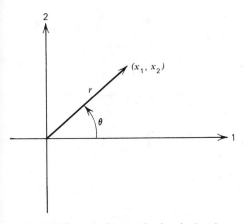

**Fig. 14** Polar coordinates of points in the plane.
$x_1 = r \cos \theta, x_2 = r \sin \theta, 0 \leqslant r, 0 \leqslant \theta < 2\pi.$

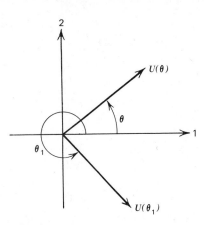

**Fig. 15**

determines a unique point in the plane with coordinates $(x_1, x_2)$ defined by (1). The radius vector $X = (x_1, x_2)$ can now be written

$$X(\theta) = (r \cos \theta, r \sin \theta) = r(\cos \theta, \sin \theta)$$
$$= f(\theta)(\cos \theta, \sin \theta).$$

We denote the vector $(\cos \theta, \sin \theta)$ by $U(\theta)$. This vector is defined for all values of $\theta$. If $0 \leqslant \theta < 2\pi$, then $U(\theta)$ is represented as in Figure 15.

As $\theta$ increases, the vector $U(\theta)$ rotates around the origin in a counterclockwise direction. If $\theta$ decreases, the direction of rotation is reversed. One revolution corresponds to $\theta = 2\pi$.

To represent the vector $X(\theta) = f(\theta)U(\theta)$, we use a directed line segment emanating from the origin of length $|f(\theta)|$. If $f(\theta) > 0$ this vector will be in the direction of $U(\theta)$. If $f(\theta) < 0$, the direction will be reversed. See Figure 16.

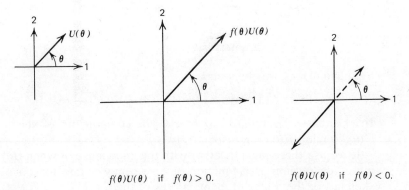

$f(\theta)U(\theta)$ if $f(\theta) > 0.$ $\qquad\qquad$ $f(\theta)U(\theta)$ if $f(\theta) < 0.$

**Fig. 16**

In addition to the vector $U(\theta)$, we shall need to consider the derivative $U'(\theta)$, which we denote by $V(\theta)$. Clearly

$$U'(\theta) = (-\sin \theta, \cos \theta) = V(\theta)$$

The vector $V(\theta)$ is orthogonal to $U(\theta)$ in the direction of positive rotation (counterclockwise) about the origin. This is illustrated in Figure 17.

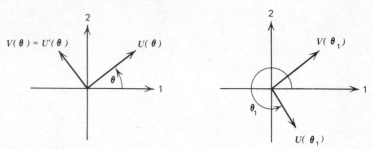

**Fig. 17**   The vectors $U(\theta)$ and $V(\theta)$.

The curves $X$ that we shall consider are those that can be written

(2) $$X(t) = f(\theta)(\cos\theta(t), \sin\theta(t)) = f(\theta)U(\theta(t)).$$

where $f(\theta)$ is some differentiable function of $\theta$ and $\theta = \theta(t)$ is a differentiable function of $t$. For the moment, let us assume that $\theta = t$, so then

(3) $$X(t) = X(\theta) = f(\theta)(\cos\theta, \sin\theta) = f(\theta)U(\theta).$$

Thus for each $\theta$, $X(\theta)$ is a scalar multiple of $U(\theta)$. In this case the velocity vector $X'(t) = X'(\theta)$. Differentiating (2) we obtain

(4)
$$\begin{aligned}
X'(\theta) &= \frac{d}{d\theta}[f(\theta)U(\theta)]\\
&= f'(\theta)U(\theta) + f(\theta)U'(\theta)\\
&= f'(\theta)U(\theta) + f(\theta)V(\theta).
\end{aligned}$$

Equation (4) expresses the velocity vector $X'(\theta)$ as a linear combination of the vectors $U(\theta)$ and $V(\theta)$.

The coefficients $f'(\theta)$ and $f(\theta)$ in (4) are called the *radial and transverse components* respectively of the velocity $X'(\theta)$. If $\theta = \theta(t)$ is an arbitrary differentiable function of $t$, then differentiating (1) by the chain rule, we obtain

$$X'(t) = \frac{d}{dt}(f(\theta(t))U(\theta(t)))$$

$$= f'(\theta(t))\theta'(t)U(\theta(t)) + f(\theta(t))\theta'(t)V(\theta(t)).$$

Setting $\theta = \theta(t)$, this becomes

$$X'(t) = f'(\theta)\theta'(t)U(\theta) + f(\theta)\theta'(t)V(\theta).$$

In this case the radial and transverse components of the velocity are $f'(\theta)\theta'(t)$ and $f(\theta)\theta'(t)$, respectively.

If we use these facts together with the knowledge of when $f(\theta)$ is increasing or decreasing as a function of $\theta$, we may obtain reasonably accurate sketches of the traces of many curves of the form

$$X(\theta) = f(\theta)U(\theta).$$

The equation $r = f(\theta)$ is called the *polar equation* for the trace of $X$. An equation in rectangular coordinates can sometimes be obtained by eliminating $\theta$ between the two equations

$$x_1 = f(\theta)\cos\theta$$

$$x_2 = f(\theta)\sin\theta.$$

However, even when possible, this calculation is often rather tedious. We assume that $\theta = t$ in the following examples.

**Example 1.** If $f(\theta) = r$ is a positive constant, then the trace of $X(\theta) = rU(\theta)$ is a circle centered at the origin with radius $r$. The velocity $X'(\theta) = rU'(\theta) = rV(\theta)$. As usual we attach the velocity vector $X'(\theta)$ to the trace of the curve as in Figure 18.

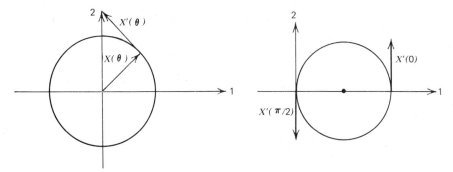

Fig. 18                                   Fig. 19

**Example 2.** Let $X(\theta) = \cos\theta U(\theta)$, then $X'(\theta) = -\sin\theta U(\theta) + \cos\theta V(\theta)$. To sketch the trace of $X$, let us compute $X$ and $X'$ for $\theta = 0, \pi/2, \pi$. Clearly $X(0) = (1, 0) = U(0)$ and $X'(0) = (0, 1) = V(0)$. Furthermore, $X(\pi/2) = (0, 0)$ and $X'(\pi/2) = -1$, $U(\pi/2) = (0, -1)$. Lastly, $X(\pi) = -U(\pi) = U(0)$, and $X'(\pi) = -V(\pi) = V(0)$. The trace of $X$ appears to be a circle centered at the point $(\frac{1}{2}, 0)$. (See Figure 19.) This can be confirmed if we compute an equation in $x_1$ and $x_2$ satisfied by the components of $X(\theta)$. Indeed, $x_1 = \cos^2\theta$ and $x_2 = \cos\theta\sin\theta$. Therefore, $x_2{}^2 = \cos^2\theta\sin^2\theta = \cos^2\theta(1 - \cos^2\theta) = x_1(1 - x_1)$ or

$$x_1{}^2 - x_1 + x_2{}^2 = \left(x_1 - \frac{1}{2}\right)^2 + x_2{}^2 - \frac{1}{4} = 0.$$

Consequently the trace is a circle centered at $(\frac{1}{2}, 0)$ with radius $\frac{1}{2}$.

**Example 3.** Let $X(\theta) = \cos 2\theta U(\theta)$. Differentiating we obtain $X'(\theta) = -2\sin 2\theta U(\theta) + \cos 2\theta V(\theta)$. To sketch the graph let us compute the points $\theta$ for which $X(\theta) = 0$ or $|X(\theta)| = 1$. These correspond to the points $\theta$ at which $\cos 2\theta = 0$ or $|\cos 2\theta| = 1$. Clearly for $0 \le \theta \le 2\pi$, $\cos 2\theta = 0$ if $\theta = \pi/4, 3\pi/4, 5\pi/4, 7\pi/4$. At $\theta = 0, \pi/2, \pi, 3\pi/2, |\cos 2\theta| = 1$. Now at $\theta = 0, X(0) = (1, 0), X'(0) = (0, 1)$. For $\theta = \pi/4, X(\pi/4) = (0, 0)$ and $X'(\pi/4) = -2U(\pi/4) = -\sqrt{2}(1, 1)$. At $\theta = \pi/2, X(\pi/2) = (0, -1)$ and $X'(\pi/2) = V(\pi/2) = (1, 0)$. If we plot these three points, together with the respective tangent vectors and sketch the curve for $0 \le \theta \le \pi/2$, we obtain Figure 20.

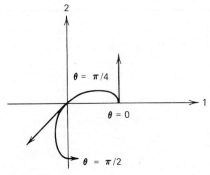

Fig. 20

Note that $\cos 2\theta$ decreases from 1 to 0 as $\theta$ traverses the interval $[0, \pi/4]$, and as $\theta$ traverses the interval $[\pi/4, \pi/2]$, $\cos 2\theta$ goes from 0 to $-1$. If we continue for $\pi/2 \le \theta \le 2\pi$, we obtain Figure 21. The trace of this curve is called a four-leafed rose.

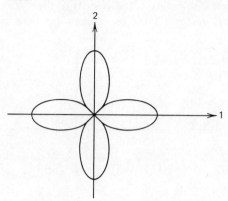

**Fig. 21**    Trace of $X(\theta) = \cos 2\theta U(\theta)$.

**Example 4.**    Let $X(\theta) = ad/(1 - d \cos \theta) U(\theta)$. We assert that the trace of this curve is a conic section with major axis along the $x_1$ axis and having eccentricity equal to $|d|$. The focus of the conic is at the origin. We shall discuss the case $|d| < 1$, leaving the others for the exercises. Let us for simplicity take $a = 1$ and $0 < d < 1$. Then since $-1 \le \cos \theta \le 1$,

$$\frac{d}{1+d} \le \frac{d}{1-d\cos\theta} \le \frac{d}{1-d}.$$

Indeed

$$X(0) = \frac{d}{1-d} U(0) = \frac{d}{1-d}(1,0) \quad \text{and} \quad X(\pi) = \frac{d}{1+d} U(\pi) = -\frac{d}{1+d}(1,0).$$

Also

$$X'(\theta) = \frac{-d^2 \sin \theta}{(1-d\cos\theta)^2} U(\theta) + \frac{d}{1-d\cos\theta} V(\theta).$$

Now $X'(0) = d/(1-d)V(0) = d/(1-d)(0,1)$ and $X'(\pi) = d/(1+d)V(\pi) = -d/(1+d)(0,1)$. Therefore for $\theta = 0, \pi$ the tangents are vertical. Clearly $f(\theta)$ decreases on the interval $[0, \pi]$ and increases on the interval $[\pi, 2\pi]$. A sketch of the trace of $X$ is Figure 22.

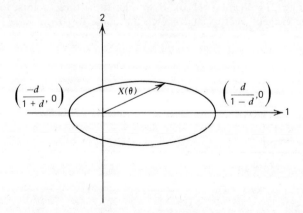

**Fig. 22**

To show that the trace of this curve satisfies the equation of an ellipse by eliminating $\theta$ between the equations

$$x_1 = \frac{d\cos\theta}{1-d\cos\theta}$$

$$x_2 = \frac{d\sin\theta}{1-d\cos\theta}$$

is rather tedious. However, we can show that this trace is an ellipse by verifying the focal-directrix property of a conic section. (See Section 5.9.)

Indeed we assert that one focus of the ellipse is at the origin and the line $x_1 = -1$ is a corresponding directrix. To see this let $(x_1, x_2)$ be a point on the trace and let $X(\theta)$ be the radius vector from the origin to this point. Let $|Y(\theta)|$ be the distance from $(x_1, x_2)$ to the directrix $x_1 = -1$. (See Figure 23.)

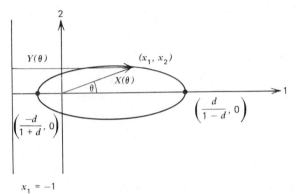

**Fig. 23**   Focal-directrix property of ellipse $\dfrac{|X(\theta)|}{|Y(\theta)|} = d$.

Then

$$|X(\theta)| = \frac{d}{1-d\cos\theta}$$

and

$$|Y(\theta)| = 1 + |X(\theta)|\cos\theta.$$

Hence

$$\frac{|X(\theta)|}{|Y(\theta)|} = \frac{\dfrac{d}{1-d\cos\theta}}{1+\dfrac{d}{1-d\cos\theta}\cos\theta} = d$$

as required.

In general, assuming $a,\ d > 0$, the conic section with equation $X(\theta) = ad/(1-d\cos\theta)U(\theta)$ has one focus at the origin and has the line $x_1 = -a$ as the corresponding directrix. If $X(\theta) = ad/(1+d\cos\theta)U(\theta)$, then the corresponding directrix has equation $x_1 = a$. Similarly, it may be verified that the trace of the curve

$$X(\theta) = a\frac{d}{1\pm d\sin\theta}U(\theta),$$

is also a conic section. Now the major axis is on the $x_2$ axis, and the directrix has equation

$$x_2 = \pm a.$$

**EXERCISES**

1. Sketch the traces of the following curves. Obtain an equation of the form $f(x_1, x_2) = c$ for the trace and identity the trace.

   (a) $X(\theta) = 2\cos\theta U(\theta)$     (c) $X(\theta) = (2, 1) - 4\cos\theta U(\theta)$
   (b) $X(\theta) = 4\sin\theta U(\theta)$     (d) $X(\theta) = (1, -3) + 2\sin\theta U(\theta)$.

2. Sketch the traces of the following curves. Determine $X'(\theta)$ when $X(\theta) = 0$ or $|X(\theta)|$ is a maximum.

   (a) $X(\theta) = \sin 2\theta U(\theta)$     (d) $X(\theta) = (1 + \sin\theta)U(\theta)$
   (b) $X(\theta) = \cos 3\theta U(\theta)$     (e) $X(\theta) = (\cos\theta/2)U(\theta)$
   (c) $X(\theta) = (1 - \cos\theta)U(\theta)$

3. Sketch the traces of the following conic sections. Determine the eccentricity and the equation of the directrix corresponding to the focus at the origin.

   (a) $X(\theta) = \dfrac{2}{2 + 2\cos\theta} U(\theta)$     (d) $X(\theta) = \dfrac{2}{3 - \sin\theta} U(\theta)$

   (b) $X(\theta) = \dfrac{4}{1 - 3\cos\theta} U(\theta)$     (e) $X(\theta) = \dfrac{4}{1 - \sin\theta} U(\theta)$

   (c) $X(\theta) = \dfrac{5}{2 + \cos\theta} U(\theta)$     (f) $X(\theta) = \dfrac{5}{2 + 3\sin\theta} U(\theta)$

4. Sketch the traces of the following curves, computing $X'(\theta)$ at selected points if necessary.

   (a) $X(\theta) = \tan\theta U(\theta)$     (f) $X(\theta) = e^\theta U(\theta)$
   (b) $X(\theta) = \sec\theta U(\theta)$     (g) $X(\theta) = \sqrt{\theta} U(\theta)$     $\theta \geqslant 0$
   (c) $X(\theta) = (2 - \csc\theta)U(\theta)$
   (d) $X(\theta) = (4 + \csc\theta)U(\theta)$     (h) $X(\theta) = \dfrac{1}{\sqrt{\theta}} U(\theta)$
   (e) $X(\theta) = \theta U(\theta)$

## 9.5   Velocity and Acceleration in Polar Coordinates

Suppose now the curve $X(t)$ is of the form $X(t) = f(\theta)U(\theta)$ where $\theta = \theta(t)$. Let $f(\theta(t)) = r(t)$. To determine the radial and transverse components of the velocity and acceleration we differentiate $X(t)$ with respect to $t$ applying the chain rule. Note that

$$\frac{d}{dt} U(\theta) = \frac{d\theta}{dt} U'(\theta) = \theta'(t)V(\theta)$$

and

$$\frac{d}{dt} V(\theta) = \frac{d\theta}{dt} V'(\theta) = -\theta'(t)U(\theta).$$

The scalar function $\theta'(t)$ is called the *angular velocity* of the curve. Now since

$$X(t) = r(t)U(\theta),$$

we have

$$X'(t) = r'(t)U(\theta) + r(t)\theta'(t)U'(\theta)$$

$$= r'(t)U(\theta) + r(t)\theta'(t)V(\theta).$$

Hence the radial and transverse components of the velocity, which we denote by $v_U$ and $v_V$, are $r'(t)$ and $r(t)\theta'(t)$, respectively. Also $|X'(t)|$, which is the speed of the curve, is given by

$$|X'(t)| = (X'(t), X'(t))^{1/2}$$

$$= [[r'(t)]^2 + r^2(t)[\theta'(t)]^2]^{1/2}$$

since $U(\theta)$ and $V(\theta)$ are orthogonal.

Suppressing $t$, this may be conveniently written

$$v = |X'| = \sqrt{r'^2 + r^2\theta'^2}.$$

To compute the acceleration we differentiate $X'(t)$ obtaining

$$X''(t) = r''(t)U(\theta) + r'(t)\theta'(t)V(\theta)$$
$$+ \{r'(t)\theta'(t) + r(t)\theta''(t)\}V(\theta) - r(t)[\theta'(t)]^2 U(\theta)$$
$$= \{r''(t) - r(t)[\theta'(t)]^2\} U(\theta)$$
$$+ \{r(t)\theta''(t) + 2r'(t)\theta'(t)\} V(\theta).$$

Letting $a_U$ and $a_V$ denote the radial and transverse components of acceleration and suppressing $t$ in the above formulas we have

$$a_U = r'' - r \cdot (\theta')^2$$

and

$$a_V = r\theta'' + 2r'\theta'.$$

If $a_V$ is identically equal to zero, the acceleration is said to be radial. The equation for $a_V$ allows us to give this a simple interpretation. Note that

$$r\theta'' + 2r'\theta' = \frac{1}{r}\frac{d}{dt}(r^2\theta').$$

Therefore $a_V(t) = 0$ as a function of $t$ if and only if $d(r^2\theta')/dt = 0$. This is equivalent to the assertion that

$$r^2\theta' = \text{constant}.$$

**Example 1.**  If $X(t) = \sin\theta U(\theta)$, where $\theta(t) = t^2$, determine radial and transverse components of velocity and acceleration as functions of time. What is the speed as a function of time? Now

$$X(t) = \sin t^2 U(\theta).$$

Therefore $X'(t) = 2t\cos t^2 U(\theta) + 2t\sin t^2 V(\theta)$. The radial and transverse components $v_U$ and $v_V$ are $2t\cos t^2$ and $2t\sin t^2$, respectively. The speed $v(t) = |X'(t)| = [(4t^2(\cos^2 t^2 + \sin^2 t^2)]^{1/2} = 2t$. For the acceleration we differentiate again and obtain

$$X''(t) = (2\cos t^2 - 4t^2\sin t^2)U(\theta) + 4t^2\cos t^2 V(\theta)$$
$$+ (2\sin t^2 + 4t^2\cos t^2)V(\theta) - 4t^2\sin t^2 U(\theta)$$
$$= (2\cos t^2 - 8t^2\sin t^2)U(\theta) + (2\sin t^2 + 8t^2\cos t^2)V(\theta).$$

Thus $a_U = 2\cos t^2 - 8t^2\sin t^2$ and $a_V = 2\sin t^2 + 8t^2\cos t^2$.

If the trace of a curve is specified, then often the curve itself may be determined in the presence of additional information.

**Example 2.**  A particle moves on the circle $r = 2\cos\theta$ in a counterclockwise direction with a constant speed of 4 units per second. If the particle was at the origin when $t = 0$, determine a curve describing the motion. $X(t) = 2\cos\theta U(\theta)$, so the curve will be determined if we determine $\theta(t)$. Now using the fact that $v = 4$ we have $16 = v^2 = r'^2(t) + r^2(t)\theta'(t)^2$. But $r(t) = 2\cos\theta(t)$. Therefore

$$r'(t) = -2\sin\theta(t)\theta'(t).$$

Hence

$$16 = v^2 = 4\sin^2\theta\theta'(t)^2 + 4\cos^2\theta\theta'(t)^2 = 4\theta'(t)^2.$$

Therefore, $\theta'(t)^2 = 4$. Since $\theta'(t) > 0$, $\theta'(t) = 2$ and $\theta(t) = 2t + k$. Now $X(0) = 0$. Hence $\cos\theta(0) = \cos k = 0$, and we may take $k = \pi/2$. Therefore the curve $X(t) = 2\cos\theta U(\theta)$, $\theta(t) = 2t + \pi/2$ describes the motion.

## EXERCISES

1. The following curves are of the form $X(t) = f(\theta)U(\theta)$, $\theta = \theta(t)$. Determine $v_U$, $v_V$ and $a_U$, $a_V$ as functions of time if
   (a) $f(\theta) = \sin \theta$ $\qquad$ $\theta(t) = 2t$
   (b) $f(\theta) = 1 - \cos \theta$ $\qquad$ $\theta(t) = t^2$
   (c) $f(\theta) = e^\theta$, $\qquad$ $\theta(t) = ct$
   (d) $f(\theta) = \dfrac{1}{2 - \cos \theta}$ $\qquad$ $\theta(t) = t^2$

2. A particle moves along the circle $r = 8 \sin \theta$ in a clockwise sense with constant speed equal to 9 units per second. Determine a curve describing the motion if at $t = 0$, $X(0) = (0, 8)$.

3. A curve traverses the spiral $r = 2\theta$ (i.e., $r = 2\theta$ is the trace of the curve). If $r'(t) = 3$ and at time $t$ the particle is at the origin, determine the speed and radial and transverse components of the accelerations at $t = 2\pi$.

4. A particle traverses the spiral $r = e^\theta$ in a counterclockwise direction with unit speed. Determine a curve describing the motion if at $t = 0$, $\theta = 0$.

5. A curve traverses the cardioid $r = 2(1 + \sin \theta)$ in a counterclockwise direction. If $\theta'(t) = \pi/4$ determine $v_U$, $v_V$, $a_U$, $a_V$ at $\theta = \pi/2$, $\pi$.

6. The angle between two curves at a point of intersection is defined to be the angle between the respective tangent vectors. If one curve traverses the cardioid $r = 4(1 - \cos \theta)$, and the other the parabola $r = 4/(1 - \cos \theta)$, show that these curves intersect at right angles.

7. A curve traverses the spiral $r = ce^{a\theta}$ where $c$ and $a$ are constants. Show that the angle between the curve and the radius vector is constant. Determine this angle.

8. A particle moves along the spiral $r = e^{-\theta}$. Show that the velocity $X'(t)$ and the acceleration are always perpendicular.

9. A particle moves along a circle centered at the origin. Show that the acceleration is radial if and only if $\theta'(t) = \text{constant}$.

10. If $X(t) = f(\theta(t))U(\theta(t))$ is a curve in the plane, show the acceleration is radial if and only if $\theta'(t) = k/[f^2(\theta(t))]$.

11. Suppose the trace of the curve $X$ in space lies in the $1, 2$ plane and has polar equation $r = f(\theta)$; show that

$$X(t) \times X'(t) = r^2 \theta'(t) U(\theta(t)) \times V(\theta(t)).$$

12. The trace of the space curve $X$ lies in the $1, 2$ plane and has polar equation $r = f(\theta)$. Suppose $X''(t) = (k/r^2(t))U(\theta)$, where $r(t) = f(\theta(t))$. Verify that

$$X''(t) \times \{X(t) \times X'(t)\} = -k\theta'(t)V(\theta(t)) = k\frac{d}{dt}U(\theta(t)).$$

13. If $X(t)$ is a curve in space, show that $X(t) \times X'(t)$ is constant if and only if the acceleration is radial.

## 9.6   Area in Polar Coordinates

If the curve $X(t)$ has equation $X(t) = r(t)U(\theta)$, where $r(t) = f(\theta(t))$, then the area $A$ of the region swept out by the radius vector as $t$ goes through the interval $[a, b]$ may easily be computed. Let us assume $X$ is regular so that $X'$ never vanishes. In particular, $\theta'$ never vanishes. We assume for definiteness that $\theta'(t) > 0$, so the direction of rotation is positive.

As might be expected, the area $A$ of the region swept out by the radius vector is a certain integral. Recall that to show that $A = \int_a^b h(t)\, dt$ and to determine $h$, we must verify two things. First we must show that our area problem is additive on subintervals of $[a, b]$. This just means that if $a = t_0 < t_1 < \cdots < t_n = b$ is a partition of $[a, b]$, then the area swept out over the entire interval is the sum of the areas swept out on each subinterval.

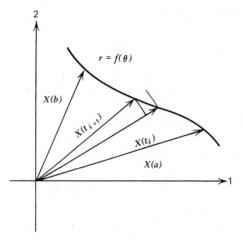

$r = f(\theta)$

$X(b)$

$X(t_{i+1})$

$X(t_i)$

$X(a)$

**Fig. 24**

This is certainly clear. (See Figure 24.)

To determine $h$ we need only to estimate the area $A_i$ swept out on a sub-interval $[t_{i-1}, t_i]$. Recall that the area of a circular sector of radius $r$ and angle $\theta$ is given by $\frac{1}{2}r^2\theta$. Therefore, if $|X(t_i')|$ is the minimum length of the radius vector to the curve on the interval $[t_{i-1}, t_i]$ and $|X(t_i'')|$ is the maximum, we have the following estimate for $A_i$. (See Figure 25.)

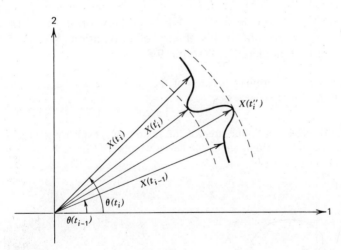

$X(t_i'')$

$X(t_i)$

$X(t_i')$

$X(t_{i-1})$

$\theta(t_i)$

$\theta(t_{i-1})$

**Fig. 25**

$$(1) \qquad \frac{1}{2}\big|X(t_i')\big|^2(\theta(t_i) - \theta(t_{i-1})) \leq A_i \leq \frac{1}{2}\big|X(t_i'')\big|^2(\theta(t_i) - \theta(t_{i-1}))$$

Next we apply the mean value theorem to the function $\theta(t)$ on the interval $[t_{i-1}, t_i]$. This asserts that there exists a point $s_i \in (t_{i-1}, t_i)$ such that

$$\theta(t_i) - \theta(t_{i-1}) = \theta'(s_i)(t_i - t_{i-1}) = \theta'(s_i)\Delta t_i.$$

Substituting this in (1) we obtain

$$\frac{1}{2}\big|X(t_i')\big|\theta'(s_i)\Delta t_i \leq A_i \leq \frac{1}{2}\big|X(t_i'')\big|^2\theta'(s_i)\Delta(t_i)$$

We conclude from Theorem 6.2 that

$$h(t) = \frac{1}{2}|X(t)|^2 \theta'(t)$$

and hence

$$A = \frac{1}{2} \int_a^b |X(t)|^2 \theta'(t)\, dt$$

If $X(t) = r(t)U(\theta)$ where $r(t) = f(\theta(t))$, then $|X(t)|^2 = r^2(t)$, and hence

(2)
$$A = \frac{1}{2} \int_a^b r^2(t)\theta'(t)\, dt.$$

If $\theta = t$ and $r = f(\theta)$, this formula has the simple form

(3)
$$A = \frac{1}{2} \int_a^b r^2\, d\theta = \frac{1}{2} \int_a^b f^2(\theta)\, d\theta.$$

We note one other fact. On p. 377 we noted that the acceleration $X''$ is radial if and only if $d(r^2(t))/dt\theta'(t)) = 0$ or $r^2(t)\theta'(t) = k$, a constant. Therefore, from (2)

$$A = \frac{1}{2} \int_a^b k\, dt = \frac{k}{2}(b-a),$$

and the area swept out by the radius vector is a fixed constant multiplied by the length of the time interval. Thus the acceleration of the curve is radial if and only if the radius vector sweeps out equal areas in equal times. This observation was made first by Sir Isaac Newton in the 17th century and formed a cornerstone in his theory of gravitation. We shall discuss this matter more completely in Section 9.8.

**Example 1.**   Determine the area of the cardioid $r = 1 + \cos\theta$. This figure is the trace of the curve

$$X(\theta) = (1 + \cos\theta)U(\theta)$$

and the desired area is the area of the region swept out by $X(\theta)$ as $\theta$ moves from 0 to $2\pi$. Sketching the trace of the curve we obtain Figure 26.

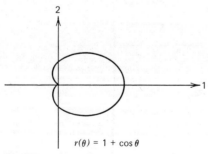

$r(\theta) = 1 + \cos\theta$

**Fig. 26**

By (3) the area of the cardioid is given by

$$A = \frac{1}{2} \int_0^{2\pi} (1 + \cos\theta)^2\, d\theta = \frac{1}{2} \int_0^{2\pi} [1 + 2\cos\theta + \cos^2\theta]\, d\theta$$

$$= \frac{1}{2} \int_0^{2\pi} \left[1 + 2\cos\theta + \frac{1}{2} + \frac{\cos 2\theta}{2}\right] d\theta = \frac{3\pi}{2}.$$

***Example 2.*** Determine the area of the region between the two loops of the limaçon $r = 1 + 2 \cos \theta$. If $X(\theta) = (1 + 2 \cos \theta) U(\theta)$, then the sketch of the trace of $X$ is Figure 27.

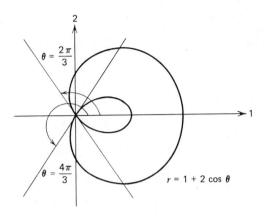

**Fig. 27**

The radius vector sweeps out the smaller loop as $\theta$ moves over the interval $[2\pi/3, 4\pi/3]$. The larger loop is swept out as $\theta$ traverses the intervals $[0, 2\pi/3]$ and $[4\pi/3, 2\pi]$. Therefore, letting $r = 1 + 2 \cos \theta$, we have from (3) that

$$A = \frac{1}{2} \int_0^{2\pi/3} r^2 \, d\theta + \frac{1}{2} \int_{4\pi/3}^{2\pi} r^2 \, d\theta - \frac{1}{2} \int_{2\pi/3}^{4\pi/3} r^2 \, d\theta.$$

Since the trace is symmetric about the first coordinate axis,

$$A = \int_0^{2\pi/3} (1 + 2 \cos \theta)^2 \, d\theta - \int_{2\pi/3}^{\pi} (1 + 2 \cos \theta)^2 \, d\theta$$

$$= \int_0^{2\pi/3} (3 + 4 \cos \theta + 2 \cos 2\theta) \, d\theta - \int_{2\pi/3}^{\pi} (3 + 4 \cos \theta + 2 \cos 2\theta) \, d\theta$$

$$= (3\theta + 4 \sin \theta + \sin 2\theta)\Big|_0^{2\pi/3} - (3\theta + 4 \sin \theta + \sin 2\theta)\Big|_{2\pi/3}^{\pi} = \pi + 3\sqrt{3}.$$

**EXERCISES**

1. Sketch the trace and determine the area of the region enclosed by the trace of the curve $X(\theta) = f(\theta) U(\theta)$ if
   (a) $f(\theta) = 2 \cos \theta$        (c) $f(\theta) = 2 + 2 \sin \theta$
   (b) $f(\theta) = 4 \sin 2\theta$         (d) $f(\theta) = 2 \cos 3\theta$
2. Determine the area of the region between the two loops of the limaçons
   (a) $r = \sqrt{2} - 2 \cos \theta$
   (b) $r = 1 + 2 \sin \theta$
   (c) $r = 2 - 3 \sin \theta$

The computation of areas of regions expressed in polar coordinates often leads to the computation of integrals of the form $\int (p(\sin \theta, \cos \theta)/q(\sin \theta, \cos \theta)) \, d\theta$, where $p$ and $q$ are polynomials in two variables. These integrals can be transformed into integrals of rational functions of $x$ by means of the substitution $x = \tan \theta/2$. Then $dx = \frac{1}{2} \sec^2 (\theta/2) \, d\theta$ or $d\theta = 2/(1 + x^2) \, dx$. Also $\sin \theta = 2x/(1 + x^2)$ and $\cos \theta = (1 - x^2)/(1 + x^2)$. For example, the integral $\int 1(a + b \cos \theta) \, d\theta$ is transformed into $2 \int 1/(a + b + (a - b)x^2) \, dx$ by this substitution. The latter integral may be evaluated by standard techniques.

3. Determine the area of the ellipse with polar equation $r = 1/(2 - \cos \theta)$.
4. Determine the area of the region in the right half plane lying to the left of the parabola $r = 1/(1 + \cos \theta)$.

5. Determine the area of that portion of the ellipse $r = 1/(2 + \cos \theta)$ lying in the left half plane.
6. Determine the area of that region in the upper half plane lying to the left of the parabola $r = 1/(1 + \cos \theta)$ and below the line $X(t) = t(1, 3)$.
7. Determine the area of the region interior to the ellipse $r = 4/(2 + \cos \theta)$ and to the left of the parabola $r = 2/(1 + \cos \theta)$.
8. Find the area of the region interior to the unit circle centered at the origin and interior to the cardioid $r = 1 + \sin \theta$.

## 9.7 Arc Length

Suppose $X(t)$ is a curve in the plane or in space. We shall now consider the problem of defining the length of the curve between two points $t = a$ and $t = b$, and then develop methods for computing this length. The problem of defining the length of a curve is similar to the problem of defining area, which we discussed in Chapter 3. However, the problems are not entirely the same as will be apparent in the following discussion.

If $P$ and $Q$ are two points on the straight line $X(t) = A + tB$, and $X(a) = P$ and $X(b) = Q$, then the length of this curve from $t = a$ to $t = b$, which we write $l(a, b)$, certainly should be defined by the formula

$$l(a, b) = |X(b) - X(a)|.$$

To generalize slightly, suppose $X(t)$ is a "polygonal" curve, that is, a curve that is composed of finitely many line segments. (See Figure 28.) To be precise we assume that there is a partition of $[a, b]$ into subintervals $[t_i, t_{i+1}]$ such that $a = t_0 < t_1 < \cdots < t_n = b$, and on each of the subintervals $X(t)$ is a straight line. In this case the length is defined by

$$l(a, b) = |X(t_1) - X(t_0)| + |X(t_2) - X(t_1)| + \cdots + |X(t_n) - X(t_{n-1})|.$$

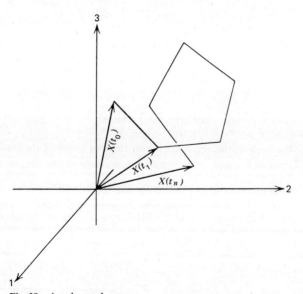

**Fig. 28** A polygonal curve.

For an arbitrary curve $X(t)$, if we partition the interval $[a, b]$ into sub-intervals $A = t_0 < t_1 < \cdots < t_n = b$, then the number $l(a, b)$ representing the length of the curve from $t = a$ to $t = b$ should exceed the length of the polygonal path obtained by connecting the points $X(t_i)$, $X(t_{i+1})$ by straight

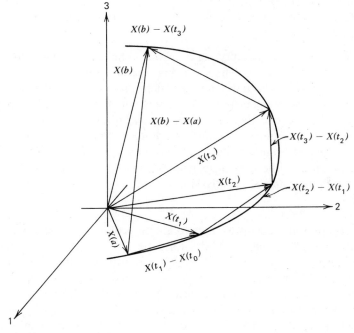

**Fig. 29**

lines as in Figure 29. This just means that

$$(1) \qquad l(a, b) \geqslant |X(t_1) - X(t_0)| + \cdots + |X(t_n) - X(t_{n-1})|.$$

The only other requirement we put on our length $l(a, b)$ is that this number should be the smallest number with property (1). That is $l(a, b)$ should be the smallest number $l$ with the property that

$$l \geqslant |X(t_1) - X(t_0)| + \cdots + |X(t_n) - X(t_{n-1})|,$$

for each choice of the partition $a = t_0 < t_1 < \cdots < t_n = b$. If there is no such number with this property, then we say that the curve does not have a length between $t = a$ and $t = b$.

If there is no length to the curve $X$ between $t = a$ and $t = b$, the problem is that there is no number $l$ at all satisfying

$$l \geqslant |X(t_1) - X(t_0)| + \cdots + |X(t_n) - X(t_{n-1})|,$$

for each choice of the partition $a = t_0 < t_1 < \cdots < t_n = b$. Indeed it is a consequence of the completeness axiom for the real numbers that if there is one number $l$ with the above property, there necessarily is a smallest such number. We formulate this result as a theorem.

**THEOREM 9.3** *Let $S$ be a nonempty set of real numbers with the property that there is a number $t$ satisfying $t \geqslant s$ for each $s \in S$. Then there is a smallest number $t_0$ satisfying $t_0 \geqslant s$ for each $s \in S$.*

**Proof.** Let $T = \{t : t \geqslant s$, for each $s \in S\}$. The set $T$ is nonempty by assumption. Certainly $S$ and $T$ are separated since if $s \in S$ and $t \in T$, then $s \leqslant t$. The completeness axiom (p. 13) implies that there is a real number $t_0$ separating $S$ and $T$. But then, $s \leq t_0 \leq t$ for each $s \in S$ and $t \in T$. Clearly, $t_0$ is the smallest real number satisfying $t_0 \geq s$ for each $s \in S$.

Whenever a nonempty set $S$ of real numbers satisfies the hypothesis of Theorem 9.3, that is, whenever there exists a number $t \geq s$ for each $s \in S$, we say that $S$ is *bounded above*. Such a number $t$ is called an *upper bound* for $S$. If $t_0$ is an upper bound for $S$ that satisfies $t_0 \leq t$ for each upper bound $t$, then $t_0$ is called the *least upper bound* for $S$. Theorem 9.3 asserts that a *nonempty set that is bounded above has a least upper bound*. The student should satisfy himself that if $t$ is a given upper bound for a set $S$, then to show that $t$ is the *least upper bound* for $S$ it suffices to show that for each $\epsilon > 0$ there is an $s \in S$ satisfying

$$s > t - \epsilon.$$

To apply this result to our arc length problem we need only take $S$ to be the set of all numbers of the form $|X(t_1) - X(t_0)| + \cdots + |X(t_n) - X(t_{n-1})|$ where $a = t_0 < t_1 < \cdots < t_n = b$ is a partition of the interval $[a, b]$. When this set $S$ is bounded above, we say the curve $X$ is *rectifiable*. The least upper bound to $S$, which we write $l(a, b)$, is then defined to be the length of $X$ from $t = a$ to $t = b$. Surprisingly enough the continuity of the function $X(t)$ on $[a, b]$ is not sufficient to imply that $X(t)$ is rectifiable. We illustrate that in Example 6. However, if the velocity $X'(t)$ is continuous on the interval $[a, b]$, then the curve $X$ is rectifiable, and indeed

$$l(a, b) = \int_a^b v(t)\, dt$$

where $v(t) = |X'(t)|$ is the speed of the curve. We state this result next.

**THEOREM 9.4**   *Let $X$ be a regular curve with continuous velocity $X'$ on the interval $[a, b]$. Then $X$ is rectifiable, and if $v(t) = |X'(t)|$ is the speed, the arc length*

$$(2) \qquad\qquad l(a, b) = \int_a^b v(t)\, dt.$$

Since the proof of 9.4 is somewhat technical we postpone the argument until after we have discussed some examples. If $X(t) = (x_1(t), x_2(t), x_3(t))$ is a curve in space, then

$$v(t) = |X'(t)| = ([x_1'(t)]^2 + [x_2'(t)]^2 + [x_3'(t)]^2)^{1/2} = \left(\left(\frac{dx_1}{dt}\right)^2 + \left(\frac{dx_2}{dt}\right)^2 + \left(\frac{dx_3}{dt}\right)^2\right)^{1/2}$$

Hence the arc length formula (2) becomes

$$l(a, b) = \int_a^b ([x_1'(t)]^2 + [x_2'(t)]^2 + [x_3'(t)]^2)^{1/2}\, dt$$

$$= \int_a^b \left(\left(\frac{dx_1}{dt}\right)^2 + \left(\frac{dx_2}{dt}\right)^2 + \left(\frac{dx_3}{dt}\right)^2\right)^{1/2} dt.$$

If the curve $X(t)$ lies in the 1, 2 plane then

$$l(a, b) = \int_a^b v(t)\, dt = \int_a^b \{[x_1'(t)]^2 + [x_2'(t)]^2\}^{1/2}\, dt.$$

When the trace of $X(t)$ is the graph of a real-valued function $s = f(t)$, and the curve $X(t) = (t, f(t))$, then $X'(t) = (1, f'(t))$ and the arc length formula takes the form

$$l(a, b) = \int_a^b \sqrt{1 + [f'(t)]^2} \, dt.$$

In polar coordinates if $X(t) = f(\theta(t))U(\theta(t))$, then

$$X'(t) = \theta'(t)[f'(\theta(t))U(\theta(t)) + f(\theta)V(\theta(t))].$$

Hence $v(t) = |\theta'(t)| \sqrt{[f'(\theta(t))]^2 + [f(\theta(t))]^2}$ and

$$l(a, b) = \int_a^b \sqrt{[f'(\theta(t))]^2 + [f(\theta(t))]^2} \, |\theta'(t)| \, dt.$$

When $t = \theta$ and $r = f(\theta)$, this may be abbreviated

$$l(a, b) = \int_a^b \sqrt{r'^2 + r^2} \, d\theta.$$

**Example 1.**   Compute the length of $X(t) = (r \cos t, r \sin t)$ from $t = 0$ to $t = a$. The trace is, of course, the circle of radius $r$ centered at the origin. (See Figure 30.)

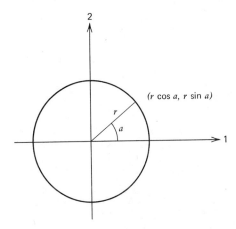

**Fig. 30**

Since $v(t) = r$, $l(0, a) = \int_0^a r \, dt = ra$. We note in passing that this proves that the radian measure $a$ of the angle is the length of arc subtended by the angle divided by the radius. The definition of this radian measure as given in Chapter 4 was twice the area of the section divided by $r^2$.

**Example 2.**   Compute the length of the helix $X(t) = (\cos t, \sin t, t)$ from $t = 0$ to $t = 2\pi$. Clearly $X'(t) = (-\sin t, \cos t, 1)$ and consequently $v(t) = \sqrt{2}$. Therefore

$$l(0, 2\pi) = \int_0^{2\pi} \sqrt{2} \, dt = 2\sqrt{2}\pi.$$

**Example 3.**   Compute the length of the curve $X(t) = (t, t^2)$ from $t = 0$ to $t = 1$. The trace of this curve is a parabola and

$$l(0, 1) = \int_0^1 \sqrt{1 + 4t^2} \, dt$$

since $v(t) = (1 + 4t^2)^{1/2}$. This integral can be evaluated by first making the substitution $2t = u$. Then

$$\int_0^1 \sqrt{1 + 4t^2} \, dt = \frac{1}{2} \int_0^2 \sqrt{1 + u^2} \, du.$$

If we now make the substitution $u = \tan \theta$, we may derive the formula

$$\int \sqrt{1+u^2}\, du = \frac{1}{2}u\sqrt{1+u^2} + \frac{1}{2}\log|u+\sqrt{1+u^2}|$$

Hence

$$\frac{1}{2}\int_0^2 \sqrt{1+u^2}\, du = \frac{1}{4}\Big[u\sqrt{1+u^2} + \log|u+\sqrt{1+u^2}|\Big]_0^2$$

$$= \frac{1}{4}[2\sqrt{5} + \log(2+\sqrt{5})].$$

**Example 4.** Compute the length of one arch of the cycloid $X(t) = (x_1(t), x_2(t))$ defined by

$$x_1(t) = a(t - \sin t)$$
$$x_2(t) = a(1 - \cos t).$$

The cycloid is the curve traced by a fixed point $P$ on the circumference of a circle of radius $a$ as that circle rolls along the first coordinate axis with unit angular velocity. Assuming the point $P$ is at the origin when $t = 0$, the motion is illustrated by Figure 31.

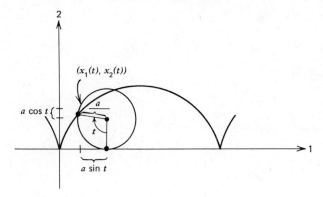

**Fig. 31** The cycloid. $x_1(t) = a(t - \sin t)$; $x_2(t) = a(1 - \cos t)$.

For the cycloid

$$x_1'(t) = a(1 - \cos t) \qquad \text{and} \qquad x_2'(t) = a \sin t.$$

Therefore

$$v(t) = a\sqrt{(1-\cos t)^2 + \sin^2 t} = a\sqrt{2 - 2\cos t} = 2a\left|\sin\frac{t}{2}\right|$$

Since $\sin^2 t/2 = (1 - \cos t)/2$. Therefore the length of one arch is

$$l(0, 2\pi) = \int_0^{2\pi} 2a\left|\sin\frac{t}{2}\right| dt = 2a \int_0^{2\pi} \sin\frac{t}{2}\, dt = -4a\cos\frac{t}{2}\Big|_0^{2\pi} = 8a.$$

**Example 5.** Arc length problems often give rise to integrals that cannot be evaluated by elementary means. For example, if the ellipse $(x^2/a^2) + (y^2/b^2) = 1$ is traced out by the curve $X(t) = (a\cos t, b\sin t)$, then

$$v(t) = \sqrt{a^2 \sin^2 t + b^2 \cos^2 t} = a\sqrt{1 - [(a^2 - b^2)/a^2]\cos^2 t}$$

Letting $k^2 = (a^2 - b^2)/a^2$ and assuming $a^2 \geqslant b^2$, we have

$$l(0, t_0) = a \int_0^{t_0} \sqrt{1 - k^2 \cos^2 t}\, dt.$$

Except for the cases $k = 0, 1$, the above integral cannot be evaluated in terms of elementary functions. These integrals, called elliptic integrals, have been extensively tabulated as functions of $k$ and $t_0$.

***Example 6.*** The following infinite polygonal curve is an example of a nonrectifiable curve.

Consider the points $(1, 0), (1/2, 0), (1/3, 0), \ldots$, on the $x$ axis and the points

$$\left(\frac{1}{2}, \frac{1}{2}\right), \left(\frac{1}{4}, \frac{1}{4}\right), \left(\frac{1}{6}, \frac{1}{6}\right) \cdots$$

on the line $x_2 = x_1$. Let $X_1(t)$ be the curve obtained by connecting the points $(1, 0)$, $(1/2, 1/2), (1/3, 0), (0, 0)$ by straight lines (Figure 32).

Let $X_2(t)$ be the curve defined by connecting $(1, 0), (1/2, 1/2), (1/3, 0), (1/4, 1/4)$, $(1/5, 0)$ by straight lines (Figure 33).

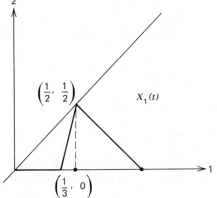

Fig. 32

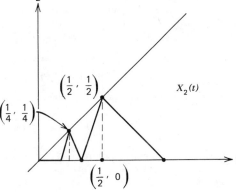

Fig. 33

For arbitrary $n$, $X_n(t)$ is the curve obtained by connecting

$$(1, 0), \left(\frac{1}{2}, \frac{1}{2}\right), \left(\frac{1}{3}, 0\right), \ldots, \left(\frac{1}{2n}, \frac{1}{2n}\right), \left(\frac{1}{2n+1}, 0\right), (0, 0)$$

by straight lines.

We now define $X(t)$ for $0 \leqslant t \leqslant 1$ by

$$X(0) = 0$$

and

$$X(t) = X_n(t) \quad \text{if} \quad \frac{1}{2n-1} \leqslant t \leqslant \frac{1}{2n+1}.$$

(See Figure 34.) $X(t)$ is clearly continuous. (Why?) To show that it is not rectifiable let $l_n(0, 1)$ be the length of the curve $X_n$. We could easily compute $l_n(0, 1)$, but all we need is to observe that $l_1(0, 1) \geqslant \frac{1}{2} =$ length of vertical line joining $(\frac{1}{2}, 0)$ to $(\frac{1}{2}, \frac{1}{2})$

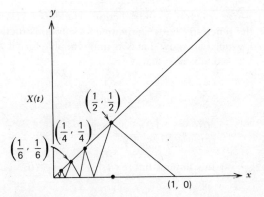

Fig. 34

and $l_2(0, 1) \geq \frac{1}{2} + \frac{1}{4}$ = sum of lengths of vertical lines joining $(\frac{1}{2}, 0)$ to $(\frac{1}{2}, \frac{1}{2})$ and $(\frac{1}{4}, 0)$ to $(\frac{1}{4}, \frac{1}{4})$. Arguing inductively we have that

$$l_n(0, 1) \geq \frac{1}{2} + \frac{1}{4} + \cdots + \frac{1}{2n} = \frac{1}{2}\left(1 + \frac{1}{2} + \frac{1}{3} + \cdots + \frac{1}{n}\right).$$

Now if $X$ is rectifiable with length $l(0, 1)$, then for each $n$, $l(0, 1) \geq l_n(0, 1)$. Hence the numbers $l_n(0, 1)$ must have an upper bound. Consequently the sums $1 + \frac{1}{2} + \frac{1}{3} + \cdots + 1/n$ must have an upper bound. This, however, is false since

$$1 + \frac{1}{2} + \cdots + \frac{1}{n} > \log(n + 1)$$

and $\lim_{n \to \infty} \log(n + 1) = \infty$.

*Example 7.*    Without too much more trouble this example may be strengthened to yield a curve $X(t)$ that is differentiable on the interval $[0, 1]$, yet the curve is still not rectifiable. Instead of connecting the points

$$\left(\frac{1}{2n - 1}, 0\right), \left(\frac{1}{2n}, \frac{1}{2n}\right), \left(\frac{1}{2n + 1}, 0\right) \qquad n = 1, 2, \ldots,$$

by straight lines as in the previous example, we connect the points

$$\left(\frac{1}{\sqrt{2n - 1}}, 0\right), \left(\frac{1}{\sqrt{2n}}, \frac{1}{2n}\right), \left(\frac{1}{\sqrt{2n + 1}}, 0\right),$$

by a nonnegative polynomial $p_n(t)$ with a maximum at $1/\sqrt{2n}$ and having zero derivative at $1/\sqrt{2n - 1}$ and $1/\sqrt{2n + 1}$. Each "arch" of the curve is illustrated by Figure 35.

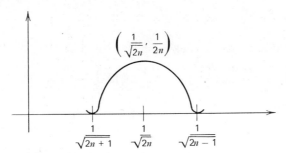

**Fig. 35**

The trace of the entire curve is sketched in Figure 36. To show this curve is not rectifiable note that if it were, the arc length $l(0, 1)$ would be greater than the sum of the altitudes to the curve at the points $1/\sqrt{2}, \ldots, 1/\sqrt{2n}$. But this sum equals

$$\frac{1}{2} + \cdots + \frac{1}{2n} \geq \frac{1}{2}\log(n + 1).$$

As $n \to \infty$, $\log(n + 1) \to \infty$, which is a contradiction. The curve $X(t) = (t, f(t))$ and at every point $t \neq 0$, the function $f$ has a derivative by our construction. We assert that at $t = 0$, $f'(0) = 0$, which will show that $X$ is differentiable for all values of $t$. To show $f'(0) = \lim_{h \to 0} (f(h)/h) = 0$, observe that if

$$\frac{1}{\sqrt{2n + 1}} \leq h \leq \frac{1}{\sqrt{2n - 1}}$$

then $f(h) \leq 1/2n$ and

$$0 \leq \frac{f(h)}{h} \leq \frac{1/2n}{1/\sqrt{2n + 1}} = \frac{\sqrt{2n + 1}}{2n}.$$

Since $\lim_{n \to \infty} (\sqrt{2n + 1}/2n) = 0$, this implies $\lim_{h \to 0} (f(h)/h) = 0 = f'(0)$.

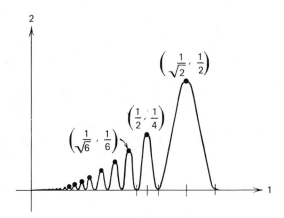

**Fig. 36**

*We turn next to the proof of Theorem 9.4. (This material may be omitted if desired.) There are two things that must be shown. First, we must show that $\int_a^b v(t)\, dt$ is an upper bound for the sums

$$|X(t_1) - X(t_0)| + \cdots + |X(t_n) - X(t_{n-1})|$$

for each partition $a = t_0 < t_1 < \cdots < t_n = b$ of the interval $[a, b]$. Second, we must show that $\int_a^b v(t)$ is the smallest number with this property. Since $\int_a^b v(t)\, dt = \int_a^{t_1} v(t)\, dt + \cdots + \int_a^b v(t)\, dt$, to prove that

$$\int_a^b v(t)\, dt \geq |X(t_1) - X(t_0)| + \cdots + |X(t_n) - X(t_{n-1})|,$$

it suffices to show that for each pair of points $c, d$ satisfying $a \leq c < d \leq b$,

(1) $$|X(d) - X(c)| \leq \int_c^d v(t)\, dt.$$

If $X(d) = X(c)$, (1) certainly holds since $v(t) \geq 0$ and $|X(d) - X(c)| = 0$. So let $C = X(d) - X(c)$ and assume $C \neq 0$. Then by the Schwarz inequality, for each $t$ in $[a, b]$

(2) $$(C, X'(t)) \leq |(C, X'(t))| \leq |C| |X'(t)| = |C| v(t).$$

But if $C = (c_1, c_2, c_3)$ and $X'(t) = (x_1'(t), x_2'(t), x_3'(t))$, equation (2) asserts that

(3) $$c_1 x_1'(t) + c_2 x_2'(t) + c_3 x_3'(t) \leq |C| v(t).$$

Now it is a fundamental property of the integral that if $f(t) \leq g(t)$ on the interval $[c, d]$, then

$$\int_c^d f(t)\, dt \leq \int_c^d g(t)\, dt$$

(Theorem 3.5, Chapter 3). Applying this to (3) yields

$$c_1 \int_c^d x_1'(t)\, dt + c_2 \int_c^d x_2'(t)\, dt + c_3 \int_c^d x_3'(t)\, dt \leq |C| \int_c^d v(t)\, dt.$$

But

$$c_1 \int_c^d x_1'(t)\, dt + c_2 \int_c^d x_2'(t)\, dt + c_3 \int_c^d x_3'(t)\, dt$$

$$= \left( C, \left( \int_c^d x_1'(t)\, dt, \int_c^d x_2'(t)\, dt, \int_c^d x_3'(t)\, dt \right) \right)$$

$$= (C, (x_1(d) - x_1(c), x_2(d) - x_2(c), x_3(d) - x_3(c)))$$

$$= (C, X(d) - X(c)) = (C, C) = |C|^2.$$

Therefore

$$|C|^2 \le |C| \int_c^d v(t)\, dt.$$

Since $|C|$ is not zero by assumption, we may divide by it obtaining

$$|C| \le \int_c^d v(t)\, dt.$$

which was to be proved.

Next we must show that $\int_a^b v(t)\, dt$ is the least upper bound of the sums

$$|X(t_1) - X(t_0)| + \cdots + |X(t_n) - X(t_{n-1})|.$$

This will be true if for each $\epsilon > 0$ there is a partition $a = t_0 < t_1 < \cdots < t_n = b$ such that

$$(4) \qquad |X(t_1) - X(t_0)| + \cdots + |X(t_n) - X(t_{n-1})| \ge \int_a^b v(t)\, dt - \epsilon.$$

To prove (4) we need one estimate on the size of $v(t)$. Since $X'(t)$ is assumed to be continuous on $[a, b]$, the component functions $x_1', x_2', x_3'$ are uniformly continuous on this interval (see Appendix 2, Chapter 5). As a consequence we may conclude that for a given $\epsilon > 0$ there is a partition $a = t_0 < t_1 < \cdots < t_n = b$ such that for each interval $[t_{i-1}, t_i]$ of this partition

$$\left| |X'(r_i)| - \{ [x_1'(s_i)]^2 + [x_2'(u_i)]^2 + [x_3'(w_i)]^2 \}^{1/2} \right| < \epsilon$$

for each choice of points $r_i, s_i, u_i, w_i$ in the interval $[t_{i-1}, t_i]$. Using this partition it follows that

$$\int_a^b v(t)\, dt - \left[ |X(t_1) - X(t_0)| + \cdots + |X(t_n) - X(t_{n-1})| \right]$$

$$\le \left| \int_{t_0}^{t_1} v(t)\, dt - |X(t_1) - X(t_0)| \right| + \cdots + \left| \int_{t_{n-1}}^{t_n} v(t)\, dt - |X(t_n) - X(t_{n-1})| \right|$$

$$= \left| |X'(r_i)| - \{ [x_1'(s_1)]^2 + [x_2'(u_1)]^2 + [x_3'(w_1)]^2 \}^{1/2} \right| \Delta t_1 + \cdots$$

$$+ \left| |X'(r_n)| - \{ [x_1'(s_n)]^2 + [x_2'(u_n)]^2 + [x_3'(w_n)]^2 \}^{1/2} \right| \Delta t_n$$

$$\le \epsilon(\Delta t_1 + \cdots + \Delta t_n) = \epsilon(b - a).$$

Since $\epsilon$ is an arbitrary positive number, this proves (4). The passage from the second line to the third is justified by applying the mean value theorem to each integral $\int_{t_{i-1}}^{t_i} v(t)\, dt$ and to the component functions in each term $|X(t_1) - X(t_{i-1})|$. (See Exercise 9.)

**EXERCISES**

1. If $X(t) = (t, f(t))$ is a curve in the plane, determine $l(a, b)$ for the following choices of $a, b, f$.

(a) $f(t) = 2t^{3/2}$      $a = 1, b = 2$

(b) $f(t) = t^2/2$      $a = 0, b = 2$

(c) $f(t) = \log t$      $a = 1, b = 2$

(d) $f(t) = \cosh t$      $a = 0, b = 1$

2. Let $X(t) = (x_1(t), x_2(t))$ be a curve in the plane. Determine $l(a, b)$ for the following choices of $x_1, x_2, a, b$.

(a) $x_1(t) = t^2$, $x_2(t) = t^3$      $a = 1, b = 4$

(b) $x_1(t) = e^t \cos t$, $x_2(t) = e^t \sin t$      $a = 0, b = \pi$

(c) $x_1(t) = \cos t + t \sin t$, $x_2(t) = \sin t - t \cos t$      $a = 0, b = 2\pi$

(d) $x_1(t) = \cos^3 t$, $x_2(t) = \sin^3 t$      $a = 0, b = 2\pi$.

(e) $x_1(t) = t + \log t$, $x_2(t) = t - \log t$      $a = 1, b = e$.

3. If $X(t) = (2t, t^2, \log t)$, determine the length of $X$ between $t = 1$ and $t = 4$.

4. If $X(t) = (\cosh t, \sinh t, t)$, determine the length of $X$ between $t = 0$ and $t = 1$.

5. If $X(t) = (2t, t^2, t^3/3)$, determine the length of $X$ between $t = 0$ and $t = 1$.

6. If the curve $X$ is given in polar coordinates by $X(\theta) = f(\theta)U(\theta)$ determine the length between $\theta = a$ and $\theta = b$ for the following choices of $a, b, f$.

(a) $f(\theta) = 4 \cos \theta$      $a = 0, b = 2\pi$

(b) $f(\theta) = \theta^2$      $a = 0, b = 1$

(c) $f(\theta) = 2e^\theta$      $a = 0, b = 4$

(d) $f(\theta) = \dfrac{2}{\sin \theta}$      $a = \pi/4, b = 3\pi/4$

(e) $f(\theta) = a(1 - \cos \theta)$      $a = 0, b = 2\pi$

(f) $f(\theta) = 4 \sin^2 \dfrac{\theta}{2}$      $a = 0, b = 2\pi$

7. Sketch the trace and compute the length of $X(\theta) = a \sin^3 (\theta/3) U(\theta)$ from $a = 0$ to $b = 2\pi$.

8. If $X(t) = (x_1(t), x_2(t), x_3(t))$ is a continuous vector-valued function for $t \in [a, b]$ we may define

$$\int_a^b X(t)\, dt = \left( \int_a^b x_1(t)\, dt, \int_a^b x_2(t)\, dt, \int_a^b x_3(t)\, dt \right).$$

The integral is called a vector-valued integral. Show that if $Y$ is continuous and $c$ and $d$ are scalars, then

$$\int_a^b [cX(t) + dY(t)]\, dt = c \int_a^b X(t)\, dt + d \int_a^b Y(t)\, dt.$$

Also show that

$$\left| \int_a^b X(t)\, dt \right| \le \int_a^b |X(t)|\, dt.$$

(A special case of this is contained in the proof of Theorem 9.4.)

9. Let the curve $X$ have continuous velocity on the interval $[c, d]$. Use the mean value theorem to show that there exist points $r, s, u, w$ in the interval $(c, d)$ satisfying

$$\int_c^d v(t)\, dt - |X(d) - X(c)| = \left[ |X'(r)| - \{ [x_1'(s)]^2 + [x_2'(u)]^2 + [x_3'(w)]^2 \}^{1/2} \right]$$

$$\times (d - c)$$

(We used this fact in the proof of (4).)

10. By examining the helix $X(t) = (\cos t, \sin t, t)$ defined on the interval $[0, 2\pi]$, show that no mean value theorem of the form

$$X(b) - X(a) = (b - a)X'(t_0)$$

$a < t_0 < b$ can possibly hold for general vector-valued functions.

11. The curve $X$ of Example 5 had infinitely many oscillations on the interval $[0, 1]$, which is a partial explanation of why it is not rectifiable. Show that if the plane curve $X$ is defined by $X(t) = (t, f(t))$ where the function $f$ is monotone on the interval $[a, b]$, then $X$ is rectifiable. Thus if the plane curve $X$ of the form $X(Y) = (t, f(t))$ has finitely many oscillations, it must be rectifiable. (Assume $f$ is monotone increasing on $[a, b]$. There is now an obvious upper bound to the sums $|X(t_1) - X(t_0)| + \cdots + |X(t_n) - X(t_{n-1})|$ where $a = t_0 < t_1 < \cdots < t_n = b$.)

12. A nonempty set $S$ of real numbers is called bounded below if there exists a number $t$ such that $t \leqslant s$ for each $s \in S$. Such a $t$ is called a lower bound for $S$. A number $t_0$ is a greatest lower bound for $S$ if $t_0$ is a lower bound for $S$ and if $t_0 \geqslant t$ for each lower bound $t$ of $S$. Using an appropriate modification of the argument in the proof of Theorem 9.3, prove that a nonempty set $S$ of real numbers that is bounded below has a greatest lower bound.

*13. Theorem 9.3 asserts that every nonempty bounded set of real numbers has a least upper bound. This condition is called the least upper bound property of the real numbers and is equivalent to the completeness axiom. Prove this by deriving the completeness axiom as a consequence of the least upper bound property.

14. A plane curve $X(t) = (t, f(t))$, where $f(t) \geqslant 0$, has the property that the area of the ordinate set for $f$ on the interval $[a, b]$ is proportional to the arc length $l(a, b)$. Determine $f$. (Set up an appropriate differential equation and solve it.)

15. A curve $X(\theta) = f(\theta)U(\theta)$, where $f(\theta) \geqslant 0$, has the property that for some constant $k > 1$ and for all $x > a$ the arc length $l(a, x) = k(f(x) - f(a))$. Determine $f$.

16. On a dark stormy night a coast guard boat was searching for a speedboat suspected of smuggling. During a flash of lightning the speedboat was sighted at a point 4 miles dead ahead fleeing at constant speed along a straight path of unknown direction. Assuming the speed of the coast guard boat was three times that of the speedboat, on what path should the coast guard boat proceed in order to be sure to apprehend the speedboat? (Determine a path $X(\theta) = f(\theta)U(\theta)$ for the coast guard boat assuming that the speedboat was at the origin when it was sighted.) (This problem is due to Lester Ford.)

## *9.8   Planetary Motion

On the basis of observations of the orbits of the planets made over long periods of time at the end of the 16th century and the beginning of the 17th, Johannes Kepler formulated the following laws describing the motion of the planets.

1. Each planet moves about the sun in an ellipse with the sun at the focus.

2. The radius vector from the sun to a planet sweeps out equal areas in equal times.

3. The square of the period of motion of a planet is proportional to the cube of the length of the semimajor axis.

These deductions were even more remarkable because the mathematical model envisaged by Kepler as describing the motions of the planets was completely fallacious.

During 1665–1666, in the midst of the great plague in England, Isaac

Newton deduced that Kepler's laws implied that the gravitational force exerted on each planet by the sun must be directed toward the sun and must have a magnitude inversely proportional to the square at the distance of the planet from the sun. Furthermore, he observed that Kepler's laws could be deduced as a consequence of this "inverse square law" of gravitation. These results were not published until 1686 when they appeared in Newton's famous monograph "Principia Mathematica." There has been much debate over the reason for this long delay in publication. It has been conjectured that one reason was that Newton had difficulty in establishing that as far as gravitational forces were concerned the masses of the sun and the planets could be assumed to be concentrated at their respective centers, a result that he needed to complete his theory.

In this section we shall derive Kepler's laws from Newton's law of gravitation. It is interesting to note that Newton does not supply a proof for the assertion that the inverse square law implies that planetary orbits are elliptical beyond saying that the orbit of a planet is uniquely determined by the gravitational force. In 1710, Johann Bernoulli took exception to this reasoning and supplied a proof of his own for this fact. The vector techniques that we shall use were invented by J. Willard Gibbs, a 19th-century American mathematician and physicist. Gibbs himself used these techniques in a paper on the determination of planetary orbits published in 1877.

To derive Kepler's laws, let $X(t)$ denote the radius vector of a planet at time $t$ with the sun assumed to be at the origin of our coordinate system. By Newton's second law of motion, assuming that the only force $F(t)$ acting on the earth is the gravitational attraction of the sun, we have

(1) $$mX''(t) = F(t)$$

where $m$ denotes the mass of the earth. We next assume that $F(t)$ is directed toward the sun and is inversely proportional to the square of the length of $X(t)$. This means that we may write

(2) $$F(t) = -\frac{GmM}{|X(t)|^2} \frac{X(t)}{|X(t)|}$$

where $M$ is the mass of the sun and $G$ is an appropriate constant. Combining (1) and (2) we see that

(3) $$X''(t) = f(t)X(t)$$

where $f(t) = -GM/|X(t)|^3$ is a scalar function of time. Therefore, $X(t)$ and $X''(t)$ are linearly dependent. Hence, if we compute the cross product we have

$$0 = X(t) \times X''(t) = \frac{d}{dt}(X(t) \times X'(t)).$$

Consequently

$$X(t) \times X'(t) = N, \qquad \text{a constant,}$$

and the vector $X(t)$ always lies in the plane through the origin perpendicular to the vector $N$. Equation (3) implies that the acceleration is radial, and we have already seen (Section 9.6) that this is equivalent to the assertion that the radius vector sweeps out equal areas in equal times. Thus we have proved the second of Kepler's laws.

To prove the first of Kepler's laws, assume the plane of the motion is the 1, 2 plane, and write the radius vector $X(t)$ in polar coordinates. Then

$$X(t) = |X(t)|U(\theta(t)) = r(t)U(\theta(t))$$

where $r(t) = |X(t)|$. It is a somewhat remarkable fact that Kepler's first law follows simply from the determination of the vector $X'(t) \times N$. We assert that

(4)
$$X'(t) \times N = GM(U(\theta(t)) + A)$$

where $A$ is a constant vector. Let us grant (4) for a moment. If we suppress the variable $t$ in (4) and take the inner product of (4) with $X$, the left-hand side is

$$(X, X' \times N) = (X \times X', N) = |N|^2$$

To compute the right-hand side, write $X = rU$. Then

$$(X, GM(U + A)) = (rU, GM(U + A)) = rGM(1 + (U, A)).$$

Therefore

$$|N|^2 = rGM(1 + (U, A)).$$

If we define the angle $\phi$ as the angle between the vectors $U$ and $A$, we have $(U, A) = |U||A| \cos \phi = |A| \cos \phi$. (See Figure 37.)

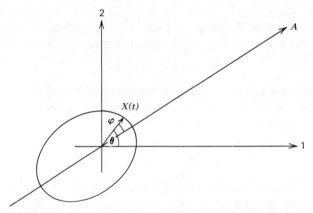

**Fig. 37**

Hence writing $r = r(\phi)$, we have

$$|N|^2 = r(\phi)GM(1 + |A| \cos \phi)$$

or

(5)
$$r(\phi) = \frac{|N|^2}{GM} \frac{1}{1 + |A| \cos \phi} = \frac{|N|^2}{GM} \frac{1}{1 + \epsilon \cos \phi}$$

where $\epsilon = |A|$. This is the equation of a conic section with eccentricity $\epsilon$ and major axis along the vector $A$, (see Example 4, Section 9.4). Now the planets are known to have periodic orbits. Since the only periodic solution to (5) is an ellipse, we must have $\epsilon < 1$, and consequently we have shown that each planet moves about the sun on an ellipse with the sun at the focus.

Equation (4), however, is easy to establish. Note first that since $X = rU$, we have

$$X' = r'U + r\theta'U' = r'U + r\theta'V.$$

Hence

(6) $$N = X \times X' = r^2\theta' U \times V,$$

since $U \times U = 0$.

Therefore

$$\frac{d}{dt}(X' \times N) = X'' \times N = -\frac{GM}{r^2} U \times r^2\theta'(U \times V)$$

$$= -GM\theta' U \times (U \times V) = GM\theta' V = \frac{d}{dt}(GMU).$$

Since the vectors $X' \times N$ and $GMU$ have the same derivative, they must differ by a constant vector, which we choose to write $GMA$. Therefore

$$X' \times N = GM(U + A)$$

which is (4).

To derive the third of Kepler's laws note first that from (6)

$$|N| = |r^2\theta' U \times V| = |r^2\theta'|$$

Assuming $\theta' > 0$, it follows that the area swept out by the radius vector during one period $T$ is

$$\frac{1}{2}\int_0^T r^2\theta'(t)\,dt = \frac{|N|}{2}\int_0^T dt = \frac{|N|}{2}T.$$

On the other hand, the area of the ellipse defined by

$$r = \frac{k\epsilon}{1 + \epsilon \cos\phi}$$

is given by $\pi ab$ where $a$ and $b$ are the lengths of the semimajor and semiminor axes, respectively. Consequently

(7) $$\frac{|N|}{2}T = \pi ab.$$

It is easily verified that

(8) $$\frac{b}{a} = \sqrt{1 - \epsilon^2}$$

and also that

(9) $$r(0) + r(\pi) = 2a.$$

But from (5)

(10) $$r(0) + r(\pi) = \frac{|N|^2}{GM}\left(\frac{1}{1+\epsilon} + \frac{1}{1-\epsilon}\right) = \frac{|N|^2}{GM}\frac{2}{1-\epsilon^2}.$$

Substituting (7), (8), and (9) in (10) we obtain

$$T^2 = \frac{4\pi^2 a^2 b^2}{|N|^2} = \frac{4\pi^2 a^4}{|N|^2}(1 - \epsilon^2) = \frac{4\pi^2}{GM}a^3.$$

Thus the square of the period of the motion is proportional to the cube of the length of the semimajor axis. This is Kepler's third law.

**EXERCISE**

1. Deduce Newton's law of gravitation from Kepler's first and second laws. (As before let $X(t)$ denote the radius vector of a planet at time $t$ and let the sun be at the origin of our coordinate system. In polar coordinates $X(t) = r(t)U(\theta(t))$.)

Applying the second of Kepler's laws we infer that the acceleration $X''$ is radial. Therefore (p. 377)

(11) $$X'' = (r'' - r(\theta')^2)U$$

and

(12) $$r^2\theta' = k,$$

where $k$ is constant (p. 377). To establish Newton's inverse square law of gravitation, we must show that $X'' = -(C/r^2)U$ where $C$ is a positive constant. In view of (11) this is equivalent to showing that the function $r$ satisfies the differential equation

(13) $$\frac{d^2r}{dt^2} - r\left(\frac{d\theta}{dt}\right)^2 = -\frac{C}{r^2}.$$

Applying Kepler's first law we infer from (5) that

(14) $$r(t) = \frac{B}{1 + \epsilon \cos \theta(t)}$$

where $B$ is constant. Now differentiate (14). Using (12) and (13) verify that

(15) $$r' = \frac{1}{B} r^2\epsilon(\sin \theta)\theta' = \frac{k\epsilon}{B} \sin \theta.$$

To complete the argument differentiate (15) and verify that $r$ satisfies (13).

# CHAPTER TEN

# IMPROPER INTEGRALS; FURTHER APPLICATIONS OF THE MEAN VALUE THEOREM

## 10.1  Improper Integrals

Up to now we have only considered the definite integral of a bounded function $f$ defined on a finite interval $[a, b]$. In this section we wish to extend the notion of a definite integral to functions defined on infinite intervals and to functions that are not bounded. Such integrals are called *improper* integrals. We assume throughout that the functions under consideration are non-negative. In both cases the extensions are quite easy. We consider the case of an infinite interval first.

*Definition.*  *Let $f$ be a nonnegative function defined for all $x \geqslant a$. Assume that $f$ is bounded and integrable on each interval $[a, t]$, $a < t$. Then if*

(1) $$\lim_{t \to \infty} \int_a^t f \quad \text{exists, we define} \quad \int_a^\infty f = \lim_{t \to \infty} \int_a^t f.$$

*If the limit in (1) exists, we say the integral $\int_a^\infty f$ converges. If the limit fails to exist, that is, $\lim_{t \to \infty} \int_a^t f = \infty$, then $\int_a^\infty f$ is not defined. We express this fact by saying that $\int_a^\infty f$ diverges.*

*Example 1.*  Show that for $a > 0 \int_a^\infty (1/x^k)\, dx$ converges if $k > 1$ and diverges if $k \leqslant 1$. We compute $\lim_{t \to \infty} \int_a^t (1/x^k)\, dx$. If $k \neq 1$,

$$\int_a^t \frac{dx}{x^k} = \frac{1}{1-k} x^{1-k} \Big|_a^t = \frac{1}{1-k} [t^{1-k} - a^{1-k}].$$

If $k > 1$, then $\lim_{t \to \infty} t^{1-k} = \lim_{t \to \infty} 1/t^{k-1} = 0$. Since $1 - k < 0$, $\lim_{t \to \infty} t^{1-k} = 0$. Hence $\int_a^\infty (1/x^k)\, dx$ converges, and indeed

$$\int_a^\infty \frac{dx}{x^k} = \frac{a^{1-k}}{k-1}.$$

For $k < 1$, $1 - k > 0$ and $\lim_{t \to \infty} t^{1-k} = \infty$. Therefore $\int_a^\infty (1/x^k)\, dx$ diverges. For $k = 1$

$$\int_a^t \frac{dx}{x} = \log t - \log a.$$

But $\lim_{t \to \infty} \log t = \infty$. Hence $\int_a^\infty (1/x)\, dx$ diverges as well.

***Example 2.*** Show that $\int_a^\infty e^{-x}\,dx$ converges. Clearly $\int_a^t e^{-x}\,dx = e^{-a} - e^{-t}$. But $\lim_{t\to\infty} e^{-t} = 0$. Hence $\int_a^\infty e^{-x}\,dx$ converges and $\int_a^\infty e^{-x}\,dx = e^{-a}$.

The technique, illustrated in the two previous examples, for showing that $\int_0^\infty f$ converges is first to compute a primitive $F$ for the function $f$. Then the integral $\int_0^\infty f$ will converge if $\lim_{t\to\infty} F(t)$ exists. Often, however, we would like to know if a given integral $\int_0^\infty f$ converges, but it is either impractical or impossible to determine a primitive for $f$. For example, what is the behavior of $\int_1^\infty e^{-x^2}\,dx$? Does this integral converge or not? We can answer these and similar questions by the use of the following comparison principle for improper integrals. It states that if $0 \leqslant f \leqslant g$ and $\int_a^\infty g$ is known to converge, then $\int_a^\infty f$ will converge also. Similarly, if $f \geqslant g \geqslant 0$ and $\int_a^\infty g$ is known to diverge, then $\int_a^\infty f$ will diverge as well. If we interpret the integrals $\int_a^\infty f$ as the area of the ordinate set for $f$, this result is intuitively clear from Figure 1.

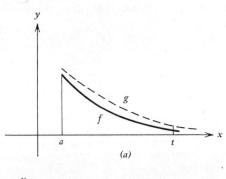

(a)

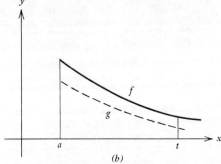

(b)

**Fig. 1** (a) If $0 \leqslant f \leqslant g$ and $\lim_{t\to\infty} \int_a^t g$ exists, then $\lim_{t\to\infty} \int_a^t f$ will exist also. (b) If $0 \leqslant g \leqslant f$ and $\lim_{t\to\infty} \int_a^t g = \infty$, then $\lim_{t\to\infty} \int_a^t f = \infty$ also.

Even though these results seem rather clear on geometrical grounds, there is a subtlety in the proof. The difficulty is, how can we say anything about $\lim_{t \to \infty} \int_a^t f$ if we cannot evaluate $F(t) = \int_a^t f$? It turns out that we cannot in general make any statement about the *value* of $\lim_{t \to \infty} \int_a^t f$. However, we can claim the limit exists by appealing, as we have done many times in the past, to the completeness axiom for the real numbers! The specific result we need follows from the least upper bound property of the real numbers, introduced in Section 9.7.

**LEMMA 10.1**   *Let f be a function defined on the interval* $[a, \infty)$. *If f is nondecreasing, that is,* $f(x) \le f(y)$ *whenever* $x \le y$, *then* $\lim_{x \to \infty} f(x)$ *exists if and only if f is bounded above.*

*Proof.*   If $f$ is bounded above, there is a constant $M$ such that

$$M \ge f(x) \qquad \text{for each } x \ge a.$$

Since the collection of numbers $\{f(x) : x \ge a\}$ is bounded above, there is a least upper bound $L$ to this collection (see p. 383). We assert $L = \lim_{x \to \infty} f(x)$. But since $L$ is the least upper bound, for each $\epsilon > 0$, there exists an $x_0$ such that

$$L \ge f(x_0) > L - \epsilon.$$

Since $f$ is increasing we may infer that for $x \ge x_0$

$$L \ge f(x) \ge f(x_0) > L - \epsilon.$$

Therefore $\lim_{x \to \infty} f(x) = L$.

Conversely, if $\lim_{x \to \infty} f(x) = L$, then since $f$ is increasing, $L \ge f(x)$ for each $x$, which means that $f$ is bounded above.

The desired result about integrals that we need now follows as a corollary.

**COROLLARY.**   *Let f be a nonnegative function defined on the infinite interval* $[a, \infty)$. *Assume that f is bounded and integrable on each interval* $[a, t]$. *Then*

$\lim_{t \to \infty} \int_a^t f$ *exists if and only if there is a constant M satisfying*

$$M \ge \int_a^t f \qquad \text{for each} \qquad t \ge a.$$

*Proof.*   Since $f \ge 0$, the integrals $F(t) = \int_a^t f$ are increasing with $t$, and the result follows from Lemma 10.1.

Next we establish our comparison principle. Actually we do not need to know that $f(x) \le g(x)$ for all $x \ge a$. It suffices to know this for all $x$ greater than some fixed $x_0 \ge a$.

**THEOREM 10.2**   *Let f, g be nonnegative functions defined for* $x \ge a$. *Assume that f and g are bounded and integrable on each interval* $[a, t]$ *and let* $x_0 \ge a$. *If* $f(x) \le g(x)$ *for all* $x \ge x_0$, *and* $\int_a^\infty g$ *converges, then* $\int_a^\infty f$ *converges. If for* $x \ge x_0$, $f(x) \ge g(x)$ *and* $\int_a^\infty g$ *diverges, then* $\int_a^\infty f$ *diverges.*

***Proof.*** Since $\int_a^t f = \int_a^{x_0} f + \int_{x_0}^t f$, the integrals $\int_a^t f$ will be bounded above if and only if the integrals $\int_{x_0}^t f$ are bounded above. If $f(x) \leqslant g(x)$ for $x \geqslant x_0$ and $\int_a^\infty g$ converges, then the integrals $\int_{x_0}^t g$ are bounded above. Since $\int_{x_0}^t f \leqslant \int_{x_0}^t g$, this implies the integrals $\int_{x_0}^t f$ are bounded above. Hence $\int_a^\infty f$ converges. The proof of the second statement is similar and is left as an exercise.

It is convenient to abbreviate the statement that the integral $\int_a^\infty f$ converges by writing $\int_a^\infty f < \infty$. Similarly, we write $\int_a^\infty f = \infty$ if the integral $\int_a^\infty f$ diverges.

***Example 3.*** Show that the integral $\int_0^\infty e^{-x^2}$ converges and the integral

$$\int_1^\infty x^{-1} \log x \, dx$$

diverges.

For $x > 1$, $x^2 > x$. Hence $e^{-x^2} < e^{-x}$. Since $\int_0^\infty e^{-x} \, dx < \infty$, this implies

$$\int_0^\infty e^{-x^2} \, dx < \infty.$$

For the second integral observe that for $x \geqslant e$, $\log x \geqslant 1$. Hence for $x \geqslant e$, $x^{-1} \log x \geqslant x^{-1}$. Since $\int_1^\infty x^{-1} \, dx = \infty$, it follows that $\int_1^\infty x^{-1} \log x \, dx = \infty$.

Next let us consider the integral of an unbounded function $f$ defined on a finite interval.

***Definition.*** *Let $f$ be a nonnegative function defined on the half-open interval $[a, b)$. Assume that $f$ is bounded and integrable on each interval $[a, t], a < t < b$. Then if*

$$(2) \qquad\qquad \lim_{t \to b^-} \int_a^t f$$

*exists, we define $\int_a^b f = \lim_{t \to b^-} \int_a^t f$. If the limit in (2) exists, the integral $\int_a^b f$ of the unbounded function $f$ is said to converge. If $\lim_{t \to b^-} \int_a^t f = \infty$, the integral $\int_a^b f$ is not defined, and we say that $\int_a^b f$ diverges. We write $\int_a^b f < \infty$, $\int_a^b f = \infty$ according as the integrals converge or diverge.*

***Example 4.*** Let $0 < a < b$. Show that $\int_a^b (b-x)^{-k} \, dx$ converges if $k < 1$ and diverges if $k \geqslant 1$. The computation is virtually the same as for Example 1. If $k \neq 1$

$$\int_a^t \frac{dx}{(b-x)^k} = \frac{-1}{1-k} (b-x)^{1-k} \Big|_a^t = \frac{-1}{1-k} \Big[ (b-t)^{1-k} - (b-a)^{1-k} \Big].$$

If $k < 1$, $1-k > 0$, and $\lim_{t \to b^-} (b-t)^{1-k} = 0$. If $k > 1$, $1-k < 0$ and $\lim_{t \to b^-} (b-t)^{1-k} =$

$\infty$. In the first case the integral converges and $\int_a^b (b-x)^{-k}\,dx = (1-k)^{-1}(b-x)^{1-k}$. In the second case the integral diverges. For $k = 1$, $\int_a^t (b-x)^{-1}\,dx = -\log(b-t) + \log(b-a)$. Since $\lim_{t\to b^-} \log(b-t) = \infty$, the integral $\int_a^b (b-x)^{-1}\,dx$ diverges.

**Example 5.** Show that the integral $\int_0^1 (1-x^2)^{-1/2}\,dx$ converges. For $0 < t < 1$, $\int_0^t (1-x^2)^{-1/2}\,dx = $ arc sin $t$. Since $\lim_{t\to 1^-}$ arc sin $t = \pi/2$, the integral converges, and $\int_0^1 (1-x^2)^{-1/2}\,dx = \pi/2$.

The same considerations apply if $f$ is unbounded at the left-hand end point of an interval $[a, b]$. Then $\int_a^b f = \lim_{t\to a^+} \int_t^b f$ if this limit exists.

We have exactly the same comparison principle for integrals of unbounded functions as for integrals over infinite intervals. We state the result next, but since the proof is virtually a repetition of Theorem 10.2 we leave it as as exercise.

**THEOREM 10.3** *Let $f$ and $g$ be nonnegative functions defined on the interval $[a, b)$. Assume that $f$ and $g$ are bounded and integrable on each interval $[a, t]$, $a < t < b$. Let $x_0$ be a fixed point in $[a, b)$. If for $x_0 \leqslant x < b$, $f(x) \leqslant g(x)$ and $\int_a^b g$ converges, then $\int_a^b f$ converges. If $f(x) \geqslant g(x)$ for $x_0 < x < b$, and $\int_a^b g$ diverges, then $\int_a^b f$ diverges.*

Figure 2 illustrates the geometry of the situation.

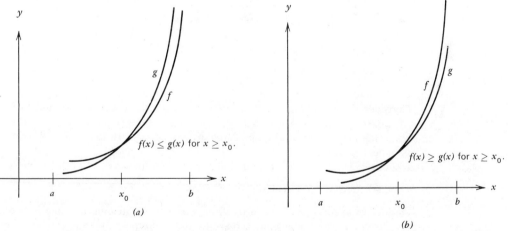

(a)

(b)

**Fig. 2** (a) If $\int_a^b g < \infty$, then $\int_a^b f < \infty$. (b) If $\int_a^b g = \infty$, then $\int_a^b f = \infty$.

**Example 6.** Show that $\int_0^1 (1/\sqrt{1-x^3})\,dx$ converges and $\int_0^1 (1-x)^{-1} \sin(\pi/2)x\,dx$ diverges. For the first integral if $0 < x < 1$, then $x^3 < x^2$. Hence $1-x^3 > 1-x^2$ and $1/\sqrt{1-x^3} < 1/\sqrt{1-x^2}$. It follows from Theorem 10.3 that since $\int_0^1 (1/\sqrt{1-x^2})\,dx$ converges, $\int_0^1 (1/\sqrt{1-x^3})\,dx$ does also. For the second integral if $\frac{1}{2} \leqslant x < 1$, then

$\sin(\pi/2)x \geq \sqrt{2}/2$. Hence by 10.3, $\int_0^1 (1-x)^{-1} \sin(\pi/2)x \, dx$ diverges since $\int_0^1 (1-x)^{-1} \, dx$ is known to diverge.

## EXERCISES

1. Determine if the following integrals converge or diverge. Give the reason for your answer.

(a) $\displaystyle\int_0^1 \frac{dx}{\sqrt{1-x}}$

(h) $\displaystyle\int_0^2 \frac{dx}{\sqrt{8-x^3}}$

(b) $\displaystyle\int_1^\infty \frac{dx}{(x+1)^2}$

(i) $\displaystyle\int_1^\infty xe^{-x^2} \, dx$

(c) $\displaystyle\int_2^\infty \frac{dx}{x+2}$

(j) $\displaystyle\int_1^\infty \frac{dx}{\sqrt{1+x^2}}$

(d) $\displaystyle\int_0^2 \frac{dx}{\sqrt{4-x^2}}$

(k) $\displaystyle\int_1^\infty \frac{dx}{x\sqrt{1+x^2}}$

(e) $\displaystyle\int_0^\infty \frac{dx}{1+x^2}$

(l) $\displaystyle\int_1^\infty \frac{dx}{x\sqrt{4x}}$

(f) $\displaystyle\int_1^\infty \frac{x \, dx}{1+x^2}$

(m) $\displaystyle\int_1^\infty xe^{-x} \, dx$

(g) $\displaystyle\int_1^\infty \frac{x^2 \, dx}{1+x^4}$

2. Determine if the following integrals are convergent or divergent.

(a) $\displaystyle\int_1^\infty (\log x)e^{-x} \, dx$

(c) $\displaystyle\int_2^\infty \frac{dx}{x(\log x)^k}, \quad k \geq 1$

(b) $\displaystyle\int_2^\infty \frac{dx}{x \log x}$

(d) $\displaystyle\int_0^{\pi/2} \sec x \, dx$

3. Show that the integral $\int_1^\infty x^k e^{-x} \, dx$ converges for $k = 1, 2, \ldots$ (Use induction.)

4. Show that for all real numbers $r$, $\int_1^\infty x^r e^{-x} \, dx$ converges.

5. Find the area of the region in the first quadrant lying below the curve $y = xe^{-x^2}$.

6. Consider the region in the first quadrant below the curve $xy = 1$ and to the right of the line $x = 1$. Show that this region has infinite area. Show on the other hand that if the region is revolved about the $x$ axis, the resulting solid has finite volume.

7. If a curve $X(t)$ is defined on an infinite interval $I$, say $(a, \infty)$, and has continuous velocity, then $X$ is rectifiable if the integral $\int_a^\infty v(t) \, dt$ converges. Determine if the following curves $X(\theta) = f(\theta)U(\theta)$ in polar coordinates are rectifiable on the given intervals and determine the arc length where possible. Sketch the traces of the curves.

(a) $X(\theta) = e^{-\theta}U(\theta), \quad I = [0, \infty)$

(b) $X(\theta) = \dfrac{1}{\theta}U(\theta), \quad I = [\pi, \infty)$

(c) $X(\theta) = \dfrac{1}{\theta^2}U(\theta), \quad I = [\pi, \infty)$.

8. Show that if $f(x) \geq g(x)$ for $x \geq x_0$ and $\int_a^\infty g$ diverges, then $\int_a^\infty f$ diverges. (Use Lemma 10.1 in a similar way as in the proof of Theorem 10.2.)

9. Let $f$ be an increasing function defined on the interval $[a, b)$. Show that if $f$ is bounded above, then $\lim_{x \to b^-} f(x)$ exists.

10. Prove Theorem 10.3. (Use Exercise 9 and the technique of the proof of Theorem 10.2.)

## 10.2 Cauchy's Mean Value Theorem and l'Hospital's Rule

The mean value theorem states that if the real-valued function $f$ is continuous on $[a, b]$ and differentiable on $(a, b)$, then for some point $x_0$ in the interval $a < x_0 < b$, it follows that

$$f'(x_0) = \frac{f(b) - f(a)}{b - a}.$$

This result is proved by examining the vertical distance $d(x)$ between the chord joining $(a, f(a))$ and $(b, f(b))$ and the function $f$. See Figure 3.

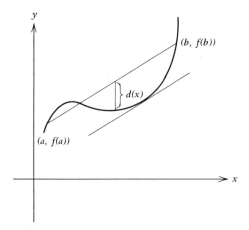

**Fig. 3**

Clearly

(1) $$d(x) = [f(b) - f(a)] \frac{x - a}{b - a} + f(a) - f(x).$$

Now $d(a) = d(b) = 0$. Hence by Rolle's theorem there is a point $x_0$ in $(a, b)$ such that $d'(x_0) = 0$. From (1) if $d'(x_0) = 0$, then

$$f'(x_0) = \frac{f(b) - f(a)}{b - a}.$$

To generalize the mean value theorem let us examine (1). Multiplying (1) by $(b - a)$, we have

(2) $$d(x)(b - a) = [f(b) - f(a)](x - a) + [f(a) - f(x)](b - a).$$

Suppose in the right-hand side of (2) we replace the function $g(x) = x$ by an arbitrary function $g$ satisfying the same conditions on $[a, b]$ that $f$ does. Then in the right-hand side of (2), $x$ becomes $g(x)$; $a$ becomes $g(a)$; and $b$ becomes $g(b)$. Calling this new function $m(x)$ we have

(3) $$m(x) = [f(b) - f(a)](g(x) - g(a)) + (f(a) - f(x))(g(b) - g(a)).$$

Just as for $d(x)$, we have $m(b) = m(a) = 0$. Hence by Rolle's theorem $m'(x_0) = 0$ for some $x_0$ in $(a, b)$.

From (3) $m'(x) = (f(b) - f(a))g'(x) - f'(x)(g(b) - g(a))$. Therefore, we have shown that for some $x_0$ in $(a, b)$

(4)    $$[f(b) - f(a)]g'(x_0) - [g(b) - g(a)]f'(x_0) = 0.$$

If we assume $g'$ never vanishes, then $g(b) \neq g(a)$ (why?), and we may rewrite (4) as

(5)    $$\frac{f(b) - f(a)}{g(b) - g(a)} = \frac{f'(x_0)}{g'(x_0)}$$

for some point $x_0$ in $(a, b)$. This extension of the mean value theorem is called *Cauchy's mean value theorem*. Formula (4) or (5) is called Cauchy's mean value formula. We state this result as our next theorem.

**THEOREM 10.4**    *Let $f$ and $g$ be continuous on $[a, b]$ and differentiable on the open interval $(a, b)$. Then for some $x_0$ in the open interval $(a, b)$*

(6)    $$[f(b) - f(a)]g'(x_0) = [g(b) - g(a)]f'(x_0).$$

This result has several interesting corollaries. The first extends the geometric interpretation of the earlier mean value theorem.

**THEOREM 10.5**    *Let $X$ be a regular curve in the plane such that $X(b) \neq X(a)$. Let $T(t)$ be the unit tangent vector to the curve at $t$. Then for some $t_0$ in the open interval $(a, b)$, the tangent vector $T(t_0)$ is parallel to $X(b) - X(a)$.*

**Proof.**    Figure 4 illustrates this result. To prove the theorem let $X(t) = (x_1(t), x_2(t))$. Applying (4) to the functions $x_1(t), x_2(t)$ in place of $f$ and $g$, we have

$$(x_1(b) - x_1(a))x_2'(t_0) + (x_2(b) - x_2(a))(-x_1'(t_0)) = 0,$$

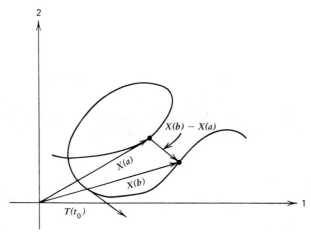

**Fig. 4**

for some value of $t_0$ between $a$ and $b$. However, this just says that the vector $U = (x_2'(t_0), -x_1'(t_0))$ is perpendicular to $X(b) - X(a)$. But $X'(t_0) = (x_1'(t_0), x_2'(t_0))$ is clearly perpendicular to $U$ since $(U, X'(t_0)) = 0$. Therefore $X'(t_0)$ must be parallel to $X(b) - X(a)$ since both vectors $X(b) - X(a)$ and $X'(t_0)$ are assumed to be different from zero. Since $T(t_0) = X'(t_0)/|X'(t_0)|$, it follows that $T(t_0)$ and $X(b) - X(a)$ are parallel.

**Example 1.**  If $f(x) = \sin x$ and $g(x) = \cos x$ are defined on the interval $[0, \pi/2]$. determine a value of $x_0$ in this interval that satisfies Cauchy's mean value theorem, that is, a value of $x$ satisfying equation (6). Since $f'(x) = \cos x$ and $g'(x) = -\sin x$, (6) implies that

$$-\sin x = -\cos x$$

or $\tan x = 1$. Therefore $x = \pi/4$.

**Example 2.**  If $X(t) = (t^2, t^3)$, determine a point $t$ in the interval $(1, 4)$ at which the tangent to this curve is parallel to $X(4) - X(1)$. Now $X(4) - X(1) = (16, 64) - (1, 1) = (15, 63)$. We seek a point $t$ such that $(x_2'(t), -x_1'(t)) = (3t^2, -2t')$ is perpendicular to $X(4) - X(1)$. But the inner product, $((15, 63), (3t^2, -2t)) = 45t^2 - 126t = 9t(5t - 14) = 0$ if $t = \frac{14}{5}$. Therefore the unit tangent $T(\frac{14}{5})$ at this point, which is perpendicular to $(x_2'(\frac{14}{5}), -x_1'(\frac{14}{5}))$, is parallel to $X(4) - X(1)$.

**Example 3.**  Cauchy's mean value formula (6) may be translated into a formula concerning integrals. If $f$ and $g$ are continuous on $[a, b]$, let $F(x) = \int_a^x f(t)\, dt$ and $G(x) = \int_a^x g(t)\, dt$. Then $F'(x) = f(x)$ and $G'(x) = g(x)$ for each $x$ in $[a, b]$. Applying (6) to $F$ and $G$ we obtain $[F(b) - F(a)]G'(x_0) = [G(b) - G(a)]F'(x_0)$ for some value of $x_0$ in $[a, b]$. But $F(b) - F(a) = \int_a^b f(t)\, dt$, and $G(b) - G(a) = \int_a^b g(t)\, dt$. Hence we have the following integral mean value formula. If $f$ and $g$ are continuous on $[a, b]$, then

$$(7) \qquad\qquad g(x_0) \int_a^b f(t)\, dt = f(x_0) \int_a^b g(t)\, dt$$

for some value of $x_0$ in $(a, b)$.

The second application of Cauchy's mean value theorem is to the calculation of limits. If $\lim_{x \to a} f(x) = 0 = \lim_{x \to a} g(x)$, then we have observed many times that nothing may be concluded from the limit theorems as to the existence of $\lim_{x \to a} f(x)/g(x)$. It may exist or it may not. However, Cauchy's mean value theorem provides a convenient tool called "l'Hospital's rule," which is quite useful in this situation. We formulate the result in terms of one-sided limits.

**THEOREM 10.6**  *Let $f$ and $g$ be continuous on $[a, b]$ and differentiable on the open interval $(a, b)$. Suppose further that $g'(x) \neq 0$ for each $x$ in the open interval $(a, b)$. If $\lim_{x \to a^+} f(x) = 0$ and if $\lim_{x \to a^+} g(x) = 0$, then*

$$\lim_{x \to a^+} \frac{f(x)}{g(x)} = \lim_{x \to a^+} \frac{f'(x)}{g'(x)}$$

*if this latter limit exists.*

**Proof.**  Let us apply the Cauchy mean value theorem to $f$ and $g$ on the interval $[a, x]$ where $a < x < b$. Then since $g'$ does not vanish on the interval $(a, b)$, we have from (5)

$$\frac{f(a) - f(x)}{g(a) - g(x)} = \frac{f'(x_0)}{g'(x_0)} \qquad \text{where} \qquad a < x_0 < x.$$

Since $f(a) = g(a) = 0$, we have that for each $x$ in $(a, b)$, there is an $x_0$ in the interval $(a, x)$ such that

$$\frac{f(x)}{g(x)} = \frac{f'(x_0)}{g'(x_0)}.$$

Therefore if $\lim_{x \to a^+} f'(x)/g'(x)$ exists and equals $L$, it must follow that

$$\lim_{x \to a^+} \frac{f(x)}{g(x)} = L.$$

The statement and proof of Theorem 10.6 may be easily extended to left-hand limits and combined to yield the following statement for two-sided limits.

**THEOREM 10.7**  *Let $f$ and $g$ be continuous on $[a, b]$ and suppose $a < c < b$. Suppose also that $f$ and $g$ are differentiable for all $x$ in the open interval $(a, b)$, except possibly at $x = c$, and assume $g'(x)$ never vanishes. Then if $\lim_{x \to c} f(x) = \lim_{x \to c} g(x) = 0$, it follows that*

$$\lim_{x \to c} \frac{f(x)}{g(x)} = \lim_{x \to c} \frac{f'(x)}{g'(x)}$$

*if this latter limit exists.*

If $\lim_{x \to \infty} f(x) = 0 = \lim_{x \to \infty} g(x)$ then l'Hospital's rule, that is, the statement that

$$\lim_{x \to \infty} \frac{f(x)}{g(x)} = \lim_{x \to \infty} \frac{f'(x)}{g'(x)}$$

if the latter limit exists, is still valid. To prove that, we need only make the change of variable $u = 1/x$. Then defining $\tilde{f}(u) = f(1/u) = f(x)$ and $\tilde{g}(u) = g(1/u) = g(x)$, we have

$$\lim_{u \to 0} \frac{\tilde{f}(u)}{\tilde{g}(u)} = \lim_{x \to \infty} \frac{\tilde{f}(x)}{\tilde{g}(x)}.$$

But differentiating $f$ and $g$ and applying the chain rule yields

$$f'(x) = \tilde{f}'(u) \frac{du}{dx}$$

and

$$g'(x) = \tilde{g}'(u) \frac{du}{dx}.$$

Therefore

$$\frac{f'(x)}{\tilde{g}'(x)} = \frac{\tilde{f}'(u)}{g'(u)} \qquad \text{where} \qquad u = \frac{1}{x}.$$

Hence if we know $\lim_{x \to \infty} f'(x)/g'(x) = L$, it follows that

$$L = \lim_{u \to 0} \frac{\tilde{f}'(u)}{\tilde{g}'(u)} = \lim_{u \to 0} \frac{\tilde{f}(u)}{\tilde{g}(u)} = \lim_{x \to \infty} \frac{f(x)}{g(x)}.$$

**Example 4.**  Compute $\lim_{x \to 1} \log x/(x - 1)$. Clearly $f(x) = \log x \to 0$ as $x \to 1$ as does $g(x) = x - 1$. But $f'(x) = 1/x$ and $g'(x) = 1$. Therefore

$$\lim_{x \to 1} \frac{\log x}{x - 1} = \lim_{x \to 1} \frac{1/x}{1} = 1.$$

**Example 5.**  Often l'Hospital's rule can be iterated if both $f'$ and $g' \to 0$ as $x \to a$ and if $g''(x)$ does not vanish for $x$ close to $a$. Compute

$$\lim_{x \to 0} \frac{\sin x - x}{x^3}.$$

Now by l'Hospital's rule

$$\lim_{x \to 0} \frac{\sin x - x}{x^3} = \lim_{x \to 0} \frac{\cos x - 1}{3x^2}.$$

Both numerator and denominator tend to zero as $x$ tends to zero. Applying l'Hospital's rule twice again, we obtain

$$\lim_{x \to 0} \frac{\cos x - 1}{3x^2} = \lim_{x \to 0} \frac{-\sin x}{6x}$$

$$= \lim_{x \to 0} -\frac{\cos x}{6} = -\frac{1}{6}.$$

When iterating l'Hospital's rule as above, care must be taken that at each stage both numerator and denominator approach zero. If either fails to tend to zero, the method is not applicable. Furthermore, at each stage the derivative of the function in the denominator cannot vanish for $x \neq a$ and close to $a$.

**Example 6.** Sometimes l'Hospital's rule yields no information if applied directly. Determine $\lim_{x \to 0^+} (e^{-1/x}/x)$. Clearly, $e^{-1/x} \to 0$ as $x \to 0^+$ so the rule appears to be applicable. However, applying the rule, we obtain

$$\lim_{x \to 0^+} \frac{e^{-1/x}}{x} = \lim_{x \to 0^+} \frac{1}{x^2} \frac{e^{-1/x}}{1} = \lim_{x \to 0^+} \frac{e^{-1/x}}{x^2}.$$

If we try again, we obtain

$$\lim_{x \to 0^+} \frac{e^{-1/x}}{x^2} = \lim_{x \to 0^+} \frac{e^{-1/x}}{2x^3},$$

and it is evident that nothing is to be gained by further application of the technique. To evaluate the limit, make the substitution $u = 1/x$. Then

$$\lim_{x \to 0^+} \frac{e^{-1/x}}{x} = \lim_{x \to 0^+} \frac{1}{xe^{1/x}} = \lim_{u \to \infty} \frac{u}{e^u}.$$

We may now apply l'Hospital's rule and conclude that $\lim_{u \to \infty} (u/e^u) = \lim_{u \to \infty} (1/e^u) = 0$. Hence

$$\lim_{x \to 0^+} \frac{e^{-1/x}}{x} = 0.$$

## EXERCISES

1. Find a value of $x$ that satisfies Cauchy's mean value formula for the following pairs of functions defined on the given intervals $[a, b]$.
   (a) $f(x) = x - 1$, $g(x) = 2x^2$, $[a, b] = [2, 5]$
   (b) $f(x) = (x + 1)^2$, $g(x) = (x - 1)^3$, $[a, b] = [0, 1]$
   (c) $f(x) = \sin x$, $g(x) = \sin^2 x$, $[a, b] = [0, \pi/2]$
   (d) $f(x) = \log x$, $g(x) = x^2$, $[a, b] = [1, e]$
   (e) $f(x) = e^{x^2}$, $g(x) = e^{-x^2}$, $[a, b] = [0, 1]$
2. Find a point $c$ in the interval $[a, b]$ at which the tangent to the curve $X(t)$ is parallel to $X(b) - X(a)$ if
   (a) $X(t) = (t^2 + 1, t^3 - 1)$, $[a, b] = [0, 1]$
   (b) $X(t) = (\sin t, \cos 2t)$, $[a, b] = [0, \pi/2]$
   (c) $X(t) = (t - \sin t, 1 - \cos t)$, $[a, b] = [0, \pi/2]$
3. Evaluate the following limits.

   (a) $\lim_{x \to 0} \dfrac{\tan x}{x}$

   (b) $\lim_{x \to \pi/2} \dfrac{\cos x}{\sin 2x}$

   (c) $\lim_{x \to 1} \dfrac{\log x}{x^2 + x - 2}$

   (d) $\lim_{x \to 0} \dfrac{\sin x - \tan x}{x^2}$

   (e) $\lim_{x \to 0} \dfrac{a^x - b^x}{x}$   $a, b > 0$

   (f) $\lim_{x \to 0} \dfrac{\log(\cos ax)}{\log(\cos bx)}$   $a, b > 0$

   (g) $\lim_{x \to 0} \dfrac{a^x - a^{\sin x}}{x^3}$   $a > 0$

   (h) $\lim_{x \to 0} \dfrac{\cosh x - \cos x}{x^2}$

   (i) $\lim_{x \to 1^+} \dfrac{x^x - x}{1 - x + \log x}$

   (j) $\lim_{x \to 0^+} \dfrac{e^{-1/x}}{x^2}$

4. Evaluate the following limits.

(a) $\lim\limits_{x\to 0} \dfrac{1}{x^3} \displaystyle\int_0^x \sin t^2\, dt$

(b) $\lim\limits_{x\to 0} \dfrac{1}{x-\sin x} \displaystyle\int_0^x \dfrac{t^2\, dt}{\sqrt{1+t}}$

(c) $\lim\limits_{x\to 0} \dfrac{1}{\sin x - \tan x} \displaystyle\int_0^x t^2\sqrt{1+\sin t}\, dt$

5. If $f$ and $g$ are continuous on $[a, b]$ and $g$ does not vanish on $(a, b)$, show on the basis of formula (7) that at some point $x_0$ in $(a, b)$

(8) $$\int_a^b f(t)g(t)\, dt = f(x_0) \int_a^b g(t)\, dt.$$

6. Use the result of Exercise (5) to derive the following inequalities.

(a) $\dfrac{1}{6\sqrt{2}} \leqslant \displaystyle\int_0^1 \dfrac{x^5}{(1+x)^{1/2}}\, dx \leqslant \tfrac{1}{6}$

(b) $\dfrac{e-1}{2} < \displaystyle\int_0^1 \dfrac{e^x}{1+x}\, dx < e-1$

(c) $\dfrac{1}{\sqrt{2}} < \displaystyle\int_0^{\pi/2} \dfrac{\sin x}{\sqrt{1+x}} < 1$

7. The second mean value theorem for integrals states that if $f$ and $g$ are continuous on $[a, b]$ and if $f'$ is continuous and never vanishes on $[a, b]$, then

(9) $$\int_a^b f(t)g(t)\, dt = f(a) \int_a^{x_0} g(t)\, dt + f(b) \int_{x_0}^b g(t)\, dt$$

for some point $x_0$ in $(a, b)$. To derive this formula, let $G(x) = \displaystyle\int_a^x g(t)\, dt$. Integrating $\displaystyle\int_a^b f(t)g(t)\, dt$ by parts yields

$$\int_a^b f(t)g(t)\, dt = f(b)G(b) - \int_a^b f'(t)G(t)\, dt.$$

Derive formula (9) by applying the integral mean value formula (8) to the integral $\displaystyle\int_a^b f'(t)G(t)\, dt$.

8. In Theorem 10.5, must the tangent $T(t)$ have the same direction as $X(b) - X(a)$ for some point $t$ in $(a, b)$? (Draw some pictures.)

## 10.3   Extensions of l'Hospital's Rule

If $f(x) \to 0$ and $g(x) \to 0$ as $x \to a$, then as we have noted nothing can be deduced from Theorem 1.10 as to the behavior of $f(x)/g(x)$. There are many other examples of this phenomenon. Suppose $f(x) \to \infty$ and $g(x) \to \infty$ as $x \to a$. What is the behavior of $f(x)/g(x)$, and $f(x) - g(x)$, as $x \to a$? Theorem 10.3 may be adapted to handle the first case. The second and other similar examples can often be reduced to the case of a quotient by appropriate algebraic manipulation.

**THEOREM 10.8**   *Suppose $f$ and $g$ are differentiable for all values of $x$ in the open interval $(a, b)$. Suppose, in addition, that $g'$ never vanishes on this interval. If $\lim\limits_{x\to a^+} f(x) = \infty = \lim\limits_{x\to a^+} g(x)$, then*

$$\lim_{x \to a^+} \frac{f(x)}{g(x)} = \lim_{x \to a^+} \frac{f'(x)}{g'(x)}$$

*if this latter limit exists. If*

$$\lim_{x \to a^+} \frac{f'(x)}{g'(x)} = \infty,$$

*then*

$$\lim_{x \to a^+} \frac{f(x)}{g(x)} = \infty,$$

*as well.*

This result has the same formulation as Theorem 10.4 for two-sided limits and applies when $a = \pm\infty$. For a sketch of the proof, see Exercise 18.

Whenever the behavior of

$$\lim_{x \to a} \frac{f(x)}{g(x)}, \ \lim_{x \to a} (f(x) - g(x)), \ \lim_{x \to a} f(x)^{g(x)}$$

cannot be determined from Theorem 1.10 or other theorems on limits, we say that the limit is "*indeterminate.*" This makes no assertion as to the existence of the limit. It just means that the existence of the limit has not been determined. As a consequence, the symbols $0/0$, $\infty/\infty$, $\infty - \infty$, etc. are often called "*indeterminate*" *forms.*

We illustrate Theorem 10.8 and other examples that can be reduced to either Theorems 10.7 or 10.8.

***Example 1.*** Determine $\lim_{x \to \infty} x^n/e^x$. We already know this limit is zero, but iterating Theorem 10.5 $n$ times we obtain

$$\lim_{x \to \infty} \frac{x^n}{e^x} = \lim_{x \to \infty} \frac{n!}{e^x} = 0.$$

***Example 2.*** Determine

$$\lim_{x \to \infty} \frac{x^2 - 1}{4x^2 + x}.$$

Applying Theorem 10.7 twice, we have

$$\lim_{x \to \infty} \frac{x^2 - 1}{4x^2 + x} = \lim_{x \to \infty} \frac{2x}{8x + 1} = \lim_{x \to \infty} = \frac{1}{4}.$$

To determine the behavior of $f(x) - g(x)$, one usually tries to rewrite $f(x) - g(x)$ as a quotient, which may then be handled by either Theorem 10.7 or 10.8.

***Example 3.*** Determine

$$\lim_{x \to 1} \left( \frac{1}{\log x} - \frac{1}{x - 1} \right).$$

Now

$$\frac{1}{\log x} - \frac{1}{x - 1} = \frac{x - 1 - \log x}{\log x(x - 1)}.$$

Also $x - 1 - \log x \to 0$ and $\log x(x - 1) \to 0$ as $x \to 1$. Therefore, applying Theorem 10.7 twice, we have

$$\lim_{x \to 1} \frac{x - 1 - \log x}{\log x(x - 1)} = \lim_{x \to 1} \frac{1 - (1/x)}{(x - 1)/x + \log x}$$

$$= \lim_{x \to 1} \frac{1/x^2}{(1/x^2) + (1/x)} = \lim_{x \to 1} \frac{1}{1 + x} = \frac{1}{2}.$$

*Example 4.*   Determine $\lim\limits_{x\to 0^+} x^x$. Now $x^x = e^{x\log x}$, and

$$\lim_{x\to 0^+} x\log x = \lim_{x\to 0^+} \frac{\log x}{1/x}.$$

We now may apply Theorem 10.8, since $\log x \to -\infty$ and $1/x \to \infty$ as $x \to 0^+$. Therefore

$$\lim_{x\to 0^+} \frac{\log x}{1/x} = \lim_{x\to 0^+} \frac{1/x}{-1/x^2} = \lim_{x\to 0^+} -x = 0.$$

Consequently

$$\lim_{x\to 0^+} x\log x = 0,$$

and

$$\lim_{x\to 0^+} x^x = \lim_{x\to 0^+} e^{x\log x} = 1.$$

**EXERCISES**

Determine the following limits.

1. $\lim\limits_{x\to\infty} \dfrac{(\log x)^2}{x}$.

2. $\lim\limits_{x\to\infty} \dfrac{(\log x)^n}{x}$.

3. $\lim\limits_{x\to\pi/2} (\sec x - \tan x)$.

4. $\lim\limits_{x\to\infty} x(e^{1/x} - 1)$.

5. $\lim\limits_{x\to 1} x^{1/(1-x)}$.

6. $\lim\limits_{x\to 0^+} \dfrac{\log\sin x}{\log\tan x}$.

7. $\lim\limits_{x\to 0} \left(\dfrac{1}{x} - \dfrac{1}{e^x - 1}\right)$.

8. $\lim\limits_{x\to 0} (\csc x - \cot x)$.

9. $\lim\limits_{x\to 0} (e^x + x)^{1/x}$.

10. $\lim\limits_{x\to 0} \dfrac{e^x - 1}{\tan 2x}$.

11. $\lim\limits_{x\to 0} (\cos x)^{1/x}$.

12. $\lim\limits_{x\to 0} \left(\dfrac{\sin x}{x}\right)^{1/x}$.

13. $\lim\limits_{x\to 0} x^{1/\log x}$.

14. $\lim\limits_{x\to 0} \dfrac{(1+x)^{1/x} - e}{x}$.

15. $\lim\limits_{x\to\pi/2} (\sin x)^{\tan x}$.

16. Show that $\lim\limits_{x\to\infty} x^p/a^x = 0$ if $a \geq 1$ and $p$ is any real number. (For $p > 0$, why is it sufficient to prove the result for $p$ an integer?)

17. Let $f(x) = e^{-1/x^2}$, $x \neq 0$, and $f(0) = 0$; show that for each value of $n$ $\lim\limits_{x\to 0} f^{(n)}(x) = 0$. (Show that $\lim\limits_{x\to 0} e^{-1/x^2}/x^n = 0$, for $n = 0, 1, 2, \ldots$)

18. To prove Theorem 10.8 we have to get around the difficulty that $f$ and $g$ cannot be evaluated at $x = a$. However, if $a < x < x_1 < b$, then by Cauchy's mean value formula

$$\frac{f'(x')}{g'(x')} = \frac{f(x) - f(x_1)}{g(x) - g(x_1)} = \frac{f(x)}{g(x)} \frac{[1 - f(x_1)/f(x)]}{[1 - g(x_1)/g(x)]}$$

for some point $x'$ in the interval $(x, x_1)$. Assuming that $f'(x)/g'(x) \to L$ as $x \to a^+$, use the above identity to show that $f(x)/g(x) \to L$ as $x \to a^+$.

## 10.4   Polynomial Approximation and Taylor's Formula

Up to now, for real-valued functions $f$, we have not been concerned with the problem of actually calculating values of the function for given values of $x$. Many times, in fact most, the rule defining the function $f$ does not provide a technique for computing values of $f$. This is certainly true for such common functions as $\sqrt{x}$, $\sin x$, $e^x$. If $f$ is defined as an integral, for example, $\log x = \int_1^x (1/t)\,dt$, we have no clear way of computing values of this function, except by appealing to the definition of the definite integral.

However, polynomials

$$p(x) = a_0 + a_1 x + \cdots + a_n x^n$$

are not subject to this difficulty. The recipe for computing $p(x)$ is present in the rule defining $p$. It makes sense then to investigate to what extent general functions may be approximated by polynomials. Also what schemes are available for carrying out this approximation? We shall consider this problem in some detail in the next two sections.

If $f$ has derivatives up to order $n$ at some point $a$, one approximation scheme would be to try to construct a polynomial having the same value as $f$ at the point $x = a$ and also having the same derivatives as $f$ at this point. A little experimentation shows how to construct such a polynomial. For $n = 1$ instead of writing $p(x) = a_0 + a_1 x$, let us write $p(x) = c_0 + c_1(x - a)$. Then the coefficients $c_0$ and $c_1$ are immediately determined by $p(a)$ and $p'(a)$. Indeed

$$c_0 = p(a) \quad \text{and} \quad c_1 = p'(a).$$

Therefore, if we are to require that $p(a) = f(a)$ and $p'(a) = f'(a)$, we need only set $c_0 = f(a)$ and $c_1 = f'(a)$.

For $n = 2$, if we write

$$p(x) = c_0 + c_1(x - a) + c_2(x - a)^2,$$

then as before $c_0 = p(a)$, $c_1 = p'(a)$. Differentiating twice, we see that $p''(a) = 2c_2$ or $c_2 = p''(a)/2$. Hence, if we wish to require that $p(a) = f(a)$, $p'(a) = f'(a)$, and $p''(a) = f''(a)$, then we should define $c_0 = f(a)$, $c_1 = f'(a)$, and $c_2 = f''(a)/2$.

It is now clear how to proceed in general. If

$$p(x) = c_0 + c_1(x - a) + \cdots + c_n(x - a)^n,$$

then arguing by induction we have for the $k$th derivative

$$p^{(k)}(a) = k \cdot (k - 1) \cdots 1 \cdot c_k.$$

Abbreviating $k \cdot (k - 1) \cdots 1$ by the factorial symbol $k!$, we have

$$p^{(k)}(a) = k! c_k$$

or

$$c_k = \frac{p^{(k)}(a)}{k!}.$$

Therefore, if we require that $p(a) = f(a)$ and $p^{(k)}(a) = f^{(k)}(a)$ for $k = 1, 2, \ldots, n$, then

$$c_k = \frac{f^{(k)}(a)}{k!}.$$

This formula for $c_k$ is valid for $k = 0$ if we agree to define $f^{(0)}(a) = f(a)$ and $0! = 1$.

The polynomial

$$(1) \qquad p_n(x) = f(a) + f'(a)(x - a) + \cdots + \frac{f^{(n)}(a)}{n!}(x - a)^n$$

is called the *Taylor polynomial for the function $f$ about the point $x = a$*. Formula (1) is called *Taylor's formula*, in honor of the English mathematician,

Brook Taylor (1685–1731). If $a = 0$, then the Taylor polynomial becomes

(2) $$p_n(x) = f(0) + f'(0)x + \cdots + \frac{f^{(n)}(0)}{n!} x^n.$$

Formula (2) is sometimes called *Maclaurin's formula*. The subscript $n$ in $p_n$ indicates that the degree of the polynomial in (1) is at most equal to $n$. We state the facts concerning Taylor polynomials as our next result.

**THEOREM 10.9** *Let the function $f$ have derivatives up to order $n$ at the point $a$. Then there exists a unique polynomial $p_n(x)$ of degree $\leq n$ satisfying $p(a) = f(a), p'(a) = f'(a), \ldots, p^{(n)}(a) = f^{(n)}(a)$. This polynomial $p_n$ is defined by the formula*

$$p_n(x) = f(a) + f'(a)(x - a) + \cdots + \frac{f^{(n)}(a)}{n!}(x - a)^n.$$

We have already demonstrated the existence of the polynomial $p_n$. We postpone the proof of uniqueness until the next section. Let us compute the Taylor polynomials for a few functions.

***Example 1.*** If $f(x) = e^x$, then $f^{(n)}(x) = e^x$ for all $n$. Hence $f^{(n)}(0) = 1$. Therefore, the Taylor polynomial for $f(x) = e^x$ about $x = 0$ is given by

$$p_n(x) = 1 + x + \frac{x^2}{2} + \cdots + \frac{x^n}{n!}.$$

***Example 2.*** If $f(x) = \sin x$, then $f'(x) = \cos x$, $f''(x) = -\sin x$, etc. Computing the polynomial about $a = 0$, we have $f(0) = 0$, $f'(0) = 1$, $f''(0) = 0$, $f'''(0) = -1$. In general

$$f^{(2n)}(0) = 0$$

and

$$f^{(2n+1)}(0) = (-1)^n.$$

Hence the Taylor polynomial of degree $2n + 1$ for $f(x) = \sin x$ is given by

$$p_{2n+1}(x) = x - \frac{x^3}{3!} + \frac{x^5}{5!} - \cdots + (-1)^n \frac{x^{2n+1}}{(2n+1)!}.$$

***Example 3.*** If $f(x) = \log x$, then

$$f'(x) = \frac{1}{x}, \quad f''(x) = -\frac{1}{x^2},$$

and in general

$$f^{(n)}(x) = (-1)^{n+1}(n-1)! \frac{1}{x^n}.$$

We cannot compute a Taylor polynomial for $\log x$ about $x = 0$, since the function is not defined at $x = 0$. However, about $x = 1$ we have

$$p_n(x) = x - 1 - \frac{(x-1)^2}{2} + \cdots + (-1)^{n+1} \frac{1}{n}(x-1)^n$$

since $f^{(n)}(1) = (-1)^{n+1}(n-1)!$ If in place of $\log x$ we take $\log(x+1)$, then the Taylor polynomial for $\log(x+1)$ about $x = 0$ is given by

$$p_n(x) = x - \frac{x^2}{2} + \frac{x^3}{3} - \cdots + (-1)^{n+1} \frac{x^n}{n}.$$

For small values of $n$, the graphs of these Taylor polynomials for the functions $e^x$, $\sin x$, and $\log(x+1)$ are given in Figures 5 to 7.

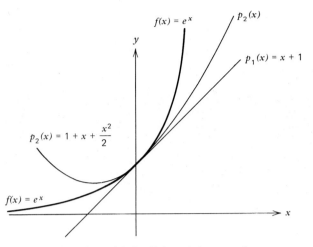

**Fig. 5** Taylor polynomials for $f(x) = e^x$ about $a = 0$.

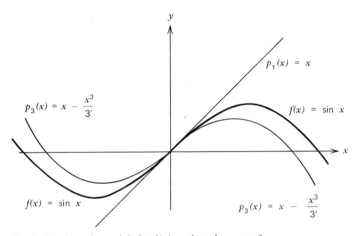

**Fig. 6** Taylor polynomials for $f(x) = \sin x$ about $a = 0$.

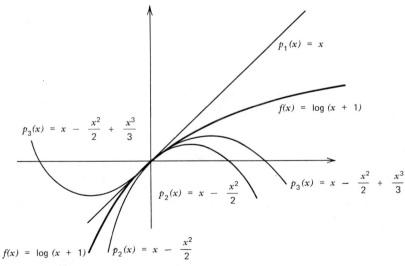

**Fig. 7** Taylor polynomials for $f(x) = \log(x+1)$ about $a = 0$.

Knowing the Taylor polynomials for certain functions enables us to compute the Taylor polynomials for other functions without laboriously calculating all the necessary derivatives. The techniques that enable us to do this are contained in the following theorems. The proofs are easy, and we leave them as exercises. We assume throughout that the functions have derivatives of all orders at zero, although in most cases this is more than we need to assume.

**THEOREM 10.10**   *If $p_n$ and $q_n$ are the Taylor polynomials of degree $n$ for $f$ and $g$ about zero, and $a$ and $b$ are constants, then $ap_n + bq_n$ is the Taylor polynomial of degree $n$ for $af + bg$.*

**THEOREM 10.11**   *If $p_n$ is the Taylor polynomial for $f$ and $k$ is constant, then $q(x) = p_n(kx)$ is the Taylor polynomial for $g(x) = f(kx)$.*

**THEOREM 10.12**   *If*

(1) $$p_n(x) = c_0 + c_1 x + \cdots + c_n x^n$$

*is the Taylor polynomial for $f$ about zero, then*

$$p_n'(x) = c_1 + 2c_2 x + \cdots + nc_n x^{n-1}$$

*is the Taylor polynomial of degree $n-1$ for $f'$. Also*

$$\int_0^x p_n(t)\, dt = c_0 x + \frac{c_1}{2} x^2 + \cdots + \frac{c_n}{n+1} x^{n+1}$$

*is the Taylor polynomial of degree $n+1$ for $F(x) = \int_0^x f(t)\, dt$.*

**Example 4.**   Determine the Taylor polynomial for $f(x) = \frac{1}{2}\log(1+x)/(1-x)$ about $x = 0$. We begin with the fact that the Taylor polynomial for $f(x) = 1/(1-x)$ about zero is $p_n(x) = 1 + x + \cdots + x^n$. This may be immediately verified from the definition.

Applying Theorem 10.11 we deduce that the Taylor polynomial $q_n$ for $g(x) = 1/(1+x)$ can be obtained from $p_n$ by replacing $x$ by $-x$. Hence $q_n(x) = 1 - x + \cdots + (-1)^n x^n$. But

$$-\log(1-x) = \int_0^x \frac{dt}{1-t}$$

and

$$\log(1+x) = \int_0^x \frac{dt}{1+t}.$$

Hence by Theorem 10.12 the Taylor polynomials for these functions can be obtained by integrating $p_n$ and $q_n$ term by term. Thus

$$r_{n+1}(x) = x + \frac{x^2}{2} + \cdots + \frac{x^{n+1}}{n+1}$$

and

$$s_{n+1}(x) = x - \frac{x^2}{2} + \cdots + (-1)^n \frac{x^{n+1}}{n+1}$$

are the Taylor polynomials for $-\log(1-x)$ and $\log(1+x)$, respectively. To finish, note that

$$\frac{1}{2}\log\left(\frac{1+x}{1-x}\right) = \frac{1}{2}[\log(1+x) - \log(1-x)].$$

Hence by Theorem 10.10 the Taylor polynomial for $\frac{1}{2}\log(1+x/1-x)$ will be $\frac{1}{2}[r_n(x) + s_n(x)]$. Since the even order terms in this sum will cancel, we determine the

Taylor polynomial of order $2n+1$. Denoting this by $t_{2n+1}$ we have

$$t_{2n+1}(x) = \frac{1}{2}[r_{2n+1}(x) + s_{2n+1}(x)]$$

$$= \frac{1}{2}\left[x + \frac{x^2}{2} + \cdots + \frac{x^{2n+1}}{2n+1} + x - \frac{x^2}{2} + \cdots + \frac{x^{2n+1}}{2n+1}\right]$$

$$= x + \frac{x^3}{3} + \cdots + \frac{x^{2n+1}}{2n+1}.$$

This is the Taylor polynomial of degree $2n+1$ for $h(x) = \frac{1}{2}\log(1+x)/(1-x)$.

Another extremely useful technique is the following substitution principle. The proof is not automatic but depends on estimates of $f(x)-p_n(x)$. We give it in the next section.

**THEOREM 10.13**  *If $p_n(x) = c_0 + c_1 x + \cdots + c_n x^n$ is the Taylor polynomial of order $n$ for $f(x)$ and $k = 2, 3, \ldots$, then the Taylor polynomial $q_{kn}$ for $g(x) = f(x^k)$ of order $kn$ is given by $q_{nk}(x) = c_0 + c_1 x^k + \cdots + c_n x^{kn} = p_n(x^k)$.*

*Example 5.*  Complete the Taylor polynomial of degree $2n+1$ for arc tan $x$. We use Theorems 10.13 and 10.12. First the Taylor polynomial of degree $n$ for $1/(1+x)$ is $1-x+\cdots+(-1)x^n$. Hence by Theorem 10.13, the Taylor polynomial of degree $2n$ for $1/(1+x^2)$ is $1-x^2+x^4-x^6+\cdots+(-1)^n x^{2n}$. But arc tan $x = \int_0^x (1+t^2)^{-1} dt$. Therefore by Theorem 10.12, the Taylor polynomial $p_{2n+1}(x)$ for arc tan $x$ is given by

$$p_{2n+1}(x) = \int_0^x (1-t^2+\cdots+(-1)^n t^{2n}) dt$$

$$= x - \frac{x^3}{3} + \frac{x^5}{5} - \cdots + (-1)^n \frac{x^{2n+1}}{2n+1}.$$

**EXERCISES**

Determine the Taylor polynomials of degree $n$ about the origin for the following functions. Exercise care with the $n$th-order term.

1. $f(x) = (1-x)^{-1}$.        4. $f(x) = (1-x)^{-3/2}$.
2. $f(x) = (1-x)^{-1/2}$.        5. $f(x) = (1+x)^{3/2}$.
3. $f(x) = (1+x)^{1/2}$.

For the following functions the Taylor polynomials about zero have either the odd-order terms or the even-order terms identically equal to zero. In the former case, determine a Taylor polynomial of degree $2n$ for $f$; in the latter, determine one of degree $2n+1$.

6. $f(x) = \cos x$.        8. $f(x) = \sinh x$.
7. $f(x) = \cosh x$.        9. $f(x) = \tan x$.

Determine the Taylor polynomials about the specified points for the following polynomials.

10. $f(x) = x^3$       about $a = 1$.
11. $f(x) = x^5$       about $a = -1$.
12. $f(x) = 4x^3 - 2x + 1$       about $a = 2$.
13. Determine Taylor polynomials for $\sin x$ and $\cos x$ of degree $n$ about the point $a = \pi/4$.

Using Theorems 10.10 to 10.13 determine Taylor polynomials about the origin of indicated degree for the following functions.

14. $f(x) = \sqrt{1+4x}$       degree $n$
15. $f(x) = (1-x^2)^{-1/2}$       degree $2n$
16. $f(x) = $ arc sin $x$       degree $2n+1$
17. $f(x) = \sinh^{-1} x$       degree $2n+1$
18. $f(x) = \int_0^x e^{t^2} dt$       degree $2n+1$

19. $g(x) = \int_0^x \sqrt{1-t^3}\, dt$     degree $3n+1$
20. Prove Theorem 10.10.
21. Prove Theorem 10.11.
22. Prove Theorem 10.12.

## 10.5   The Error in Taylor's Formula

Figures 5 to 7 in the previous section indicate that the approximation to $f$ by the Taylor polynomial $p_n(x)$ is quite good around the point $a$, but away from this point, the approximation deteriorates. To give an estimate of the error involved in this approximation is our next task. Let us set

$$R_n(x) = f(x) - p_n(x)$$

$$= f(x) - \left( f(a) + \cdots + \frac{f^{(n)}(a)}{n!}(x-a)^n \right).$$

We call $R_n(x)$ the remainder or error at $x$ in Taylor's formula. To estimate the size of $R_n(x)$, we derive the formula for the Taylor polynomial by successively integrating by parts. This procedure keeps track of the magnitude of $R_n(x)$ for each value of $n$.

**THEOREM 10.14**   *If $f^{(n+1)}(x)$ is continuous, then*

(1)     $f(x) = f(a) + f'(a)(x-a) + \cdots + \dfrac{1}{n!} f^{(n)}(a)(x-a)^n +$

$$\frac{1}{n!} \int_a^x f^{(n+1)}(t)(x-t)^n\, dt.$$

*Hence, if $R_n(x) - f(x) - p_n(x)$ where $p_n(x)$ is the Taylor polynomial for $f$ about $x = a$, then*

(2)                     $R_n(x) = \dfrac{1}{n!} \int_a^x f^{(n+1)}(t)(x-t)^n\, dt.$

Formula (1) is often called *Taylor's formula with integral remainder*. To derive formula (1) we successively integrate by parts. Note that

$$f(x) - f(a) = \int_a^x f'(t)\, dt = f'(t)(x-t)\, \Big|_a^x + \int_a^x f''(t)(x-t)\, dt.$$

Hence

(3)                     $f(x) = f(a) + f'(a)(x-a) + \displaystyle\int_a^x f''(t)(x-t)\, dt.$

In the integration by parts formula

$$\int_a^x u(t)v'(t)\, dt = u(t)v(t)\, \Big|_a^x - \int_a^x v(t)u'(t)\, dt$$

we have chosen $u(t) = f'(t)$ and $v'(t) = 1$. Hence, $u'(t) = f''(t)$ and we may take $v(t) = (t-x)$.

Integrating by parts again, we have

$$\int_a^x f''(t)(x-t) = -\frac{1}{2} f''(t)(x-t)^2\, \Big|_a^x + \frac{1}{2}\int_a^x f'''(t)(x-t)^2\, dt$$

$$= \frac{1}{2} f''(a)(x-a)^2 + \frac{1}{2}\int_a^x f'''(t)(x-t)^2\, dt.$$

Substitution in (3) yields

$$f(x) = f(a) + f'(a)(x-a) + \frac{1}{2}f''(a)(x-a)^2 + \frac{1}{2}\int_a^x f'''(t)(x-t)^2\,dt.$$

Formula (1) now follows by induction.

Another useful formula for $R_n(x)$ can be obtained by applying formula (8), p. 408, of Section 10.2. This formula states that if $g$ and $h$ are continuous on the interval $[a, b]$ and $h$ does not vanish on the open interval $(a, b)$, then for some point $c$, $a < c < b$,

$$\int_a^b g(t)h(t)\,dt = g(c)\int_a^b h(t)\,dt.$$

If we let $g(t) = f^{(n+1)}(t)$ and $h(t) = (x-t)^n$, then we may apply this formula to (2) obtaining

$$R_n(x) = \frac{1}{n!}\int_a^x f^{(n+1)}(t)(x-t)^n\,dt$$

$$= \frac{f^{n+1}(c)}{n!}\int_a^x (x-t)^n\,dt \qquad a < c < x$$

$$= \frac{f^{n+1}(c)}{(n+1)!}(x-a)^{n+1}.$$

We state this result as our next theorem.

**THEOREM 10.15** *Let $p_n(x)$ be the Taylor polynomial for $f$ of degree $n$ about $x = a$, and let $R_n(x) = f(x) - p_n(x)$ be the remainder. If $f^{(n+1)}$ is continuous on the interval $[a, x]$, then*

(4)
$$R_n(x) = \frac{f^{(n+1)}(c)}{(n+1)!}(x-a)^{n+1}$$

*for some point $c$ satisfying $a < c < x$.*

Let us use this formula to estimate the error in Taylor's formula for the three examples considered in the previous section.

**Example 1.** If $f(x) = e^x$, then the Taylor polynomial about zero for this function is

$$p_n(x) = 1 + x + \cdots + \frac{x^n}{n!}.$$

The remainder $R_n(x)$ by Theorem 10.15 is given by

$$R_n(x) = \frac{e^c}{(n+1)!}x^{n+1} \qquad \text{where} \qquad 0 < c < x.$$

If $x \leqslant 1$, then $R(x) \leqslant e/(n+1)!$ since $c < x \leqslant 1$. To achieve an accuracy of 0.001 in an approximation of $e^x$ by $p_n(x)$ for $0 \leqslant x \leqslant 1$, we would have to choose $n$ so that $e/(n+1)! \leqslant 0.001$. Letting $e = 2.718$, we see that this is achieved if $1/(n+1)! \leqslant 1/2718$ or $n \geqslant 6$. Thus the approximation $p_n(x) = 1 + x + \cdots + x^6/6!$ gives an approximation to $e^x$ on the interval $[0, 1]$ with an error less than 0.001.

**Example 2.** For $f(x) = \sin x$, the remainder $R_n(x)$ for the Taylor polynomial about zero is

$$R_n(x) = \frac{f^{n+1}(c)}{(n+1)!}x^{n+1}.$$

However, the derivatives of $\sin x$ are all bounded in absolute value by one. Hence

$$|R_n(x)| \leq \frac{|x^{n+1}|}{(n+1)!}.$$

Thus to achieve an accuracy of 0.001 on the interval $|x| \leq \pi$, we must choose $n$ so large that

$$\frac{\pi^{n+1}}{(n+1)!} \leq 0.001.$$

This occurs if $n \geq 13$. Hence for $|x| \leq \pi$, the Taylor polynomial $x - (x^3/3!) + \cdots + (x^{13}/13!)$ approximates $\sin x$ to within 0.001.

**Example 3.**   If $f(x) = \log(x+1)$, then the Taylor polynomial about zero is

$$p_n(x) = x - \frac{x^2}{2} + \cdots + (-1)^{n+1}\frac{x^n}{n}.$$

To compute $R_n(x)$ for $0 \leq x \leq 1$, we note that $|f^{(n)}(x)| = (n-1)!(x+1)^{-n}$. Therefore

$$|R_n(x)| \leq \frac{n!}{(n+1)!}(c+1)^{-n-1}x^{n+1} \leq \frac{1}{n+1}.$$

Now to achieve an accuracy of 0.001, we must choose $1/(n+1) \leq 0.001$ or $n+1 \geq 1000$. To compute $\log 2$ to an accuracy of 0.001 by a sum of the form $1 - \frac{1}{2} + \frac{1}{3} - \cdots + (-1)^{n+1}(1/n)$ takes approximately 1000 terms!

Taylor polynomials are only effective as methods for approximating functions when the remainders $R_n(x)$ become small rapidly as $n$ becomes large. We say that $R_n(x)$ tends to zero as $n$ tends to $\infty$ if, for each positive $\epsilon$, there exists a positive integer $N$ such that $|R_n(x)| < \epsilon$ whenever $n > N$. We abbreviate this by writing $R_n(x) \to 0$ as $n \to \infty$. When this is true we say that the Taylor polynomials $p_n(x)$ converge to the function $f(x)$ as $n \to \infty$. For $e^x$ and $\sin x$, the above discussion shows that this is true for all values of $x$. For $\log(x+1)$, $R_n(x)$ does not tend to zero if $x > 1$ although this is not obvious from the formula for $R_n(x)$. There are functions $f$ for which $R_n(x)$ fails to tend to zero whenever $x \neq a$. (See Exercise 14.)

In certain cases, more explicit information on the behavior of $R_n(x)$ may be obtained.

**Example 4.**   If $f(x) = \arctan x$, then we assert

$$R_{2n+1}(x) = (-1)^{n+1}\int_0^x \frac{t^{2n+2}}{1+t^2}\,dt.$$

To derive this formula we proceed exactly as in the previous section when we derived the formula for the Taylor polynomial for $\arctan x$ except that we keep track of the remainder at each stage. We start with the formula

(5) $$\frac{1}{1-x} = 1 + x + \cdots + x^n + \frac{x^{n+1}}{1-x}$$

valid for all $x \neq 1$. Formula (5) shows incidentally that the remainder in Taylor's formula for $1/(1-x)$ is exactly $x^{n+1}/(1-x)$. Replacing $x$ by $-x^2$ we have

(6) $$\frac{1}{1+x^2} = 1 - x^2 + x^4 - \cdots + (-1)^n x^{2n} + (-1)^{n+1}\frac{x^{2n+2}}{1+x^2}$$

and

(7)
$$\text{arc tan } x = \int_0^x \frac{dt}{1+t^2}$$

$$= \int_0^x (1 - t^2 + \cdots + (-1)^n t^{2n}) dt + (-1)^{n+1} \int_0^x \frac{t^{2n+2}}{1+t^2} dt$$

$$= x - \frac{x^3}{3} + \cdots + (-1)^n \frac{x^{2n+1}}{2n+1} + (-1)^{n+1} \int_0^x \frac{t^{2n+2}}{1+t^2} dt.$$

Since we know the Taylor polynomial of degree $2n+1$ for $f(x) = \text{arc tan } x$ is

(8)
$$p_{2n+1}(x) = x - \frac{x^3}{3} + \cdots + (-1)^n \frac{x^{2n+1}}{2n+1},$$

we conclude that the remainder $R_{2n+1}(x)$ is given by

(9)
$$R_{2n+1}(x) = (-1)^{n+1} \int_0^x \frac{t^{2n+2}}{1+t^2} dt.$$

To estimate $|R_{2n+1}(x)|$, we observe that if $|x| \leqslant 1$, and $0 \leqslant |t| \leqslant |x|$, then $1 + t^2 \geqslant 1$. Hence

$$\left| \int_0^x \frac{t^{2n+2} dt}{1+t^2} \right| \leqslant \left| \int_0^x t^{2n+2} dt \right| = \frac{1}{2n+3} |x|^{2n+3} \leqslant \frac{1}{2n+3}.$$

As $n \to \infty$, $1/(2n+3) \to 0$. Hence for $|x| \leqslant 1$ the Taylor polynomials, given by (8), converge to $\text{arc tan } x$ as $n \to \infty$. If we take $x = 1$, then $\text{arc tan } 1 = \pi/4$. Hence we have the following formula:

(10)
$$\frac{\pi}{4} = \lim_{n \to \infty} \left( 1 - \frac{1}{3} + \frac{1}{5} - \cdots + (-1)^n \frac{1}{2n+1} \right).$$

Evaluating the sum on the right of (10) is not a practical way to compute $\pi/4$ because the convergence is too slow. Indeed after $n$ terms we have an error greater than $1/(4n+12)$ (Exercise 5). However, the Taylor polynomial for $\text{arc tan } x$ can be used to give very accurate estimates of $\pi/4$ if we make use of the following trick. The addition formula for the tangent states that

$$\tan(\alpha + \beta) = \frac{\tan \alpha + \tan \beta}{1 - \tan \alpha \tan \beta}$$

If we choose $\alpha$, $\beta$ so that $\tan \alpha = \frac{1}{2}$ and $\tan \beta = \frac{1}{3}$, we have

$$\tan(\alpha + \beta) = \frac{\frac{1}{2} + \frac{1}{3}}{1 - \frac{1}{2} \frac{1}{3}} = 1.$$

Therefore $\alpha + \beta = \pi/4$, or

$$\frac{\pi}{4} = \text{arc tan } \frac{1}{2} + \text{arc tan } \frac{1}{3}$$

$$= \lim_{n \to \infty} \left( \frac{1}{2} - \frac{1}{3} \frac{1}{2^3} + \frac{+(-1)^n}{2n+1} \frac{1}{2^{2n+1}} \right)$$

$$+ \lim_{n \to \infty} \left( \frac{1}{3} - \frac{1}{3} \frac{1}{3^3} + \cdots + \frac{(-1)^n}{2n+1} \frac{1}{3^{2n+1}} \right).$$

The convergence now is very rapid. Exercise 6 shows that after only 5 terms in each of the sums the error is less than $5 \times 10^{-5}$.

We conclude this section with a result that characterizes the Taylor polynomial for a function $f$. This result is commonly called *Taylor's theorem*.

**THEOREM 10.16**   *Suppose $f^{(n+1)}(x)$ is continuous and let $p_n(x)$ be the Taylor polynomial of degree n for f about zero. Then $p_n(x)$ is the unique polynomial of degree n satisfying*

(11)
$$\lim_{x \to 0} \frac{f(x) - p_n(x)}{x^n} = 0.$$

**Proof.**   Note that (11) asserts that $f(x) - p_n(x) \to 0$ as $x \to 0$ faster than $x^n$. Formula (11) is a simple consequence of the derivative form (4) of the error in Taylor's formula. That is,

$$f(x) - p_n(x) = \frac{f^{(n+1)}(c)x^{n+1}}{(n+1)!}$$

or

$$\frac{f(x) - p_n(x)}{x^n} = \frac{f^{(n+1)}(c)x}{(n+1)!}.$$

Since $0 < c < x$, if $x \to 0$, then $c \to 0$. Hence

$$\lim_{x \to 0} \frac{f(x) - p_n(x)}{x^n} = \lim_{c \to 0} \frac{f^{n+1}(c)}{(n+1)!} \lim_{x \to 0} x = 0.$$

This shows that $p_n$ satisfies (11). To show that $p_n$ is the only polynomial of degree $n$ with this property, suppose

(12)
$$\lim_{x \to 0} \frac{f(x) - q_n(x)}{x^n} = 0$$

where $q_n(x)$ is another polynomial of degree $n$. Subtracting (12) from (11), we have

$$0 = \lim_{x \to 0} \left[ \frac{f(x) - p_n(x)}{x^n} - \frac{f(x) - q_n(x)}{x^n} \right] = \lim_{x \to 0} \frac{q_n(x) - p_n(x)}{x^n}.$$

Now $q_n(x) - p_n(x) = r_n(x)$ is a polynomial of degree $n$. We shall be finished if we can show that

$$\lim_{x \to 0} \frac{r_n(x)}{x^n} = 0$$

implies that $r_n$ is the zero polynomial. But

$$\lim_{x \to 0} \frac{r_n(x)}{x^n} = 0$$

implies

$$\lim_{x \to 0} \frac{r_n(x)}{x^k} = 0$$

for $k = 0, 1, \ldots, n$ since

$$\frac{r_n(x)}{x^k} = \frac{r_n(x)}{x^n} x^{n-k}.$$

Now $r_n(x) = c_0 + c_1 x + \cdots + c_n x^n$, and if $r_n(x) \to 0$ as $x \to 0$ then $c_0 = 0$. If $r_n(x)/x \to 0$ as $x \to 0$, then $c_1 = 0$. Continuing inductively, we see that all the coefficients $c_0 = c_1 = \cdots = c_n = 0$. Therefore, $r_n$ is the zero polynomial, and the proof is complete.

The uniqueness statement in Theorem 10.9 now follows as a corollary. We leave the verification of this as Exercise 11. We may use Theorem 10.16 to prove the substitution theorem, 10.13.

To establish Theorem 10.13 we must show that if $p_n$ is the Taylor polynomial of degree $n$ for $f$ and $k = 2, 3, \ldots$, then $p_n(x^k)$ is the Taylor polynomial of degree $kn$ for $f(x^k)$. But by Theorem 10.16

$$\lim_{x \to 0} \frac{f(x) - p_n(x)}{x^n} = 0.$$

But as $x \to 0$, $x^k \to 0$, hence

(13) $$\lim_{x \to 0} \frac{f(x^k) - p_n(x^k)}{x^{kn}} = 0.$$

Now $p_n(x^k)$ is a polynomial of degree $kn$, and Theorem 10.16 asserts that the Taylor polynomial of degree $kn$, that is, $p_{nk}(x)$ is the only polynomial satisfying (13). Therefore $p_{nk}(x) = p_n(x^k)$ for all $x$, and the proof of Theorem 10.13 is complete.

### EXERCISES

1. Write Taylor's formula with integral remainder about $a = 0$ for $f(x) = e^x$ and $f(x) = \sin x$.
2. If $f(x) = e^x$, estimate the remainder $R_n(x)$ in Taylor's formula and show that $R_n(x)$ tends to zero as $n$ tends to infinity.
3. Obtain Taylor's formula for $\log (1 + x)$ by integrating the expression

$$\frac{1}{1 + t} = 1 - t + t^2 - \cdots + (-1)^{n-1} t^{n-1} + (-1)^n \frac{t^n}{1 + t}.$$

   (Follow the same technique as for arc tan $x$.)
4. In Exercise 3 give upper and lower estimates for $|R_n(1)|$ by using formula (8), p. 408, of Section 10.2.
5. Give upper and lower estimates on the absolute value of the remainder $R_{2n+1}(1)$ in Taylor's formula for the arc tan $x$ by using formula (8), p. 408, of Section 10.2.
6. Show that

$$\frac{1}{2} - \frac{1}{3}\frac{1}{2^3} + \frac{1}{5}\frac{1}{2^5} - \frac{1}{7}\frac{1}{2^7} + \frac{1}{9}\frac{1}{2^9} + \frac{1}{3} - \frac{1}{3}\frac{1}{3^3} + \frac{1}{5}\frac{1}{3^5} - \frac{1}{7}\frac{1}{3^7} + \frac{1}{9}\frac{1}{3^9}$$

   approximates $\pi/4$ with an error of no more than $5 \times 10^{-5}$.
7. Determine the Taylor polynomial for $\log (1+x)/(1-x)$ about $x = 0$ by using the result of Exercise 3 and the fact that the Taylor polynomial for $\log (1-x)$ may be determined by exploiting the identity

$$\frac{1}{1-x} = 1 + x + \cdots + x^{n-1} + \frac{x^n}{1-x}.$$

   Write the remainder as an integral.
8. Determine a Taylor approximation to $e$ accurate to 0.0001.
9. Determine a Taylor approximation to $\log 2$ accurate to 0.01 using the Taylor polynomial in Exercise 7. (If $f(x) = \frac{1}{2} \log (1+x)/(1-x)$, then $p_{2n+1}(x) = x + \frac{1}{3}x^3 + \cdots + 1/(2n+1)x^{2n+1}$.)
10. Determine a Taylor polynomial approximation to $f(x) = \sin x$ on the interval $|x| \leqslant \pi/4$ that is accurate to 0.001.
11. Prove that if $p(x)$ is a polynomial of degree $n$, then

$$p(x) = p(a) + p'(a)(x - a) + \cdots + \frac{p^{(n)}(a)}{n!} (x - a)^n.$$

This proves the uniqueness part of Theorem 10.9. (Examine the remainder.)

12. Obtain a Taylor polynomial approximation to the integral $\int_0^x e^{t^2}\, dt$. How many terms must be taken if the error is to be less than 0.001 when $0 \leqslant x \leqslant 1$?

13. If $f(t) = e^{-t^2}$, show that the Taylor polynomial $p_8(t) = 1 - t^2 + t^4/2! - t^6/3! + t^8/4!$. For $0 \leqslant t \leqslant \frac{1}{2}$, show that the remainder $R_8(t)$ satisfies $|R_8(t)| \leqslant (\frac{1}{2})^{10}/5!$. Using this show that

$$\int_0^{1/2} e^{-t^2}\, dt = \frac{1}{2} - \frac{1}{3 \cdot 2^3} + \frac{1}{5 \cdot 2^5 \cdot 2!} - \frac{1}{7 \cdot 2^7 \cdot 3!} + \frac{1}{9 \cdot 2^9 \cdot 5!} + \epsilon$$

where $|\epsilon| \leqslant 4.5 \times 10^{-6}$.

14. Determine the Taylor polynomial about zero for the function $f(x) = e^{-1/x^2}$, $x \neq 0$, $f(0) = 0$. Does $R_n(x) \to 0$ as $n \to \infty$?

15. Prove that the Taylor polynomials for $f(x) = \sin x$ converges to $\sin x$ for each value of $x$ as $n \to \infty$.

16. Show that the Taylor polynomial for $f(x) = \frac{1}{2}\log (1+x)/(1-x)$ converges to $f(x)$ for each $x$, $0 < |x| < 1$, as $n \to \infty$.

17. Use Theorem 10.16 and equation (5), p. 418 to conclude that $p_n(x) = 1 + x + \cdots + x^n$ is the Taylor polynomial of degree $n$ for $f(x) = 1/(1-x)$.

# CHAPTER ELEVEN

# INFINITE SERIES

**11.1 Preliminaries — Summation Notation**

In the last chapter we derived many limit relations such as

$$e = \lim_{n \to \infty} \left( 1 + 1 + \frac{1}{2!} + \frac{1}{3!} + \cdots + \frac{1}{n!} \right)$$

$$\log 2 = \lim_{n \to \infty} \left( 1 - \frac{1}{2} + \frac{1}{3} - \cdots + (-1)^{n-1} \frac{1}{n} \right)$$

$$\frac{\pi}{4} = \lim_{n \to \infty} \left( 1 - \frac{1}{3} + \frac{1}{5} - \cdots + \frac{(-1)^{n-1}}{2n-1} \right).$$

We also showed that many functions could be written as the limit of certain sums. For example we showed that

$$\sin x = \lim_{n \to \infty} \left( x - \frac{x^3}{3!} + \cdots + (-1)^n \frac{x^{2n+1}}{(2n+1)!} \right)$$

for all values of $x$, and

$$\text{arc tan } x = \lim_{n \to \infty} \left( x - \frac{x^3}{3} + \cdots + (-1)^n \frac{x^{2n+1}}{2n+1} \right)$$

if $|x| \leq 1$. In many cases, these limit relations give important methods for computing the values of the function $f$.

In this chapter we shall consider the problem of evaluating limits of the form $\lim_{n \to \infty} (a_1 + a_2 + \cdots + a_n)$ where $a_1, a_2, \ldots, a_n \ldots$ are real numbers. In the usual way we say that $\lim_{n \to \infty} (a_1 + \cdots + a_n) = s$, if for each $\epsilon > 0$ there exists a corresponding integer $N$ such that $|a_1 + \cdots + a_n - s| < \epsilon$ whenever $n > N$. Indeed, if $\lim_{n \to \infty} (a_i + \cdots + a_n)$ exists and equals $s$, then we say that the number $s$ is the *sum* of the infinite collection of numbers $a_1, a_2, \ldots, a_n, \ldots$ and we write

$$s = a_1 + a_2 + \cdots + a_n + \cdots.$$

A formal expression $a_1 + a_2 + \cdots + a_n + \cdots$ is called an *infinite series*. If $\lim_{n \to \infty} (a_1 + \cdots + a_n)$ exists, then this series is said to *converge*. If the limit fails to exist, then the series is said to *diverge*. When the limit $s = \lim_{n \to \infty} (a_1 + \cdots + a_n)$ does exist, $s$ is called the *sum* of the series. The numbers $a_1, a_2, \ldots$ are called the *terms* of the series, and the sum $s_n$ of the first $n$ terms, $a_1 + \cdots + a_n$, is called the *nth partial sum* of the series. Thus the infinite series $a_1 + \cdots + a_n + \cdots$ converges if the nth partial sums

$$s_n = a_1 + \cdots + a_n$$

possess a limit as $n \to \infty$.

When we write the numbers $a_1, a_2, \ldots$, we are asserting, although we have not precisely said so, that to each positive integer $n$ is associated a well-defined real number $a_n$. Thus, we are in reality defining a function $f$ with domain the set of positive integers such that for each $n$

$$f(n) = a_n.$$

Such a function is called a *sequence*. However, we do not use the function, or $f(x)$, notation to describe sequences. The notation $a_1, a_2, \ldots, a_n, \ldots$ indicates clearly enough that associated with each integer $n$ is the unique number $a_n$. Thus the sequence $a_1, a_2, \ldots$ is called the *sequence* of *terms* of the series $a_1 + a_2 + \cdots + a_n + \cdots$. If $s_1 = a_1$, $s_2 = a_1 + a_2, \ldots, s_n = a_1 + \cdots + a_n$, then $s_1, s_2, \ldots$ is called the *sequence* of *partial sums* of the series.

When the expression for $a_n$ is complicated, the notation

(1) $$a_1 + \cdots + a_n + \cdots$$

may become awkward. It is convenient to introduce the summation notation as a tool for expressing (1). We use the capital Greek letter $\Sigma$ (sigma) to indicate a sum and write

(2) $$\sum_{n=1}^{\infty} a_n = a_1 + a_2 + \cdots + a_n + \cdots.$$

When writing (2) we are not asserting that the series converges, that is, that the partial sums have a limit. The symbol $\sum_{n=1}^{\infty} a_n$ is just an abbreviation for $a_1 + \cdots + a_n + \cdots$. However, if independently it is known that the series does converge and $s$ is the sum, then we write

$$s = \sum_{n=1}^{\infty} a_n.$$

We use the summation notation for finite sums as well. Indeed

$$\sum_{n=1}^{N} a_n = a_1 + a_2 + \cdots + a_N$$

The letter $n$ in the summation symbol is called the *index* of summation. When it is clear which letter denotes the index of summation, the latter is often omitted from the symbol $\Sigma$. Thus

$$\sum_{n=1}^{\infty} a_n = \sum_{1}^{\infty} a_n$$

are both abbreviations for the same thing. Indeed, if it is clear that we are talking about an infinite rather than a finite series, we may write

$$\sum a_n \quad \text{for} \quad \sum_{n=1}^{\infty} a_n.$$

In the summation sign

$$\sum_{n=1}^{\infty} \quad \text{or} \quad \sum_{n=3}^{N},$$

the symbols $1, \infty$ and $3, N$ are called the *limits of summation*. For example,

$$\sum_{n=1}^{\infty} \frac{1}{n} = 1 + \frac{1}{2} + \frac{1}{3} + \cdots.$$

$$\sum_{n=0}^{\infty} \frac{1}{n!} = 1 + 1 + \frac{1}{2!} + \frac{1}{3!} + \cdots. \qquad \text{(remember } 0! = 1\text{)}$$

$$\sum_{n=1}^{\infty} \frac{(-1)^{n+1}}{1+n^2} = \frac{1}{2} - \frac{1}{5} + \frac{1}{10} - \cdots.$$

$$\sum_{n=0}^{\infty} (-1)^n \frac{x^{2n}}{(2n)!} = 1 - \frac{x^2}{2!} + \frac{x^4}{4!} - \frac{x^6}{6!} + \cdots.$$

In the previous chapter we met many examples of convergent infinite series. Indeed, we showed that

$$\frac{1}{1-x} = \sum_{n=0}^{\infty} x^n \qquad \text{if} \qquad |x| < 1$$

$$e^x = \sum_{n=0}^{\infty} \frac{x^n}{n!} \qquad \text{for all} \qquad x$$

$$\log(1+x) = \sum_{n=1}^{\infty} (-1)^{n+1} \frac{x^n}{n} \qquad \text{if} \qquad -1 < x \leqslant 1$$

Each of these is an example of a convergent infinite series. However,

$$\sum_{n=1}^{\infty} \frac{1}{n}$$

does not converge since by Exercise 13, p. 147 we know that

$$\sum_{n=1}^{N} \frac{1}{n} \geqslant \log(N+1)$$

and $\lim_{N \to \infty} \log(N+1) = \infty$.

The calculus of limits of sequences is exactly the same as that for limits of functions which we have discussed earlier. Since sequences are special cases of functions, this is to be expected. The precise definition of $\lim_{n \to \infty} s_n$ is as follows.

**Definition.** $\lim_{n \to \infty} s_n = s$ *if to each $\epsilon > 0$ there corresponds an integer $N$ such that whenever $n > N$ it follows that*

$$|s_n - s| < \epsilon.$$

The fundamental rules for operating with limits of sequences are contained in the following.

**THEOREM 11.1** *If $\lim_{n \to \infty} s_n = s$ and $\lim_{n \to \infty} t_n = t$, then*

1. $\lim_{n \to \infty} (as_n + bt_n) = as + bt$ *for arbitrary real numbers $a$ and $b$.*
2. $\lim_{n \to \infty} s_n t_n = st$.
3. $\lim_{n \to \infty} \dfrac{s_n}{t_n} = \dfrac{s}{t}$ *if $t \neq 0$.*

The proofs of these facts are exactly the same as those for the fundamental theorem on limits, Theorem 1.10. They are left as exercises. Note in part 3 that if $t \neq 0$, then $t_n \neq 0$ for $n$ sufficiently large. (Why?) Hence for $n$ large enough $s_n/t_n$ always may be defined.

We shall use the first statement in Theorem 11.1 repeatedly in operations with series. It states that if $\sum\limits_{n=1}^{\infty} a_n$ and $\sum\limits_{n=1}^{\infty} b_n$ converge, then for each pair of real numbers $c$ and $d$ the series $\sum\limits_{n=1}^{\infty} (ca_n + db_n)$ converges. Furthermore

$$\sum_{n=1}^{\infty} (ca_n + db_n) = c \sum_{n=1}^{\infty} a_n + d \sum_{n=1}^{\infty} b_n.$$

It also follows immediately from Theorem 11.1 that finitely many terms of a series $\sum\limits_{n=1}^{\infty} a_n$ may be assigned arbitrarily without affecting the convergence or divergence of the series. We leave the proof of this fact to Exercise 17.

We close this section with a necessary condition for a series to converge. That is, if the series converges, then the condition must hold.

**THEOREM 11.2** *If $\sum\limits_{n=1}^{\infty} a_n$ converges, then $\lim\limits_{n \to \infty} a_n = 0$.*

**Proof.** Write $s_n = a_1 + \cdots + a_n$. If the series converges, $s = \lim\limits_{n \to \infty} s_n$ exists. But by the definition of a limit, $s = \lim\limits_{n \to \infty} s_{n-1}$. Hence by Theorem 11.1, part 1

$$0 = s - s = \lim_{n \to \infty} s_n - \lim_{n \to \infty} s_{n-1}$$
$$= \lim_{n \to \infty} (s_n - s_{n-1})$$
$$= \lim_{n \to \infty} a_n.$$

Therefore, if a series converges, the $n$th term tends to zero. Equivalently, if the $n$th term fails to tend to zero as $n$ tends to $\infty$, the series *does not converge*. However, if the $n$th term does tend to zero, this implies nothing about the convergence. For example, $\sum\limits_{n=1}^{\infty} 1/n$ diverges yet $\sum\limits_{n=1}^{\infty} 1/2^n$ converges. In both cases the $n$th term tends to zero.

**Example 1.** $\sum\limits_{n=1}^{\infty} n/(n+1)$ does not converge since $\lim\limits_{n \to \infty} n/(n+1) = 1$. The $n$th term of $\sum\limits_{n=2}^{\infty} n/(n^2 - 1)$ tends to zero, as does the $n$th term of

$$\sum_{n=1}^{\infty} \frac{n^2}{4 - n^4}.$$

We shall verify presently that the first series diverges but the second converges. If for the series $\sum a_n$, $a_n = f(n)$ for a familiar function $f$, then often l'Hospital's rule is useful in determining if $\lim\limits_{n \to \infty} a_n = \lim\limits_{n \to \infty} f(n) = 0$. Indeed, to show that the $n$th term of $\sum\limits_{n=1}^{\infty} n^2/e^n$ tends to zero, set $f(x) = x^2/e^x$. If we apply l'Hospital's rule twice to $f$, we deduce that $\lim\limits_{x \to \infty} f(x) = 0$.

**EXERCISES**

1. Determine the $n$th term of each of the following series and abbreviate the series by using the summation notation. It may be more convenient to denote the first term of the series by $a_0$, or $a_2$, or $a_{-1}$ rather than $a_1$.

   (a) $\dfrac{1}{5} + \dfrac{1}{\sqrt{26}} + \dfrac{1}{\sqrt{29}} + \dfrac{1}{\sqrt{34}} + \cdots$

   (b) $1 + \dfrac{1}{9} + \dfrac{1}{25} + \cdots$

(c) $1 + \dfrac{2}{9} + \dfrac{3}{25} + \dfrac{4}{49} + \cdots$

(d) $1 + 1 + \dfrac{4}{6} + \dfrac{8}{24} + \dfrac{16}{120} + \cdots$

(e) $1 - \dfrac{1}{\sqrt{3}} + \dfrac{1}{\sqrt{5}} - \dfrac{1}{\sqrt{7}} + \cdots$

(f) $\dfrac{2}{3} - \dfrac{3}{8} + \dfrac{4}{15} - \dfrac{5}{24} + \cdots$

2. Write each of the following series in the form $c \, \Sigma \, a_n + d \, \Sigma \, b_n$ with appropriate limits of summation.

(a) $5 + \left(\dfrac{2}{3} + \dfrac{3}{2}\right) + \left(\dfrac{2}{9} + \dfrac{3}{4}\right) + \left(\dfrac{2}{27} + \dfrac{3}{8}\right) + \cdots$

(b) $\dfrac{1}{4} - \dfrac{3}{5} + \dfrac{1}{16} - \dfrac{3}{25} - \dfrac{1}{64} - \dfrac{3}{125} + \cdots$

(c) $1 + \dfrac{1}{2} + \dfrac{1}{4} - \dfrac{2}{25} + \dfrac{1}{8} - \dfrac{2}{125} + \dfrac{1}{16} - \dfrac{2}{625} + \cdots$

(d) $\dfrac{1}{e^2} + \dfrac{1}{e^3} + \dfrac{1}{e^4} - \dfrac{1}{16} + \dfrac{1}{e^5} - \dfrac{1}{32} + \dfrac{1}{e^6} - \dfrac{1}{64} + \cdots$

Determine the limits of the following sequences if these limits exist.

3. $x_n = \dfrac{n}{n+1} - \dfrac{n+1}{n}$.

4. $x_n = \dfrac{n^2}{n+1} - \dfrac{n^2+1}{n}$.

5. $x_n = \dfrac{n}{2^n}$.

6. $x_n = 1 + (-1)^n$.

7. $x_n = \dfrac{n^2 + 3n - 4}{5n^2}$.

8. $x_n = 1 + \dfrac{1}{n} \cos \dfrac{n\pi}{2}$.

9. $x_n = \sin \dfrac{n\pi}{2}$.

10. $x_n = n^{(-1)^n}$.

11. $x_n = \dfrac{\log n}{n}$.

12. On the basis of Theorem 11.2, for which of the Exercises 3 to 11 would $\Sigma \, x_n$ diverge?

13. Prove Theorem 11.1.

14. If $a_n \geqslant 1/n$, does $\displaystyle\sum_{n=1}^{\infty} a_n$ converge?

15. If $\displaystyle\sum_{n=1}^{\infty} a_n$ converges and $\displaystyle\sum_{n=1}^{\infty} b_n$ diverges, what can be said about the convergence

of $\displaystyle\sum_{n=1}^{\infty} (a_n + b_n)$?

16. Does $\displaystyle\sum_{n=1}^{\infty} (1/n - 1/2^n)$ converge?

17. Show that finitely many terms of the series $\displaystyle\sum_{n=1}^{\infty} a_n$ may be assigned arbitrarily without affecting the convergence or divergence of the series.

## 11.2 Geometric and Telescoping Series

Usually it is much easier to determine if a given series $\Sigma \, a_n$ converges than it is to determine the explicit value of the sum. In this section we discuss two situations where the sum of the series can be explicitly computed. The first is the familiar and important geometric series.

**THEOREM 11.3**    *The series*

$$\sum_{n=0}^{\infty} x^n$$

*converges if and only if $|x| < 1$. If $|x| < 1$ then*

$$\frac{1}{1-x} = \sum_{n=0}^{\infty} x^n.$$

**Proof.**    If $|x| \geq 1$, then $x^n$ does not tend to zero as $n \to \infty$. Hence for these values of $x$ the series does not converge. To show the series converges to $1/(1-x)$ if $|x| < 1$, note that

$$\frac{1}{1-x} - (1 + x + \cdots + x^n) = \frac{x^{n+1}}{1-x}.$$

Hence as $n \to \infty$, $\lim_{n\to\infty} x^{n+1}/(1-x) = 0$ since $|x| < 1$. The second part of this result has, of course, been observed before.

Another class of series for which the sum can be explicitly computed is the so-called telescoping series. The series $\Sigma\, a_n$ is called a telescoping series if we can write

$$a_1 = b_1 - b_2$$
$$a_2 = b_2 - b_3$$
$$.$$
$$.$$
$$.$$
$$a_n = b_n - b_{n+1}.$$

where $\lim_{n\to\infty} b_n$ is known to exist. Then $a_1 + \cdots + a_n = (b_1 - b_2) + (b_2 - b_3) + \cdots + (b_n - b_{n+1}) = b_1 - b_{n+1}$. Hence

$$\sum_{n=1}^{\infty} a_n = b_1 - \lim_{n\to\infty} b_{n+1},$$

and the series $\sum_{n=1}^{\infty} a_n$ converges if and only if $\lim_{n\to\infty} b_n$ exists.

**Example 1.**    Show that the series $\sum_{n=1}^{\infty} 1/n(n+1)$ converges and compute the sum. We note that

$$\frac{1}{n(n+1)} = \frac{1}{n} - \frac{1}{n+1}.$$

Hence

$$\sum_{n=1}^{m} \frac{1}{n(n+1)} = 1 - \frac{1}{2} + \frac{1}{2} - \frac{1}{3} + \cdots - \frac{1}{m} + \frac{1}{m} - \frac{1}{m+1} = 1 - \frac{1}{m+1}.$$

Therefore

$$\sum_{n=1}^{\infty} \frac{1}{n(n+1)} = \lim_{m\to\infty} \sum_{1}^{m} \frac{1}{n(n+1)} = \lim_{m\to\infty} \left(1 - \frac{1}{m+1}\right) = 1.$$

**EXERCISES**

Abbreviate the following series using the summation symbol and compute the sum if the series converges.

1. $1 - \dfrac{1}{2} + \dfrac{1}{4} - \dfrac{1}{8} + \dfrac{1}{16} - \cdots$

2. $9 - 3 + 1 - \frac{1}{3} + \frac{1}{9} - \frac{1}{27} + \cdots$

3. $25 - 16 + 9 - 1 + \frac{1}{4} - \frac{1}{3} + \frac{1}{16} - \frac{1}{9} + \frac{1}{64} - \frac{1}{27} + \cdots$.

4. $1 - 2 + 4 - 8 + 16 - \cdots$.

5. $1 - \frac{1}{2} - 1 + \frac{1}{4} + 1 - \frac{1}{8} - 1 + \frac{1}{16} + 1 - \cdots$.

For what values of $x$ do the following geometric series converge? Abbreviate the series using the summation symbol and compute the sum.

6. $1 + 4x + 16x^2 + \cdots$.

7. $\frac{1}{2}x - \frac{1}{4}x^2 + \frac{1}{8}x^3 + \cdots$.

8. $\frac{1}{3}x^2 - \frac{1}{9}x^3 + \frac{1}{27}x^4 - \cdots$.

9. $\frac{1}{5}x^3 + \frac{1}{25}x^4 + \frac{1}{125}x^5 + \cdots$.

Decompose the $n$th term of the following series using partial fractions, and verify that the series is a telescoping series and converges to the given sum.

10. $\displaystyle\sum_{n=1}^{\infty} \frac{1}{(2n-1)(2n+1)} = \frac{1}{2}$.

11. $\displaystyle\sum_{n=1}^{\infty} \frac{2n+1}{n^2(n+1)^2} = 1$.

12. $\displaystyle\sum_{n=1}^{\infty} \frac{3}{(3n+2)(3n+5)} = \frac{1}{5}$.

13. $\displaystyle\sum_{n=2}^{\infty} \frac{1}{n^2-1} = \frac{3}{4}$.

14. $\displaystyle\sum_{n=1}^{\infty} \frac{\sqrt{n+1}-\sqrt{n}}{\sqrt{n^2+n}} = 1$.

15. $\displaystyle\sum_{n=1}^{\infty} \frac{n}{(n+1)(n+2)(n+3)} = \frac{1}{4}$.

16. One of the paradoxes of Zeno states the following. A runner can never finish a race because when he has gone one-half the distance he still has one-half to go. When he runs one-half of the remaining distance he has one-fourth of the distance to run and so on. Therefore he never completes the race. Prove or disprove.

## 11.3 Series with Positive Terms; The Comparison Principle

Let us suppose that $a_n \geq 0$ for each term in the series $\Sigma\, a_n$. If $s_n$ denotes the $n$th partial sum, then

$$s_n = a_1 + \cdots + a_n,$$

and $s_n \leq s_{n+1}$. Therefore, the partial sums are increasing and, using the least upper bound property of bounded sets of real numbers, it is easy to show that the series will converge if and only if the partial sums $s_n$ are bounded. We formulate this as our next result.

**THEOREM 11.4**   *If for each k, $a_k \geq 0$, then the series $\sum_{k=1}^{\infty} a_k$ converges if and only if the partial sums*

$$s_n = \sum_{k=1}^{n} a_k$$

*are bounded.*

**Proof.**   If the series converges, then $s = \lim_{n \to \infty} s_n$ exists. Since $s_n \leq s_{n+1}$ for each $n$, it is clear from the definition of a limit that $s \geq s_n$ for each $n$. Hence, the partial sums are bounded. Conversely, if the partial sums $s_n$ are bounded, that is, if there is a number $M$ satisfying

$$(1) \qquad\qquad M \geq s_n, \qquad \text{for each } n,$$

then, by the least upper bound property of bounded sets of real numbers, there is a smallest number $M_0$ satisfying (1). We assert $M_0 = \lim_{n \to \infty} s_n$. Since $M_0$ is the least upper bound of the sequence $s_n$, for each $\epsilon > 0$ there is an $n_0$ such that

$$s_{n_0} > M_0 - \epsilon.$$

(Why?) It then follows since $s_{n+1} \geq s_n$, that for all $n \geq n_0$

$$M_0 \geq s_n \geq s_{n_0} > M_0 - \epsilon.$$

Hence by definition of the limit

$$M_0 = \lim_{n \to \infty} s_n.$$

This theorem forms the basis for numerous tests to determine the convergence or divergence of a given series with positive term. The first of these is the *comparison test*. Suppose we wish to determine the convergence or divergence of the series $\Sigma \, a_n, a_n \geq 0$. (For convenience we omit the limits on the summation sign.) Suppose also that we know that another series $\Sigma \, b_n$ with positive term converges. For example, $\Sigma \, b_n$ might be the geometric series. Then if there is a constant $C$ such that for all values of $n$

$$a_n \leq Cb_n,$$

then the series $\Sigma \, a_n$ will converge. The proof of this follows easily from the previous theorem. To see this let $s_n = a_1 + \cdots + a_n$, $t_n = b_1 + \cdots + b_n$ be the partial sums of the series $\Sigma \, a_n$ and $\Sigma \, b_n$, respectively. Since we are assuming that $\Sigma \, b_n$ converges, the theorem implies that there is a fixed number $M$ such that for all values of $n$

$$t_n \leq M.$$

But $a_n \leq Cb_n$ for each integer $n$. Therefore, $s_n \leq Ct_n \leq CM$. By Theorem 11.4 we may conclude that $\Sigma \, a_n$ converges. By precisely the same reasoning if instead the series $\Sigma \, b_n$ is known to diverge, and there exists a constant $C > 0$ such that for all values of $n$,

$$a_n \geq Cb_n,$$

then the series $\Sigma \, a_n$ diverges. This is clear, since if the series $\Sigma \, b_n$ diverges, then by the theorem the partial sums are unbounded. Since $a_n \geq Cb_n$ for all $n$, and $C > 0$, it follows that the partial sums of the series $\Sigma \, a_n$ are unbounded. Hence the series $\Sigma \, a_n$ diverges. Let us illustrate this test with some examples.

Consider

$$\sum_{n=1}^{\infty} \frac{n-1}{n} \frac{1}{2^n} = \frac{1}{4} + \frac{2}{3} \cdot \frac{1}{4} + \frac{3}{4} \cdot \frac{1}{8} + \cdots.$$

The series $\sum_{n=1}^{\infty} (\frac{1}{2})^n$ is known to converge.

For each value of $n$, $(n-1)/n \leq 1$. Therefore $(n-1)/n2^n \leq 1/2^n$ for all $n$, and by the comparison test $\sum_{n=1}^{\infty} (n-1)/n2^n$ converges.

Consider $\sum_{n=2}^{\infty} (n-1)/n^2$. For each $n \geq 2$, $(n-1)/n \geq \frac{1}{2}$. Therefore $(n-1)/n^2 \geq \frac{1}{2}(1/n)$, and since $\sum_{n=1}^{\infty} 1/n$ is known to diverge, the series $\sum_{n=2}^{\infty} (n-1)/n^2$ diverges as well.

A few words of caution are in order here. If $\Sigma b_n$ converges and $a_n > Cb_n$, then nothing can be said about the convergence of $\Sigma a_n$. Similarly, if $\Sigma b_n$ diverges, and $a_n \leq Cb_n$, then the test yields no information about the behavior of $\Sigma a_n$. Also to apply the comparison test it is essential that both series $\Sigma a_n$ and $\Sigma b_n$ have only positive terms.

The comparison between series $\Sigma a_n$ and $\Sigma b_n$, both having only positive terms, can often be made by evaluating $\lim_{n \to \infty} a_n/b_n$. If the limit exists, and the series $\Sigma b_n$ is known to converge, then $\Sigma a_n$ converges. To prove this, let $L = \lim a_n/b_n$. Choose $n_0$ so that for all values of $n > n_0$

$$\left| \frac{a_n}{b_n} - L \right| < 1.$$

Then for $n > n_0$, $a_n/b_n < L+1$ or $a_n < (L+1)b_n$. Since it is always true that for finitely many terms $a_1, \ldots, a_m$ and $b_1, \ldots, b_m$ there exists a constant $C$ satisfying $a_n \leq Cb_n$ (why?), the series converges by the comparison test. Similarly if $\Sigma b_n$ is known to diverge and $L = \lim_{n \to \infty} a_n/b_n$ exists *and is different from zero*, then $\Sigma a_n$ will diverge.

We summarize these facts in the following theorem.

**THEOREM 11.5 (Comparison Test).** *Assume for each $n$ that $a_n \geq 0$ and $b_n \geq 0$. If the series $\Sigma b_n$ is known to converge, then the series $\Sigma a_n$ will converge if either of the following conditions are satisfied.*
(i) *There is a constant $C$ such that for each integer $n$*

$$a_n \leq Cb_n$$

(ii) $\lim_{n \to \infty} a_n/b_n$ *exists.*

*If the series $\Sigma b_n$ is known to diverge, then the series $\Sigma a_n$ will diverge if either of the following conditions are satisfied.*
(iii) *There is a constant $C > 0$ such that for each integer $n$*

$$a_n \geq Cb_n$$

(iv) $\lim_{n \to \infty} a_n/b_n$ *exists and is different from zero.*

***Example 1.*** Determine the convergence or divergence of the series

$$\sum_{n=1}^{\infty} n2^{-n} \sin 1/n.$$

We compare this series with $\Sigma\, 2^{-n}$ which is known to converge. If $a_n = n2^{-n} \sin 1/n$ and $b_n = 2^{-n}$, then

$$\frac{a_n}{b_n} = n \sin \frac{1}{n} = \frac{\sin 1/n}{1/n}.$$

Since $(\sin 1/n)/(1/n) \to 1$ as $n \to \infty$, we may apply Theorem 11.5 (ii) and conclude that the series $\sum\limits_{n=1}^{\infty} n2^{-n} \sin 1/n$ converges.

**Example 2.** The series $\sum\limits_{n=2}^{\infty} (n-1)/(2n^2 + 4n + 5)$ diverges. This follows by part (iv) of Theorem 11.5 since, if $a_n = (n-1)/(2n^2 + 4n + 5)$ and $b_n = 1/n$, then

$$\lim_{n\to\infty} \frac{a_n}{b_n} = \lim_{n\to\infty} \frac{n^2 - n}{2n^2 + 4n + 5} = \frac{1}{2}.$$

The student should note that Theorem 11.4 for series and Lemma 10.1 for functions are in reality the same theorem. Furthermore, the comparison principle for series is exactly the same as the comparison principle for improper integrals discussed in Section 10.1. Often, the convergence of a series may be ascertained by comparing it to an improper integral. This test for convergence is called the *integral test*.

**THEOREM 11.6   (Integral Test).** *Let $f$ be a decreasing positive function defined on the interval $[0, \infty)$. Let $a_n = f(n)$. If $\int_0^{\infty} f(x)\, dx$ converges, then the series $\sum\limits_{n=1}^{\infty} a_n$ converges. If $\int_0^{\infty} f(x)\, dx$ diverges, then $\sum\limits_{n=1}^{\infty} a_n$ diverges.*

**Proof.** If $\int_0^{\infty} f(x)\, dx$ exists, then by definition of the improper integral

$$\int_0^{\infty} f(x)\, dx = \lim_{n\to\infty} \int_0^{n} f(x)\, dx.$$

If we approximate $\int_0^{n} f(x)\, dx$ by Riemann sums, and use the fact that $f$ is decreasing, we obtain

$$\int_0^{n} f(x)\, dx \geqslant f(1) + \cdots + f(n) = a_1 + \cdots + a_n = s_n.$$

This inequality is illustrated in Figure 1. Therefore the partial sum $\sum\limits_{k=1}^{n} a_k$ of the series are bounded. Hence by Theorem 11.4, the series $\Sigma\, a_n$ converges.

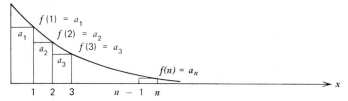

**Fig. 1**   $a_1 + \cdots + a_n = s_n \leqslant \int_0^{n} f(x)\, dx.$

Conversely, if $\int_0^{\infty} f(x)\, dx$ fails to exist, then

$$\lim_{n\to\infty} \int_0^{n} f(x)\, dx = \lim_{n\to\infty} \int_1^{n+1} f(x)\, dx = \infty.$$

and

$$\int_1^{n+1} f(x)\,dx \leqslant f(1) + \cdots + f(n) = a_1 + \cdots + a_n.$$

Therefore the partial sums of $\Sigma\, a_n$ are unbounded, and by Theorem 11.4, $\Sigma\, a_n$ diverges. The above inequality is illustrated in Figure 2.

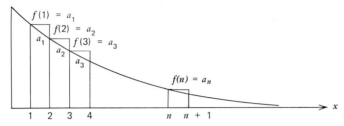

**Fig. 2**    $s_n = a_1 + \cdots + a_n \geqslant \int_1^{n+1} f(x)\,dx.$

In the above theorem it is unimportant what the lower limit of integration is. If we evaluate $\int_1^\infty f$ or $\int_k^\infty f$, this just corresponds to omitting a certain finite number of terms from the series $\Sigma\, a_n$. This, of course, does not affect the convergence.

**Example 3.**    Show that the series $\sum\limits_{n=1}^\infty 1/n^k$ converges if and only if $k > 1$. If $f(x) = 1/x^k$, then $f(n) = 1/n^k$. If $k \leqslant 0$, the series diverges since the $n$th term does not tend to zero. If $k > 0$, then $f(x) = 1/x^k$ decreases as $x$ increases. Hence we may compare the series with the integral $\int_1^\infty 1/x^k\,dx$. We have already verified that this integral converges if and only if $k > 1$. Thus $\sum\limits_{n=1}^\infty 1/n^k$ converges if and only if $k > 1$.

**EXERCISES**

Establish the convergence or divergence of the following series by using the comparison test.

1. $\displaystyle\sum_{n=1}^\infty \frac{1}{n!}.$

2. $\displaystyle\sum_{n=1}^\infty \frac{1}{n^n}.$

3. $\displaystyle\sum_{n=1}^\infty \frac{n+2}{n}\,\frac{1}{3^n}.$

4. $\displaystyle\sum_{n=1}^\infty \frac{n+2}{n^2}.$

5. $\displaystyle\sum_{n=1}^\infty \left(1+\frac{1}{n}\right)^n \frac{1}{n^2}.$

6. $\displaystyle\sum_{n=2}^\infty \frac{n+2}{n^2-2n+1}.$

7. $\displaystyle\sum_{n=1}^\infty \frac{1}{n^2+n}.$

8. $\displaystyle\sum_{n=1}^\infty \frac{1}{n^2} \sin \frac{1}{n}.$

Establish the convergence or divergence of the following series by the use of the integral test.

9. $\displaystyle\sum_{n=1}^\infty \frac{1}{\sqrt{n^2+1}}.$

10. $\displaystyle\sum_{n=2}^\infty \frac{1}{n \log n}.$

11. $\displaystyle\sum_{n=1}^\infty \frac{n}{e^n}.$

12. $\displaystyle\sum_{n=1}^\infty n^2 e^{-n}.$

13. $\displaystyle\sum_{n=2}^\infty \frac{1}{n \log n} (\log \log n).$

14. $\displaystyle\sum_{n=2}^\infty \frac{1}{(\log n)^2}.$

15. $\displaystyle\sum_{n=1}^{\infty} \frac{n}{n^3+1}.$        16. $\displaystyle\sum_{n=1}^{\infty} \frac{n^2}{3n^3+1}.$

Establish the convergence or divergence of the following series by any method.

17. $\displaystyle\sum_{n=1}^{\infty} \frac{n^2}{n^2+1}\, 2^{-n}.$        20. $\displaystyle\sum_{n=1}^{\infty} \frac{1}{n(\log n)^{3/2}}.$

18. $\displaystyle\sum_{n=1}^{\infty} \frac{1}{\sqrt{n(n+1)}}.$        21. $\displaystyle\sum_{n=1}^{\infty} \frac{1}{(n!)^{1/n}}$        (Compare $n!$ with $2^n$).

19. $\displaystyle\sum_{n=1}^{\infty} \frac{e^{-1/n}}{n!}.$

22. Let $p$ and $q$ be polynomials such that for $n = 1, 2, \ldots, p(n) > 0$ and $q(n) > 0$. Show that $\displaystyle\sum_{n=1}^{\infty} p(n)/q(n)$ converges if and only if degree $q \geqslant 2 +$ degree $p$.

23. Assuming $p$ and $q$ the same as in Exercise 22, guess a necessary and sufficient condition for the convergence of $\displaystyle\sum_{n=1}^{\infty} \sqrt{p(n)/q(n)}$, and prove it.

## 11.4   The Ratio and Root Tests

We may infer from the comparison test that the series $\Sigma\, a_n, a_n > 0$, will converge if there is a constant $C$ such that for $n = 1, 2, \ldots$

$$a_n \leqslant C x^n$$

where $0 < x < 1$. Furthermore, we need only to verify this inequality for those values of $n$ greater then some fixed integer $n_0$. Similarly, if

$$a_n \geqslant C x^n$$

where $x > 1$, the series $\Sigma\, a_n$ will diverge. These two facts are the basis for two convergence tests, called the *ratio test* and the *root test*.

**THEOREM 11.7 (Ratio Test).** *Let* $a_n > 0$ *and assume that* $L = \lim_{n \to \infty} (a_{n+1}/a_n)$ *exists.*

*If* $L < 1$, *the series* $\Sigma\, a_n$ *converges.*
*If* $L > 1$, *the series* $\Sigma\, a_n$ *diverges.*
*If* $L = 1$, *the series may either converge or diverge.*

**Proof.**   To establish these tests we proceed as follows. Let $M = (L+1)/2$. If $L < 1$, then $L < M < 1$. Furthermore, $\lim_{n \to \infty} a_{n+1}/a_n = L$ implies that for all integers $n$ greater than some fixed integer $n_0$,

$$\frac{a_{n+1}}{a_n} \leqslant M.$$

In particular

$$\frac{a_{n_0+1}}{a_{n_0}} \leqslant M, \qquad \frac{a_{n_0+2}}{a_{n_0}} = \frac{a_{n_0+2}}{a_{n_0+1}} \cdot \frac{a_{n_0+1}}{a_n} \leqslant M^2,$$

and

$$\frac{a_{n_0+k}}{a_{n_0}} \leqslant M^k.$$

Therefore if $n \geqslant n_0$,

$$\frac{a_n}{a_{n_0}} \leqslant M^{n-n_0} = \left(\frac{1}{M^{n_0}}\right) M^n,$$

or

$$a_n \leqslant \left(\frac{a_{n_0}}{M^{n_0}}\right) M^n = C M^n.$$

Since $M < 1$, the series $\Sigma\ M^n$ converges. Hence by the comparison principle $\Sigma\ a_n$ converges. The same argument shows that if $M > 1$, then for all integers $n \geqslant n_0$

$$a_n \geqslant \left(\frac{a_{n_0}}{M^{n_0}}\right) M^n.$$

Since $M > 1$, the series $\Sigma\ M^n$ diverges. By the comparison test, the series $\Sigma\ a_n$ diverges as well.

To show that if $L = 1$ the series may either converge or diverge, consider the two series

$$\Sigma \frac{1}{n^2} \quad \text{and} \quad \Sigma \frac{1}{n}.$$

In the first case

$$\frac{a_{n+1}}{a_n} = \frac{1/(n+1)^2}{1/n^2} = \frac{n^2}{(n+1)^2},$$

which tends to 1 as $n \to \infty$. For the second series $(a_{n+1}/a_n) = (1/(n+1)/(1/n)) = n/(n+1)$, and $\lim_{n\to\infty} n/(n+1) = 1$. However, we may conclude from the integral test, p. 432, that the first series converges and the second diverges.

A second useful criterion for convergence is the following *root test* which is derived from the comparison test in much the same way as Theorem 11.7.

**THEOREM 11.8 (Root Test).**  *Let  $a_n > 0$,  and  assume  that  $L = \lim_{n\to\infty} (a_n)^{1/n}$ exists.*

*If $L < 1$, the series $\Sigma\ a_n$ converges.*
*If $L > 1$, the series $\Sigma\ a_n$ diverges.*
*If $L = 1$, the series may either converge or diverge.*

To prove these results, we argue just as in the case of the ratio test. Let $M = (L+1)/2$. If $0 \leqslant L < 1$, then $M < 1$, and for all integers $n$ beyond some fixed integer $n_0$

$$(a_n)^{1/n} \leqslant M.$$

Therefore $a_n \leqslant M^n$. Since $\Sigma\ M^n$ converges, it follows from the comparison test that the series $\Sigma\ a_n$ must converge. If $L > 1$, then for $n \geqslant n_0$, $a_n^{1/n} \geqslant M > 1$. Hence

$$a_n \geqslant M^n,$$

Since $\Sigma\ M^n$ diverges, $\Sigma\ a_n$ diverges as well.

To show that no conclusions can be drawn if $L = 1$, consider $\Sigma\ (1/n^2)$, and $\Sigma\ (1/n)$. Observe first that $\lim_{n\to\infty} (1/n)^{1/n} = 1$ and $\lim_{n\to\infty} (1/n^2)^{1/n} = 1$. (It follows from l'Hospital's rule that $\lim_{x\to0} x^x = 1 = \lim_{x\to0} x^{2x}$. Letting $x = 1/n$ the two statements follow.) But $\Sigma\ (1/n^2)$ converges and $\Sigma\ (1/n)$ diverges. Hence in neither case does the test yield any information.

The utility of the ratio and root tests depends on how easy it is to evaluate $\lim_{n\to\infty} (a_{n+1}/a_n)$ or $\lim_{n\to\infty} a_n^{1/n}$. One should expect to use l'Hospital's rule frequently.

*Example 1.*   Determine the convergence or divergence of the series

$$\Sigma\ (\log n)^k/2^n \quad \text{and} \quad \Sigma\ (e^n/n^k)$$

For the first series

$$\frac{a_{n+1}}{a_n} = \left(\frac{\log (n+1)}{\log n}\right)^k \frac{\frac{1}{2}^{n+1}}{\frac{1}{2}^n} = \left(\frac{\log (n+1)}{\log n}\right)^k \frac{1}{2} \to \frac{1}{2}$$

as $n \to \infty$ since $(\log(n+1)/\log n) \to 1$. Applying the ratio test, we conclude that for any real number $k$, $\Sigma \, ((\log n)^k/2^n)$ converges. For the second series

$$(a_n)^{1/n} = \left(\frac{e^n}{n^k}\right)^{1/n} = \frac{e}{(n^{1/n})^k}.$$

Therefore, as $n \to \infty$, $(e/(n^{1/n})^k) \to e > 1$. By the root test the series diverges.

### EXERCISES

Determine the convergence or divergence of the following series by the use of the root or ratio test.

1. $\sum\limits_{n=1}^{\infty} n \, \dfrac{2^{n+1}}{3^n}.$

7. $\sum\limits_{n=1}^{\infty} \left(\dfrac{2}{n}\right)^n.$

2. $\sum\limits_{n=1}^{\infty} \dfrac{2^{2n}}{(2n)!}.$

8. $\sum\limits_{n=1}^{\infty} \dfrac{1}{(2+1/n)^n}.$

3. $\sum\limits_{n=1}^{\infty} \dfrac{n^2}{n!}.$

9. $\sum\limits_{n=1}^{\infty} \dfrac{2^n n!}{n^n}.$

4. $\sum\limits_{n=1}^{\infty} \dfrac{1}{n(n+1)} \left(\dfrac{3}{2}\right)^n.$

10. $\sum\limits_{n=1}^{\infty} \dfrac{n!}{2^n}.$

5. $\sum\limits_{n=1}^{\infty} \dfrac{n^n}{n!}.$

11. $\sum\limits_{n=1}^{\infty} \left(\sqrt[n]{n} - \dfrac{1}{2}\right)^n.$

6. $\sum\limits_{n=1}^{\infty} \dfrac{n^2}{2^n}.$

12. $\sum\limits_{n=2}^{\infty} \dfrac{1}{(\log n)^{1/n}}.$

Determine for which values of $x > 0$ the following series converge.

13. $\sum\limits_{n=1}^{\infty} \dfrac{x^{2n}}{(2n)!}.$

16. $\sum\limits_{n=1}^{\infty} n! x^n.$

14. $\sum\limits_{n=1}^{\infty} n(n+1) x^n.$

17. $\sum\limits_{n=1}^{\infty} \dfrac{\log n}{n} x^n.$

15. $\sum\limits_{n=1}^{\infty} \dfrac{n+1}{2^n} x^n.$

18. $\sum\limits_{n=1}^{\infty} \dfrac{n!}{n^n} x^n.$

### 11.5  Alternating Series; Conditional and Absolute Convergence

We have observed several times that the *harmonic* series $\sum\limits_{n=1}^{\infty} (1/n)$ diverges, yet the *alternating harmonic* series $\sum\limits_{n=1}^{\infty} (-1)^{n-1} 1/n$ converges. Indeed, in Chapter 10 we established that

(1) $$\log 2 = \sum_{n=1}^{\infty} (-1)^{n-1} \frac{1}{n}.$$

Equation (1) was verified by estimating the remainder in Taylor's formula for $\log(1+x)$. It is quite easy to show that the alternating harmonic series converges without appealing to Taylor's formula. This method does not tell us what the sum is, however. To demonstrate convergence of this series, note that

$$\sum_{n=1}^{\infty} (-1)^{n-1} \frac{1}{n} = 1 - \frac{1}{2} + \frac{1}{3} - \frac{1}{4} + \cdots$$

$$= 1 - \left(\frac{1}{2} - \frac{1}{3} + \frac{1}{4} - \frac{1}{5} + \cdots\right)$$

$$= 1 - \sum_{n=1}^{\infty} \left( \frac{1}{2n} - \frac{1}{2n+1} \right)$$

$$= 1 - \sum_{n=1}^{\infty} \frac{1}{2n(2n+1)}.$$

Since $\sum_{n=1}^{\infty} (1/n^2)$ is known to converge, $\sum_{n=1}^{\infty} (1/2n(2n+1))$ will converge by the comparison test, and therefore $\sum_{n=1}^{\infty} (-1)^{n-1}1/n$ converges.

The series $\sum_{n=1}^{\infty} (-1)^{n-1}(1/n)$ is an example of an *alternating series*, that is, a series of the form $\sum_{n=1}^{\infty} (-1)^{n-1}a_n$ where $a_n \geq 0$. For such series we have an extremely simple test for convergence, which is due to Leibniz.

**THEOREM 11.9 (Leibniz' Rule).** *If $a_n \geq 0$, then the alternating series $\sum_{n=1}^{\infty} (-1)^{n-1}a_n$ will converge if*

$$\text{(i) } \lim a_n = 0$$

*and if*

$$\text{(ii) } a_n \geq a_{n+1} \text{ for each integer } n.$$

**Proof.** To establish this rule, let us examine the partial sums $s_n$,

$$s_n = a_1 - a_2 + a_3 - \cdots + (-1)^{n-1}a_n.$$

We assert that

(2) $$s_2 \leq s_4 \leq s_6 \leq \cdots \leq s_5 \leq s_3 \leq s_1.$$

If we write the even partial sums in the form

$$s_{2n} = (a_1 - a_2) + (a_3 - a_4) + \cdots + (a_{2n-1} - a_{2n})$$

each term in the parentheses is greater than or equal to zero since $a_n \geq a_{n+1}$. Hence $s_2 \leq s_4 \leq s_6 \leq \cdots$. Similarly the odd partial sums can be written

$$s_{2n+1} = a_1 - (a_2 - a_3) - (a_4 - a_5) - \cdots - (a_{2n} - a_{2n+1}).$$

Again, since each term in a parenthesis is nonnegative, we have

$$s_1 \geq s_3 \geq s_5 \geq \cdots.$$

But

$$s_{2n+1} = (a_1 - a_2) + \cdots + (a_{2n-1} - a_{2n}) + a_{2n+1} = s_{2n} + a_{2n+1}.$$

Therefore $s_{2n} \leq s_{2n+1}$ for all $n$ and consequently (2) holds. The oscillating behavior of the partial sums is illustrated in Figure 3.

**Fig. 3** Relative order of the partial sums $s_n$ for an alternating series $\sum_{n=1}^{\infty} (-1)^{n-1}a_n$ if $a_n \geq a_{n+1} \geq 0$.

If $S = \{s_2, s_4, \ldots\}$ and $T = \{s_1, s_3, s_5, \ldots\}$, then (2) asserts that $S$ and $T$ are nonempty separated sets of real numbers. However, by the completeness axiom for the real numbers (p. 13), we know that $S$ and $T$ are separated by a real number $s$. In other words, there exists a real number $s$ satisfying

$$s_2 \leqslant s_4 \leqslant \cdots \leqslant s \leqslant \cdots \leqslant s_3 \leqslant s_1.$$

To show that $s = \lim_{n \to \infty} s_n$, notice that

$$0 \leqslant s - s_{2n} \leqslant s_{2n+1} - s_{2n} = a_{2n+1}$$

and $0 \leqslant s_{2n+1} - s \leqslant s_{2n+1} - s_{2n+2} = a_{2n+2}$. Therefore, for each $n$

$$(3) \qquad\qquad |s - s_n| \leqslant a_{n+1}$$

Since $\lim_{n \to \infty} a_{n+1} = 0$, it follows that $\lim_{n \to \infty} |s - s_n| = 0$, or $\lim_{n \to \infty} s_n = s$.

Notice that (3) gives us an estimate on the remainder of an alternating series which satisfies (i) and (ii). If we estimate $s$ by the $n$th partial sum $s_n$, then the absolute value of the error committed is always less than the first term $a_{n+1}$ which is omitted.

The series $\sum_{n=1}^{\infty} (-1)^{n-1} 1/n$ is an example of a series $\sum a_n$ with the property that $\sum a_n$ converges but $\sum |a_n|$ diverges. Such a series is called *conditionally convergent*. However, if the series $\sum |a_n|$ is known to converge, then this forces the series $\sum a_n$ to converge as well. We establish this result next.

**THEOREM 11.10**   *If $a_n$ are real numbers and the series $\sum |a_n|$ converges, then the series $\sum a_n$ converges.*

To prove this fact we argue as follows. Introduce

$$b_n = a_n + |a_n|.$$

Now if $a_n \geqslant 0$, then $b_n = 2|a_n|$, and if $a_n < 0$, then $b_n = 0$. In any case, for each integer $n$

$$0 \leqslant b_n \leqslant 2|a_n|.$$

So if $\sum |a_n|$ converges, $\sum 2|a_n|$ converges, and by the comparison test $\sum b_n$ converges. But

$$a_n = b_n - |a_n|.$$

Hence $\sum a_n$ must converge since the series is the difference of two convergent series. Namely $\sum a_n = \sum b_n - \sum |a_n|$.

A series $\sum a_n$ for which $\sum |a_n|$ converges is called *absolutely convergent*. The importance of absolutely convergent series derives from the fact that if the order of the terms in the series is rearranged, then the sum of the series is unaltered. However, this is no longer true for a conditionally convergent series. Indeed, if a real number $t$ is specified, then the terms in the series may be rearranged in such a way that the series converges to $t$. In fact, the terms may be rearranged in such a way that the series will diverge. This is illustrated in Exercise 14.

*Example 1.*   The series $\sum_{n=1}^{\infty} ((-1)^n/\log n)$ converges by Leibnitz' rule since it is an alternating series and $\log n$ tends monotonically to zero. The series is only conditionally convergent, however, since the series $\sum (1/\log n)$ diverges. (The fact that $\sum (1/\log n)$ diverges follows immediately from the comparison test since $1/\log n \geqslant 1/n$.) By the Leibnitz rule we may infer that $\sum ((-1)^n/n^2)$ converges. But $\sum (1/n^2)$ is known to converge as well hence $\sum ((-1)^n/n^2)$ converges absolutely.

**EXERCISES**

Determine by any method the convergence or divergence of the following series. If convergent specify if absolutely or conditionally convergent.

1. $\sum_{n=1}^{\infty} (-1)^n \dfrac{n+1}{n^2}.$

2. $\sum_{n=1}^{\infty} (-1)^n \dfrac{n!}{10^n}.$

3. $\sum_{n=1}^{\infty} \dfrac{\sin n\pi/2}{n^2}.$

4. $\sum_{n=1}^{\infty} (-1)^{n-1} \log \dfrac{n}{n+1}.$

5. $\sum_{n=1}^{\infty} (-1)^{n-1} \dfrac{n^2}{2^{n-1}}.$

6. $\sum_{n=1}^{\infty} (-1)^n \dfrac{1}{\sqrt[3]{n}}.$

7. $\sum_{n=1}^{\infty} \dfrac{(-2)^n}{n!}.$

8. $\sum_{n=1}^{\infty} (-1)^n \dfrac{\log n}{n}.$

9. $\sum_{n=1}^{\infty} (-1)^{n+1} \dfrac{n^3}{(n+1)^n}$

10. $\sum_{n=1}^{\infty} \sin(\log n).$

11. $\sum_{n=1}^{\infty} \dfrac{(-1)^n}{n^{1/n}}.$

12. $\sum_{n=1}^{\infty} (-1)^n \left(1 - \cos \dfrac{1}{n}\right).$

13. Suppose the series $\Sigma\, a_n$ converges absolutely. If $s = \Sigma\, a_n$, show that

$$|s| \le \Sigma\, |a_n|.$$

14. Rearrange the order of the terms in the series $\sum_{n=2}^{\infty} (-1)^n (1/n)$ so that the series diverges. (Verify that

$$\left(\frac{1}{2} + \frac{1}{4} - \frac{1}{3}\right) + \left(\frac{1}{6} + \frac{1}{8} + \frac{1}{10} + \cdots + \frac{1}{16} - \frac{1}{5}\right)$$

$$+ \left(\frac{1}{18} + \frac{1}{20} + \cdots + \frac{1}{32} + \frac{1}{34} + \frac{1}{36} + \cdots + \frac{1}{64} - \frac{1}{7}\right) + \cdots$$

is a rearrangement of the given series. Yet each term in parenthesis has sum greater than $\frac{1}{4}$.)

15. Rearrange the alternating harmonic series so that its sum is $\frac{3}{2} \log 2$. (Note that

$$1 - \frac{1}{2} + \frac{1}{3} - \frac{1}{4} + \cdots = \log 2.$$

Therefore

$$\frac{1}{2} - \frac{1}{4} + \frac{1}{6} - \frac{1}{8} + \cdots = \frac{1}{2} \log 2.$$

But

$$1 - \frac{1}{2} + \frac{1}{3} - \frac{1}{4} + \frac{1}{5} - \frac{1}{6} + \frac{1}{7} - \frac{1}{8} + \cdots = \log 2$$

$$+ \frac{1}{2} \qquad - \frac{1}{4} \qquad + \frac{1}{6} \qquad - \frac{1}{8} + \cdots = \frac{1}{2} \log 2.$$

Adding these two equations we see that

$$1 + \frac{1}{3} - \frac{1}{2} + \frac{1}{5} + \frac{1}{7} - \frac{1}{4} + \cdots = \frac{3}{2} \log 2.$$

Verify that the series obtained is a rearrangement of the alternating harmonic series.)

## 11.6  Power Series

Next we turn to the study of series of the form

(1) $$\sum a_n x^n.$$

Such an infinite series is called a *power series*. We shall show by examples that such a series may converge for all values of $x$ or it may only converge for $x = 0$. If, however, there is a number $y \neq 0$ such that $\Sigma \, a_n y^n$ converges and another number $z$ for which $\Sigma \, a_n z^n$ diverges, then there is an *interval of convergence* for the series (1). Specifically, there exists a number $r > 0$ such that if $|x| < r$ then the series $\Sigma \, a_n x^n$ converges absolutely, and if $|x| > r$ the series $\Sigma \, a_n x^n$ diverges. If $x = \pm r$, then the series may either converge or diverge.

To show that such an interval of convergence exists for the power series $\Sigma \, a_n x^n$ we proceed as follows. We say that the terms $a_n x^n$ of the series are *bounded* if there exists a constant $M$ such that for all positive integers $n$

$$|a_n x^n| \leq M.$$

If no such $M$ exists, then the terms $a_n x^n$ are said to be *unbounded*. Assuming that there is a number $y$ such that $\Sigma \, a_n y^n$ diverges, the number $r$ we seek will be the least upper bound of those $x \geq 0$ such that the terms $a_n x^n$ are bounded. To show this we must verify a few facts about power series. First we show that if for some $y \neq 0$ the terms $a_n y^n$ are bounded, then for all $x$, $|x| < |y|$, the series $\Sigma \, a_n x^n$ converges absolutely. We prove this by the comparison test. Note that

$$|a_n x^n| = |a_n y^n| \left|\frac{x^n}{y^n}\right| \leq M \left|\left(\frac{x}{y}\right)^n\right|.$$

But since $|x/y| < 1$, $\Sigma \, |(x/y)|^n$ is a convergent geometric series. Therefore $\Sigma \, a_n x^n$ converges absolutely. Second, if for some $z$ the terms $a_n z^n$ of the series are unbounded, then the series $\Sigma \, a_n z^n$ certainly diverges since the $n$th term does not tend to zero. Furthermore if $|x| > |z|$, then the terms $a_n x^n$ are unbounded. To see this note that $|x/z| > 1$ implies

$$|a_n x^n| = |a_n z^n| \left|\frac{x^n}{z^n}\right| > |a_n z^n|.$$

Since the numbers $|a_n z^n|$ are assumed to be unbounded, the numbers $|a_n x^n|$, which are even larger, must also be unbounded.

Let $S$ be the set of all nonnegative numbers $x$ such that the terms $a_n x^n$ are bounded. If each $x \geq 0$ belongs to $S$, the above discussion shows that the series $\Sigma \, a_n x^n$ converges absolutely for all real numbers $x$. If, on the other hand, there is a number $y \geq 0$ that does not belong to $S$, then for $z \geq y$ the terms $a_n z^n$ are unbounded. Hence $S$ is bounded above by $y$. Since $0 \in S$, the set $S$ is nonempty and bounded. Hence, $S$ has a least upper bound, which we denote by $r$. It now follows by the above remarks that if $|x| < r$, then $\Sigma \, a_n x^n$ converges absolutely. If $|x| > r$, then $\Sigma \, a_n x^n$ diverges. The verification of these two facts is left as Exercise 13.

The root and ratio tests provide the most useful techniques for computing $r$. Suppose as $n \to \infty$, $|a_n|^{1/n} \to L$. Then $|a_n x^n|^{1/n} = |a_n|^{1/n}|x| \to L|x|$. Applying the root test we see that the series converges absolutely if $L|x| < 1$ or if $|x| < 1/L$. The series diverges if $L|x| > 1$, or $|x| > 1/L$. Therefore

$$r = \frac{1}{L}.$$

Similarly if

$$\lim_{n \to \infty} \left|\frac{a_{n+1}}{a_n}\right| = L,$$

then

$$\left|\frac{a_{n+1}x^{n+1}}{a_n x^n}\right| = \left|\frac{a_{n+1}}{a_n}\right| |x| \to |x|L.$$

The ratio test asserts that the series $\Sigma\, a_n x^n$ converges absolutely if $|x|L < 1$ and diverges if $|x|L > 1$. Thus, just as before

$$r = \frac{1}{L}.$$

We summarize these results in the following theorem.

**THEOREM 11.11** *If $-r < x < r$ is the interval of convergence for the power series $\Sigma\, a_n x^n$, then*

$$\frac{1}{r} = \lim_{n\to\infty} |a_n|^{1/n} \ \text{if this limit exists}$$

*and*

$$\frac{1}{r} = \lim_{n\to\infty} \left|\frac{a_{n+1}}{a_n}\right| \ \text{if this limit exists.}$$

Thus if both $\lim_{n\to\infty} |a_n|^{1/n}$ and $\lim_{n\to\infty} |a_{n+1}/a_n|$ exist, these limits must be equal.

However, these are power series for which neither of these limits exist. Hence we cannot define the interval $[-r, r]$ of convergence in terms of these limits. Furthermore, one limit may exist and the other may not.

**Example 1.** The geometric series $\Sigma\, x^n$ is, of course, an example of a power series. Since the series converges absolutely if $|x| < 1$ and diverges if $|x| > 1$ this shows that $r = 1$. If $x = \pm 1$, then the series diverges since the $n$th term does not tend to zero.

**Example 2.** The interval of convergence of the power series $\Sigma\, x^n/n$ is $-1 < x < 1$ since

$$\frac{1}{r} = \lim_{n\to\infty} \sqrt[n]{(a_n)} = \lim_{n\to\infty} \sqrt[n]{1/n} = 1.$$

We determine the convergence or divergence at the end points of the interval by using other information. Indeed, we know that for $x = 1$, $\Sigma\,(1/n)$ diverges and for $x = -1$ the series $\Sigma\,((-1)^n/n)$ conditionally converges.

**Example 3.** The interval of convergence of the series $\Sigma\,(1/n^2)x^n$ is also the interval $-1 < x < 1$ since

$$\frac{1}{r} = \lim_{n\to\infty} \left|\frac{a_{n+1}}{a_n}\right| = \lim_{n\to\infty} \left|\frac{1/(n+1)^2}{1/n^2}\right| = 1.$$

However, the series $\Sigma\,(1/n^2)$ is known to converge. Hence $\Sigma\,((-1)^n/n^2)$ converges as well.

These three examples show that the interval of convergence can often be determined quite readily. The determination of the convergence or divergence of the series at the end points of the interval always requires other information and can be quite involved.

**Example 4.** The power series $\Sigma\,(x^n/n!)$ converges absolutely for all values of $x$. This follows readily from the ratio test since if $a_n = 1/n!$

$$\left|\frac{a_{n+1}x^{n+1}}{a_n x^n}\right| = \frac{1/(n+1)!}{1/n!}|x| = \frac{|x|}{n+1} \to 0,$$

as $n \to \infty$. Therefore the interval of convergence for this series is the entire real line. On the other hand, the series $\Sigma\, n! x^n$ does not converge for any value of $x \neq 0$.

### EXERCISES

Determine the intervals of convergence of the following power series.

1. $\displaystyle\sum_{n=1}^{\infty} n4^n x^n.$

2. $\displaystyle\sum_{n=1}^{\infty} \frac{\log n}{2^n} x^n.$

3. $\displaystyle\sum_{n=1}^{\infty} \frac{x^n}{\sqrt{n}}.$

4. $\displaystyle\sum (-1)^{n+1} \frac{x^n}{n \cdot 2^n}.$

5. $\displaystyle\sum_{n=1}^{\infty} n^n x^n.$

6. $\displaystyle\sum_{n=1}^{\infty} \frac{2^{2n}}{2 \cdot 4 \cdots 2n} x^n.$

Determine for what values of $x$ the following power series converge. (In addition to finding the interval of convergence, you must determine the convergence behavior at $x = \pm r$.)

7. $\displaystyle\sum_{n=0}^{\infty} \frac{x^n}{(n+1)2^n}.$

8. $\displaystyle\sum_{n=1}^{\infty} \frac{x^n}{n(n+1)}.$

9. $\displaystyle\sum_{n=1}^{\infty} \frac{2n-1}{2^n} x^n.$

10. $\displaystyle\sum_{n=1}^{\infty} \frac{n}{n+1} \frac{1}{3^n} x^n.$

11. $\displaystyle\sum_{n=1}^{\infty} \frac{2^{2n}}{2n} x^n.$

12. $\displaystyle\sum_{n=1}^{\infty} \frac{2^{\sqrt{n}}}{\sqrt{n^2+1}} x^n.$

13. Verify that it follows from the definition of $r$ (p. 440) that $\Sigma\, a_n x^n$ converges absolutely if $|x| < r$ and $\Sigma\, a_n x^n$ diverges if $|x| > r$.

   Assume for Exercise 14 and 15 that

$$\frac{1}{r} = \lim_{n\to\infty} \left|\frac{a_{n+1}}{a_n}\right| \qquad \text{or} \qquad \frac{1}{r} = \lim_{n\to\infty} |a_n|^{1/n}.$$

14. Show that if the power series $\Sigma\, a_n x^n$ is differentiated term by term to obtain $\Sigma\, na_n x^{n-1}$, then the interval of convergence is unchanged.

15. Show that if the power series $\Sigma\, a_n x^n$ is integrated term by term to obtain $\Sigma\, a_n x^{n-1}/(n+1)$, then the interval of convergence is unchanged.

16. If $(-r, r)$ is the interval of convergence of $\Sigma\, a_n x^n$, what is the interval of convergence of $\Sigma\, a_n x^{n+k}$ for a fixed positive integer $k$?

17. If $(-r, r)$ is the interval of convergence for the power series $\Sigma\, a_n x^n$, show that $(-r^{1/k}, r^{1/k})$ is the interval of convergence for $\displaystyle\sum_{n=0}^{\infty} a_n x^{kn}.$

## 11.7   Taylor Series

In our discussion of Taylor polynomials in Chapter 10 we verified the formulas

$$(1) \qquad\qquad \sin x = \lim_{n\to\infty} \left(x - \frac{x^3}{3!} + \cdots + (-1)^{n-1} \frac{x^{2n-1}}{(2n-1)!}\right)$$

for each $x$ and

$$(2) \qquad\qquad \log(1+x) = \lim_{n\to\infty} \left(x - \frac{x^2}{2} + \cdots + (-1)^{n+1} \frac{x^n}{n}\right)$$

for $|x| < 1$. Therefore

$$(3) \qquad\qquad \sin x = \sum_{n=1}^{\infty} (-1)^{n-1} \frac{x^{2n-1}}{(2n-1)!} \qquad \text{for all } x$$

and

$$(4) \qquad\qquad \log(1+x) = \sum_{n=1}^{\infty} (-1)^{n+1} \frac{x^n}{n} \qquad \text{for } |x| < 1.$$

Furthermore if we differentiate the series in (3) and (4) term by term it follows by what we did in Chapter 10 that the new series converge to $\cos x$ and $1/(1+x)$, respectively. That is

$$\cos x = \lim_{n \to \infty} \left(1 - \frac{x^2}{2!} + \cdots + (-1)^n \frac{x^{2n}}{(2n)!}\right) = \sum_{n=0}^{\infty} (-1)^n \frac{x^{2n}}{(2n)!} \qquad \text{for each } x$$

and

$$\frac{1}{1+x} = \lim_{n \to \infty} (1 - x + x^2 - \cdots + (-1)^n x^n) = \sum_{n=0}^{\infty} (-1)^n x^n \qquad \text{for } -1 < x < 1$$

Each of the series in (3) and (4) is a power series $\Sigma\, a_n x^n$. If we denote the sum of the series by $f(x)$, then, at least in these two cases, the derivative $f'(x)$ is the sum of the power series $\Sigma\, n a_n x^{n-1}$, which is obtained by differentiating the original series term by term. This differentiation property is a very important property of power series that we now wish to discuss.

Let us suppose that the power series $\sum_{n=0}^{\infty} a_n x^n$ has $(-r, r)$ as interval of convergence and assume that $r > 0$. Define the function $f$ to be the sum of this series for $-r < x < r$. That is,

$$(5) \qquad f(x) = \sum_{n=0}^{\infty} a_n x^n.$$

Then it follows that $f$ is differentiable for each $x$ in the interval $(-r, r)$, and

$$f'(x) = \sum_{n=1}^{\infty} n a_n x^{n-1}.$$

The interval of convergence of the power series $\sum_{n=1}^{\infty} n a_n x^{n-1}$ is also the interval $(-r, r)$. Hence we may differentiate again and obtain

$$f''(x) = \sum_{n=2}^{\infty} n(n-1) a_n x^{n-2}$$

Indeed, the function $f$ has derivatives of all orders, and for each integer $k$

$$(6) \qquad f^{(k)}(x) = \sum_{n=k}^{\infty} n(n-1)\ldots(n-k+1) a_n x^{n-k}.$$

If we evaluate (6) at $x = 0$, we see that $f^{(k)}(0)$ is just the constant term of the series on the right-hand side. Indeed

$$f^{(k)}(0) = k! a_k$$

Hence $a_n = (f^{(n)}(0)/n!)$, and the function $f(x) = \sum_{n=0}^{\infty} a_n x^n$ may be written

$$(7) \qquad f(x) = \sum_{n=0}^{\infty} \frac{f^{(n)}(0)}{n!} x^n.$$

The series in (7) is called the *Taylor series* for the function $f$. The $m$th partial sum of the series $\sum_{n=0}^{m} f^{(n)}(0)/n!$ is, of course, the Taylor polynomial of order $m$ for the function $f$.

In addition to being able to differentiate the series (5) term by term, we may integrate it as well. Indeed, for each $x$, $-r < x < r$, it follows that

$$\int_0^x f(t)\, dt = \sum_{n=0}^{\infty} \frac{a_n x^{n+1}}{n+1}.$$

To demonstrate these facts there are several things that have to be shown. First, we must show that if $I = (-r, r)$ is the interval of convergence for the power series $\Sigma a_n x^n$, then the differentiated and integrated series,

$$\sum n a_n x^{n-1}, \sum \frac{a_n x^{n+1}}{n+1}$$

respectively, also have $I$ as their interval of convergence. This is easy if

$$\frac{1}{r} = \lim_{n \to \infty} \left| \frac{a_{n+1}}{a_n} \right|$$

or

$$\frac{1}{r} = \lim_{n \to \infty} \left| a_n \right|^{1/n}.$$

(Exercises 14 and 15, p. 442). However, we need a proof without this restriction. This we do next.

**THEOREM 11.12** *Assume $I = (-r, r)$ is the interval of convergence for the series $\Sigma a_n x^n$. Then $I = (-r, r)$ is also the interval of convergence for differentiated series $\Sigma n a_n x^{n-1}$ and the integrated series $\Sigma a_n x^{n+1}/(n+1)$.*

**Proof.** We shall prove the case of the differentiated series, leaving the demonstration for the integrated series as an exercise. Now by definition of $r$, the terms $a_n x^n$ are bounded for $|x| < r$ and unbounded for $|x| > r$. We must show that the same is true for the terms $n a_n x^{n-1}$. If $|x| > r$ then

$$\left| n a_n x^{n-1} \right| = \frac{n}{|x|} \left| a_n x^n \right|.$$

Clearly $(n/|x|) |a_n x^n| > |a_n x^n|$ if $n$ is large enough. Hence the terms $n a_n x^{n-1}$ must be unbounded if $|x| > r$. On the other hand if $0 < |x| < r$, choose $z$ so that $|x| < |z| < r$. Then

$$\left| n a_n x^{n-1} \right| = \frac{n}{|x|} \left| a_n x^n \cdot \frac{z^n}{z^n} \right| = \frac{n}{|x|} \left| \frac{x}{z} \right|^n \left| a_n z^n \right|.$$

Now it is not hard to show that as $n \to \infty$, $(n/|x|)|x/z|^n \to 0$ since $|x/z| < 1$. The verification of this fact we leave as Exercise 12. Hence, for $n$ large enough

$$\left| n a_n x^{n-1} \right| < \left| a_n z^n \right|.$$

Since the numbers $|a_n z^n|$ are bounded, the numbers $|n a_n x^{n-1}|$ must be bounded also. This completes the proof.

Next we must show that if

$$f(x) = \sum_{n=0}^{\infty} a_n x^n \qquad \text{for} \qquad |x| < r$$

then

$$f'(x) = \sum_{n=1}^{\infty} n a_n x^{n-1}.$$

The proof is straightforward but involves the following algebraic identity which may easily be verified by induction (Exercise 8). If $h \neq 0$ and $n$ is a positive integer $\geq 2$, then

$$(8) \quad h[(x+h)^{n-2} + 2x(x+h)^{n-3} + 3x^2(x+h)^{n-4} + \cdots + (n-1)x^{n-2}] =$$

$$\frac{(x+h)^n - x^n}{h} - n x^{n-1}.$$

**THEOREM 11.13**  *If $I = (-r, r)$ is the interval of convergence for the power series $\Sigma\, a_n x^n$, and we define*

$$f(x) = \sum_{n=0}^{\infty} a_n x^n \qquad \text{for} \qquad |x| < r.$$

*then for $|x| < r$*

$$f'(x) = \sum_{n=1}^{\infty} n a_n x^{n-1}.$$

**Proof.**  If $|x| < r$ and $h$ is small enough, we may assume that for some fixed positive number $r_1 < r$ we have $|x|, |x+h| < r_1$. We must prove that

$$\lim_{h \to \infty} \frac{f(x+h) - f(x)}{h} = \sum_{n=1}^{\infty} n a_n x^{n-1}.$$

But applying the definition of f we have

$$\frac{f(x+h) - f(x)}{h} - \sum_{n=1}^{\infty} n a_n x^{n-1} = \frac{1}{h}\left[ \sum_{n=0}^{\infty} a_n (x+h)^n - \sum_{n=0}^{\infty} a_n x^n \right] - \sum_{n=1}^{\infty} n a_n x^{n-1}$$

$$= a_0 - a_0 + a_1 - a_1 + \sum_{n=2}^{\infty} a_n \left[ \frac{(x+h)^n - x^n}{h} - n x^{n-1} \right]$$

$$= h \sum_{n=2}^{\infty} a_n [(x+h)^{n-2} + 2x(x+h)^{n-3} + \cdots + (n-1)x^{n-2}].$$

The last equality holds by virtue of equation (8). Now since $|x|, |x+h| < r_1$ by assumption, it follows that

$$|(x+h)^{n-2} + 2x(x+h)^{n-3} + \cdots + (n-1)x^{n-2}|$$

$$\leqslant |(x+h)^{n-2}| + 2|x||(x+h)^{n-3}| + \cdots + (n-1)|x^{n-2}|$$

$$\leqslant r_1^{n-2}(1 + 2 + \cdots + n - 1) = \frac{1}{2} n(n-1) r_1^{n-2}.$$

Consequently

(9)  $$\left| \frac{f(x+h) - f(x)}{h} - \sum_{n=1}^{\infty} n a_n x^{n-1} \right| \leqslant \frac{1}{2} |h| \sum_{n=2}^{\infty} n(n-1) |a_n| r_1^{n-2}.$$

This latter series converges since by Theorem 11.12 the interval of convergence for the series

$$\sum_{n=2}^{\infty} n(n-1) a_n x^{n-2}$$

is the same as for the series $\Sigma\, a_n x^n$. Therefore, as $h \to 0$

$$\frac{1}{2}|h| \sum_{n=2}^{\infty} n(n-1) |a_n| r_1^{n-2} \to 0.$$

Hence by (9)

$$\lim_{h \to 0} \left| \frac{f(x+h) - f(x)}{h} - \sum_{n=1}^{\infty} n a_n x^{n-1} \right| = 0.$$

This shows that

$$f'(x) = \lim_{h \to 0} \frac{f(x+h) - f(x)}{h} = \sum_{n=1}^{\infty} n a_n x^{n-1}$$

which was to be proved.

This theorem shows that the function $f(x) = \sum_{n=0}^{\infty} a_n x^n$ has derivatives of all orders. Furthermore

$$f^{(k)}(x) = \sum_{n=k}^{\infty} n(n-1) \cdots (n-k) a_n x^{n-k}.$$

Since the value of the series at $x = 0$ is clearly $k(k-1) \cdots 1 \, a_k$, we have

$$f^{(k)}(0) = k! \, a_k.$$

Therefore

$$f(x) = \sum_{n=0}^{\infty} \frac{f^{(n)}(0)}{n!} x^n.$$

We have already remarked that this series is called the Taylor series for the function $f$. On the basis of this result, it follows readily that if both $a$, $x \in (-r, r)$, then

$$f(x) = \sum_{n=0}^{\infty} \frac{f^{(n)}(a)}{n!} (x-a)^n.$$

This series is called the Taylor series for $f$ about the point $x = a$.

If we apply the fundamental theorem of calculus to the result of Theorem 11.13, we see that we can integrate the power series $f(x) = \sum a_n x^n$ term by term and the resulting integrated series converges to the integral of $f$.

**THEOREM 11.14**  *If $(-r, r)$ is the interval of convergence for the series $\sum a_n x^n$, and we set $f(x) = \sum a_n x^n$, $-r < x < r$, then*

$$\int_0^x f(t) \, dt = \sum_{n=0}^{\infty} \frac{a_n}{n+1} x^{n+1} \qquad -r < x < r.$$

The verification of Theorem 11.14 we leave as an exercise. These theorems allow us to represent many functions as infinite series.

*Example 1.*  We know that for all $x$

(10) $$\cos x = \sum_{n=0}^{\infty} (-1)^n \frac{x^{2n}}{(2n)!}.$$

Therefore for all $x$

$$\cos x^2 = \sum_{n=0}^{\infty} (-1)^n \frac{x^{4n}}{(2n)!}$$

and

$$\int_0^x \cos t^2 \, dt = \sum_{n=0}^{\infty} (-1)^n \frac{x^{4n+1}}{(4n+1)(2n)!}.$$

The integral $\int_0^x \cos t^2 \, dt$ is called the Fresnel cosine integral and occurs in problems of the diffraction of light.

Similarly for all $x$

(11) $$\sin x = \sum_{n=0}^{\infty} (-1)^{n-1} \frac{x^{2n+1}}{(2n+1)!}.$$

Hence

$$\frac{\sin x}{x} = \sum_{n=0}^{\infty} (-1)^{n-1} \frac{x^{2n}}{(2n+1)!}$$

and

$$\int_0^x \frac{\sin t}{t} \, dt = \sum_{n=0}^{\infty} (-1)^{n-1} \frac{x^{2n+1}}{(2n+1)(2n+1)!}.$$

It is perhaps worthwhile to note that formulas (10) and (11) can be used to define the trigonometric functions $\sin x$ and $\cos x$ without any reference to the geometry of the unit circle $x^2 + y^2 = 1$.

### EXERCISES
Determine the Taylor series for the following functions. Specify the interval of convergence.

1. $f(x) = e^{x^2}$.

2. $f(x) = \int_0^x e^{-t^2}\, dt$.

3. $f(x) = \int_0^x \dfrac{1 - \cos t}{t^2}\, dt$.

4. $f(x) = \dfrac{1}{1 - x^3}$.

5. $f(x) = \int_0^x \dfrac{dt}{1 + t^3}$.

6. $f(x) = \int_0^x \log\,(1 + t^2)\, dt$.

7. Show that if $(-r, r) = I$ is the interval of convergence for the power series $\Sigma\, a_n x^n$, then $I$ is the interval of convergence for the series $\Sigma\, a_n/(n+1)x^{n+1}$. (Examine the argument used to prove Theorem 11.12.)

8. Show that for $h \neq 0$, and $n \geq 2$

$$\frac{(x+h)^n - x^n}{h} - nx^{n-1}$$
$$= h[(x+h)^{n-2} + 2x(x+h)^{n-3} + 3x^2(x+h)^{n-4} + \cdots + (n-1)x^{n-2}].$$

$\left(\text{Expand } \dfrac{(x+h)^n - x^n}{h} \text{ and derive the identity}\right.$

$$\frac{(x+h)^n - x^n}{h} = (x+h)\left[\frac{(x+h)^{n-1} - x^{n-1}}{h}\right] + x^{n-1}.$$

Now use induction.$\Big)$

9. If $(-r, r)$ is the interval of convergence for the series $\Sigma\, a_n x^n$, show that $f(x) = \sum\limits_{n=0}^{\infty} a_n x^n$ is continous for each $x$, $-r < x < r$.

10. Prove Theorem 11.14. That is, show that if $f(x) = \sum\limits_{n=0}^{\infty} a_n x^n$ for $-r < x < r$ then

$$\int_0^x f(t)\, dt = \sum_{n=0}^{\infty} \frac{a_n}{n+1} x^{n+1}.$$

(Use the fundamental theorem of calculus and Theorems 11.12 and 11.13.)

11. Show that the result of Theorem 11.13 does not necessarily hold if $x$ is the end point of the interval of convergence of a power series. (Examine the Taylor series for the functions $\log\,(1+x)$ and $1/(1+x)$ at $x = 1$.)

12. Show by the ratio test that the series $\sum\limits_{n=1}^{\infty} nr^n$ converges for each $r$, $0 \leq r < 1$.

Conclude from this that $\lim\limits_{n \to \infty} nr^n = 0$ if $0 \leq r < 1$.

# ANSWERS TO EXERCISES

## CHAPTER 1

**Section 1.1 – p. 3**   **1.** (a) $\{2, 7\}$   (b) $\{2, 4, 7, 9\}$   (c) $\{1, 2, 3, 4, 7, 8\}$   (d) $\{1, 4\}$
**2.** (a) yes   (b) no   **3.** $\emptyset, \{1\}, \{2\}, \{3\}, \{1, 2\}, \{2, 3\}, \{1, 3\}, \{1, 2, 3\}$   **4.** (a) $\{1, 2, 3, 4\}$

(b) $\emptyset$   (c) $\left\{ -\dfrac{\sqrt{\pi}}{\pi}, \dfrac{\sqrt{\pi}}{\pi} \right\}$   (d) $\emptyset$   **5.** $\cup, \cap$ are associative and commutative by definition.

**6.** $A_1 \cap \cdots \cap A_k$ is the set of elements common to all of the sets $A_1, A_2, \ldots, A_k$.   **7.** (a) Let
$X = A \cap (A \cup B)$. If $x \in X, x \in A$ and $x \in A \cup B$. Hence, $x \in A$ and $X \subset A$. On the other
hand, if $x \in A$, then $x \in A \cup B$. Hence, $x \in A \cap (A \cup B)$. Consequently, $A \subset X$, which
proves $X = A$.   **8.** (a) false   (b) false   (c) true   (d) false   (e) false

**Section 1.2 – p. 7**   **1.** prop (1): If $x > y$ and $y > z$, then $x - y > 0$ and $y - z > 0$. Hence by (ii)
$(x - y) + (y - z) = x - z > 0$. Therefore, $x > z$.   **5.** (a) $\{x : x > 4\} \cup \{x : x < 3\}$
(b) $\{x : 1 < x < 2\}$   (c) $\emptyset$   (d) $\{x : x > 1\} \cup \{x : -1 < x < 0\}$   (e) $\{x : 2 < x < 7/2\}$
(f) $\{x : 3/2 \geq x \geq 5/4\}$   g) $\{x : 2 - \sqrt{5} < x < 0\} \cup \{x : 4 < x < 2 + \sqrt{5}\}$
(h) $\{x : -2/3 < x < -1/2\}$   (i) $\{x : x < 7/4\} \cup \{x : x > 45/21\}$

(j) $\{x : x < -2\} \cup \left\{ x : \dfrac{3 - \sqrt{65}}{4} < x < \dfrac{3 + \sqrt{65}}{4} \right\}$
**6.** (c) $2x^2 - x + 2 = 2[(x - 1/4)^2 + 15/16] \geq 2(15/16) > 0$   **7.** $x = 1$   **8.** $x = 1/2$
**9.** $4b > a^2$   **10.** $4ac > b^2, a > 0$   **11.** no   **12.** $a = 0, b < 1$

**Section 1.4 – p. 11**   (b) Apply Theorem 1.1.   (d) Write $x = x + y - y$. Hence $|x| = |x + y - y|$
$\leq |x + y| + |y|$ by Theorem 1.4. Hence $|x| - |y| \leq |x + y|$   **5.** (a) $(1/3, 1)$   (b) all $x$
(c) $\{0, 1\}$   (d) $\emptyset$   (e) $\{x : x \leq -3\} \cup \{x : x \geq 3\}$   (f) $\left( \dfrac{-3 + \sqrt{17}}{4}, \dfrac{-3 + \sqrt{33}}{4} \right) \cup$

$\left( \dfrac{-3 - \sqrt{33}}{4}, \dfrac{-3 - \sqrt{17}}{4} \right)$   (g) $\left( \dfrac{3 - \sqrt{21}}{3}, \dfrac{3 + \sqrt{21}}{3} \right)$.   (h) $\{x : x \leq -1\} \cup \{x : x \geq 1\}$
(i) $(\sqrt[4]{3}, \sqrt[4]{5}) \cup (-\sqrt[4]{5}, -\sqrt[4]{3})$   (j) $\{-5, -1/2, -3/2, 3\}$   (k) $\{5/3, 13/3\}$   (l) $\emptyset$
**6.** (a) $|x - 2| < 1$ implies $-1 < x - 2 < 1 \therefore -3 < 3x - 6 < 3$, i.e., $|3x - 6| < 3$   (c) $|x - 1| < 2$
implies $-2 < x - 1 < 2 \therefore 1 < x + 2 < 5$, so $-5 < 1 < x + 2 < 5 \therefore |x + 2| < 5$. Hence
$|x^2 + x - 2| = |x + 2| |x - 1| < 5 \cdot 2 = 10$   **7.** (b) $|x - 1| < \delta < 1$ implies $-1 < x - 1 < 1$, so
$1 < x + 1 < 3 \therefore |x + 1| < 3 \therefore |x^2 - 1| = |x + 1||x - 1| < 3\delta$   **8.** (a) $\delta \leq 1/4$
(b) If $|x + 1| < \delta < 1$, then $x - 1 < x + 1 < 1$ and $-3 < x - 1 < 1$. Hence $|x - 1| < 3$.
Therefore $|x^2 - 1| = |x - 1||x + 1| < 3\delta$. Hence, if $\delta \leq 1/6, |x^2 - 1| < 1/2$.   (c) $\delta \leq \sqrt{1/20}$
(d) $\delta \leq \epsilon/6$

**Section 1.5 – p. 16**   **1.** not separated   **2.** separated, $0 \leq m \leq 1$   **3.** separated, $m = 3$
**4.** not separated   **5.** not separated   **6.** separated, $m = 1$   **7.** separated, $\sqrt{5} \leq m \leq 5$
**8.** separated, $m = \sqrt{3}$   **9.** If $0 < \delta < \dfrac{m^2 - 2}{2m}$, then $(m - \delta)^2 > 2$.
**10.** $S = \{x : x > 0 \text{ and } x^2 < 3\}$   $T = \{x : x > 0 \text{ and } x^2 > 3\}$

**Section 1.6 – p. 20**   **1.** (a) yes   (b) no   (c) no   (d) no   **2.** (a) $R$   (b) $\{x : x \neq 4\}$
(c) $\{x : -2 \leq x \leq 2\}$   (d) $\{u : u \geq a\} \cup \{u : u \leq -a\}$   **3.** (a) $R$   (b) $\{y : y \geq -1\}$
(c) $\{t : -\sqrt{2} \leq t \leq \sqrt{2}\}$   **4.** (a) Domain $f = R$, Range $f = R, f(2) = 9, f(a) = a^3 + 1$,
$f[(y + a)/(y - a)] = (2y^3 + 6a^2 y)/(y - a)^3$   (b) Domain $g = \{x : x \geq 1\}$, Range $g = \{x : x \geq 0\}$,
$g(2) = 1, g(a) = \sqrt{a - 1}, g[(y + a)/(y - a)] = \sqrt{2a(y - a)}/(y - a)$   (c) Domain
$h = \{x : x \neq \pm 1\}$, Range $h = \{x : x \leq -1\} \cup \{x : x > 0\}, h(2) = 1/3, h(a) = 1/(a^2 - 1)$,
$h[(y + a)/(y - a)] = (y - a)^2/4ay$   **5.** Domain $f = \{x : x \geq 1\}$, Range $f = \{x : x \geq 1\}$,
$f(0) = 1, f(4) = 16 + \sqrt{3}, f(a + b) = (a + b)^2 + \sqrt{a + b - 1}, f(1 + h) = (1 + h)^2 + \sqrt{h}$.

$[f(1+h) - f(1)]/h = 2 + h + 1/\sqrt{h}, [f(y^2+k) - f(y^2)]/k = 1 + 1/\sqrt{y^2+k-1} + \sqrt{y^2-1})$

**6.** Domain $g = \{x: x > 1\} \cup \{x: x < -1\}$, Range $g = [0, 1) \cup \{x: x > 1\}, g(2) = \sqrt{3}$,
$g[(x+1)/(x-1)] = \sqrt{x}, g[(x-1)/(x+1)] = \sqrt{-x}, g(2+h) = \sqrt{3+4h+h^2}/(1+h)$,
$[g(2+h) - g(2)]/h = -2/[\sqrt{3+4h+h^2} + \sqrt{3}(1+h)]$,
$[g(a+h) - g(a)]/h = -2/[(a-1)\sqrt{(a+h)^2-1} + (a+h-1)\sqrt{a^2-1}]$

**7.** (a) 1　　(b) 1　　(c) undefined　　(d) 1/7　　(e) undefined　　(f) undefined　　(g) 1/3

**Section 1.7 – p. 23**　　**1.** function; Domain $f =$ Range $f = R$　　**2.** function; Domain $f =$ Range $f = R$　　**3.** not a function　　**4.** function; Domain $f = [-1, 1]$ Range $f = [0, 1]$　　**5.** function; Domain $f = R$, Range $f = [0, \infty)$　　**6.** not a function　　**7.** function; Domain $f = R$, Range $f = [0, \infty)$　　**8.** function; Domain $f =$ Range $f = R$　　**9.** function; Domain $f = [-2, 2]$, Range $f = [0, 2]$　　**10.** function; Domain $f =$ Range $f = R$　　**11.** function; Domain $f = R$, Range $f = \{x: x \le 1\}$　　**12.** not a function　　**13.** $\emptyset$　　**14.** function; Domain $f = R - \{4\}$ Range $f = R - \{0\}$　　**15.** $(f+g)(x) = 2x-1$ if $x > 1$; $= 1$ if $x \le 1$. $(f-g)(x) = -1$ if $x > 1$; $= 1 - 2x$ if $x \le 1$. $(2f-g)(x) = x-2$ if $x > 1$; $= 2 - x$ if $x \le 1$.　　**16.** $(f+g)(x) = 2-x$ if $x > -1/2$; $= -5x$ if $x \le -1/2$. $(f-g)(x) = 5x$ if $x > -1/2$; $= x - 2$ if $x \le -1/2$. $(2f-3g)(x) = 13x-1$ if $x > -1/2$; $= 5(x-1)$ if $x \le -1/2$.　　**17.** $(f+g)(x) = x^2 + \sqrt{x+1}, D = \{x: x \ge -1\}$ $(g \cdot h)(x) = (x+1)^{3/2}/(x-1)$, $D = [-1, 1] \cup \{x: x > 1\}$; $(f-h)(x) = (x^3 - x^2 - x - 1)/(x-1), D = R - \{1\}$, $(h/g)(x) = (x+1)^{1/2}/(x-1), D = (-1, 1) \cup \{x: x > 1\}$.　　**18.** yes　　**19.** no, $f$ is not defined when $x = 0$　　**21.** (a) $(2x+1)(\sqrt{x+2} + \sqrt{x+1})$　　(b) 4　　(c) $h(1)$ undefined　　(d) $(x+1)^4 - 2\sqrt{x}(x+1)^2 + x$　　(e) $2\sqrt{3}$

**Section 1.8 – p. 27**　　**1.** $f(x) = 6x - 8$　　**2.** $f(x) = \frac{4}{3}x + \frac{5}{3}$　　**3.** $f(x) = 2x - 4$

**4.** $f(x) = x + 1$　　**5.** $f(x) = -\frac{2}{3}x + \frac{14}{3}$　　**6.** $f(x) = -\frac{3}{2}x + 7$　　**7.** $f(x) = \frac{1}{2}x^2 - \frac{7}{2}x + 4$

**8.** $f(x) = -\frac{5}{6}x^2 - \frac{5}{6}x + 4$　　**9.** $f(x) = 2x^2 - 8x + 13$　　**10.** $f(x) = 8x^2 - 16x + 7$

**11.** (a) $\left(-\frac{1}{6}, -\frac{13}{12}\right)$　　(b) $(2, 11)$　　**12.** Completing the square in $f$, we obtain

$$a_2 x^2 + a_1 x + a_0 = a_2\left(x + \frac{a_1}{2a_2}\right)^2 + a_0 - \frac{a_1^2}{4a_2}$$　　**13.** $y = -\frac{1}{4}x + \frac{11}{8}$　　**14.** $24y - 16x - 103 = 0$

**15.** $y = 4x - 4$　　**16.** $y = -\frac{3}{4}x + 8$　　**17.** $t = \frac{5}{2}\sqrt{10}$ sec.

**Section 1.9 – p. 33**　　**7.** (a) 2　　(b) 1　　(c) 2　　(d) 2　　**8.** 2　　**9.** 8　　**12.** (a) 12　　(b) 23　　(c) −6　　**18.** 1 if $a > 0$, −1 if $a < 0$.　　**20.** 1 if $a > 0$, −1 if $a < 0$.　　**22.** no　　**23.** $\lim_{x \to 0} f(x)$ does not exist　　**24.** $\lim_{x \to 0} f(x) = 0$.

**Section 1.10 – p. 37**　　**1.** limit does not exist　　**2.** limit does not exist　　**4.** (a) 6　　(b) 5/6　　(c) 1　　(d) 3　　(e) does not exist　　(f) does not exist　　(g) 1/2　　(h) does not exist　　(i) does not exist　　(j) 3/8　　(k) 0　　**5.** (a) continuous everywhere　　(b) discontinuous at $x = -1$, not removable　　(c) discontinuous at $x = 1$, removable　　(d) discontinuous at $x = \pm 1$, −1 is removable but +1 is not　　(e) continuous everywhere　　(f) discontinuous at $x = \pm 2$, not removable　　(g) discontinuous at $x = 1$, not removable　　(h) discontinuous at $x = 2$, removable　　(i) discontinuous at $x = 2$, removable　　(j) discontinuous at $x = \pm 1$, not removable　　**6.** If $a = n$, then $f$ is continuous at $x = 1$.　　**7.** (a) all values of $a$; $g(0) = 2a$　　(b) all values of $a$; $g(0) = 6a - 1$　　(c) all values of $a$ except $a = 0$; $g(0) = -2/a^2$　　(d) all values of $a$ except $a = 0$; $g(0) = -6/a^3$　　(e) all values of $a$ except $a = -1$; $g(0) = 2/(a+1)^2$　　(f) all values of $a$ except $a = 1/3$; $g(0) = -7/(3a-1)^2$　　**8.** Yes, use Theorem 1.11.

**Section 1.11 – p.40**　　**1.** (b) Let $F = \{n \in W: (1+x)^n < 1+nx, x > -1\}$. $1 \notin F$ since $1 + x = 1 + x$. Assume $F \ne \emptyset$. Let $n$ be the smallest integer in $F$. Then $n - 1 \notin F$ and hence $(1+x)^{n-1} \ge 1 + (n-1)x$. Multiply both sides of this inequality by $(1+x)$ to obtain a contradiction.　　**2.** (b) If $n = 1$, the result is clear. Suppose $1^2 + 2^2 + \cdots + (n-1)^2 < n^3/3$. Add $n^2$ to both sides of this inequality and show that $n^2 + \frac{n^3}{3} < \frac{(n+1)^3}{3}$.　　**3.** $\frac{1}{2} + \frac{2}{4} + \cdots + \frac{n}{2^n} = 2 - \frac{n+2}{2^n}$.　　**4.** $(1 - 1/2)(1 - 1/3) \cdots (1 - 1/n) = 1/n$

# CHAPTER 2

**Section 2.1 – p. 46**    **1.** $-8, -14, -20$, $v(1) = -5$, $v(2) = -11$, $v(3) = -17$, $v(4) = -23$,

$v(t) = 1 - 6t$    **2.** $\frac{1}{2}, \frac{5}{6}$, $a(a+h) - 1$, $v(t) = 1 - \frac{1}{t^2}$    **3.** $2a - h$; $v(t) = 2t$; $t = 0$

**4.** $v(t) = 3 - 3t^2$, $t = \pm 1$    **6.** (a) $t = 1/2$    (b) does not change sign    (c) $t = \pm 1$

**7.** (a) $a(t) = 0$    (b) $a(t) = 6$    (c) $a(t) = \frac{2}{t^3}$    (d) $a(t) = \frac{4}{(t-1)^3}$    **9.** $v(t) = -32t$,

$-80$ ft/sec.    **10.** $v(t) = 2at + b$, $t = \frac{-25 + 5\sqrt{65}}{4}$, $s = 2250$ ft    **11.** $+1$ for $t > 0$,

$-1$ for $t < 0$, not defined at $t = 0$

**Section 2.2 – p. 51**    **1.** $y = 16x - 12$    **2.** $y = -4x + 5$    **3.** $x = 1$    **4.** $x = \pm 1$

**5.** $x = \pm 2$    **7.** (a) $(2, 3)$    (b) $\left(-\frac{1}{2}, -\frac{17}{4}\right)$    (c) $\left(-\frac{3}{2}, -\frac{13}{4}\right)$    **8.** $-\frac{1}{4}$    **11.** no, no

**Section 2.3 – p. 56**    **1.** (a) $f'(x) = -3$, $f''(x) = 0$    (b) $f'(x) = 4x + 1$, $f''(x) = 4$

(c) $f'(t) = 1 - 6t$, $f''(t) = -6$    (d) $g'(s) = 2(s-1)$, $g''(s) = 2$    (e) $h'(s) = \frac{1}{(4-s)^2}$,

$h''(s) = \frac{2}{(4-s)^3}$    (f) $g'(s) = \frac{-2}{(2s-1)^2} + 1$, $g''(s) = \frac{8}{(2s-1)^3}$    (g) $h'(t) = 3t^2$, $h''(t) = 6t$

(h) $p'(u) = 3(u-1)^2$, $p''(u) = 6(u-1)$    (i) $p'(r) = \frac{-2r}{(r^2-1)^2}$, $p''(r) = \frac{4r^2+2}{(r^2-1)^3}$

(j) $f'(t) = \frac{-2}{(t-1)^2}$, $f''(t) = \frac{4}{(t-1)^3}$    (k) $g'(s) = \frac{1-s^2}{(s^2+1)^2}$, $g''(s) = \frac{2s(s^2-3)}{(s^2+1)^3}$

(l) $h'(u) = \frac{-2}{(1+t)^3}$, $h''(u) = \frac{6}{(1+t)^4}$    **2.** $y = 5x - 2$    **3.** $y = t$    **4.** $x_0 = \frac{3}{4}$    **5.** $x_0 = \pm 2$

**6.** (a) $8y - x + 54 = 0$    (b) $y = s + 2$    (c) $12y + t - 97 = 0$    **9.** (a) not cont. ∴ not diff.
(b) not cont. ∴ not diff.    (c) cont. and diff.    (d) cont. but not diff.    (e) cont. and diff.

**Section 2.4 – p. 60**    **1.** $3$    **2.** $6x + 4$    **3.** $20t^4 + \frac{6}{t^4}$    **4.** $8x + 20$    **5.** $-\frac{14}{(3p-4)^2}$

**6.** $2s(3s-1)$    **7.** $\frac{-4}{(4t-7)^2}$    **8.** $\frac{2y^4 + 24y^2 + 2y}{(1-2y^3)^2}$    **9.** $\frac{-2x}{(x^2-1)^2}$    **10.** $\frac{-2}{(y-1)^2}$

**11.** $\frac{1-4s}{(2s^2-s)^2}$    **12.** $\frac{-4t}{(t^2+1)^2}$    **13.** $\frac{t^2(28t^3 + 35t^2 - 2t - 3)}{(t+1)^2}$    **14.** $6t^2 + 4t - 4$

**15.** $\frac{y^2 + 2y + 7}{[(2-y)(3+y)]^2}$    **16.** $(2-t)(-3t-4)$    **17.** $\frac{24z^4 - 8z^3 - 21z^2 + 12z}{(6z^2 - z - 2)^2}$    **18.** $\frac{3(s^2 + 4s + 1)}{(s+2)^2}$

**19.** $\frac{4u^5 - 15u^4 - 12}{(u-3)^2}$    **20.** $6x^2 + 8x - 10$    **21.** $5s^4 - 6s^2 + 1$    **22.** $\frac{5 - 9t^2}{4t^6}$

**23.** $\frac{2(-8x^2 + 10x - 3)}{x^7}$    **24.** $\lim_{h \to 0} \frac{cf(x+h) - cf(x)}{h} = c \lim_{h \to 0} \frac{f(x+h) - f(x)}{h} = cf'(x)$

**29.** (a) $15(3x+1)^4$    (b) $24x(2x^3-1)^5$    (c) $8(t^2 + 3t)^7(2t+3)$    (d) $\frac{-10}{(2s+3)^6}$

(e) $\frac{-96x}{(2x^2+3)^7}$    (f) $\frac{(2x-3)^3(6x+43)}{(3x+2)^4}$    (g) $(4t^2 + 3t)^4 \left[\frac{5t(8t+3)(1+1/t) + 4(4t+3)}{t(1+1/t)^5}\right]$

(h) $\left(1 - \frac{1}{t}\right)^2 \left[\frac{-6t^5 + 3t^4 + 72t - 84}{t^5(t + 4/t^3)^7}\right]$    (i) $\frac{2(4x-1)^4(2x+3)^3(-48x^2 + 158x + 121)}{(5-2x)^4}$

(j) $\frac{(2-3t)}{(t+1)^5(4-t)^4}(-15t^2 + 35t - 50)$

**Section 2.5 – p. 64**    **1.** $f(t) = \sqrt{1 + (2t-4)^2}$, $g(y) = \sqrt{1 + (y-3)^2}$, $h(t) = 2t - 1$

**2.** $d(t) = 2|3t^2 - 5|$, $f(t) = 2t^2 - 4$, $g(x) = |3x + 2|$    **3.** $d(t) = \frac{1}{64}|3(t^2 + 4)^2 - 6(t^2 + 4) + 2|^3$,

$f(u) = |u|^3$, $g(x) = \frac{3x^2}{4} - \frac{3x}{2} + 1/2$, $h(t) = t^2 + 4$    **4.** $(f \circ g)(x) = 2x^2 - 5$,

$(g \circ f)(x) = 4x^2 + 12x + 5$, $(f \circ f)(x) = 4x + 9$, $(g \circ g)(x) = x^2 - 8x + 12$;
$(f \circ g)(1) = -3$, $(g \circ f)(1) = 21$, $(f \circ f)(1) = 13$, $(g \circ g)(1) = 5$

**5.** $(f \circ g)(x) = \frac{1}{2x^2 - 1}$, domain $= \{x : x \neq \pm 1/2\}$, $(g \circ f)(x) = \frac{1}{(2x-1)^2}$, domain $= \{x : x \neq \pm 1/2\}$

**6.** $(f \circ f)(x) = \frac{3x - 4}{19 - 12x}$, domain $= \{x : x \neq 19/12\}$    **7.** $(f \circ g)(x) = \frac{1}{\sqrt{1 - x} - 1}$, domain $= \{x : x \leq 1, x \neq 0\}$, $(g \circ f)(x) = \sqrt{1 - \dfrac{1}{x-1}}$, domain $= \{x : x < 1, x > 2\}$

8. $(f \circ g)(x) = \dfrac{1}{1-x^4}$, $(g \circ f)(x) = \left(\dfrac{1}{1-x^2}\right)^2$, $(f \circ g)(2) = -\dfrac{1}{15}$, $(g \circ f)(2) = \dfrac{1}{9}$,

domain of $f \circ g = \{x : x \neq \pm 1\}$, domain of $g \circ f = \{x : x \neq \pm 1\}$,

$(f \circ g \circ f)(x) = \dfrac{(1-x^2)^2}{x^4 - 2x^2}$, $(g \circ f \circ g)(x) = \left(\dfrac{1}{1-x^4}\right)^2$    9. $(f \circ g)(x) = \sqrt{1 - |x-1|}$,

$(g \circ f)(x) = |\sqrt{1-x} - 1|$, domain of $f \circ g = [0, 2]$, domain of $g \circ f = \{x : x \leq 1\}$.

10. $(f \circ g)(x) = \dfrac{1}{3x}$, domain $= \{x : x \neq 0\}$; $-\dfrac{3}{x^2}$    11. $(f \circ g)(x) = \dfrac{1}{4x(x-1)}$; $\dfrac{2x-1}{4x^2(x-1)^2}$

**Section 2.6 – p. 70**    1. $\dfrac{3}{2\sqrt{1+3x}}$    2. $10x(3+x^2)^4$    3. $\dfrac{4(1+t)}{3\sqrt[3]{(3+4t+2t^2)^2}}$    4. $15(1+3s)^4$

5. $\dfrac{1}{\sqrt{x-1}\sqrt{x+1}(x+1)}$    6. $\dfrac{3(2x+\sqrt{x})^2(4\sqrt{x}+1)}{2\sqrt{x}}$    7. $\dfrac{5\sqrt{x}(1+\sqrt{x})^9}{x}$

8. $\dfrac{2}{\sqrt{2t-1}\sqrt{(2t+1)^3}}$    9. $3(4-x+3x^2)^2(6x-1)$    10. $\dfrac{-15}{(3x-5)^6}$    11. $\dfrac{3x^3+x}{\sqrt{1+x^2}}$

12. $-\dfrac{6}{(3x-2)^3}$    13. $\dfrac{-x(5+x^2)}{3\sqrt{1-x^2}\sqrt[3]{(1+x^2)^4}}$    14. $\dfrac{1}{5}(1+x^2-x^4)^{-4/5}(2x-4x^3)$

15. $\dfrac{1}{2}\left[-1+\dfrac{\sqrt{x}-\sqrt{1+x}}{\sqrt{x^3}+\sqrt{(1+x)^3}}\right]$    16. $\dfrac{2x}{(48-x^2)^2}$    17. $-3(2+3x^2-\sqrt{x})^{-4}\left(\dfrac{12x^2-\sqrt{x}}{2x}\right)$

18. $\dfrac{1}{8\sqrt{1+\sqrt{1+\sqrt{x}}}\sqrt{1+\sqrt{x}}\sqrt{x}}$    19. $3\dfrac{x}{(x^2-1)^{1/4}(x^2+1)^{7/4}}$

20. $\dfrac{3}{2}(2+1/x-\sqrt{x})^{-5/2}\left(\dfrac{2+x\sqrt{x}}{2x^2}\right)$    21. $\dfrac{3}{2}[(1+\sqrt{x})^2+1]^{1/2}\left(\dfrac{x+\sqrt{x}}{x}\right)$    22. $\dfrac{26+2x-19x^2}{(2x+1)^4\sqrt{x^2-1}}$

23. $\dfrac{1}{2}\dfrac{3x^{3/2}-5x^{1/2}+2x-2x^2+2}{(1-x)^{3/2}\sqrt{1-2x^{1/2}+x}}$    24. $-\dfrac{1}{2}\left[\left(x-\dfrac{1}{x}\right)^3+\sqrt{x}\right]^{-3/2}\left[3\left(x-\dfrac{1}{x}\right)^2\left(1+\dfrac{1}{x^2}\right)+\dfrac{\sqrt{x}}{2x}\right]$

**Section 2.7 – p. 73**    1. $\dfrac{dy}{dx} = 1+\dfrac{1}{2}(x-1)^{-3/2}$, $\dfrac{d^2y}{dx^2} = -\dfrac{3}{4}(x-1)^{-5/2}$    2. $\dfrac{ds}{dt} = -t(1+t^2)^{-3/2}$,

$\dfrac{d^2s}{dt^2} = \dfrac{2t^2-1}{(1+t^2)^{5/2}}$    3. $(Df)(x) = \dfrac{2x(1+x^2)}{\sqrt{x^4+2x^2+2}}$, $(D^2f)(x) = \dfrac{2x^6+6x^4+12x^2+4}{(x^4+2x^2+2)^{3/2}}$, $D^2f(0) = \sqrt{2}$

4. $\dfrac{du}{dv} = \dfrac{3w-v}{\sqrt{v-w(v+w)^2}}$, $\dfrac{du}{dw} = \dfrac{w-3v}{\sqrt{v-w(v+w)^2}}$

**Section 2.8 – p. 77**    1. $y' = -\dfrac{3x}{4y}$    2. $y' = \dfrac{2x}{y}$    3. $y' = \dfrac{1}{8y+1}$    4. $y' = \dfrac{2x+y^3}{1-3xy^2}$

5. $y' = \dfrac{2\sqrt{x+y^2}-1}{2y-2\sqrt{x+y^2}}$    6. $y' = \dfrac{1-18x^2y^3}{18x^3y^2-5y^4}$    7. $y' = \dfrac{(x-y)^2+2y}{2x}$    8. $y' = -\left(\dfrac{y+a}{x-a}\right)^2$

10. $y' = \dfrac{2\sqrt{x^2-y^2}+\sqrt{x+y}-\sqrt{x-y}}{\sqrt{x-y}+\sqrt{x+y}}$    11. $y' = -\dfrac{2x+y}{x+2y}$, $y'' = \dfrac{6(x^2+xy+y^2)}{(x+2y)^3}$

12. $y' = \dfrac{x}{2y}$, $y'' = \dfrac{2y^2-x^2}{4y^3}$    13. $y' = \dfrac{1-y}{x+2y+1}$, $y'' = \dfrac{2(y^2+xy+y-x-2)}{(x+2y+1)^3}$    14. $y' = \dfrac{1}{16}$

15. $y' = -\dfrac{x_0}{y_0}\cdot\dfrac{b^2}{a^2}$    16. $y = \dfrac{\sqrt{5}}{2}x-\dfrac{1}{2}$    17. $11y-x-16 = 0$

21. (a) meaningful    (b) not meaningful    (c) meaningful

**Section 2.9 – p. 81**    1. (a) $x^3+x+C$    (b) $x^5+x^3-2x+C$    (c) $\dfrac{x^6}{6}-\dfrac{7}{3}x^3+3x+C$

(d) $\dfrac{x^5}{5}+\dfrac{5}{2}x^4+9x^3+5x^2+x+C$    (e) $-\dfrac{1}{x^2}+\dfrac{1}{x^4}+C$    (f) $x^4+\dfrac{7}{2x^2}+2\sqrt{x^3}+C$

(g) $\dfrac{8\sqrt{x^3}}{3}-\dfrac{2\sqrt{x^5}}{5}+C$    (h) $\dfrac{6\sqrt[3]{x^5}}{5}-\dfrac{4x^3}{3}+\dfrac{25\sqrt[5]{x^4}}{4}+C$    (i) $5x-\dfrac{8}{\sqrt{x^3}}+\dfrac{3}{2\sqrt[3]{x^4}}+C$

(j) $3x^2-\dfrac{7x^4}{4}+20\sqrt{x}-\dfrac{1}{4x^4}+C$    2. (a) $x^2-x+4$    (b) $\dfrac{x^3}{3}-2x^2+x+\dfrac{8}{3}$

(c) $\dfrac{3}{5}x^5+\dfrac{5}{3}x^2-6x+7$    (d) $\dfrac{2}{x}-\dfrac{x^2}{2}-\dfrac{1}{x^4}+\dfrac{1}{2}$    (e) $2\sqrt{x^3}+\dfrac{1}{x^2}+\dfrac{15-16\sqrt{2}}{4}$    (f) $\dfrac{2\sqrt{x^5}}{5}-8\sqrt{x}+\dfrac{33}{5}$

(g) $\dfrac{x^3}{3}+\dfrac{4}{\sqrt{x}}-\dfrac{1}{3}$    (h) $x^3+\dfrac{12}{\sqrt[3]{x}}-(7+6\sqrt[3]{4})$    3. $\dfrac{x^4}{4}-x^3+x^2+1$    4. $\dfrac{x^3}{3}+2x-\dfrac{1}{x}-\dfrac{4}{3}$

**5.** $\dfrac{x^2}{2}-\dfrac{1}{x}+\dfrac{1}{6x^2}-\dfrac{5x}{3}-2$    **6.** $\dfrac{2x^5}{5}+\dfrac{16\sqrt{x^9}}{21}+\dfrac{x^4}{2}+\dfrac{4\sqrt{x^7}}{35}-\dfrac{234x}{35}+1$    **7.** $-2x^3+6x^2-6x+4$

**8.** $\sqrt[4]{264}$    **9.** 5/2 sec    **10.** 625/16 ft, 25/8 sec    **11.** 84 ft/sec    **13.** $k=2$ yards/sec$^2$

**14.** (a) $\dfrac{(2x+1)^6}{6}+C$    (b) $\dfrac{(1-x)^4}{4}+C$    (c) $\dfrac{2\sqrt{(1+x)^3}}{3}+C$    (d) $-\dfrac{1}{2(2x+1)^2}+C$

(e) $-\dfrac{2}{\sqrt{1+2x}}+C$    (f) $\dfrac{\sqrt{(1+2x)^3}}{3}+C$    (g) $\dfrac{1}{2(1-x)^2}+C$    (h) $\dfrac{\sqrt{(1+x^2)^3}}{3}+C$

**Section 2.10 – p. 84**    **1.** $\dfrac{8}{3}$    **2.** $-\dfrac{32}{3}(10\sqrt{10}-8)^{-1/3}$    **3.** $y'=\dfrac{3}{4},\,y''=-\dfrac{173}{64}$

**5.** $\dfrac{\sqrt{2}}{2}$ ft/sec, $\dfrac{\sqrt{2}}{4}$ ft/sec, no    **6.** 22 knots    **7.** $\dfrac{375\sqrt{29}}{29}$ ft/sec

**8.** $25\sqrt{2}$ ft/sec, $25\sqrt{2}$ ft/sec    **9.** $-8\sqrt{\sqrt{2}+1}$    **10.** (a) $\dfrac{1}{8\pi}$ ft/sec    (b) $\dfrac{1}{6\pi}$ ft/sec

**11.** $\dfrac{10}{3}\sqrt{13}$    **12.** $\dfrac{r}{3}$

**Section 2.11 – p. 88**    **1.** (a) $df(h)=13h$    (b) $df(h)=\dfrac{3\sqrt{10}}{10}h$    (c) $df(h)=-3\sqrt{2h}$

(d) $df(h)=\dfrac{1}{\sqrt{a-1}\sqrt{(a+1)^3}}h$    (e) $df(h)=\dfrac{3a^2+4a+1}{2\sqrt{(a+1)^3}}h$

(f) $df(h)=\dfrac{\sqrt{a^2+1}(2a^2-3a-1)}{(a-1)^2}h$    **2.** (a) $l(x)=16x-20$    (b) $l(x)=25x-27$

(c) $l(x)=\dfrac{\sqrt{6}(5x+2)}{12}$    (d) $l(x)=\dfrac{7\sqrt{3}}{3}-\dfrac{2\sqrt{6}x}{3}$    (e) $l(x)=\dfrac{\sqrt{2}}{2}$

**3.** (a) 105,000    (b) 3.0019    (c) 2.0125    (d) 4.987    (e) 4.035    **4.** all values of $x_0$ except $x_0=\pm1$    **5.** all $x\neq0$    **6.** .12 in.$^3$, .24 in.$^2$    **7.** $5\times10^{-7}$ in.

# CHAPTER 3

**Section 3.1 – p. 94**    **1.** 7/2, 9/2; 15/4, 17/4; 31/8, 33/8    **2.** 3/4, 5/4; 7/8, 9/8; 15/16, 17/16
**3.** 3/16, 9/16; 16/64, 30/64; 74/256, 102/256    **4.** 14/16, 26/16, 68/64, 96/64; 308/256,
364/256    **5.** 41/4, 56/4; 183/16, 214/16; 752/64, 819/64

**Section 3.2 – p. 98**    **1.** $6[x+1],12[x+1],=18[x+1]+1,111/10,180/7,733/6$
**2.** $26-2(\sqrt{2}+\sqrt{3})-(\sqrt{5}+\sqrt{6}+\sqrt{7}+\sqrt{8})$,    **3.** $7/2+2\sqrt{2}+\sqrt{3}$    **4.** $30=\int_{-3}^{9}f$

**5.** $4(\sqrt{12}-\sqrt{6})+2(\sqrt{8}+\sqrt{7}-\sqrt{10})=\int_{-\sqrt{12}}^{-\sqrt{6}}f+\int_{\sqrt{6}}^{\sqrt{12}}f$    **6.** 26/3    **7.** 28, 122, 25/2

**8.** $\int_m^n g=\left(\dfrac{n-m}{2}\right)(m+n-1)$

**Section 3.4 – p. 107**    **3.** 124/3, 9    **4.** 208/3    **7.** 255/4    **8.** 1/6, 9, 3/2
**12.** Use Exercise 11.    **13.** yes

**Section 3.5 – p. 114**    **1.** 7/2    **2.** 7/3    **3.** 242/3    **4.** 7/9    **5.** $\dfrac{68\sqrt{2}-22}{15}$    **6.** 81

**7.** $40+4a$    **8.** $\dfrac{21b}{2}+\dfrac{3}{10a^2}$    **9.** $2(1+3t^2)/3$    **10.** $x^2-x+1/3$    **11.** $(7^{11}-5^{11})/11$

**12.** $(3^7+1)/14$    **13.** $2(\sqrt{5}-\sqrt{3})$    **14.** $\dfrac{(2-a)^5+(2+a)^5}{10}$    **15.** $(2a^3-a^4)/4$    **16.** 57/4

**17.** 81/5    **18.** $20\sqrt{5}/3$    **19.** 64/3    **20.** 36    **21.** 18    **22.** $\sqrt{1+x^2}$, $\sqrt{5}$
**23.** $\sqrt{1+x^3}$    **24.** $(1+s)^3$    **25.** $-1/\sqrt{3}$    **26.** $1/\sqrt{2}$

**Section 3.6 – p. 123**    **1.** $-11/2$    **2.** $-2/3$    **3.** 0    **4.** 5/2    **5.** $-139/15$

**6.** $\dfrac{1}{2n+1}[2^{2n+1}+3^{2n+1}]$    **7.** 3/2    **8.** 5/2    **9.** 11/6    **10.** 123/4    **11.** 8

**12.** $6-\sqrt{3}-\sqrt{2}$    **13.** 125/6    **14.** 45/6    **15.** 32/3    **16.** 8    **17.** 11

**18.** $\dfrac{28\sqrt{7}+12\sqrt{3}-52}{3}$    **19.** 37/12    **22.** (a) $A=2\sqrt{2},B=2\sqrt{10}$,    (b) $A=2,B=6$

(c) $A=2/3,B=2$    **23.** (a) $1/x$    (b) $\sqrt{1+x^2}$,    (c) $-1/x$,    (d) $3\sqrt{1+9x^2}$,
(e) $1/(1+5x+6x^2)$    (f) $2x\sqrt{1+x^4}$

**Section 3.7 – p. 130**   1. (a) $2(2\sqrt{2}-1)$   (b) $2(\sqrt{3}-\sqrt{2})+\frac{9}{2}(\sqrt[3]{4}-\sqrt[3]{9})$   (c) $-\frac{548}{7}$

(d) $\frac{9}{2}\sqrt[3]{9}+\frac{2}{27}\sqrt{3}-\frac{13}{2}$   (e) $[9(2\sqrt[3]{2}-\sqrt[3]{4})-32]/3$   (f) $30-15\sqrt[3]{4}$   (g) $(46+6\sqrt{3})/3$

(h) $39c+3d(\sqrt[3]{5}-\sqrt[3]{2})$   2. $16/3$   3. $26/15$   4. $10/3$   5. $7/15$

6. (a) $\frac{2}{3}(3\sqrt{3}-2\sqrt{2})$   (b) $\frac{1}{11}(3^{11}-2^{11})$   (c) $\frac{3}{4}(3\sqrt[3]{3}-1)$   (d) $-\frac{5}{72}$   (e) $\frac{1}{4}$   (f) $0$

(g) $\frac{5}{12}(3\sqrt[5]{6}-\sqrt[5]{2})$   (h) $-\frac{36}{49}$   (i) $40+(11^{3/2}-3^{3/2})/6$   (j) $\frac{1}{3}(3^{3/2}-1)+\frac{1}{2}(10^{2/3}-7^{2/3})$

(k) $\frac{1}{12}[1-5^6]+\frac{1}{4}[5^{-4}-2^{-4}]$   (l) $\frac{2}{3}[8-\sqrt{8}]$   (m) $\frac{2}{3}[\sqrt{125}-\sqrt{8}]$   (n) $\frac{3}{4}[\sqrt[3]{36}-\sqrt[3]{9}]$

(o) $\frac{1}{33}[2^{11}-1]$   7. $2\sqrt{3}+14/3\sqrt{7}-2$   8. If $a\leqslant t\leqslant b$, then $1/b\leqslant 1/t\leqslant 1/a$. Now use

(10), p. 121.   9. Use same estimate as in Problem 8.   11. Compute upper and lower

estimates for $\displaystyle\int_{k-1}^{k}\frac{dt}{t}$ and then add.

**Section 3.8 – p. 137**   1. $4/3$   2. $4$   3. $3/4$   4. $81/4$   5. $6$   6. $\displaystyle\int_{1}^{4}(5-x-1/x)\,dx$

7. $128\pi/3$   8. $\pi h^2\left(r-\dfrac{h}{3}\right)$   9. $64\pi$   10. $8\sqrt{3}\pi$   11. $128\pi/5$   12. $41\pi/2$

13. $4\pi\sqrt{3}$   14. $128/3$   15. $32\sqrt{3}/2$   16. $32,\dfrac{1024\sqrt{2}}{15}$   17. $125\pi/1152$   18. $\pi abh/3$

19. $480$ ft-lb   20. $600$ ft-lb   21. $10,000, 7,500$ ft-lb   22. $15,000, 10,000$ ft-lb

23. $k$ mile-lb   24. $W=k\displaystyle\int_{4000}^{4010}\frac{(ax+b)}{x^2}\,dx+20,000\,k\displaystyle\int_{4010}^{5000}\frac{dx}{x^2}$ where $a,b$ are constants

determined by the equations $30,000=4000a+b$, $20,000=4010a+b$. No, since we cannot at

this point evaluate $\displaystyle\int_{4000}^{4010}\frac{dx}{x}$   25. $12,480\pi$ ft-lb   26. $13,000\pi, 39,000\pi$ ft-lb

# CHAPTER 4

**Section 4.2 – p. 146**   4. (a) $\log x+1$   (b) $\log_{10}(x^2+1)+2x^2/(x^2+1)\log 10$
(c) $2(x+1)\log x/x+(\log x)^2$   (d) $-2/(x^2-1)\log 2$   (e) $x/(1+x^2)$
(f) $8x[\log (x^2+4)]^3/(x^2+4)$   (g) $\dfrac{1}{x\log x}$   (h) $\sqrt{1+x^2}/\displaystyle\int_{1}^{x}\sqrt{1+t^2}\,dt$
(i) $+2+\log (x^2\log x)+1/\log x$   (j) $-\sqrt{1+x^3}/\displaystyle\int_{x}^{2}\sqrt{1+t^3}\,dt$   (k) $4x/(x^4-1)$
5. (a) $\log |x+4|+C$   (b) $\log |x^2-4|+C$   (c) $\log |x^3+a^3|+C$   (d) $1/3(x^2+1)^3+C$
(e) $\sqrt{x^2+1}+C$   (f) $\log |\log x|+C$   6. (a) $1/3(x^2+1)^{3/2}+C$   (b) $-\sqrt{1-x^2}+C$
(c) $1/2\log |2x+4|+C$   (d) $1/9(3x-2)^3+C$   (e) $1/15(x^3+a^3)^5+C$
(f) $1/4(4x^2-1)^{1/2}+C$   (g) $1/4\log (x^4+1)+C$   (h) $1/2\log |\log (2x+1)|+C$
(i) $x-2\log |x+1|+C$   (j) $1/2\log\log (x^2+1)+C$   (k) $2(\log 10)\log |\log_{10} x|+C$
7. $a=-1/3$   8. $63/4-12\log 2$   9. $2-3\log\sqrt{3}$   10. $\pi\log 4$   11. $\pi[3/2-2\sqrt{2}+\log 4]$
17. (a) $\left(\dfrac{2}{x-1}+\dfrac{6}{2x+3}-\dfrac{1}{x-4}\right)\dfrac{(x-1)^2(2x+3)^3}{(x-4)}$   (b) $\left(\dfrac{1}{2x-3}-\dfrac{3}{(x+4)}\right)\dfrac{\sqrt{2x-3}}{(x+4)^3}$
(c) $\left(\dfrac{6}{2x-1}+\dfrac{2}{4x-3}-\dfrac{70}{7x+3}-\dfrac{2}{3(x-1)}\right)\left(\dfrac{(2x-1)^3\sqrt{4x-3}}{(7x+3)^{10}(x-1)^{2/3}}\right)$
(d) $\left(\dfrac{3}{x-1}+\dfrac{2}{x-2}+\dfrac{1}{x-3}-\dfrac{3}{x}-\dfrac{2}{x+1}-\dfrac{1}{x+2}\right)\left(\dfrac{(x-1)^3(x-2)^2(x-3)}{x^3(x+1)^2(x+2)}\right)$
(e) $\dfrac{1}{n}\left[\dfrac{1}{(x-a_1)}+\dfrac{1}{(x-a_2)}+\cdots+\dfrac{1}{(x-a_n)}\right]\sqrt[n]{(x-a_1)(x-a_2)\cdots(x-a_n)}$

**Section 4.3 – p. 153**   3. (a) $y'=2xe^{(x^2+1)}, y''=e^{(x^2+1)}[4x^2+2]$   (b) $y'=(x+1)e^x$,
$y''=(x+2)e^x$   (c) $y'=2x+1, y''=2$   (d) $y'=\dfrac{-e^{-x}(1+x)}{x^2}, y''=\dfrac{e^{-x}(x^2+2x+2)}{x^3}$
(e) $y'=\dfrac{e^x+1}{e^x+x}, y''=\dfrac{e^x(e^x+x)-(e^x+1)^2}{(e^x+x)^2}$   4. (a) $10^x(x\log 10+1)$   (b) $\dfrac{1-x\log x\log 2}{x2^x\log 2}$
(c) $\dfrac{4}{(e^x+e^{-x})^2}$   (d) $\dfrac{(\log 10)\,10^{\log x}}{x}$   (e) $\dfrac{1}{x(\log x)[\log (\log x)]}$   (f) $\dfrac{x^{1/x}(1-\log x)}{x^2}$

(g) $\dfrac{2(\log x)x^{\log x}}{x}$  (h) $\log x+1$  5. (a) $-1$  (b) $\left(\dfrac{e^{x+y}-e^x}{e^y-e^{x+y}}\right)$  (c) $\dfrac{y\log y}{(y^{(y+1)}[\log y+1]-x)}$

8. $p(x)=1+x+x^2/2$  9. $p(x)=1+x+x^2/2+x^3/6+\cdots+x^n/n!$  10. (a) $\dfrac{2}{\log 2}$

(b) $\left(\dfrac{10^3-10^{-2}}{\log 10}\right)$  11. $1/e-e^{-x}$, area approaches $1/e$  12. $e+1/e-3$  13. $\pi/4[4+e^2-e^{-2}]$

14. (a) $1/3e^{x^3}+C$  (b) $\dfrac{1}{\log 100}10^{x^2}+C$  (c) $e^{x+\log x}+C$  (d) $\log (e^x+1)+C$

(e) $\dfrac{1}{\log 4}2^{-x^2}+C$  (f) $\dfrac{1}{2}\log (e^{x^2}+1)+C$  (g) $x^x+C$  (h) $-\dfrac{1}{\log 5}\dfrac{2^x}{10^x}+C$

15. (a) $2e^{x^4}(4x^4+1)$  (b) $\dfrac{e^x(1-x\log x)}{x(\log x)^2}$  (c) $8x^4e^{x^4}+2e^{x^4}-2xe^{x^2}$  23. $1061.00$, $e^{0.06}-1$

**Section 4.4 – p. 159**  1. $1:1$; $f^{-1}(y)=2/3y+4/3$, range $f^{-1}=$ domain $f^{-1}=R$.

2. $1:1$; $f^{-1}(y)=+\sqrt{\dfrac{4+y}{3}}$, domain $f^{-1}=\{x:x\geqslant -4\}$, range $f^{-1}=\{x:x\geqslant 0\}$

3. not $1:1$  4. $1:1$ $f^{-1}(y)=1/y$, domain $f^{-1}=$ range $f^{-1}=R-\{0\}$
5. $1:1$ $f^{-1}(y)=1/y^{1/n}$, domain $f^{-1}=$ range $f^{-1}=R-\{0\}$  6. not $1:1$
7. $1:1$; $f^{-1}(y)=-1/y-2$, domain $f^{-1}=R-\{0\}$, range $f^{-1}=R-\{-2\}$
8. $1:1$; $f^{-1}(y)=y+2$, domain $f^{-1}=R-\{-4\}$, range $f^{-1}=R-\{-2\}$.
9. $1:1$; $f^{-1}(y)=\dfrac{2(y+1)}{(y-1)}$, domain $f^{-1}=R-\{1\}$, range $f^{-1}=R-\{2\}$.  10. not $1:1$
11. not $1:1$  12. $a=0$, $f_1^{-1}(x)=x^{1/4}$ $f_2^{-1}(x)=-x^{1/4}$  13. $a=0$ $f_1^{-1}(x)=1\sqrt{x}$
$f_2^{-1}(x)=-1/\sqrt{x}$  14. $a=1$, $f_1^{-1}(x)=2-x$, $f_2^{-1}(x)=x$  15. $a=1$, $f_1^{-1}(x)=1+\sqrt{x}$,
$f_2^{-1}(x)=1-\sqrt{x}$  16. $a=3/2$, $f_1^{-1}(x)=3/2+\sqrt{x+1/4}$, $f_2^{-1}(x)=3/2-\sqrt{x+1/4}$
17. $a=1$, $f_1^{-1}(x)=1+\sqrt{x-1}$, $f_2^{-1}(x)=1-\sqrt{x-1}$

**Section 4.5 – p. 161**  1. no  2. Differentiate equation $e^{\log x}=x$.  3. If $g(x)=f^{-1}(x)$,
then $[g(x)]^n=x$. Differentiate and apply the chain rule.
7. $f'(x)=\dfrac{1}{\sqrt{1-x^2}}$ $F(y)=\sin y$, $G(y)=\cos y$

**Section 4.6 – p. 166**  2. *Hint.* Let $x=2y$ and use double angle formula.  4. $a=\sqrt{2}$,
$b=-\sqrt{2}$  5. $a=\alpha\cos\beta$, $b=-\alpha\sin\beta$  7. (a) $\{k\pi:k=0,\pm 1,\pm 2,\ldots\}$

(b) $\left\{\dfrac{(2k+1)\pi}{2}:k=0,\pm 1,\pm 2,\ldots\right\}$  (c) $\left\{\dfrac{(4k+1)\pi}{2}:k=0,\pm 1,\pm 2,\ldots\right\}$

(d) $\{2k\pi:k=0,\pm 1,\pm 2,\ldots\}$  (e) $\left\{\dfrac{(4k+3)\pi}{2}:k=0,\pm 1,\pm 2,\ldots\right\}$

(f) $\{(2k+1)\pi:k=0,\pm 1,\pm 2,\ldots\}$  8. (a) $\left\{\dfrac{(4k+1)\pi}{4}:k=0,\pm 1,\pm 2,\ldots\right\}$

(b) $\{2k\pi:k=0,\pm 1,\pm 2,\ldots\}$  (c) $\{\pi/2+2\pi k, \pi+2\pi k:k=0,\pm 1,\pm 2,\ldots\}$
9. Period of $|\tan x|$ is $\pi$  12. $f(x)=(x^2+1)+(-x)$ where $f_1(x)=x^2+1$ is even,
$f_2(x)=-x$ is odd, $g(x)=\dfrac{e^x+e^{-x}}{2}+\dfrac{e^x-e^{-x}}{2}$ where $g_1(x)=\dfrac{e^x+e^{-x}}{2}$ is even and
$g_2(x)=\dfrac{e^x-e^{-x}}{2}$ is odd.

**Section 4.7 – p. 172**  1. (a) $3\cos (3x+2)$  (b) $-2x\sec^2 (2-x^2)$  (c) $-5\cos^4 x\sin x$

(d) $6x\sin^2 (x^2-1)\cos (x^2-1)$.  (e) $\dfrac{-2}{(1-x)^2}\sin\left(\dfrac{1+x}{1-x}\right)$  (f) $e^x(\sin 2x+2\cos 2x)$

(g) $\dfrac{\tan x\sec^2 x}{(\tan^2 x+1)^{1/2}}$  (h) $\dfrac{e^x(\sin x+\cos x+1)}{(1+\cos x)^2}$  (i) $\dfrac{\cos x-\sin x-1}{(1+\sin x)(1-\cos x)}$
(j) $\sin^2 x\sqrt{1-x^2}+2\sin x\cos x\displaystyle\int_0^x \sqrt{1+t^2}\,dt$  (k) $-\sin x\,|\sin x|$
4. (a) $\dfrac{(x+1)(\sec^2 x+\sec x\tan x)-(\tan x+\sec x)}{(x+1)^2}$  (b) $-4e^{2x}\csc^2 (1+e^{2x})\cot (1+e^{2x})$

(c) $\dfrac{\sec x(\tan x-\sec x-1)}{\sqrt[3]{1+\sec x}(1+\tan x)^{3/2}}$  (d) $\dfrac{2\csc 2x(2x\csc^2 (1+x^2)-\cot 2x)}{\cot (1+x^2)}$

5. (a) $\dfrac{-\sin (y+x)}{x\cos y+\sin (x+y)}$  (b) $\cos^2 y\sin x$  (c) $\dfrac{\sec x\tan x+\csc^2 (x+y)}{[\csc y\tan y-\csc^2 (x+y)]}$

**7.** (a) $1/2 \sin x^2 + C$  (b) $-\cos e^x + C$  (c) $1/2 \tan x^2 + C$  (d) $-1/3 \cos(x^3 + 1) + C$
(e) $-\log|\cos x| + C$  (f) $e^{\sin x} + C$  (g) $\log|\sin x| + C$  (h) $-1/2 e^{\cos x^2} + C$
(i) $1/2 \sec x^2 + C$  (j) $\tan e^x + C$  **8.** (a) $1/2$  (b) $2$  (c) $0$  (d) $1/2$

**9.** $\dfrac{x}{2} + \dfrac{1}{4}\sin 2x + C$  **11.** $2$  **12.** $\pi/4 + 1/2$  **15.** $12{,}000\,\pi$ ft/min, $\dfrac{100{,}000\,\pi}{3}$ ft/min

**16.** $\dfrac{25\,\pi}{3}$ ft/hour  **17.** $\pi^2/2$

**Section 4.8 – p. 180**  **1.** (a) $\pi/4$  (b) $-2/3$  (c) $0$  **2.** (a) $2$  (b) $0$

**3.** (a) $\dfrac{1}{\sqrt{(5-x)(x-1)}}$  (b) $\dfrac{-3\cos 3x}{\sqrt{-\sin 3x(2 + \sin 3x)}}$  (c) $\dfrac{-xe^x}{(x+1)^2 + e^{2x}}$

(d) $\dfrac{-\sin(3 + \arctan x)}{x^2 + 1}$  (e) $\dfrac{\sqrt{1-x^2} - 1}{\sqrt{1-x^2}(x - \arcsin x)}$  (f) $1$  (g) $\dfrac{-e^{(1/2)\sin x}\cos x}{\sqrt{2 - e^{\sin x}}}$

**4.** (a) $2xe^{x^2}[(x+y)^2 + 1] - 1$  (b) $\dfrac{1 + (x+y)^2 - \sqrt{1-(x-y)^2}}{1 + (x+y)^2 + \sqrt{1-(x-y)^2}}$

(c) $\dfrac{-y}{\sqrt{1-(x+\sin y)^2} + y\cos y}$  (d) $\dfrac{x+y}{x-y}$  **6.** $\pi/4$  **7.** $\pi/2$  **8.** (a) $1/2 \arcsin 2x + C$
(b) $1/2 \arccos x^2 + C$  (c) $1/a \arctan(x/a) + C$  (d) $\arcsin e^x + C$
(e) $\arcsin(\log x) + C$  (f) $\arctan(\sin x) + C$  **9.** $(2k+1)\pi/2, k = 0, \pm 1, \pm 2, \ldots$

**10.** $F(x) = \pi/2 - \displaystyle\int_0^x \dfrac{dt}{1+t^2}$  **11.** $\dfrac{-1}{x\sqrt{x^2-1}}$  **12.** $\dfrac{1}{x\sqrt{x^2-1}}$  **13.** $\dfrac{-1}{x\sqrt{x^2-1}}$

**Section 4.9 – p. 182**  **5.** $F'(x) = \dfrac{1}{\sqrt{x^2-1}}, G'(x) = \dfrac{-1}{\sqrt{x^2-1}}, F(x) = \displaystyle\int_1^x \dfrac{dt}{\sqrt{t^2-1}},$

$G(x) = -\displaystyle\int_1^x \dfrac{dt}{\sqrt{t^2-1}}$  **6.** $a^2\left(e - \dfrac{1}{3}\right)$  **7.** $A(x) = \dfrac{1}{2}x\sqrt{x^2-1} - \displaystyle\int_1^x \sqrt{t^2-1}\,dt,$

$A(y) = \dfrac{1}{2}y\sqrt{y^2+1} - \displaystyle\int_0^y \sqrt{t^2+1}\,dt$

## CHAPTER 5

**Section 5.1 – p. 188**  **1.** (a) No maximum, Minimum is $-2$, intermediate value property
satisfied  (b) Maximum is 1, Minimum is $-7$, intermediate value property not satisfied
(c) Maximum is 1, Minimum is $-1$, intermediate value property satisfied  (d) Minimum is 1/6,
No maximum, intermediate value property satisfied  **2.** (a) No. Let $f(x) = x$ for
$0 \le x < 1, f(x) = x - 1$ for $1 \le x \le 2$.  (b) No. Consider (1a) above.  **3.** (a) No.
Consider (1b) above  (b) No. Consider (1d) above.  **4.** (a) Yes (both)  (b) Yes (both)
**5.** Yes (both)  **6.** Intermediate value property satisfied. Extreme Value Property may not
be: for example, take $f(x) = \dfrac{1}{x}$ on $(0, 1)$.

**Section 5.2 – p. 194**  **1.** (a) min $= f(1) = -1$, max $= f(3) = 5$  (b) min $= f(1) = 3$,
max $= f(0) = f(2) = 5$  (c) min $= f(1) = 6$, max $= f(3) = 10$  (d) min $= f(3) = -11$,
max $= f(0) = 4$  (e) min $= f(3) = 0$, max $= f(1) = 2$  (f) min $= f(-1/2) = -3/2$,
max $= f(-3) = 4$  (g) min $= f(1) = f(-2) = -3$, max $= f(-1) = f(2) = 5$
(h) min $= f(-1) = -15$, max $= f(0) = 2$  (i) min $= f(2) = -7$, max $= f(-2) = 25$
(j) min $= f(2) = 12$, max $= f(1) = 17$  **2.** (a) $x_0 = 2$  (b) $x_0 = (1 + \sqrt{37})/3$
(c) $x_0 = \sqrt{2}/2$  (d) $x_0 = (t-1)/\log t$  **4.** $f'(0)$ does not exist.  **7.** (a) Apply mean value
theorem to $f(t) = e^t$ on $[0, x]$.  (b) $f(t) = \log t$ on $[1, x]$  (c) $f(t) = \sin t$ on $[x, y]$
(d) $f(t) = \arctan t$ on $[0, x]$  (e) $f(t) = \arctan t$ on $[x, y]$.  **8.** Apply mean value theorem
to  (a) $f(x) = \sqrt{x}$ on $[25, 26]$,  (b) $f(x) = \sqrt[3]{x}$ on $[27, 28]$,  (c) $f(x) = \sqrt[5]{x}$ on $[32, 33]$.

**Section 5.3 – p. 701**  **1.** (a) $x = 0$, rel max; $x = 1/2$, rel min; inc. for $x \le 0$ and $x \ge 1/2$;
dec. for $0 \le x \le 1/2$.  (b) $x = 1$, rel. max.; $x = 4$, rel. min.; inc. for $x \le 1$ and $x \ge 4$;
dec. for $1 \le x \le 4$.  (c) $x = 0$, rel. max.; $x = \pm 1$ rel. min.; inc. if $-1 \le x \le 0$ or $x \ge 1$;
dec. if $x \le -1$ or $0 \le x \le 1$.  (d) $x = 1$, not rel. extreme, inc. for all $x$.  (e) $x = 0$, rel. min.;
$x = 2$, not rel. extreme; inc. if $x \ge 0$, dec. if $x \le 0$.  (f) $x = -2$, rel. min.; inc. if $x \ge -2$;

dec. if $x \le -2$.    (g) $x = 0$, rel. max.; $x = -1/2$, 2 rel. min.; inc. if $-1/2 \le x \le 0$ or $x \ge 2$; dec. if $0 \le x \le 2$ or $x \le -1/2$.    (h) $x = -1$ not rel. extreme; inc. for all $x$.
(i) $x = -3$ rel. max.; $x = -1$ rel. min.; $x = 0$ not rel. extreme; inc. if $x \le -3$ or $x \ge -1$; dec. if $-3 \le x \le -1$.    (j) $x = 1, -\sqrt{2}$ rel. max.; $x = -1, \sqrt{2}$ rel. min. $x = 0$ not rel. extreme; inc. if $x \le -2$ or $-1 \le x \le 1$ or $x \ge \sqrt{2}$; dec. if $-\sqrt{2} \le x \le -1$ or $1 \le x \le \sqrt{2}$.
**2.** (a) $x = \sqrt[3]{2}$ rel. min.; inc. if $x \ge \sqrt[3]{2}$; dec. if $x < 0$ or $0 < x \le \sqrt[3]{2}$.    (b) No critical points; inc. if $x > 0$; dec. if $x < 0$.    (c) $x = 1$ rel. max.; inc. if $x \le 1$; dec. if $x \ge 1$.
(d) $x = 2/3$ rel. max., $x = 1$ rel. min.; inc. if $x \le 2/3$ or if $x \ge 1$; dec. if $2/3 \le x \le 1$.
(e) $x = 1/4$ rel. max.; $x = 0$ not rel. ext.; inc. if $x \ge 1/4$; dec. if $x \le 1/4$.    (f) $x = 1$ rel. max.; inc. if $x \le 1$, dec. if $x \ge 1$.    **3.** (a) $(-1, 1/3)$ abs. min.; $(2, e^2)$ abs. max.
(b) $(-1, -e)$ abs. min., $(1, 1/e)$ abs. max.    (c) $(-1, 1/e), (1, 1/e)$ abs. min.; $(0, 1)$ abs. max.
(d) $(0, 0)$ abs. min. $(-2, 4e^{-2})$ rel. max., $(3, 9e^3)$ abs. max.    (e) $(7\pi/4, -\sqrt{2})$ abs. min., $(3\pi/4, \sqrt{2})$ abs. max.    (f) $(0, 0), (\pi, 0), (2\pi, 0)$ abs. min., $(\pi/2, 1), (3\pi/2, 1)$ abs. max.
(g) $\left(7\pi/4, \dfrac{-\sqrt{2}}{2} e^{7\pi/4}\right)$ abs. min. $(3\pi/4, \sqrt{2}/2 e^{3\pi/4})$ abs. max.    (h) $(1, 0)$ abs. min.,
$(4, (\log 2)^2)$ abs. max.    **7.** No, $f'$ may vanish.

**Section 5.4 – p. 205**    **2.** (a) $x = y = 5$    (b) $x = y = 5$    **3.** $6/5$    **4.** $\sqrt{6}/2$    **6.** $\sqrt{2}r/2$, $\sqrt{2}r$    **7.** $4r/\sqrt{5}$, $r/\sqrt{5}$. There is no minimum if both dimensions are positive.    **8.** $8/\sqrt{3}$, $32/3$    **9.** 2, 15    **10.** $l/4, l/2$    **11.** $\sqrt{A/2}, \sqrt{2A}$    **12.** $\sqrt{15}/15$ miles from point on the shore closest to the boat.    **13.** $h/r = 2$    **14.** length of each side $= r\sqrt{3}$    **15.** 11.1 miles    **19.** $5\sqrt{5}$ ft    **20.** no    **21.** 125    **22.** If $l/2 \le b$, then the dimensions are $l/4$ by $l/2$. If $l/3 > b$, then the dimensions are $(l + b)/4$ by $(l + b)/4$.    **23.** Minimum crease is $(3\sqrt{3}w)/4$ where $w$ is the width of the paper. A right triangle should be formed whose legs are $3w/4$, $(3\sqrt{2}w)/4$ and whose hypotenuse is $(3\sqrt{3}w)/4$.

**Section 5.5 – p. 210**    (a) up if $x \ge 1/2$, down if $x \le 1/2$    (b) up if $-\sqrt{2}/2 \le x \le 0$ or $\sqrt{2}/2 \le x$, down if $x \le -\sqrt{2}/2$ or $0 \le x \le \sqrt{2}/2$    (c) up if $x \le (-4 + \sqrt{7})/30$ or $x \ge 0$, down if $0 \le x \le (-4 + \sqrt{7})/30$.    (d) up if $0 \le x \le -1$ or $0 \le x \le 1$, down if $-1 \le x \le 0$ or $x \ge 1$.    (e) up if $x \le -\sqrt{2}/2$ or $x \ge \sqrt{2}/2$, down if $-\sqrt{2}/2 \le x \le \sqrt{2}/2$    (f) up for all $x$    (g) up if $-1 \le x \le 1$, down if $x \le -1$ or $x \ge 1$

**Section 5.6 – p. 218**    **1.** (a) Always increasing; no extreme points; intercept at $x = 0$; concave down if $x \le 0$, up if $x \ge 0$; antisymmetric about $y$-axis.    (b) Always decreasing; no extreme points; intercept at $x = 1$; concave down if $1 \le x$, up if $x \le 1$.
(c) Increasing if $-2\sqrt{3}/3 \le x \le 2\sqrt{3}/3$; decreasing if $x \le -2\sqrt{3}/3$ or $2\sqrt{3}/3 \le x$; Relative maximum at $(2\sqrt{3}/3, 16\sqrt{3}/9)$; relative minimum at $(-2\sqrt{3}/3, -32\sqrt{3}/9)$; intercepts at $x = 2, x = -2, x = 0$; concave down if $0 \le x$, up if $x \le 0$; antisymmetric about $y$-axis.    (d) Increasing if $x \le -1$ or $0 \le x \le 1$; decreasing if $-1 \le x \le 0$ or $1 \le x$; relative maximum at $(1, 0)$ and $(-1, 0)$; relative minimum at $(0, -1)$; intercepts at $x = 1, x = -1$; concave down if $x \le -\sqrt{3}/3$ or $\sqrt{3}/3 \le x$; concave up if $-\sqrt{3}/3 \le x \le \sqrt{3}/3$, symmetric about $y$-axis.    (e) Increasing if $x \le 0$; decreasing if $x \ge 0$; absolute maximum at $(0, 0)$; intercepts at $x = 0$; concave down if $x \le 2/3$ or $x \ge 2$; concave up if $2/3 \le x \le 2$.    (f) Increasing if $x \le -1$ or $x \ge 1$; decreasing if $-1 \le x \le 1$; relative maximum at $(-1, 2)$; relative minimum at $(1, -2)$; intercepts at $x = 0, \pm\sqrt{2}$; concave down if $x \le -\sqrt{3/10}$ or $0 \le x \le \sqrt{3/10}$, up if $-\sqrt{3/10} \le x \le 0$ or $x \ge \sqrt{3/10}$; antisymmetric about $y$-axis.    (g) Intercept at $x = 0$; no extreme points; increasing for all $x$, concave down if $x \le 0$ or $2 \le x \le 4$; concave up if $0 \le x \le 2$ or $x \ge 4$.    (h) Intercepts at $x = 0, x = \sqrt{15}, x = -\sqrt{15}$; relative maximum at $x = 3$; relative minimum at $x = -3$; stationary point at $x = 0$; increasing if $-3 \le x \le 3$; decreasing if $x \le -3$ or $x \ge 3$; concave down if $-3\sqrt{2}/2 \le x \le 0$ or $x \ge 3\sqrt{2}/2$; concave up if $x \le -3\sqrt{2}/2$ or $0 \le x \le 3\sqrt{2}/2$; antisymmetric about $y$-axis.
**2.** (a) No stationary or extreme points. Decreasing if $x < 1$ or $-1 < x < 1$ or $x > 1$.
(b) Absolute minimum at $x = -5$; increasing if $x \ge -5$; decreasing if $x \le -5$; concave down if $x \le -8$; concave up if $x \ge -8$.    (c) Absolute maximum at $x = 1$; absolute minimum at $x = -1$; increasing if $-1 \le x \le 1$; decreasing if $x \le -1$ or $x \ge 1$; concave down if $x \le -\sqrt{3}$ or $0 \le x \le \sqrt{3}$; concave up if $-\sqrt{3} \le x \le 0$ or $x \ge \sqrt{3}$.
(d) Absolute maximum at $x = -4 - 2\sqrt{5}$; absolute minimum at $x = -4 + 2\sqrt{5}$; increasing if $x \le -4 - 2\sqrt{5}$ or $x \ge -4 + 2\sqrt{5}$; decreasing if $-4 - 2\sqrt{5} \le x \le -4 + 2\sqrt{5}$.
(e) Relative maximum at $x = 0$; increasing if $x < -2$ or $-2 < x \le 0$; decreasing if $0 \le x < 2$ or $x > 2$; concave down if $-2 < x < 2$;

concave up if $x < -2$ or $x > 2$; symmetric with respect to $y$-axis.
(f) Absolute maximum at $x = 0$; concave down if $-\sqrt{3}/3 \leqslant x \leqslant \sqrt{3}/3$;
concave up if $x \leqslant -\sqrt{3}/3$ or $x \geqslant \sqrt{3}/3$; increasing if $x \leqslant 0$; decreasing if $x \geqslant 0$;
symmetric with respect to $y$-axis.　　(g) No extreme points. Increasing if $x < -1$ or $x > -1$;
concave up if $x < -1$; concave down if $x > -1$.　　(h) Relative maximum at $x = -1$;
increasing if $-2 \leqslant x \leqslant -1$; decreasing if $x \leqslant -2$ or $-1 \leqslant x < 1$ or $x > 1$;
concave up if $x > 1$; concave down if $x < -2$ or $-2 < x < 1$.
(i) Relative maximum at $x = -1/\sqrt[3]{2}$; increasing if $x \leqslant -1/\sqrt[3]{2}$;
decreasing if $-1/\sqrt[3]{2} \leqslant x < 1$ or $x > 1$; concave up if $x \leqslant -\sqrt[3]{2}$ or $x > 1$;
concave down if $-\sqrt[3]{2} \leqslant x < 1$; antisymmetric with respect to $y$-axis.

(j) Relative maximum at $\left(-\dfrac{1}{\sqrt[3]{2}}, \dfrac{2}{3 \cdot \sqrt[3]{2}}\right)$; increasing if $x \leqslant -1/\sqrt[3]{2}$;

decreasing if $-1/\sqrt[3]{2} \leqslant x < 1$ or if $x > 1$; concave down if $-\sqrt[3]{2} \leqslant x < 1$;
concave up if $x \leqslant \sqrt[3]{2}$ or if $x > 1$　　**3.** (a) $y = x + 1, x = 1$　　(b) $y = 1, y = -1$
(c) $y = 2x + 2, x = 1$　　(d) $y = 3x + 7, x = 2$　　(e) $y = 1, x = \pm 1$　　(f) $y = 0, x = -4$
**10.** For $n$ odd, $\lim\limits_{x \to a^-} f(x) = -\infty$, $\lim\limits_{x \to a^+} = +\infty$; for $n$ even, $\lim\limits_{x \to a^-} f(x) = +\infty$, $\lim\limits_{x \to a^+} = +\infty$.
**13.** no　　**14.** $(f \circ g)(x) \to \infty$

**Section 5.7 – p. 223**　　**1.** (a) Absolute maximum at $x = \pi/4$; absolute minimum at $x = 5\pi/4$;
inflection points at $x = 3\pi/4, x = 7\pi/4$.　　(b) Absolute maximum at $x = 0, x = 2\pi$; relative
maximum at $x = \pi$; absolute minimum at $x = 2\pi/3$ and $x = 4\pi/3$; inflection points at $x = $ arc
cos $5/8$ and at $x = $ arc cos $(-7/8)$.　　(c) Absolute maximum at $x = 0, x = 2\pi$; absolute
minimum at $x = \pi$; relative maximum at $x = $ arc cos $(-\sqrt{5}/5)$; relative minimum at
$x = $ arc cos $\sqrt{5}/5$.　　(d) Absolute maximum at $x = $ arc sin $\sqrt{3}/3$; absolute minimum at
$x = $ arc sin $(-\sqrt{3}/3)$; relative maximum at $x = 3\pi/2$; relative minimum at $x = \pi/2$; inflection
points at $x = 0, \pi, 2\pi$ and at $x = $ arc sin $\sqrt{7}/3$ and $x = $ arc sin $(-\sqrt{7}/3)$.　　(e) Absolute
minimum at $x = -1$; inflection point at $x = -2$.　　(f) No extreme points and no inflection
points, decreasing for all $x$.　　(g) Absolute minimum at $x = 0$, absolute maximum at $x = 1$,
$x = -1$; inflection points at $x = \pm\sqrt{\dfrac{5 \pm \sqrt{17}}{4}}$.　　(h) Absolute minimum at $x = 0$; inflection

point at $x = \log 2$.　　(i) Absolute maximum at $x = 0$; relative extremes solutions of $x = \tan x$,
$\lim\limits_{x \to \infty} f(x) = 0$.　　(j) Absolute minimum at $x = 1/e$.　　(k) Relative maximum at $x = \pi/4 + 2\pi k$;
relative minimum at $x = 5\pi/4 + 2\pi k$; inflection points at $x = 0, \pi, 2\pi + 2\pi k$.　　(l) Absolute

maximum at $x = 0$.　　**2.** *Note.* We define $f'_-(a) = \lim\limits_{h \to 0^-} \dfrac{f(a+h) - f(a)}{h}$ and

$f'_+(a) = \lim\limits_{h \to 0^+} \dfrac{f(a+h) - f(a)}{h}$.　　(a) Absolute minimum at $x = -1/4$; $f'(0)$ does not exist

(vertical tangent).　　(b) Relative maximum at $x = \pm\sqrt{3}/3$; absolute minimum at $x = 0, \pm 1$,
$f'_-(1) = -2, f'_-(0) = -1, f'_-(-1) = -2, f'_+(1) = 2, f'_+(0) = 1, f'_+(-1) = 2$.　　(c) Absolute
maximum at $x = 0$; absolute minimum at $x = 1, -1$.　　(d) Absolute maximum at $x = \pi/2 + k\pi$,
$f'_-(\pi + k\pi) = -1, f'_+(\pi + k\pi) = 1$.　　(e) Not defined for $-1 < x \leqslant 0$ $f'(-1) = \infty$ (vertical
tangent at $x = -1$); decreasing if $x \leqslant -1$ or if $x > 0$.　　**4.** yes; no, no

**Section 5.8\* – p. 230**　　**1.** (a) $y^2 = -12x$　　(b) $y^2 = 8x$　　(c) $x^2 = 4y$　　(d) $x^2 = -12y$
(e) $2y^2 = 9x$　　(f) $y = 4x^2$　　**2.** (a) $(y - 1)^2 = -2(x - 3/2)$　　(b) $(y - 2)^2 = 8(x - 3)$
(c) $(x - 2)^2 = -6(y - 5/2)$　　(d) $(y - 2)^2 = 9(x - 3)$　　(e) $(y - 3) = -7(x - 1)^2$
(f) $(x - 2) = 3(y + 2)^2$　　**3.** $-8$　　**4.** $7$　　**5.** $8c^2/3$　　**6.** $2\pi c^3$　　**7.** $16\pi c^3/5$

**Section 5.9\* – p. 235**　　**1.** (a) $\dfrac{x^2}{25} + \dfrac{y^2}{9} = 1$　　(b) $\dfrac{x^2}{25} + \dfrac{y^2}{16} = 1$　　(c) $\dfrac{x^2}{41} + \dfrac{y^2}{25} = 1$

(d) $\dfrac{x^2}{25} + \dfrac{y^2}{16} = 1$　　(e) $\dfrac{x^2}{25} + \dfrac{y^2}{9} = 1$　　(f) $25x^2 + 9y^2 = 81$　　**2.** (a) $\dfrac{(x-1)^2}{25} + \dfrac{(y+2)^2}{16} = 1$

(b) $\dfrac{(x-2)^2}{81} + \dfrac{(y-1)^2}{72} = 1$　　(c) $\dfrac{(x-5)^2}{63} + \dfrac{(y+2)^2}{144} = 1$　　(d) $\dfrac{(x-1)^2}{25} + \dfrac{(y-3)^2}{16} = 1$

**3.** $y = 1 - 3x$　　**4.** $3y + 8x - 41 = 0, y = 3$　　**5.** no　　**6.** $y = 4x - 5, y = 3$　　**7.** $4ab^2\pi/3$
**8.** $4ba^2\pi/3$

**Section 5.10\* – p. 242**　　**1.** (a) $(\pm\sqrt{41}, 0), (\pm 5, 0), x = \pm\dfrac{25}{\sqrt{41}}, y = \pm 4x/5, \sqrt{41}/5$

(b) $(\pm 2\sqrt{3}, 0), (\pm 2\sqrt{2}, 0)$ $x = \pm 4y/3\sqrt{3} = \pm x/\sqrt{2}, \sqrt{3}/2$　　(c) $(\pm 5, 0), (\pm 3, 0), x = \pm 9/5$,

$y = \pm 4x/3, 5/3$   (d) $(\pm\sqrt{3}/2, 0), (\pm 1/2, 0)$  $x = \pm\sqrt{3}/6, y = \pm\sqrt{2}x, \sqrt{3}$   (e) $(0, \pm\sqrt{3}/2),$
$(0, \pm 1/\sqrt{2}), y = \pm 3/\sqrt{2}, y = \pm x/\sqrt{2}, \sqrt{3}$   (f) $(0, \pm\sqrt{7}/12), (0, \pm\sqrt{1/3})$  $y = \pm 4\sqrt{21/7}$

$y = \pm 2x/\sqrt{3}, \sqrt{7}/6$   **2.** (a) $\dfrac{x^2}{16} - \dfrac{y^2}{144} = 1$   (b) $\dfrac{4x^2}{25} - \dfrac{y^2}{25} = -1$   (c) $\dfrac{x^2}{15} - \dfrac{y^2}{10} = 1$

(d) $x^2 - y^2 = -8$   (e) $\dfrac{y^2}{64} - \dfrac{x^2}{36} = 1$   (f) $x^2 - y^2 = 4$   **3.** (a) $\dfrac{27}{40}\left(x + \dfrac{2}{3}\right)^2 - \dfrac{9}{40}(y-1)^2 = 1$

(b) $\dfrac{3}{16}(x-2)^2 - \dfrac{9}{16}\left(y + \dfrac{5}{3}\right)^2 = -1$   (c) $\dfrac{4}{9}\left(x - \dfrac{3}{2}\right)^2 - \dfrac{(y-4)^2}{18} = 1$   (d) $\dfrac{x^2}{45} - \dfrac{(y-9)^2}{36} = -1$

**4.** $4\sqrt{2}y - 2x = 2, \sqrt{2}y + 4x = 5$   **6.** $y = 2 \pm \dfrac{1}{2}\sqrt{17}x$, no

**11.** $\dfrac{\pi b^2}{3a^2}[2a^3 + (b^2 - 2a^2)\sqrt{a^2 + b^2}], \dfrac{4\pi}{3}\dfrac{b^4}{a}$

**Section 5.11\* – p. 246**   **1.** parabola   **2.** parabola   **3.** ellipse   **4.** graph is empty
**5.** circle   **6.** hyperbola,   **7.** hyperbola,   **8.** graph is empty   **9.** pair of vertical lines
**10.** hyperbola,   **11.** the point $(-1, -1)$   **12.** pair of crossed lines
**13.** If $a, b > 0, c \leqslant \dfrac{1}{4}\left(\dfrac{c^2}{a} + \dfrac{d^2}{b}\right)$.

**Section 5.12\* – p. 251**   **1.** $I_2 = [1, 3/2], I_3 = [7/4, 3/2], I_4 = [11/8, 3/2], I_5 = [11/8, 23/16]$
**2.** $I_2 = [3/2, 2], I_3 = [3/2, 7/4], I_4 = [13/8, 7/4], I_5 = [27/16, 7/4]$   **3.** $1/5 = 00110$,
$2/3 = 10101, 1/7 = 00100$   **4.** Divide $[0, 1]$ into 10 equal intervals, and number them
$0, 1, \ldots, 9$. If $x \in I_k$ where $0 \leqslant k \leqslant 9$, then $k$ is the first number in the expansion. Divide
$I_k$ into 10 equal intervals and repeat the process.

**Section 5.13\* – p. 254**   **1.** no

# CHAPTER 6

**Section 6.1 – p. 258**   **1.** (a) $\dfrac{1}{5}(2x-1)^5$   (b) $\dfrac{-1}{(x^3+x)}$   (c) $\log|x^2+1|$   (d) $\dfrac{1}{6}(x^4+x^3)^6$

(e) $e^{-x^2}$   (f) $e^{\sin x}$   (g) $\dfrac{1}{4}\sin^4 x$   (h) $\dfrac{1}{2}(\tan x)^2$   (i) arc tan $(e^x)$   (j) $\dfrac{1}{2}(\text{arc tan } x)^2$

(k) $\dfrac{1}{2}(\log x)^2$   (l) tan $(e^x)$   (m) arc sin $(x-1)$   (n) tan $(2x+1)$   **2.** (a) $\dfrac{1}{4}\log|4x+1|$

(b) $-\dfrac{1}{4} \cdot \dfrac{1}{(x^2+1)^2}$   (c) $\dfrac{1}{8}(\log x^2)^4$   (d) $\dfrac{1}{2}e^{\sin x^2}$   (e) $-\dfrac{1}{16}(\cos^4 4x)$   (f) $\dfrac{1}{2}$ arc tan $(x^2)$

(g) $-\dfrac{1}{2}\log|\cos(x^2)|$   (h) $-\dfrac{2}{9}(1-3x^2)^{3/2}$   (i) $-\dfrac{1}{3}(1-x^2)^{3/2}$   (j) $\dfrac{2^x}{\log 2}$   (k) $\dfrac{1}{4}\log|\sin 4x|$

(l) $-\dfrac{1}{8}(\cos 2x)^2$   (m) $-\dfrac{2}{3}(1-x^3)^{1/2}$   (n) $\dfrac{\sqrt{3}}{12}$ arc tan $\left(1 + \dfrac{\sqrt{3}}{2}x^2\right)$   **3.** (a) $\dfrac{1}{4}\sin 2x + \dfrac{1}{2}x$

(b) $\dfrac{x}{2} - \dfrac{1}{8}\sin 4x$   (c) $\sin x - \dfrac{1}{3}\sin^3 x$   (d) $-\dfrac{1}{3}\cos^3 x + \dfrac{1}{5}\cos^5 x$   (e) $\dfrac{1}{2}\tan^2 x + \dfrac{1}{4}\tan^4 x$

(f) $\dfrac{1}{3}\log|\sin 3x|$   (g) $\dfrac{1}{5}\sin^5 x - \dfrac{1}{7}\sin^7 x$   (h) $-\dfrac{1}{3}\cos^3 x + \dfrac{2}{5}\cos^5 x - \dfrac{1}{7}\cos^7 x$

(i) $-\dfrac{1}{4}\cos 2x - \dfrac{1}{8}(\cos 2x)^2$   (j) $-\dfrac{1}{14}\cos 7x - \dfrac{1}{2}\cos x$   (k) $\dfrac{1}{6}\sin 3x - \dfrac{1}{10}\sin 5x$

(l) $\dfrac{1}{8}\sin 4x + \dfrac{1}{4}\sin 2x$   **4.** (a) $\dfrac{-\sin^3 x \cos x}{4} + \dfrac{3}{8}x - \dfrac{3}{16}\sin 2x$   (b) $\dfrac{1}{8}x - \dfrac{1}{32}\sin 4x$

(c) $\dfrac{-\sin^2 x \cos x}{3} - \dfrac{2}{3}\cos x$   (d) $\dfrac{1}{8}\sin 2x + \dfrac{15}{16}x - \dfrac{1}{192}\sin 4x + \dfrac{\cos^5 x \sin x}{6}$

(e) $\dfrac{\cos^5 x \sin x}{6} + \dfrac{5\sin 4x}{192} + \dfrac{5}{24}\sin 2x + \dfrac{5}{16}x$   (f) $\dfrac{1}{16}x - \dfrac{1}{32}\sin 2x + \dfrac{1}{24}\sin^3 x$

**Section 6.2 – p. 266**   **1.** (a) $\dfrac{4}{15}(\sqrt{2}+1)$   (b) $\dfrac{184}{105}$   (c) $\dfrac{93}{10}$   (d) $-\dfrac{1}{28}(12\sqrt[3]{4}+1)$

(e) $\dfrac{46}{15}$   (f) 4   **2.** (a) $\dfrac{\pi}{4}$   (b) $\dfrac{1}{3}$   (c) $4[\pi/3 - \sqrt{3}/2]$   (d) $\sqrt{2}/2$   (e) $\dfrac{1}{108}[1 + \pi/2]$
(f) $3(\log(2+\sqrt{3}) - \sqrt{3}/2)$   **3.** (a) $-2/3 (1-x)^{3/2} + 2/5 (1-x)^{5/2}$
(b) $2/189 (3x+1)^{7/2} - 4/135 (3x+1)^{5/2} + 2/81 (3x+1)^{3/2}$
(c) $3/10 (x+4)^{10/3} - 24/7 (x+4)^{7/3} + 12 (x+4)^{4/3}$   (d) $3/8 (x-1)^{8/3} + 3/5 (x-1)^{5/3}$

(e) $2/5 (x+1)^{5/2} - 4/3 (x+1)^{3/2} + 2 (x+1)^{1/2}$

(f) $1/224 (4x-1)^{7/2} + 1/80 (4x-1)^{5/2} - 15/96 (4x-1)^{3/2}$    (g) $-2/9 (1-x^3)^{3/2} + 2/15 (1 \quad x^3)^{5/2}$

(h) $2/3 (1+x)^{3/2} - 2(1+x)^{-1/2}$    (i) $2 \log (\sqrt{x}+1)$    (j) $2(1+\sqrt{x}) - 2 \log (1+\sqrt{2})$

**4.** (a) $x\sqrt{4-9x^2} - 2/3 \arcsin (3x/2)$    (b) $4x\sqrt{1-4x^2} - 1/64 \arcsin (2x) - 1/16 x(1-4x^2)^{3/2}$

(c) $\log \left| \dfrac{x}{5} + \dfrac{\sqrt{x^2-25}}{5} \right| - \dfrac{\sqrt{x^2-25}}{x}$    (d) $1/2\sqrt{2x-x^2} - \sqrt{2x-x^2} [1-(x/2)] - \arcsin (\sqrt{x}/2)$

(e) $\dfrac{x}{2(x^2+1)} + \dfrac{\arctan x}{2}$    (f) $1/25 \dfrac{x}{\sqrt{x^2+25}}$    (g) $\log \left| \dfrac{\sqrt{x^2-a^2}}{a} + \dfrac{x}{a} \right| - \dfrac{\sqrt{x^2-a^2}}{x}$

(h) $2 \log \left| \dfrac{1}{2}\sqrt{x-4} + \dfrac{\sqrt{x}}{2} \right|$    (i) $\log \left| \dfrac{\sqrt{x^2+1}}{x} - \dfrac{1}{x} \right|$    **6.** $\dfrac{\sqrt{3}}{2} + \dfrac{2\pi}{3}$    **7.** $4\pi - 3\sqrt{3}$

**8.** $4\sqrt{15} + \log (\sqrt{15}+4)$    **9.** $3\pi$

**Section 6.3 – p. 269**    **1.** (a) $1/2\, x \sin 2x + 1/4 \cos 2x$    (b) $-1/2\, e^{-2x} (x+1/2)$

(c) $-1/3\, x^2 \cos 3x + 2/9\, x \sin 3x + 2/27 \cos 3x$    (d) $e^x (x^2 - 2x + 2)$    (e) $1/2\, x^2 (\log x - 1/2)$

(f) $1/2\, x^2 [(\log x)^2 - \log x + 1/2]$    (g) $2/3\, x^{3/2} (\log x - 2/3)$

(h) $\dfrac{b}{a^2+b^2} e^{ax} \sin bx + \dfrac{a}{a^2+b^2} e^{ax} \cos bx$    (i) $x(\arcsin x) + \sqrt{1-x^2}$

(j) $x(\arctan x) - 1/2 \log |1+x^2|$    **2.** (a) $1/4\, x^2 - 1/4\, x \sin 2x - 1/8 \cos 2x$

(b) $1/3\, x^3 \arctan x - 1/6\, x^2 + 1/3 \log |\sqrt{x^2+1}|$    (c) $1/2 \sec x \tan x + 1/2 \log |\tan x + \sec x|$

(d) $1/2 (\tan x)^2 + \log |\cos x|$    (e) $\sec x \tan^2 x - 2/3 \sec^3 x$

**3.** (a) $f(x) = \cos^{n-1} x, g'(x) = \cos x$    (b) $f(x) = x^n, g'(x) = e^{ax}$

(c) $f(x) = (\log x)^n, g'(x) = x^m$    **4.** (a) $1$    (b) $2 \log 2 - 1$    (c) $1/2\, e^{\pi/2} + 1/2$    (d) $\pi/4$

(e) $\pi^2/4 - 2$    (f) $3\pi/16$    (g) $15\pi/96$    (h) $\pi/2 - 1$

**Section 6.4 – p. 274**    **1.** (a) $1/2 \log \left| \dfrac{x-1}{x+1} \right|$    (b) $1/2 \log |(x-1/2)^2 + 23/4| + \dfrac{3\sqrt{23}}{23} \arc$

$\tan \left[ \dfrac{2\sqrt{23}}{23} \left( x - \dfrac{1}{2} \right) \right]$    (c) $1/3 \log \left( \left| \dfrac{x-2}{x+1} \right|^4 \right) + x$    (d) $1/5 \log \left| \dfrac{x-1}{2x+3} \right|$

(e) $1/8 \log \left| \dfrac{(x-2)(x+2)}{x^2} \right|$    (f) $-\dfrac{1}{x} - \dfrac{1}{2x^2}$    (g) $1/3 \log |x(x+3)^2|$    (h) $1/2x^2 + 1/2 \log |x^2-1|$

(i) $\log \left( \left| \dfrac{x}{x-1} \right|^{10} \right) - \dfrac{4}{x} - \dfrac{6}{x-1}$    (j) $\log \left| \dfrac{x}{\sqrt{x^2-1}} \right| - \dfrac{3}{2} \cdot \dfrac{1}{(x^2-1)}$

(k) $x - \log (|x-1|^3) - \dfrac{3}{x+1} + \dfrac{1}{(x+1)^2}$    (l) $\dfrac{1}{20} \log \left( \left| \dfrac{x-2}{x+2} \right| \right) - \dfrac{1}{5} \arctan x$

(m) $\dfrac{1}{3} \log |x-1| - \dfrac{1}{6} \log |x^2+x+1| - \dfrac{\sqrt{3}}{3} \arctan \left[ \dfrac{2\sqrt{3}}{3} \left( x + \dfrac{1}{2} \right) \right]$

(n) $\dfrac{\sqrt{5}}{4} \arctan \left( \dfrac{\sqrt{5}}{5} x \right) - \dfrac{1}{4} \arctan x$    (o) $\dfrac{1}{4} \log \left| \dfrac{x^2-1}{x^2+1} \right|$    (p) $\dfrac{1}{2}x^2 - 2 \log |x^2+4|$

(q) $\log x - \dfrac{1}{2} \log |x^2+1| + \dfrac{1}{x} - \arctan x$    (r) $\dfrac{-1}{2(x^2+1)}$    **2.** (a) $\arcsin (\sqrt{x}) - \sqrt{x-x^2}$

(b) $\dfrac{1}{3} \arccos \left( \dfrac{3}{x+1} \right)$    (c) $\log |x| - \dfrac{1}{2} \log |x^2+1| - \dfrac{\arctan x}{x}$    (d) $x \log |x + \sqrt{x^2+1}| - \sqrt{x^2+1}$

(e) $-\dfrac{1}{2} \csc x \cot x + \dfrac{1}{2} \log |\csc x - \cot x|$    (f) $\dfrac{1}{3} \tan^3 x - \tan x + x$

(g) $\log |x + \sqrt{x^2+1}| - \dfrac{x}{\sqrt{x^2+1}}$    (h) $\left( \dfrac{1}{2}x^2 - \dfrac{1}{4} \right) \arcsin x + \dfrac{1}{4} x\sqrt{1-x^2}$

(i) $\dfrac{\sqrt{2}}{2} \log \left| \sqrt{\dfrac{1+x^2}{1-x^2}} + \sqrt{\dfrac{2}{1-x^2}} \right|$    (j) $\log \left| \dfrac{\sqrt{1+x^2}-1}{x} \right|$

(k) $-2x^{3/2}e^{-\sqrt{x}} - 6xe^{-\sqrt{x}} - 12e^{-\sqrt{x}}(\sqrt{x}+1)$    (l) $(x+1) + 4\sqrt{x+1} + \log (|\sqrt{x+1}-1|^4)$

**Section 6.5 – p. 283**    **3.** $16\pi$    **4.** $2\pi^2$    **5.** $\pi \left( \dfrac{69}{8} - 8 \log 2 \right)$    **6.** $4\pi e^2$    **7.** $\dfrac{-284}{15} \pi$

**8.** $2R\pi^2 r^2$    **10.** $\bar{y} = h/3$    **11.** $\bar{x} = 0, \bar{y} = 4/3\pi$    **12.** $\bar{x} = 5/8, \bar{y} - 1/3$; yes

**13.** $\bar{x} = \pi/2, \bar{y} = \pi/8$    **14.** $\bar{x} = \dfrac{1}{2} \dfrac{e^2+1}{e^2-1}, \bar{y} = \dfrac{e^2+1}{4}$    **16.** $\bar{x} = 0, \bar{y} = 4b/3\pi$

**CHAPTER 7**

**Section 7.1 – p. 287**    **13.** $y' = \dfrac{x}{y}$    **14.** $y'' - y' - 2y = 0$

**15.** $y'' - 2y'(x+1) + 2y\left( 1 + \dfrac{1}{x} \right) - xe^x = 0$    **16.** $(y')^2 + (y'')^2 = 1$    **17.** $y' + (y')^3 - xy'' = 0$

**18.** $y''(x-y) - (y')^3 - (y')^2 - y' - 1 = 0$    **19.** $xy'' - y' = 0$    **20.** $(y')^2 + (y-1)y'' = 0$

**24.** (a) $y = \dfrac{x^2}{2} - e^{-x} + c$    (b) $y = -\sin x + c_1 x + c_2$    (c) $y = \dfrac{e^{2x}}{8} + c_1 x^2 + c_2 x + c_3$

**25.** $f(x) = \dfrac{1}{2}(1 + e^{2(1-x)}),\ f(x) = \dfrac{1}{2} + \left(y_0 - \dfrac{1}{2}\right)e^{2(x_0-x)}$, yes

**Section 7.2 – p. 290**    **1.** $y = ce^{-10x}$    **2.** $y = ce^{x^3/3}$    **3.** $\log y = \dfrac{1}{2}e^{-x^2} + c$    **4.** $y = A \csc x$

**5.** $y = Ae^{-x} - 1$    **6.** $y = cx^a$    **7.** $y = c$    **8.** $y^3 = \dfrac{3}{2}\left(\dfrac{1}{2}\sin 2x + x\right) + c$

**9.** $\log y = 2\cos 2x + c$    **10.** arc tan $y = -$arc tan $x + c$    **11.** $y = 2x$    **12.** $y = 2\log x$
**13.** $y = 2\sin^2 x$    **14.** $y^2 = 2\log|x-1|$

**Section 7.3 – p. 292**    **1.** $e^{-\log 2/15.9}$    **2.** $T = (\log 10/\log 2)3.05$    **3.** about 15,540 years ago
**4.** about 1000 A.D.    **5.** (a) about every 34.2 years    (b) 208 million, assuming the
population in 1968 is 200 million    **6.** 60°F.    **7.** (a) $V(p) = c/p$.    (b) volume is halved
**8.** (a) $y(t) = v_0 \sin \alpha t - 16t^2, x(t) = v_0 \cos \alpha t$    (b) $\alpha = \pi/4$
**9.** $x = \log\left|\dfrac{100+y}{\sqrt{(100)^2 - y^2}}\right| + \dfrac{y}{100} + 100$    **10.** 11:23    **11.** 1160

**Section 7.4 – p. 295**    **1.** $y = ce^{x^2/2}$    **2.** $y = ce^{-2/3 x^3}$    **3.** $y = ce^{\cos x}$    **4.** $y = -\dfrac{1}{2} + ce^{2x}$

**5.** $y = \dfrac{1}{2} + ce^{-2x^2}$    **6.** $y = 1 + ce^{-1/2x^2}$    **7.** $y = -\dfrac{1}{2} + ce^{-4/3 x^3}$    **8.** $y = -(x^2 + 2x + 2) + ce^x$

**9.** $y = -\dfrac{1}{2}e^{2x} + ce^{4x}$    **10.** $y = 1 + ce^{-\sin x}$    **11.** $y = \dfrac{1}{2}x^3 + cx$

**12.** $y = \dfrac{4}{3}x - \dfrac{1}{x} + cx^{-1/2}$    **13.** $y = \dfrac{1}{5}x^3 + x + cx^{1/2}$    **14.** $y = -\dfrac{1}{2}\left(\dfrac{x}{x+1}\right)e^{-x^2} + \dfrac{cx}{x+1}$
**15.** (a) $I(t) = E_0/R(1 - e^{-(R/L)t})$    (b) $\lim\limits_{t \to \infty} I(t) = E_0/R$
(c) $I(t) = \dfrac{L}{R^2\omega + L^2}(R\omega \cos \omega t + L \sin \omega t - R\omega e^{-(R/L)t})$    **16.** $100(1 - (9/10)^8)$, no

**17.** $\dfrac{dy}{dt} = -\alpha y + m,\ \alpha = \dfrac{\log 2}{70} = .01;\ y(100) = 100\,m\left(1 - \dfrac{1}{e}\right) = 63\,m$, maximum $= 100\,m$, 6 ounces.
**18.** $f(x) = 2e^{x-1} - 1$    **19.** $f(x) = x$    **20.** $f(x) = 2\log x + 2$    **22.** $y = -x\log x + cx$
**23.** $\dfrac{x^2 + y^2}{x^4} = c$    **24.** Arc tan $(y/x) - \log y = c$    **25.** $\dfrac{x+y}{x^2 y} = c$    **26.** $x^2(x^2 - 2y^2) = c$

**27.** $\dfrac{y}{x} - \log y = c$

**Section 7.5 – p. 301**    **1.** $y(x) = \sin 2x - \dfrac{1}{2}\cos 2x$; amplitude $= \dfrac{\sqrt{5}}{2}$;
initial phase $=$ arc cos $\dfrac{2}{\sqrt{5}} = $ arc sin $\left(-\dfrac{1}{\sqrt{5}}\right)$    **2.** $y(x) = 10\sin(\sqrt{5}x + \pi/4)$.

**3.** $y(x) = 4\sin 6(x + \pi/72)$    **4.** period $= 2\pi$    **5.** $y = 2\cos 3x + 2\sin 3x$
**6.** $y = \sin x - \cos x$    **7.** $y = -\sin 2x + \cos 2x$    **8.** $y(1) = -\cos \sqrt{g}, y'(1) = \sin \sqrt{g}$
**9.** amplitude $= \dfrac{1}{\sqrt{28}}$, period $= \sqrt{\dfrac{2}{g}}\pi$    **13.** $y(x) = \dfrac{(2+\sqrt{2})e^{-\sqrt{2}}}{4}e^{\sqrt{2}x} + \dfrac{(1-\sqrt{2})e^{\sqrt{2}}}{4}e^{-\sqrt{2}x}$
**14.** $y(t) = 2e^{-t}$    **15.** yes    **16.** $y = e^{-\sqrt{2}t}$    **17.** $y = 3/4e^{2t} + 1/4e^{-2t}$
**18.** $(e^{t+1} + e^{2-t})/(e+1)$

# CHAPTER 8

**Section 8.1 – p. 306**    **1.** $(1, 7), (-2, 5), (1, -1), (10, 26), \sqrt{601}$
**2.** $(-1, 1, 0), (27, 3, -15), (0, 0, -2), \sqrt{74}$    **6.** $(2, -4) = 14(1, 1) - 6(2, 3)$
**7.** $(1, -1, 2) = -\dfrac{1}{13}(1, 0, -1) - \dfrac{2}{13}(2, 2, 1) + \dfrac{9}{13}(2, -1, 3)$    **8.** no    **9.** no

**Section 8.2 – p. 312**    **1.** (a) $1/\sqrt{27}, 1/\sqrt{27}, 5/\sqrt{27}$    (b) $3/\sqrt{14}, 2/\sqrt{14}, -1/\sqrt{14}$
(c) $-5/\sqrt{33}, 2/\sqrt{33}, 2/\sqrt{33}$    **2.** (a) $1/\sqrt{14}$ $(3, 2, -1)$    (b) $1/\sqrt{21}$ $(4, -1, 2)$
(c) $1/\sqrt{5}$ $(0, -1, 2)$    **3.** $1/\sqrt{5}$ $(2, -1, 0)$; $1/3$ $\sqrt{3}$ $(1, 1, 5)$; $1/\sqrt{117}$ $(4, 1, 10)$;
$1/\sqrt{30}$ $(1, -2, -5)$    **4.** $1/\sqrt{34}$ $(5, 3)$; $1/\sqrt{26}$ $(-1, -5)$; $1/13$ $(12, 5)$.    **5.** (a) $(4, 2, 1)$
(b) $(5, 3, -1)$    (c) $(6, 4, 4)$    **6.** (c)    **7.** (a) $(-2, -3, 4)$    (b) $(-5, -3, 3)$
(c) $(-7, 0, 2)$

**Section 8.3 – p. 317**    **3.** (a) $1/\sqrt{21}\,(2,1,4)$    (b) $1/\sqrt{51}\,(1,5,5)$    (c) $(0,0,1)$
**4.** Straight line passing through the tip of $A$ in the direction of $B$.    **6.** Straight line segment passing through the tips of vectors $A$ and $B$.    **7.** (a) Straight line passing through the tip of $B$ in the direction of $A$.    (b) Straight line segment connecting the tips of $A$ and $B$.

**8.** (a) $(0,0) = OA + OB$    (b) $(2,4) = \dfrac{7}{5}A + \dfrac{1}{5}B$    **9.** (a) $(0,0) = OA + OB$

(b) $(2,4) = \dfrac{2}{7}A + \dfrac{8}{7}B$    **11.** A plane through $(1,2,3)$ no; $d = 0$    **12.** A plane through

$(1,0,-1)$; no values of $d$.    **13.** yes; $(-1,0,-1) = \dfrac{1}{2}A - \dfrac{5}{2}B - C$; $(0,0,0) = OA + OB + OC$.

**14.** yes; $(-1,0,-1) = -A + 2C$; $(0,0,0) = OA + OB + OC$    **15.** $50(-1-5\sqrt{2}, 5\sqrt{2})$,
$1/(101 + 10\sqrt{2})(-1-5\sqrt{2}, 5\sqrt{2})$    **17.** $F_1 = 50(-2/\sqrt{21}, 1)$, $F_2 = 50(2/\sqrt{21}, 1)$ in both cases

**Section 8.4 – p. 321**    **1.** $6, 0, 2$    **2.** $1$    **3.** $-1/7$    **4.** $2$    **5.** $3$    **6.** no; consider
$A = (0,0,1)$, $B = (0,1,0)$, $C = (1,0,0)$    **7.** $(1,1,2)$; yes    **8.** $(2,3,-1)$

**9.** $\pm 1/\sqrt{2}(1,1,0)$.    **10.** $\pm 1/\sqrt{2}(-1,0,1)$    **11.** $x \neq 0, y = -5x$: $x = \pm \dfrac{1}{3\sqrt{5}}$

**12.** $x \neq 0, y = 2x$: $x = \pm 1/\sqrt{6}$    **17.** The diagonals must be equal in length.    **18.** yes

**Section 8.5 – p. 324**    **1.** $A$ and $B$ parallel; $A$ and $D$ orthogonal; $B$ and $D$ orthogonal
**2.** $0, 1/2, \sqrt{3}/2$    **3.** (a) $x = -1$    (b) For no $x$    **4.** (a) $x = 13$    (b) $x = -1$
**8.** $\sqrt{23}/6, \sqrt{13}/6, 0$    **9.** $-\sqrt{2}/8$    **10.** $t = -|A|$    **12.** $x = \dfrac{-(A,B)}{|B|^2}$

**13.** no    **14.** a diamond, a square

**Section 8.6 – p. 331**    **1.** (a) $X = (1,-1) + t(2,3)$; $3x_1 - 2x_2 - 5 = 0$
(b) $X = (2,-3) + t(-3,5)$; $5x_1 + 3x_2 - 1 = 0$    (c) $X = (4,1) + t(2,1)$; $x_1 - 2x_2 - 2 = 0$
(d) $X = (1,3) + t(-3,5)$; $5x_1 + 3x_2 - 14 = 0$    **2.** $X = (0,3) + t(2,-4)$,
$X = (3/2, 1/2) + t(5/2, -7/2)$, $X = (1/2, 3/2) + t(1/2, 1/2)$    **3.** $(1,3) + t(1/2, -1/2)$
**4.** $(-1,1)$, $(5,5)$, $(4,6)$    **5.** $D = \sqrt{5}$    **6.** (a) $X = (1,-1,1) + t(2,0,-3)$;
$\dfrac{x_1 - 1}{2} = \dfrac{x_3 - 1}{-3}$    (b) $X = (2,1,-2) + t(1,-1,-5)$; $\dfrac{x_1 - 2}{1} = \dfrac{x_2 - 1}{-1} = \dfrac{x_3 + 2}{-5}$

(c) $X = (1,-1,1) + t(0,3,2)$; $\dfrac{x_2 + 1}{3} = \dfrac{x_3 - 1}{2}$    **10.** (a) $X = (5,-3,1) + t(2,-4,6)$
(b) $X = (0,-1,1/2) + t(-2,2,-1)$    **11.** $D = \sqrt{62/21}$    **14.** $D = (8\sqrt{5})/5$
**15.** (a) $(3\sqrt{30})/5$    (b) $\sqrt{3}$    (c) $(3\sqrt{66})/11$    **16.** $\sqrt{182/7}$    **17.** $\sqrt{3}$

**Section 8.7 – p. 339**    **1.** (a) $\{5/2, -5, 5/3\}$    (b) $\dfrac{1}{\sqrt{14}}(2,-1,3)$    (c) $X = (0,-5,0)$
$+ s(1,-1,-1) + t(1,2,0)$    (d) $(1,-1,-1)$ and $(1,2,0)$    **2.** (a) $x_1 - x_2 + x_3 = 3$
(b) $(1,1,0)$ and $(0,1,1)$    (c) $\{3,-3,3\}$    (d) $D = \sqrt{3}$    **3.** $2x_1 + 2x_2 + x_3 = 6$;
$N = (2/3, 2/3, 1/3)$, $A = (-3,1,4)$, $B = (-1,2,-2)$    (d) $D = 5/3$    **4.** (b) and (d) are in $\pi$
**5.** $x_1 + 3x_2 + x_3 = 18$    **6.** $x_1 + 9x_2 - 2x_3 + 12 + 5\sqrt{86} = 0$, $x_1 + 9x_2 - 2x_3 + 12 - 5\sqrt{86} = 0$.
**8.** $D = \sqrt{389/62}$    **9.** $2x_1 + x_2 - x_3 = 7$    **10.** (a) $D = 9/(2\sqrt{14})$    (b) $D = 8/\sqrt{11}$
**13.** $41x_1 + 3x_2 - 18x_3 = 17$    **14.** $n_1x_1 + n_2x_2 + n_3x_3 + n_4x_4 = n_1c_1 + n_2c_2 + n_4c_4$; $D = \sqrt{3}/3$
**15.** $x_1 \neq 0, x_1 = x_2 = x_3$    **16.** $x_1 \neq 0, x_2 = 3x_1, x_3 = -4x_1, x_4 = -3x_1$.

**Section 8.8 – p. 346**    **1.** (a) $(-3,3,3)$    (b) $(-3,3,3)$    (c) $(0,0,0)$    (d) $(-14, 14, 14)$
(e) $(12,0,-3)$    (f) $15$    (g) $15$    (h) $0$    (i) $0$    **2.** (a) $(-1,1,1)$    (b) $(1,5,3)$
(c) $(-1,0,1)$    **3.** (a) $(2,0,1)$    (b) $(4,0,-1)$    (c) $(20,-3,8)$

**4.** $A = \dfrac{1}{2}|(Q - P) \times (R - P)|$    **5.** (a) $\dfrac{1}{2}\sqrt{53}$    (b) $\dfrac{1}{2}\sqrt{174}$    (c) $\dfrac{3}{2}\sqrt{11}$

**7.** $A$ is perpendicular to $B$    **8.** *Hint.* The altitude is the distance from $C$ to the plane

spanned by $A$ and $B$.    **9.** $t = \pm \dfrac{1}{28}$

**Section 8.9\* – p. 352**    **1.** (a) independent    (b) dependent    (c) independent    (d)
(d) dependent    **2.** (a) $t = -7/3$    (b) no $t$ exists    (c) all $t$    **3.** (a) linearly dependent;
no solution    (b) linearly dependent; no solution    (c) $x_1 = -7/13, x_2 = 4/13, x_3 = -1/13$
(d) $x_1 = 1/2, x_2 = -1, x_3 = 3/2$    **4.** (a) $x_1 = 1/5, x_2 = -4/5, x_3 = 0$    (b) $A, B, A \times B$ are

linearly dependent; no solution.     (c) $x_1 = 23/43$, $x_2 = 65/43$, $x_3 = 15/43$

(d) $x_1 = 2$, $x_2 = 0$, $x_3 = 0$

**12.**

$$x_2 = \frac{\begin{vmatrix} a_1 & d_1 & c_1 \\ a_2 & d_2 & c_2 \\ a_3 & d_3 & c_3 \end{vmatrix}}{\begin{vmatrix} a_1 & b_1 & c_1 \\ a_2 & b_2 & c_2 \\ a_3 & b_3 & c_3 \end{vmatrix}} \qquad x_3 = \frac{\begin{vmatrix} a_1 & b_1 & d_1 \\ a_2 & b_2 & d_2 \\ a_3 & b_3 & d_3 \end{vmatrix}}{\begin{vmatrix} a_1 & b_1 & c_1 \\ a_2 & b_2 & c_2 \\ a_3 & b_3 & c_3 \end{vmatrix}}$$

## CHAPTER 9

**Section 9.1 – p. 362**     **1.** $Y(t) = (4, 0) + t(0, 2)$, $Y(t) = (2\sqrt{2}, \sqrt{2}) + t(-2\sqrt{2}, \sqrt{2})$; ellipse
**2.** (a) $Y(t) = (-2, -3) + t(1, 2)$     (b) $t = \pm 2$     **3.** $X'(t) = (2t, \cos t, -\sin t)$,
$Y(t) = \left(\frac{\pi^2}{16}, \frac{\sqrt{2}}{2}, \frac{\sqrt{2}}{2}\right) + t\left(\frac{\pi}{2}, \frac{\sqrt{2}}{2}, \frac{-\sqrt{2}}{2}\right)$     **4.** $X'(t) = (e^t, -e^t, t)$, $Y(t) = (1, 1, 0) + t(1, -1, 1)$
**5.** $\cos\theta = \dfrac{t}{\sqrt{2} + 2t^2}$     **6.** $Y(t) = \left(\dfrac{e^2 + 1}{e}, \dfrac{e^2 - 1}{2e}\right) + t\left(\dfrac{e^2 - 1}{e}, \dfrac{e^2 + 1}{2e}\right)$;
trace is a hyperbola     **7.** $X(t)$ is differentiable at $t = 0$. Both curves have a direction at $t = 0$.
**14.** The acceleration is orthogonal to the velocity.

**Section 9.2 – p. 364**     **1.** (a) $(X, Y)'(t) = t \sin t + (t^2 + 1) \cos t + 4t^3$,
$(fX)'(t) = e^t(\cos t - \sin t, \cos t + \sin t, t + 1)$, $(fY)'(t) = e^t(t + 1, t^2 + 2t, t^3 + 3t^2)$ $(X \times Y)'(t)$
$= (3t^2 \sin t + t^3 \cos t - 3t^2, t^3 \sin t - 3t^2 \cos t + 2t, -t^2 \sin t + t \cos t - \sin t)$
(b) $(X, Y)'(t) = \left[\dfrac{-e^t(1 + t^2)}{(1 + t)^2} + e^{-t} - te^{-t} + \dfrac{2}{(1 - t)^2}\right]$,
$(fX)'(t) = \left[-2t, (3t + 1)(t + 1), \dfrac{2(2 - t)(1 + t)^2}{(1 - t)^2}\right]$, $(fY)'(t) = (1 + t)(3e^t + te^t, e^{-t} - te^{-t}, 2)$,
$(X \times Y)'(t) = \left[\dfrac{1 - t + e^{-t}(1 + t)}{1 - t}, \dfrac{e^t(3 - t^2)}{(1 - t)^2} + \dfrac{2}{(1 + t)^2}, \dfrac{e^{-t}(t^2 - 3)}{(1 + t)^2} - te^t - e^t\right]$
(c) $(X, Y)'(t) = \dfrac{1 - \log t}{t^2} + 2t \, e^{t^2} + 1$ $(fX)'(t) = \left(\dfrac{\sin t}{t} + \cos t \log t, \cos t, t \cos t + \sin t\right)$
$(fY)'(t) = \left[\dfrac{t \cos t - \sin t}{t^2}, e^{t^2}(2t \sin t + \cos t), \cos t\right]$
$(X \times Y)'(t) = \left[-e^{t^2}(2t^2 - 1), -\dfrac{1}{t}, 2t \, e^{t^2} \log t + \dfrac{e^{t^2}}{t} + \dfrac{1}{t^2}\right]$     **4.** $x_1^2 + x_2^2 + x_3^2 = r^2$

**Section 9.3 – p. 369**     **1.** (a) $x_1 - 4x_2 + 3x_3 = -18$     (b) $A = (1, 2, -3)$, $B = (2, 1, -2)$
(c) at $t = 5/8$, $64x_1 + 160x_2 + 192x_3 = 525$     **2.** (a) $\pi x_2 + 4x_3 = 12$     (b) $1, -1, 0$
(c) $\pi x_2 - 2x_3 = 0$, $\pi x_2 + 2x_3 = 0$, $x_2 = 0$     **5.** $x_1 = 0$, $x_1 + 4x_2 + 12x_3 = 114$, $x_1 - 2x_2 + 3x_3 = -6$
**6.** (a) $x''(2) = (2, 0)$, $\left\{\dfrac{4}{\sqrt{5}}, \dfrac{2}{\sqrt{5}}\right\}$     (b) $x''(3\pi/4) = (-\sqrt{2}, \sqrt{2})$, $\{0, 2\}$
(c) $x''(0) = (1, 1)$, $\{0, \sqrt{2}\}$     (d) $x''(\pi/2) = (-3, 0)$, $\{0, 3\}$     **7.** $X(t) = (t + 2, \sqrt{8}(t + 2)^{1/2})$,
$v'(t) = -\dfrac{(t + 2)^{-3/2}}{(t + 4)^{1/2}}$     **8.** $X(t) = (2t^{1/3}, 2t)$, $X'(1) = (2/3, 2)$, $X''(1) = (-4/9, 0)$,
$\left\{\dfrac{-4}{3\sqrt{10}}, \dfrac{16\sqrt{10}}{3}\right\}$     **9.** $X(t) = (3t, 3\sqrt{25} - t^2)$, $\left\{\dfrac{54}{25\sqrt{5}}, \dfrac{27}{25}\right\}$     **10.** (a) $3x_1 - 3x_2 + x_3 = 1$
(b) $2x_1 - x_2 = 0$     (c) $x_2 - x_3 = -2$     (d) $2\pi x_1 - 8x_2 + (\pi^2 + 8)x_3 = \dfrac{\pi^3}{2}$

**11.** (a) $T(t) = \dfrac{1}{\sqrt{2}}(-\sin t, \cos t, 1)$, $N(t) = (-\cos t, -\sin t, 0)$, $B(t) = \dfrac{1}{\sqrt{2}}(\sin t, -\cos t, 1)$
(b) $T(t) = (e^{2t} + e^{-2t} + 1)^{-(1/2)}(e^t, -e^{-t}, 1)$, $N(t) = (5e^{-2t} + 5e^{2t} + e^{4t} + e^{-4t} + 6)^{-(1/2)}$
$(2e^{-t} + e^t, 2e^t + e^{-t}, e^{-2t} - e^{2t})$, $B(t) = [(e^{2t} + e^{-2t} + 1)(5e^{-2t} + 5e^{2t} + e^{4t} + e^{-4t} + 6)]^{-(1/2)}$
$(-e^{-3t} - e^t - e^{-t}, e^{-t} + e^t + e^{3t}, 2e^{2t} + 2e^{-2t} + 2)$     **13.** (a) $x_1 - x_3 = 0$     (b) $x_1 + x_2 - 2x_3 = 0$
(c) $x_1 - x_2 - 2x_3 = 0$     **15.** $x_1 - x_3 = 0$     **17.** (a) $3x_1 - x_2 - 2x_3 = 3$     (b) $x_1 - x_2 - 2x_3 = 1$

**Section 9.4 – p. 376**     **1.** (a) circle with radius 1 centered at $(1, 0)$     (b) circle with radius 2
centered at $(0, 2)$     (c) circle with radius 2 centered at $(0, 1)$     (d) circle with radius 1
centered at $(1, -4)$     **3.** (a) parabola, 1, $x_1 = 1$     (b) hyperbola, 3, $x_1 = -4/3$     (c) ellipse,
$1/2, x_1 = 5$     (d) ellipse, $1/3, x_2 = -2$     (e) parabola, 1, $x_2 = -4$     (f) hyperbola,
$3/2, x_2 = 5/3$

**Section 9.5 – p. 378**    **1.** (a) $a_U = 2\cos 2t$, $v_V = 2\sin 2t$, $a_U = -8\sin 2t$, $a_V = 8\cos 2t$

(b) $v_U = 2t\sin t^2$, $v_V = 2t(1-\cos t^2)$, $a_U = 8t^2\cos t^2 + 2\sin t^2 - 4t^2$,

$a_V = 2(1-\cos t^2) + 8t^2\sin t^2$    (c) $v_U = ce^{ct}$, $v_V = ce^{ct}$, $a_U = 0$, $a_V = 2c^2 e^{ct}$

(d) $v_U = \dfrac{-2t\sin t^2}{(2-\cos t^2)^2}$, $v_V = \dfrac{2t}{(2-\cos t^2)}$, $a_U = \dfrac{8t^2(\sin t^2)^2}{(2-\cos t^2)^3} - \dfrac{2(2t^2\cos t^2 + \sin t^2)}{(2-2\cos t^2)^2} - \dfrac{4t^2}{(2-\cos t^2)}$,

$a_V = \dfrac{2}{2-\cos t^2} - \dfrac{8t^2\sin t^2}{(2-\cos t^2)^2}$    **2.** $X(t) = 8\sin\theta U(\theta)$, $\theta(t) = -\dfrac{9}{8}t + \pi/2$

**3.** $v = 3\sqrt{1+9\pi^2}$, $a_U = \dfrac{-27\pi}{2}$, $a_V = 9$    **4.** $X(t) = \left(\dfrac{t}{\sqrt 2}+1\right)U(\theta)$, $\theta(t) = \log\left(\dfrac{t}{\sqrt 2}+1\right)$

**5.** at $t = \pi/2$, $v_U = 0$, $v_V = \pi$, $a_U = \dfrac{-3\pi^2}{8}$, $a_V = 0$; at $t = \pi$, $v_U = -\pi/2$, $v_V = \pi/2$, $a_U = \dfrac{-\pi^2}{8}$,

$a_V = \dfrac{-\pi^2}{8}$

**Section 9.6 – p. 381**    **1.** (a) $A = \pi$    (b) $A = 16\pi$    (c) $A = 6\pi$    (d) $A = \pi$

**2.** (a) $A = 2\pi + 8\sqrt 2 + 2$    (b) $A = \pi + 3\sqrt 3$    (c) $A = 17\arcsin(2/3) + 6\sqrt 5$    **3.** $3\pi\sqrt 2/8$

**4.** $A = 8/3$    **5.** $\dfrac{2\pi}{9\sqrt 3} - \dfrac{1}{6}$    **6.** $\dfrac{5}{18\sqrt 3}$    **7.** $\dfrac{4(\pi+1)}{3}$

**Section 9.7 – p. 391**    **1.** (a) $2/27\,(19^{3/2} - 10^{3/2})$    (b) $\log(2+\sqrt 5) + 2\sqrt 5$

(c) $(\sqrt 5 - \sqrt 2) + \log\left(\dfrac{2+2\sqrt 2}{1+\sqrt 5}\right)$    (d) $\dfrac{e^2-1}{2e}$    **2.** (a) $[8\cdot(37)^{3/2} - (13)^{3/2}]/27$

(b) $\sqrt 2(e^\pi - 1)$    (c) $2\pi^2$    (d) $6$    (e) $\sqrt 2\log\left|\dfrac{e(1+\sqrt 2)}{2(1+\sqrt{e^2+1})}\right| + \sqrt{2(e^2+1)} - 2$

**3.** $15 + \log 4$    **4.** $\dfrac{\sqrt{2(e^2-1)}}{2e}$    **5.** $7/3$    **6.** (a) $8\pi$    (b) $1/3\,(5^{3/2} - 8)$

(c) $2\sqrt 2\,(e^4 - 1)$    (d) $4$    (e) $8a$    (f) $16$    **7.** $\pi + \dfrac{3\sqrt 3}{8}$    **14.** $f(t) + \sqrt{[f(t)]^2 + 1} = ke^t$,

where $k$ is some real constant.    **15.** $f(\theta) = Ae^{(\theta/\sqrt{k^2-1})}$, where $A, k$ are real constants.

**16.** $f(\theta) = 4e^{\theta/(2\sqrt 2)}$

# CHAPTER 10

**Section 10.1 – p. 402**    **1.** (a) conv.    (b) conv.    (c) div.    (d) conv.    (e) conv.

(f) div.    (g) conv.    (h) conv.    (i) conv.    (j) div.    (k) conv.    (l) conv.

(m) conv.    **2.** (a) conv.    (b) div.    (c) div.    (d) div.    **5.** $1/2$    **7.** (a) yes, $l = \sqrt 2$

(b) no    (c) yes, $= \dfrac{\sqrt{\pi^2+4}}{2\pi^2} + \dfrac{1}{4}\log\left(\dfrac{2+\sqrt{\pi^2+4}}{\pi}\right)$

**Section 10.2 – p. 407**    **1.** (a) $x_0 = 7/2$    (b) $x_0 = \dfrac{20 - \sqrt{148}}{18}$    (c) $x_0 = \pi/6$

(d) $x_0 = \sqrt{\dfrac{e^2-1}{2}}$    (e) $x_0 = \dfrac{\sqrt 2}{2}$    **2.** (a) $t = 2/3$    (b) $t = \pi/6$    (c) $t = \pi$    **3.** (a) $1$

(b) $1/2$    (c) $1/3$    (d) $0$    (e) $\log a - \log b$    (f) $a^2/b^2$    (g) $\log a/6$    (h) $1$    (i) $-2$

(j) $0$    **4.** (a) $1/3$    (b) $2$    (c) $-2/3$    **8.** no

**Section 10.3 – p. 410**    **1.** $0$    **2.** $0$    **3.** $0$    **4.** $1$    **5.** $1/e$    **6.** $1$    **7.** $1/2$    **8.** $0$

**9.** $e^2$    **10.** $1/2$    **11.** $1$    **12.** $1$    **13.** $e$    **14.** $-e/2$    **15.** $1$

**Section 10.4 – p. 415**    **1.** $p_n(x) = 1 + x + x^2 + \cdots + x^n$

**2.** $p_n(x) = 1 + \dfrac{1}{2}x + \dfrac{3}{2\cdot 2^2}x^2 + \dfrac{5\cdot 3}{3!2^3}x^3 + \cdots + \dfrac{(2n-1)(2n-3)(2n-5)\cdots 3}{n!2^n}x^n$

**3.** $p_n(x) = 1 + \dfrac{1}{2}x - \dfrac{1}{2^2}x^2 + \cdots + \dfrac{(-1)^{n+1}(2n-3)(2n-5)\cdots 3\,x^n}{2^n n!}$

**4.** $p_n(x) = 1 + \dfrac{3}{2}x + \dfrac{5\cdot 3}{2\cdot 2^2}x^2 + \cdots + \dfrac{(2n+1)(2n-1)(2n-3)\cdots 3x^n}{n!2^n}$

**5.** $p_n(x) = 1 + \dfrac{3}{2}x + \dfrac{3}{2\cdot 2^2}x^2 - \dfrac{3\cdot 3}{3!2^3}x^3 + \cdots + \dfrac{(-1)^n(2n-3)(2n-5)\cdots 3\cdot 3}{n!2^n}x^n$

**6.** $1 - \dfrac{x^2}{2!} + \dfrac{x^4}{4!} + \cdots + \dfrac{(-1)^n x^{2n}}{(2n)!}$    **7.** $1 + \dfrac{x^2}{2!} + \dfrac{x^4}{4!} + \cdots + \dfrac{x^{2n}}{(2n)!}$

**8.** $x+\dfrac{x^3}{3!}+\dfrac{x^5}{5!}+\cdots+\dfrac{x^{2n+1}}{(2n+1)!}$   **9.** $p(x)=2x+\dfrac{2x^3}{3}+\dfrac{2x^5}{5}+\cdots+\dfrac{2x^{2n+1}}{2n+1}$

**10.** $p(x)=1+3(x-1)+3(x-1)^2+(x-1)^3$

**11.** $p(x)=-1+5(x+1)-10(x+1)^2+10(x+1)^3-5(x+1)^4+(x+1)^5$

**12.** $p(x)=29+46(x-2)+24(x-2)^2+4(x-2)^3$   **13.** $\dfrac{\sqrt{2}}{2}\Big[1+(x-\pi/4)-\dfrac{(x-\pi/4)^2}{2}-$

$\dfrac{(x-\pi/4)^3}{3!}+\dfrac{(x-\pi/4)^4}{4!}+\cdots\pm\dfrac{(x-\pi/4)^n}{n!}\Big];\ \dfrac{\sqrt{2}}{2}\Big[1-(x-\pi/4)-\dfrac{(x-\pi/4)^2}{2!}+\dfrac{(x-\pi/4)^3}{3!}$

$+\dfrac{(x-\pi/4)^4}{4!}-\cdots\pm\dfrac{(x-\pi/4)^n}{n!}\Big]$

**14.** $p_n(x)=1+x-x^2+\cdots+(-1)^{n+1}\dfrac{(2n-3)(2n-5)\cdots 3x^n}{n!}$

**15.** $p_{2n}(x)=1+\dfrac{1}{2}x^2+\dfrac{3}{2\cdot 2^2}x^4+\dfrac{5\cdot 3}{3!2^3}x^6+\cdots+\dfrac{(2n-1)(2n-3)(2n-5)\cdots 3x^{2n}}{n!2^n}$

**16.** $p_{2n+1}(x)=x+\dfrac{1}{2\cdot 3}x^3+\dfrac{3}{2\cdot 2^2\cdot 5}x^5+\dfrac{5\cdot 3x^7}{3!2^3\cdot 7}+\dfrac{(2n-1)(2n-3)(2n-5)\cdots 3x^{2n+1}}{n!2^n(2n+1)}$

**17.** $p_{2n+1}(x)=x+\dfrac{x^3}{3!}+\dfrac{x^5}{5!}+\cdots+\dfrac{x^{2n+1}}{(2n+1)!}$   **18.** $p_{2n+1}(x)=x+\dfrac{x^3}{3!}+\dfrac{x^5}{5\cdot 2!}+\dfrac{x^7}{7\cdot 3!}+\cdots$

$+\dfrac{x^{2n+1}}{(2n+1)n!}$   **19.** $p_{3n+1}(x)=x-\dfrac{x^4}{8}-\dfrac{1}{7\cdot 2^2}x^7-\cdots-\dfrac{(2n-3)(2n-5)\cdots 3x^{3n+1}}{(3n+1)2^n n!}$

**Section 10.5 – p. 421**   **1.** $e^x=1+x+\dfrac{x^2}{2!}+\dfrac{x^3}{3!}+\cdots+\dfrac{x^n}{n!}+\dfrac{1}{n!}\displaystyle\int_0^x e^t(x-t)^n\,dt,$

$\sin x=x-\dfrac{x^3}{3!}+\dfrac{x^5}{5!}+\cdots+(-1)^n\dfrac{x^{2n+1}}{(2n+1)!}+\dfrac{1}{(2n)!}\displaystyle\int_0^x(-1)^n\cos t(x-t)^{2n}\,dt$

**2.** $R_n(x)\le\dfrac{(x-x_0)^n}{n!}e^x$ where $0<x_0<x$   **3.** $\log(1+x)=$

$x-\dfrac{x^2}{2}+\dfrac{x^3}{3}-\cdots+(-1)^{n-1}\dfrac{x^n}{n}+(-1)^n\displaystyle\int_0^x\dfrac{t^n}{1+t}\,dt$   **4.** $\dfrac{1}{2(n+1)}<|R_n(1)|<\dfrac{1}{n+1}$

**5.** $\dfrac{1}{2(2n+3)}<R_n(1)<\dfrac{1}{(2n+3)}$   **7.** $\log\Big(\dfrac{1+x}{1-x}\Big)=2\Big(x+\dfrac{x^3}{3}+\dfrac{x^5}{5}+\cdots+\dfrac{x^{2n+1}}{2n+1}+$

$\displaystyle\int_0^x\dfrac{t^{n+1}}{t^2-1}\,dt\Big)$   **8.** $1+1+\dfrac{1}{2!}+\dfrac{1}{3!}+\cdots+\dfrac{1}{6!}$   **9.** $\dfrac{3}{5}+\dfrac{1}{3}\Big(\dfrac{3}{5}\Big)^3+\dfrac{1}{5}\Big(\dfrac{3}{5}\Big)^5$

**10.** $p_5(x)=x-\dfrac{x^3}{3!}+\dfrac{x^5}{5!}$   **14.** no

# CHAPTER 11

**Section 11.1 – p. 426**   **1.** (a) $\displaystyle\sum_{n=0}^{\infty}\dfrac{1}{\sqrt{5+n^2}}$   (b) $\displaystyle\sum_{n=0}^{\infty}\dfrac{1}{(2n+1)^2}$   (c) $\displaystyle\sum_{n=0}^{\infty}\dfrac{n+1}{(2n+1)^2}$

(d) $\displaystyle\sum_{n=0}^{\infty}\dfrac{2^n}{n!}$   (e) $\displaystyle\sum_{n=0}^{\infty}(-1)^n\dfrac{1}{\sqrt{2n+1}}$   (f) $\displaystyle\sum_{n=2}^{\infty}(-1)^n\dfrac{n}{(n^2-1)}$   **2.** (a) $2\displaystyle\sum_{n=0}^{\infty}\dfrac{1}{3^n}+3\displaystyle\sum_{n=0}^{\infty}\dfrac{1}{2^n}$

(b) $\displaystyle\sum_{n=1}^{\infty}\dfrac{1}{4^n}-3\displaystyle\sum_{n=1}^{\infty}\dfrac{1}{5^n}$   (c) $\displaystyle\sum_{n=0}^{\infty}\dfrac{1}{2^n}-2\displaystyle\sum_{n=2}^{\infty}\dfrac{1}{5^n}$   (d) $\displaystyle\sum_{n=3}^{\infty}\dfrac{1}{e^n}-\displaystyle\sum_{n=4}^{\infty}\dfrac{1}{2^n}$   **3.** 0   **4.** $-1$   **5.** 0

**6.** no limit   **7.** 1/5   **8.** 1   **9.** no limit   **10.** no limit   **11.** 0   **12.** 4, 6, 7, 8, 9, 10
**14.** no   **15.** diverges   **16.** no

**Section 11.2 – p. 429**   **1.** $\displaystyle\sum_{n=0}^{\infty}(-1/2)^n,\ 2/3$   **2.** $\displaystyle\sum_{n=-2}^{\infty}(-1/3)^n,\ 6\dfrac{3}{4}$

**3.** $\displaystyle\sum_{n=1}^{\infty}(1/4)^n-\displaystyle\sum_{n=0}^{\infty}(1/3)^n+18,\ 16\dfrac{5}{6}$   **4.** $\displaystyle\sum_{n=0}^{\infty}(-2)^n,$ div.   **5.** $\displaystyle\sum_{n=0}^{\infty}(-1)^n+(-1/2)^{n+1},$ div.

**6.** $|x|<1/4$   **7.** $|x|<2$   **8.** $|x|<3$   **9.** $|x|<5$

**Section 11.3 – p. 433**   **1.** conv.   **2.** conv.   **3.** conv.   **4.** div.   **5.** conv.   **6.** div.
**7.** conv.   **8.** conv.   **9.** div.   **10.** div.   **11.** conv.   **12.** conv.   **13.** div.
**14.** div.   **15.** conv.   **16.** div.   **17.** conv.   **18.** div.   **19.** conv.   **20.** conv.
**21.** div.

**Section 11.4 – p. 436**   **1.** conv.   **2.** conv.   **3.** conv.   **4.** div.   **5.** div.   **6.** conv.

**7.** conv.　　**8.** conv.　　**9.** conv.　　**10.** div.　　**11.** conv.　　**12.** div.　　**13.** for *all x*
**14.** $|x| < 1$　　**15.** $|x| < 2$　　**16.** for *no x* $\neq 0$　　**17.** $|x| < 1$　　**18.** $|x| < 1$

**Section 11.5 – p. 439**　　**1.** cond. conv.　　**2.** div.　　**3.** abs. conv.　　**4.** cond. conv.
**5.** abs. conv.　　**6.** cond. conv.　　**7.** cond. conv.　　**8.** cond. conv.　　**9.** abs. conv.
**10.** div.　　**11.** div.　　**12.** cond. conv.

**Section 11.6 – p. 442**　　**1.** $r = 1/4$　　**2.** $r = 2$　　**3.** $r = 1$　　**4.** $r = 2$
**5.** no interval of convergence　　**6.** the whole real line　　**7.** $[-2, 2)$　　**8.** $[-1, 1]$
**9.** $(-2, 2)$　　**10.** $(-3, 3)$　　**11.** $[-1/4, 1/4)$　　**12.** $(-1, 1)$　　**16.** $(-r, r)$

**Section 11.7 – p. 447**　　**1.** $\displaystyle\sum_0^\infty \frac{x^{2n}}{n!}, (-\infty, \infty)$　　**2.** $\displaystyle\sum_0^\infty (-1)^n \frac{x^{2n+1}}{n!(2n+1)}, (-\infty, \infty)$

**3.** $\displaystyle\sum_1^\infty (-1)^n \frac{x^{2n-1}}{(2n)!(2n-1)}, (-\infty, \infty)$　　**4.** $\displaystyle\sum_0^\infty x^{3n}, (-1, 1)$　　**5.** $\displaystyle\sum_0^\infty (-1)^n \frac{(-1)^n}{3n+1} x^{3n+1}, (-1, 1)$

**6.** $\displaystyle\sum_0^\infty (-1)^n \frac{x^{2n+3}}{(2n+3)(n+1)}, (-1, 1)$

# INDEX